Insect Pests in Tropical Forestry, 2nd Edition

We dedicate this book to our families, Trish, Richard, and Gev, and Ang, Cate, Nick, Richard, and Toby, for their patience and support during the writing of both editions of this book.

Insect Pests in Tropical Forestry, 2nd Edition

Dr F. Ross Wylie

*Department of Agriculture, Fisheries and Forestry,
Queensland Government, Australia*

Dr Martin R. Speight

Zoology Department, University of Oxford, and St Anne's College, Oxford, UK

CABI is a trading name of CAB International

CABI	CABI
Nosworthy Way	875 Massachusetts Avenue
Wallingford	7th Floor
Oxfordshire OX10 8DE	Cambridge, MA 02139
UK	USA
Tel: +44 (0)1491 832111	Tel: +1 617 395 4056
Fax: +44 (0)1491 833508	Fax: +1 617 354 6875
E-mail: cabi@cabi.org	E-mail: cabi-nao@cabi.org
Website: www.cabi.org	

© F.R. Wylie and M.R. Speight 2012. All rights reserved. No part of this publication may be reproduced in any form or by any means, electronically, mechanically, by photocopying, recording or otherwise, without the prior permission of the copyright owners.

A catalogue record for this book is available from the British Library, London, UK.

Library of Congress Cataloging-in-Publication Data

Wylie, F.R.
   Insect pests in tropical forestry / F. Ross Wylie, Martin R. Speight. -- 2nd ed.
     p. cm.
   Includes bibliographical references and index.
   ISBN 978-1-84593-635-8 (pbk : alk. paper) -- ISBN 978-1-84593-635-5 (hardback : alk. paper)
1. Forest insects--Tropics. 2. Forest insects--Control--Tropics.
I. Speight, Martin R. II. Title.

  SB764.T73W95 2012
  595.71734--dc23
                                      2011042626

ISBN-13: 978 1 84593 636 5 (Hbk)
          978 1 84593 635 8 (Pbk)

Commissioning editor: Rachel Cutts
Editorial assistant: Alexandra Lainsbury
Production editor: Simon Hill

Typeset by SPi, Pondicherry, India.
Printed and bound in the UK by CPI Group (UK) Ltd, Croydon, CR0 4YY.

Insect Pests in Tropical Forestry, 2nd Edition

Dr F. Ross Wylie

*Department of Agriculture, Fisheries and Forestry,
Queensland Government, Australia*

Dr Martin R. Speight

Zoology Department, University of Oxford, and St Anne's College, Oxford, UK

CABI is a trading name of CAB International

CABI	CABI
Nosworthy Way	875 Massachusetts Avenue
Wallingford	7th Floor
Oxfordshire OX10 8DE	Cambridge, MA 02139
UK	USA
Tel: +44 (0)1491 832111	Tel: +1 617 395 4056
Fax: +44 (0)1491 833508	Fax: +1 617 354 6875
E-mail: cabi@cabi.org	E-mail: cabi-nao@cabi.org
Website: www.cabi.org	

© F.R. Wylie and M.R. Speight 2012. All rights reserved. No part of this publication may be reproduced in any form or by any means, electronically, mechanically, by photocopying, recording or otherwise, without the prior permission of the copyright owners.

A catalogue record for this book is available from the British Library, London, UK.

Library of Congress Cataloging-in-Publication Data

Wylie, F.R.
  Insect pests in tropical forestry / F. Ross Wylie, Martin R. Speight. -- 2nd ed.
    p. cm.
  Includes bibliographical references and index.
  ISBN 978-1-84593-635-8 (pbk : alk. paper) -- ISBN 978-1-84593-635-5 (hardback : alk. paper)
 1. Forest insects--Tropics. 2. Forest insects--Control--Tropics.
I. Speight, Martin R. II. Title.

  SB764.T73W95 2012
  595.71734--dc23
                                  2011042626

ISBN-13: 978 1 84593 636 5 (Hbk)
          978 1 84593 635 8 (Pbk)

Commissioning editor: Rachel Cutts
Editorial assistant: Alexandra Lainsbury
Production editor: Simon Hill

Typeset by SPi, Pondicherry, India.
Printed and bound in the UK by CPI Group (UK) Ltd, Croydon, CR0 4YY.

Contents

Preface vii

1 Tropical Forests 1
2 Tropical Forests and Insect Biodiversity 22
3 Abiotic and Biotic Effects 49
4 Insect–Host Tree Interactions 75
5 Tropical Forest Pests: Ecology, Biology and Impact 91
6 Management Systems I: Planning Stage 153
7 Management Systems II: Nursery Stage 186
8 Management Systems III: Plantation Stage 204
9 Management Systems IV: Forest Health Surveillance, Invasive Species and Quarantine 240
10 Integrated Pest Management (IPM) 275

Bibliography 297

Index 357

Preface

Since the publication of the first edition of this book in 2001, there have been several significant developments in forestry and forest entomology in the tropics that warranted inclusion in a second edition. One is the emergence of new pest species associated either with the expansion of plantation forestry into new regions or marginal areas, or a gradual adaptation by indigenous species to exotic hosts or the rapid spread of new invasives. The South American carpenterworm, *Chilecomadia valdiviana*, in Chile, and the South African goat moth, *Coryphodema tristis*, are examples of emerging indigenous pests associated with the establishment and expansion of exotic *Eucalyptus* spp. plantations in these countries. The blue gum chalcid, *Leptocybe invasa*, and the erythrina gall wasp, *Quadrastichus erythrinae*, are recent examples of the rapid worldwide spread of an invasive insect, similar to that which occurred for the leucaena psyllid, *Heteropsylla cubana*, in the 1980s and 1990s. The biology, ecology and impact of these, and many additional pest species, are discussed in an expanded Chapter 5.

Another development has been the growing awareness of the 'true' global impact of forest invasive species as losses have been better quantified. Damage worldwide has been estimated at several billion US dollars per year, even without taking into account the loss of non-market value, which may well exceed that figure. This has prompted a range of international responses, including the formation of networks and surveillance programmes to provide early warning of incursions of forest invasives and the development of international standards for phytosanitary measures, such as that for wood packaging material, as discussed in Chapter 9. The global increase in self-help schemes such as plant clinics and field schools is also discussed. A range of new technologies has enhanced the accuracy and efficiency of forest health surveys over the past decade. These include the use of geographic information system–global positioning system interface tools and handheld computers to assist navigation and data collection in the field, and the application of digital, remotely sensed imagery to detect and classify damaged forest canopies. There have also been significant advances in the use of semiochemicals, including pheromones, for the detection and monitoring of forest pests and for controlling pest populations by means of mass trapping, lure and kill, lure and infect and mating disruption.

One of the fundamental principles underlying this book is that the prevention of problems is far better than attempts at cures. The book is not in any way designed to be a manual of pest management. Instead, it presents concepts and examples of tropical forest insects, their ecology, impact and control approaches where appropriate, and we hope that individual readers will make the connection between the generalities in the book and their own situation, socio-economic position and appropriate technologies. Some situations or examples have been revisited several times in different parts of the book, and some amount of repetition has been inevitable, and indeed vital. So, for example, the eucalyptus longicorn beetle, *Phoracantha semipunctata*, is discussed in chapters on ecology, host-plant relationships, plant resistance, biological control and integrated pest management, while the mahogany shoot borer, *Hypsipyla* spp., crops up almost as often.

As can be seen from Chapter 1, we have been fairly broad in our definition of 'tropical'. Clearly, insects do not take any notice of atlases and an example from the subtropics (however defined) may have just as much relevance to equatorial regions. In some cases, it has been necessary to use even temperate examples where none are available from more tropical regions. We have kept this to a minimum and though there may well have been several examples of concepts or techniques that are well established from temperate situations, we have strived to provide a tropical one instead. Our aim has also been to provide a complete global perspective on tropical forest pest ecology and management; hence the rather general nature of the contents. We have attempted to paint as broad a picture as possible, using examples from as many tropical countries as we have experienced personally or are well represented in the literature. Clearly, there will be omissions. Not all countries and all major pests have been treated exhaustively. We hope that interested readers, whether from the Pacific, Australasia, South and South-east Asia, Africa and both Central and South America, will find something of relevance to them personally and will also discover experiences outside their regions to be of significance to their own situations. Finally, we had to decide which forest types to focus on and have been sparing in our mention of native or natural forests. Essentially, this is a book about plantation forestry, either grown intensively on a large scale or at a local, small-scale level as part of village or agroforestry systems.

The layout of this second edition follows that of the first, though the content has been substantially rewritten to reflect 10 years of research and development, as well as the emergence of new pest species. Chapter 1 presents an overview, from a somewhat entomological perspective, of tropical forestry in its many guises. Chapters 2, 3 and 4 then discuss the 'pure' biology and ecology of tropical insects and their co-evolved relationships with the trees and forests in which they live. We discuss both abiotic (climatic) and biotic (food and enemies) factors, and together we hope that some of the ways in which 'neutral' forest insects become pests and, in turn, are subsequently reduced to low levels will become clear. The later, more 'applied' chapters base their management tactics on this earlier, pure science. It must not be forgotten that one person's pest may be another's rarity. Insect biodiversity in tropical forests is an extremely popular and important area for both discussion and research and we have touched on this aspect of insects in forests in Chapter 2.

Chapter 5 is necessarily the largest chapter in the book, looking in detail at a selection of major pest species from all over the tropical world. As mentioned above, these examples are not in any way exhaustive – that would be a quite impossible task – but, instead, provide case histories to illustrate a whole range of insect types, from defoliators and sap feeders, to shoot, bark and wood borers. Chapters 6, 7, 8 and 9 then discuss the theory and practice of insect pest management, starting at the fundamental planning stage, before any seeds hit the soil. If basic concepts and problems are not tackled and handled correctly at this first stage, then, in many cases, no amount of effort or expenditure in the future will fix problems to come. We then consider, in turn, nursery management and stand management in Chapters 7 and 8. The reader will notice a certain amount of 'pigeonholing' here. For example, the vital topics of biological and chemical control have relevance to all stages of tree growth and forest management, but we have chosen to put them in Chapter 8 since, in the majority of cases, this is where the techniques and concepts described find most practical use. Chapter 9 covers the topics of forest health surveillance, quarantine and forest invasive species, topics which again have significance at all stages of forestry but for convenience are presented after nursery and forest management. Ideally, readers should study all four chapters together to provide an integrated overview of all available concepts and practices. This, in fact, we attempt to do in the final chapter, Chapter 10, which combines most of the previous nine chapters in examples illustrating the concept of integrated pest management (IPM).

We are extremely indebted to numerous colleagues and friends who have helped, supported, read chapter drafts, supplied photographs or information, etc. – we gratefully acknowledge (in no particular order) Paul Embden, Clive Hambler, Peter Savill, Hugh Evans, Eric Boa, Jeff Burley, Malcolm Ryder, John Fryer, Mike Ivory, Russell Haines, Tim Hardwick, Ian Bevege, Brenton Peters, Jock Kennedy, Simon Lawson, Judy King, Murdoch DeBaar, Manon Griffiths, Janet McDonald, Helen Nahrung, Brendan Murphy, Geoff Pegg, Michael Kennedy, Chris Fitzgerald, Michael Ramsden, Norm Hartley, Trent Hemmens, Humphrey Elliott, Cliff Ohmart, Allan Watt, Mark Hunter, Rob Floyd, Carrie Hauxwell, Stephen Harris, Sean Murphy, Michael Cole, Bill Crowe, Sankaran, Zvi Mendel and Chuck Hodges. We also thank all our colleagues in the many countries where we have worked for their help and hospitality and our employing organizations for their support during the writing of this book.

Without the excellent services provided by the Queensland Department of Employment, Economic Development and Innovation library (particularly Dianne Langford, Mel Kippen, Helen Chamberlin and Kerrie McLaren), the University of Queensland, St Lucia, library and the Bodleian Library and St Anne's College Library in Oxford, the book could not have been written.

Most of the photographs have been taken by one or other of us, and those colleagues and agencies who kindly supplied the others are acknowledged in the photo credits. Many of the graphs and tables have been redrawn or reproduced from the literature. All the publishers who gave permission to do this are gratefully acknowledged here or, where requested, on the figure legends.

Finally, we are most grateful indeed to ACIAR (Australian Centre for International Agricultural Research), who provided a most generous grant to fund the provision of colour plates in the book. When all is said and done, the subject of the book is primarily a visual one and colour makes an immeasurable improvement to the whole work.

Ross Wylie and Martin Speight
Brisbane and Oxford
September 2011

1

Tropical Forests

1.1 Forests

Global land use in 2000, and estimated additional requirements by 2030, are presented in Table 1.1 (Lambin and Meyfroidt, 2011). As the table shows, considerably more land will be required by 2030 for the almost unavoidable advance in cropland, grazing land and highly controversial biofuel production. The authors predict that over 100 million hectares (Mha) of industrial forestry will be required by then.

At the moment, forests account for almost 30% of the earth's total land area, somewhere in the region of 4 bn ha or about 39.5 Mkm^2 (Freer-Smith and Carnus, 2008). If we extrapolate to include the oceans as well, this comes to around 10% of the entire globe. Most people can conjure up a tropical forest easily in their minds: hot and humid jungles, with luxuriant creepers, enormous trees and mysterious wild animals. Romantic, no doubt, but in reality only a small and fanciful part of the truth! It should be easy enough to define a forest, though Putz and Redford (2010) suggest that doing so may be 'implemented in fogs of ambiguity'. According to the Food and Agriculture Organization of the United Nations (FAO), land with tree crown cover (or stand density) of more than about 20% of the total area is defined as 'forest' in 'developed' regions, whereas in 'developing' countries, ecosystems with a minimum of only 10% tree (and/or bamboo) crown cover can be called 'forest' (FAO, 1997). These systems usually are characterized by wild flora and fauna, with natural soil conditions, and they are not associated with agricultural practices. As we shall see, though, they may well be influenced dramatically by other anthropocentric activities, whereupon they are split into two categories, natural forests, both primary and secondary, and plantation forests. Note that of the nearly 4 bn ha mentioned above, only about 3.8% is plantation forestry (Freer-Smith and Carnus, 2008). Furthermore, a large variety of natural and artificial systems, while not falling comfortably under the conventional umbrella of 'forest', contain relatively high proportions of woody vegetation. These systems are of great ecological and economic importance. Scrub, shrub or bushland are general vegetation types where the dominant woody elements are at maximum 5–7 m tall. Forest fallow consists of many types of woody vegetation derived from the clearing of natural forest for shifting agriculture, where ecological succession proceeds in patches. These woody areas encompass about 1.7 bn ha on top of the 4 bn or so of true forest already mentioned (Sharma et al., 1992). Individual trees may also be of great significance; the planting of arid-tolerant

© F.R. Wylie and M.R. Speight 2012. *Insect Pests in Tropical Forestry*, 2nd Edition
(F.R. Wylie and M.R. Speight)

Table 1.1. Estimates of global land use derived from the literature in 2000 plus additional demand by 2030. Low estimates are a conservative view, while high estimates are 'bolder' (data from Lambin and Meyfroidt, 2011).

Land-use category	Low (Mha)	High (Mha)
Land use in 2000		
Cropland	1510	1611
Pastures	2500	3410
Natural forests	3143	3871
Planted forests	126	215
Urban built-up areas	66	351
Unused, productive land	356	445
Projected land use in 2030		
Additional cropland	81	147
Additional biofuel crops	44	118
Additional grazing land	0	151
Urban expansion	48	100
Industrial forestry expansion	56	109
Protected area expansion	26	80
Land lost to degradation	30	87
Total land demand in 2030	*285*	*792*
Clearing of natural forests, 2000–2030	152	303

trees in dry regions, or the establishment of woody species in villages or small farms for fuel, shelter or browse can hardly be termed forestry, at least in the strict industrial sense of the word, but various types of agroforestry, shamba or taungya are vital to local economies. The various types of forest, from intensive industrial plantations to social and community plantings and agroforestry, will be considered in more detail below.

We should also note that forest cover in its wide sense varies considerably according to country and region, as Colour Plate 129 shows (UNEP, 2009). Countries which still have 50% or more of their land area forested include much of Brazil, Guyana, Peru, Republic of Congo, Malaysia and Papua New Guinea.

1.2 The Tropics

Geographically, the 'tropics' is a region of the earth's surface bounded by the tropic of Cancer on the 23°27'N parallel and the tropic of Capricorn on the 23°27'S parallel; the equator bisects the region (Evans and Turnbull, 2004). The tropic of Cancer runs roughly (east to west) from Hong Kong to Calcutta, Aswan, Cuba and the tip of Baja California, while the tropic of Capricorn runs from north of Brisbane to southern Madagascar and eventually Rio de Janeiro (see Colour Plate 130). The width of this geographical belt is by definition 2820 nautical miles, roughly 5200 km.

Though precise geographically, the tropics cannot be delimited so easily in terms of climate or ecology. A dry season, monsoon wind or even a swarm of locusts do not recognize lines on maps, so it is much more realistic to consider the tropics as a region of the earth characterized by a certain range of abiotic and biotic features. It must also be remembered that even though a region or locality occurs within these boundaries, it may in fact experience temperate if not arctic conditions – anyone who has reached the summit of Mount Kinabalu in Sabah, Malaysia, or Kilimanjaro in Tanzania, East Africa, will testify readily to this! Some of the conditions considered to be 'tropical' are provided by Evans (1992) and are presented in Table 1.2. Clearly, some parts (around 25%) of the tropics are stereotypically hot and wet for most, if not all, of the year;

Table 1.2. Climatic characteristics of tropical regions (data from Evans, 1992).

Temperature	
Average air temperature	Over 20°C
Mean temperature of coldest month	Over 18°C
Rainfall	
Variability	0–10,000 mm/year
Duration of moisture limitation, arable land	28% of land, no limitation
	42% of land, 4–7 months' limitation
	30% of land, over 8 months' limitation
Solar radiation	
Comparison with temperate regions	Roughly double
Day length	
	10.5–13.5 h

seasonal variations are hard to detect and, for most months, rainfall exceeds losses by evapotranspiration.

Unsurprisingly, the natural vegetation is dominated by rainforest. In other regions (about 50% of the total), some parts of the year have much less rain than others, producing distinct wet and dry seasons. Dry seasons where rainfall is low or even absent for months on end are typified by water deficits and mild to severe droughts, followed by heavy monsoon rain. In these regions, rainforests give way to deciduous or semi-deciduous forest (Colour Plate 130). In both these regions, forestry is a widespread and vital industry. The remaining 25% of tropical regions consists of mountainous areas or dry scrub and savannah or desert, where rain tends to be in short supply and crop production is difficult. Even so, some form of forestry may still be attempted, especially where local people are in need of fodder for their livestock, mulch for their crops and/or wood to burn.

1.3 Socio-economics in the Tropics

Around 40% of the earth's land area and almost half of its human population (Evans, 1992) can be found in the tropics, in its broadest sense. Indeed, the term 'forest' may need to be defined not in ecological terms as much as in cultural ones (Putz and Redford, 2009). The economic prosperity and development status of the region is much less equitable, however. Table 1.3 looks at some basic statistics of several major countries in the tropics (World Bank, 2010). Various conclusions can be derived from these data. For example, it is clear that the majority of countries show a decline in forest area between 2000 and 2005. Only one or two in the table, such as India and Vietnam, show a small increase. Of course, some tropical countries are as developed as any in temperate regions, so that sophisticated management of production and industry is perfectly possible at the highest levels, but it does seem that the lower the GDP or the higher the population density, the less forested land there is. Forestry in tropical Australia, Malaysia, Thailand, parts of India and Africa and some countries in South America is thus very different to that in most countries of the tropics, the so-called 'less developed countries', or LDCs. Not only that, but traditions of land use, tenure, subsistence and so on vary tremendously from one culture to the next; it is quite impossible to imagine that a management scheme for a forest crop (and indeed the pests within it) would work equally well and be equally acceptable to local people in Papua New Guinea, Nigeria and Nicaragua. Whatever plans are made must always take into account the needs and abilities of local communities, social structures, levels of development and economics. It seems clear that people affect forests, mainly deleteriously. As can be seen in Fig. 1.1, there are very significant declines in forest cover as human populations increase (Laurance, 2007). These relationships appear to hold true for Asia, Africa and the Americas (though note the non-linear x-axis). In addition, Laurance points out that poor countries seem to have less surviving forest cover than do richer countries, and that endemic corruption may also play a part in reducing forest cover still further.

Table 1.3. Country statistics including forest area and per cent change from 2000 to 2005 (data from the World Bank, 2010).

Country	GDP (US$bn in 2008)	Population density/km²	Annual population growth (%)	Forest area (1000 km² in 2000)	Forest area (1000 km² in 2005)	Forests total land area (%)	Change in forests 2000–2005 (%)
Africa							
Ghana	5.0	97.6	2.1	60.9	55.2	23.2	−9.4
Kenya	12.7	53.4	2.6	35.8	35.2	6.1	−1.7
Malawi	1.7	93.3	2.9	35.7	34.0	28.8	−4.8
Nigeria	46.0	135.3	2.6	131.4	110.9	12.0	−15.6
South-east Asia							
Indonesia	165.0	108.2	1.3	978.5	885.0	46.5	−9.6
Malaysia	93.8	70.6	2.3	215.9	208.9	63.3	−3.2
Philippines	75.9	259.0	2.0	79.5	71.6	23.9	−9.9
Thailand	122.7	120.9	0.8	148.1	145.2	28.3	−2.0
Vietnam	31.2	234.1	0.2	117.3	129.3	39.3	+10.2
South Asia							
Bangladesh	47.1	979.2	1.8	8.8	8.7	6.0	−1.1
India	460.2	309.1	1.7	675.5	677.0	20.6	+0.2
Nepal	5.5	163.3	2.3	39.0	36.4	24.8	−6.7
Oceania							
Fiji	1.7	44.4	0.7	10.0	10.0	55.6	0.0
Papua New Guinea	3.5	11.7	2.6	301.3	294.4	63.6	−2.3
Central America							
Costa Rica	16.0	78.4	2.3	23.8	23.9	46.9	+0.4
Honduras	7.1	55.4	2.0	54.3	46.5	41.5	−14.4
Mexico	581.4	49.9	1.4	655.4	642.4	32.7	−2.0
Panama	11.6	39.5	1.9	43.1	42.9	56.4	−0.5
South America							
Brazil	644.7	20.4	1.4	4932.1	4777.0	56.1	−3.1
Colombia	94.1	35.0	1.5	609.6	607.3	43.0	−0.4
Guyana	0.7	3.5	0.0	151.0	151.0	70.2	0.0
Peru	53.3	91.3	1.5	692.1	687.4	53.5	−0.7
Venezuela	117.5	26.3	1.8	491.5	477.1	52.3	−2.9
To compare							
Australia	405.1	2.5	1.2	1646.5	1636.8	21.1	−0.6
UK	1451.0	241.8	0.4	27.9	28.5	11.7	+2.2
USA	9764.8	29.3	1.1	3022.9	3030.9	31.5	+0.3

1.4 Forest Products in the Tropics

For most countries in the world, temperate or tropical, developed or developing, forests are a vital resource for the economy at the local, national and international level. Forest products have been defined by FAO (1997) (Table 1.4). In this context, the term 'forest product' thus includes roundwood and sawwood, chipboard, MDF, pulpwood, veneer, fuelwood, plywood and charcoal. Table 1.5 subdivides the regions of the tropical world and shows the magnitude of the value of forest products in 2008 (FAO, 2010). Tropical Asia is, of course, a vast area, but production from its forests is worth well over US$100bn per annum, a colossal figure by any standards.

The type of forest product varies according to the forests that produce them, the markets both nationally and internationally, the technological development of the local

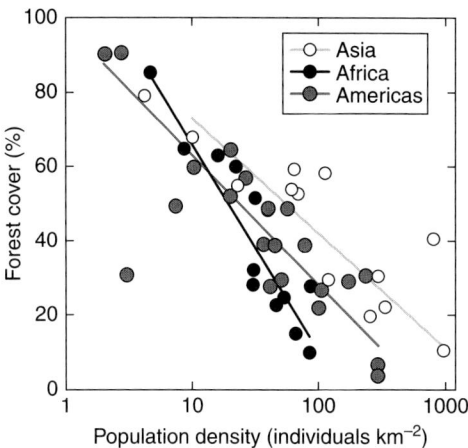

Fig. 1.1. Relationship between human population density and remaining forest cover (including logged or regenerating forests and plantations) for 45 tropical countries in three global regions (from Laurance, 2007, courtesy Elsevier Publishers).

Table 1.4. FAO definitions of forest products (from FAO, 1997).

Category	Definition
Roundwood	Wood in the rough; in its natural state, with or without bark, split or squared. It includes all wood from removals from forests and areas outside forests. It comprises saw and veneer logs, pulp and fuelwood, chips, particles, residues and charcoal.
Fuelwood	Wood in the rough (trunks and branches) intended as fuel for cooking, heating or power production.
Charcoal	Wood carbonized by partial combustion, intended as fuel for cooking, heating or power production.
Sawlogs	Logs to be cut lengthways for construction, railway sleepers, general lumber.
Veneer logs	Logs of high-quality timber mainly for peeling or slicing.
Pulp or chipwood	Wood in the rough rather than logs, may be reduced to small pieces (chips) intended to make paper, particleboard or fibreboard.
Other industrial roundwood	Roundwood used for tanning, distillation, poles, pitprops, etc.

industries and the subsistence needs of indigenous people. Some of these major categories vary considerably in their magnitudes and values depending on the region of the tropical world. Thus, for example, fuelwood and charcoal are most valuable in sub-Saharan Africa, whereas saw and veneer logs are most important in South America and South-east Asia. As well as trade, tropical forestry employs a very large number of people, all of whom are totally dependent on active forest exploitation for their livelihood. In Malaysia, for example, over 50,000 people were employed in the wood-products industry in the early 1990s (Sharma et al., 1992).

There is also a suggestion that subsistence use, such as fuelwood, is of greater value than the more common perception of forest products such as saw and veneer logs and pulpwood. In fact, Sharma et al. (1992) predicted that by 2025, when the world population will be nearing 8500 m, the demand for fuelwood and house-building materials in the developing world will be in the region of 3660 million m^3 of roundwood, whereas industrial wood (much of which is for export) will account for a 'mere' 780 million m^3. There is an inescapable link between the number of people in the world and the fundamental conflict between producing more food and managing sustainable forests. The single, most important cause of rainforest clearance is still shifting agriculture carried out by indigenous people (Brady, 1996); Sanchez and Bandy (1992) suggest, for example, that 60–80% of rainforest deforestation is for (mainly temporary) agricultural settlement.

In general, then, it can be seen that in order to meet the myriad increasing demands of humans on forest ecosystems, while still maintaining a substantial forest cover on the earth's surface, we need to increase the productivity of our forests and the efficiency of their exploitation. These ideals may be approached by considering various different types of forest separately, since their values and ecologies will vary such that management considerations and goals will have to be different. This approach may be aided by the concept of compositional

Table 1.5. Value of forest products in various tropical regions of the world in 2008 (data from FAO, 2010).

	Import value (US$1000)	Export value (US$1000)	Net difference in timber products
Africa			
Eastern	806,907	263,830	−543,140
Middle	155,923	1,659,903	+1,503,980
Southern	999,121	1,660,944	+661,823
America			
Central	3,012,209	640,860	−2,371,349
South	6,685,338	15,058,577	+8,373,239
Asia			
South	5,254,212	534,765	−4,719,447
South-east	8,638,291	13,724,291	+5,086,000
Oceania	2,799,580	4,547,355	+1,747,775

change, which can also be thought of as intervention level, where the amount of intervention (or, if preferred, interference) in a forest is used to describe its ecology and potential productivity. Figure 1.2 presents the ecological changes which occur as natural (old growth) forests become modified all the way to industrial plantations (Putz and Redford, 2010).

Some examples of the major types of tropical forest are now presented. Further details and examples may be found in Chapter 6.

1.4.1 Natural forests – primary

Any schoolchild these days will tell you willingly that tropical forests (meaning, of course, rainforests) are the most diverse and productive systems on the planet. Coral reef biologists may argue at least the former point, but there is no doubt that in terms of biodiversity (see Chapter 2), moist tropical forests support extremely rich floras and faunas. Conservation value aside, however, natural forests in the tropics, both humid rainforests and dry deciduous ones, have vital economic values too and it is impossible to consider them in isolation. Idealism must give way to pragmatism in the real world. Furthermore, in reality, it is rather difficult to recognize a truly natural forest, if we imply an area of woodland which has *never* been influenced by humans, even in historical times. It is more convenient to consider a natural forest as one which has had no *significant* anthropocentric influence for many years, perhaps centuries, and is as close to a state produced by millions of years of evolution as is possible to be. This type of forest we would consider as 'primary'; in simple terms, it has not been logged or cleared during historical time, as far as we know (Plates 1 and 2). The disappearance of primary rainforests across the world has been very well documented and is undoubtedly due to human exploitation of one sort or another, with conversion to some form of agriculture probably the most significant. Take, for example, Papua New Guinea. As Fig. 1.3 shows, there is a distinct (asymptotic) relationship between the population density in different provinces and the amount of primary forest clearance between 1972 and 2002 (Shearman *et al.*, 2009). Notice that these relationships do vary according to locality, so that lowland areas seem to have been affected more intensely by people pressure than those in the highlands.

It seems reasonable to retain the description of a tract of forest as primary if it is exploited carefully and sustainably for various valuable products. Some are likely to have more impact on the forest ecosystems than others by their harvesting; even with the most careful low-impact logging technology, it is very difficult to avoid considerable physical, and hence ecological, damage to primary forest. In addition, it will be

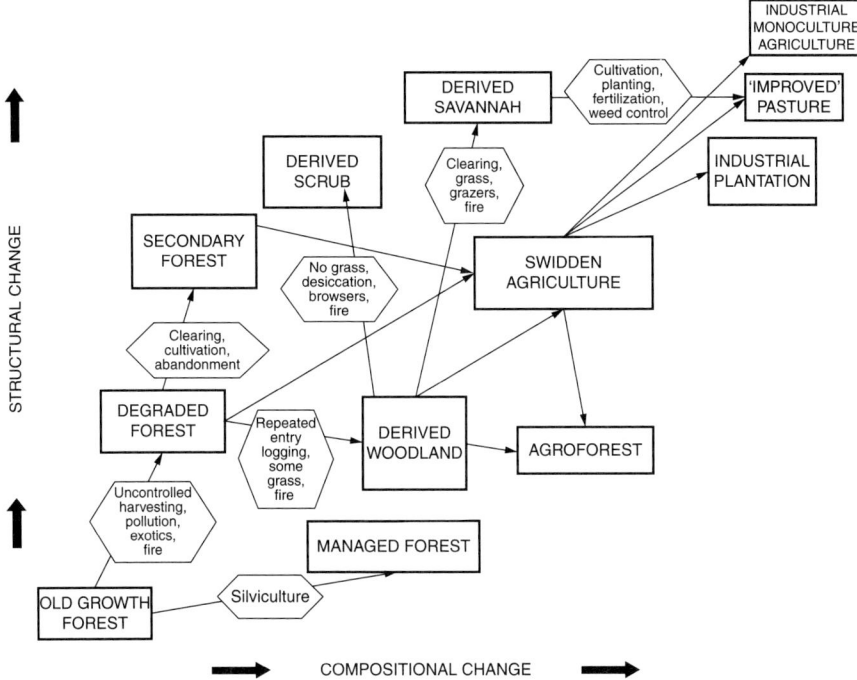

Fig. 1.2. Ecological state changes from tropical old growth forests. Black squares = forest states; hexagons = principal drivers (from Putz and Redford, 2010, courtesy Wiley-Blackwell Publishers).

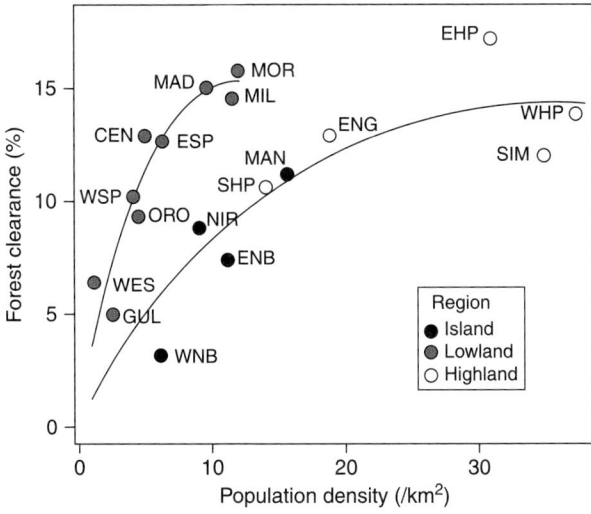

Fig. 1.3. Percentage of primary rainforest cleared due to subsistence agriculture between 1972 and 2002 according to population density in each province in 2002. Each provincial data point is labelled according to its geographic region – Islands, Highlands or Lowland coastal. Provincial abbreviations: CEN = Central; EHY = Eastern Highlands; ENB = East New Britain; ENG = Enga; ESK = East Sepik; GUL = Gulf; MAD = Madang; MAN = Manus; MIL = Milne Bay; MOR = Morobe; NIR = New Ireland; ORO = Oro; SHY = Southern Highlands; SIM = Chimbu; WES = Western; WHY = Western Highlands; WNB = West New Britain; WSK = West Sepik (from Shearman et al., 2009, courtesy Wiley-Blackwell Publishers).

several decades at least before the same operation can be performed in that area, so that this type of exploitation, lucrative though it might be in the short term, must be considered to have a limited future, assuming that alternatives such as plantation forestry are available. Other products are much easier to harvest sustainably; rattans and bamboos can be removed from natural forests without too much damage to the surrounding ecosystem. In Sumatra, for example, species of small rattan can be harvested from the natural forests on a 4-year, sustainable-yield cycle. Though this extra source of income will not remove the possibility of converting the forest into farmland, it will certainly reduce the problem (Belsky and Siebert, 1995). As long as maximum sustainable yields are first derived by sound research and then adhered to, plant products such as seeds and medicines, and animals from game to butterflies may be collected. The discovery of new medicines from tropical forests is quoted frequently as one good reason for the latter's conservation; somewhat 'throw-away' statements intensify the fervour of primary forest conservationists. In fact, Mendelsohn and Balick (1995) estimate that about one in eight plant species with pharmaceutical value have already been discovered, but a complete collection and screening of all possible plants could be worth well over US$150bn to society as a whole. Whatever the value of these naturally derived resources, it is crucial that pharmaceutical companies return some of the benefits, at least to local governments and especially native populations who provided the original research material. Shidamallayya et al. (2010), for example, point out that the development of folklore and the collection of natural medicines from tropical forests is vitally important to promote the livelihoods of forest dwellers living in poverty. In contrast, many people who are fortunate enough to live in developed countries may argue that primary forests must be left untouched at all costs. Far too many have now been reduced to the status of secondary forests already, and it is to the latter group of natural tropical forests that we should turn for exploitation.

1.4.2 Natural forests – secondary

It has often been advocated that the loss of primary tropical forests is irreversible. This would seem to be a particularly ill-informed and short-sighted approach to tropical ecology in general and the value of forest exploitation in particular. A secondary forest can be defined simply as that formed as a consequence of human intervention on forestlands (see Hashim and Hughes, 2010). However, as suggested in the previous section, the extent of the intervention is vital. Heavy commercial selective logging, clear felling or shifting cultivation (slash-and-burn agriculture) may all be precursors to natural regeneration of local flora (and subsequent recolonization of fauna). The bad news seems to be that vast areas of forest have been damaged and degraded by these sorts of changing land use (Bradley and Millington, 2008) but, looking on the bright side, regeneration, if left to its own devices, will proceed as ecological succession. In the fullness of time, between 60 and 80 years depending on circumstances, these secondary forests may reach a stage in development where they become difficult to distinguish from their primary, untouched counterparts, though subtle elements such as highly specialized plants and animals may take much longer to reappear, if they ever do.

More than 600 Mha of secondary forests (Plate 3) were in existence in the tropics in 1990 (Brown and Lugo, 1990), comprising over 30% of forestland. These widely variable areas are viewed by some as being of debatable value for production and to have a very low ecological esteem among purists; biodiversity is often reported as being significantly reduced compared with primary systems (Silver et al., 1996). On the contrary, however, even a casual glance may indicate that these areas are very floristically and faunistically species rich (Chey et al., 1997) and they have some distinct advantages over primary forests, as well as being perfectly adequate at providing most of the goods and services summarized above. These advantages may include higher rates of productivity of a few valuable tree species, which may regenerate naturally from seed banks in the soil, or more likely from seedling banks derived from regeneration present before felling. However, one of the problems with secondary forests is that the valuable tree species have often been removed by logging, especially if forest practices have been poorly controlled (Savill, personal communication). Line or enrichment plantings of native tree species established by reforestation programmes may, in fact, be the best way to achieve regeneration (Paquette et al., 2009). In general, secondary forests have higher biomass, they may serve as foster ecosystems for late successional stage species, they are often easier to access and manage because of their locality and pre-established roadways and it has even been suggested that they are able to reduce pest populations. In fact, secondary forests might be considered as rather ephemeral. Left alone, they will normally regenerate as mentioned above, and indeed, a short-rotation exotic tree species on the site may aid in this by providing shade and reducing the risk of fire and grass competition (Kuusipalo et al., 1995).

Primary forests over a large area have a level of production effectively equal to zero, since growth roughly equates to death and decay every year. Disturbed secondary forests do, however, have a net increase in production due to removals and subsequent regrowth, but they are still not the real answer to providing all the tropical forest products required in the 21st century. Plantation forests are where major efforts have been and will be directed.

1.4.3 Plantation forests

Plantations have been defined loosely as an arrangement of trees which are generated artificially and are the result of conscious management (Sedjo, 1987). In size, they may range from under 1 ha, as with a farm windbreak or a small stand in a village, to thousands of hectares (Sawyer, 1993). Such forests are usually of regular shape and have clearly defined boundaries, but perhaps their most distinctive and common characteristic is that plantation blocks within them consist in the main of one species of tree only, all individuals of which are the same age. In addition, more often than not, the species of tree is not native to the country wherein the plantation has been established. It is, instead, an exotic and hence the species richness and numbers of plants and animals indigenous to the area prior to planting are reduced (Armstrong et al., 1996).

Table 1.6 partitions the tropics into three major regions – Africa, Asia and Pacific, and Latin America plus the Caribbean – and shows the total land area and the area of forest plantations in each in the mid-2000s (ITTO, 2009); the quantities of forest plantations are really very small, especially in Africa and the Americas. Figure 1.4 breaks the world down into more detail, with India clearly leading the field (Brown, 2001). On the whole, it is significant that the rates of planting of new forests fall well short of the rates of disappearance of natural ones, though data suggest that the Asia/Pacific region is striving to balance natural forest clearance with plantations. Curiously, it is difficult to obtain accurate estimates of the extent of tropical forest plantations (Plates 4, 5 and 6). It is hard to delimit the tropics themselves, of course, and the term 'plantation' covers a multitude of scenarios, but figures suggest that the global areas of tropical

Table 1.6. Forest plantation areas in the tropics (from ITTO, 2009).

Tropical region	Land area (thousand ha)	Plantation area (thousand ha)	Land area (%)
Asia/Pacific	1,104,000	54,073	4.9
Africa	1,652,000	4,620	0.28
Latin America and Caribbean	2,250,000	8,805	0.39
Total	*5,006,000*	*67,498*	*1.35*

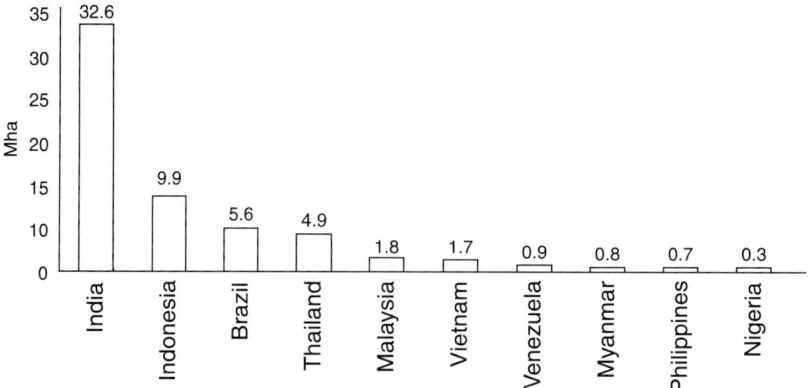

Fig. 1.4. Main tropical countries with forest plantations in 2005 (courtesy ITTO, 2009).

hardwood species occupy around 31 Mha (about 57% of plantation area), while tropical softwood species cover 24 Mha (the remaining 43%) (Brown, 2001).

Forest plantations become established in the tropics for various reasons, linked mainly to longer-term economic development (Farley, 2007). Farley provides one example of steps in the development of forest plantations in Ecuador from the 1880s until the 1990s (Table 1.5). Clearly, there is sound sense in establishing a plantation of trees rather than relying on nature to provide timber and other products. Figure 1.5 compares the productivities of various tropical forest systems and clearly shows that intensively managed tropical forests of exotic tree species provide much higher yields than do natural stands of indigenous species, even when managed. If we ignore the conservation ethic or the naturalness lobby and think purely in terms of timber production economics, it makes very good sense to grow tropical eucalypts or pines instead of natural forests. The timber produced by each system is, of course, not equally suitable for the same job; *Eucalyptus* is not a good veneer log, while tropical hardwoods should not be used as pulp raw material. However, if markets can be modified, a eucalypt plantation on a 10-year rotation could provide orders of magnitude greater yields than a natural forest. Brown (2001) discusses the major plantation tree species in the world, showing the domination of just a few species and genera, with *Pinus* being the most abundant, followed by *Eucalyptus*, *Picea* and *Abies*, *Acacia* and then teak (*Tectona grandis*). It must be noted that some tropical tree plantations might seem to the layman, or perhaps politician, to be perfectly good 'forests', so oil palm plantations, for instance, might seem to be green and luxuriant when viewed from the air. Koh *et al.* (2011) show the very significant ratios of oil palm to natural peatswamp forest in Malaysia (peatswamp forests are primary rainforest growing on peatland) (Fig. 1.6). In some parts of the region, such as Peninsula Malaysia, Sarawak and South Sumatra in particular, relatively little primary forest remains.

In any event, the future of tropical forestry must lie in some form of plantation system, but it is difficult to be sure that the areas being established are increasing. In Colour Plate 131, it can be seen that parts of India, Asia and Australia are showing some increases, whereas most of tropical Africa and South America are showing declines (UNEP, 2009). The specific type of plantation may vary with country or region, and 'plantation' can include artificial restoration of natural forests, large uniform stands of native or exotic trees, small woodlots and individual trees on a farm, or even urban landscape.

Table 1.7. Initiatives or phases in which the planting of pine was promoted in the Ecuadorian Andes (from Farley, 2007, courtesy of Wiley Blackwell Publishing).

Initiative/phase	Time period	Origin	Goals
First plantations	Late 1800s–early 1900s	National	Meet fuel and timber needs; erosion control
Forestry trials	1920s	National	Timber production
First MAG-sponsored programmes	1970s	National	(Not specified)
Convenios de Participatación	1980s	National	Increase renewable resources; production of goods and services
FONAFOR/Plan Bosque	1980s	National	Timber for export
IDB	1980s	International	Utilize low-productivity land; revitalize government forestry programmes
PLANFOR	1990s	National	Promote reforestation
PROFAFOR	1990s–	International	Carbon sequestration

Notes: MAG = Ministry of Agriculture; FONAFOR = Fondo Nacional para la Forestación; IDB = Inter-American Development Bank; PLANFOR = Plan Maestro de Forestación; PROFAFOR = Programa FACE (Forests Absorbing Carbon Dioxide) de Forestación.

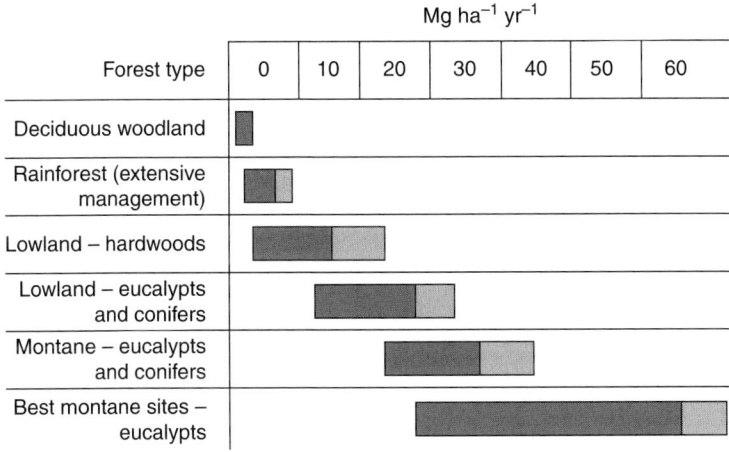

Fig. 1.5. Productivity of different types of tropical forest. Dark grey = normal range; light grey = maximum range (from Peter Savill, unpublished). Values in Mg ha^{-1} yr^{-1}.

1.4.4 Agroforestry and silvo-pastoralism

Agroforestry has been practised for centuries by local people, in somewhat informal ways. The Maya of Belize in Central America cultivated 'forest gardens' over 1000 years ago, which provided communities with a wide variety of products (Ross and Rangel, 2011). These days, agroforestry and its ally, silvo-pastoralism, are rapidly expanding strategies in the tropical world. Both systems use small-scale plantings of trees or shrubs in close association with agricultural crops or livestock. Many of the tree species used are fast-growing legumes, which not only shelter the food crops from climatic extremes but also provide fuelwood and fodder, while enriching the soil with nitrogen. Because of these various roles, they have been termed multi-purpose trees, or MPTs (Plates 7 and 8).

Organized research in agroforestry is only a few decades old (Nair, 1993) and, so far, the research and development that has taken place has attempted to screen the

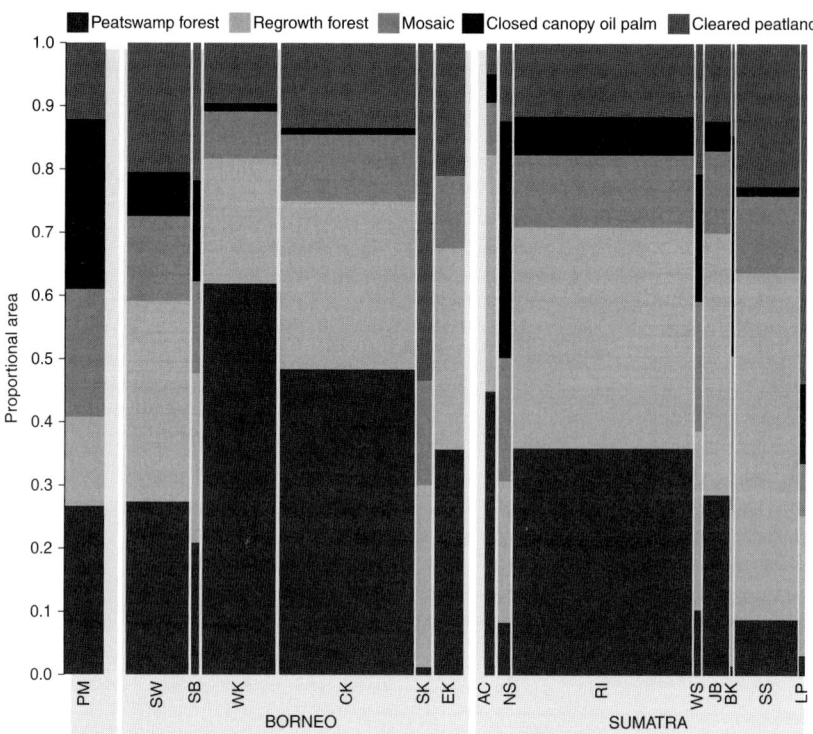

Fig. 1.6. Peatland composition in the lowlands of Peninsular Malaysia, Borneo, and Sumatra. Column widths reflect relative areas of peatland among subregions. PM = Peninsular Malaysia; SW = Sarawak; SB = Sabah; WK = West Kalimantan; CK = Central Kalimantan; SK = South Kalimantan; EK = East Kalimantan; AC = Aceh; NS = North Sumatra; RI = Riau; WS = West Sumatra; JB = Jambi; BK = Bengkulu; SS = South Sumatra; LP = Lampung (from Koh et al., 2011, courtesy National Academy of Sciences).

most suitable MPT species and to determine the best ways of deploying them. The trees may be planted in rows or alleys, separated by food crops as diverse as maize, cowpea or cocoa (Evans and Turnbull, 2004), or in small blocks adjacent to fields in a village system. The next vital step in the large-scale adoption of these systems is the education and training of local farmers and villagers (Nair, 1993).

Taungya was a system of reforestation that arose in the main in response to the need for tree plantations producing high-value timber in areas where local social conditions (such as the simple need to grow food) precluded the establishment of large-scale industrial plantations (Moses, 2009). It is thought to have originated in Burma (now Myanmar) in the mid-1800s and involves the planting by subsistence farmers of trees in mixtures with a whole host of agricultural crops including rice, cassava, maize, peppers, bananas and groundnuts. One of the incentives to establish taungya is the attempt to reforest lands denuded of trees by shifting cultivation and to provide, admittedly on a fairly long timescale, high-quality timber from individual trees (rather than whole plantations). Many commercial tree species seem suitable for taungya, but most establishments in Thailand and Indonesia, for instance, have involved teak. Taungya is similar to a system known as 'shamba' in Kenya, but it is distinguished from agroforestry in that in the latter system emphasis is placed on the food crops rather than the trees. The latter tend to be planted around the boundaries of the farm and may be used for fuel or fodder. The major ecological or entomological interest involves the relationships between the trees and the agricultural crops and the likely balance between annual (or seasonal) and perennial food sources for potential pests and diseases.

1.5 Forest Productivity and Carbon Storage

It could be argued that any sort of forest is better than no forest at all (unless you are a farmer, of course). Compared with grasslands, for example, even plantation forests are more productive, support more biodiversity, reduce soil erosion, promote natural regeneration and stabilize the microclimate (Lee et al., 2006). Figure 1.7 depicts the above-ground storage of various types of land use in the humid tropics (Paquette et al., 2009), showing that secondary forests, teak plantations and even agroforestry are capable of sequestering substantial amounts of carbon, at least until they are processed. Note that land-use changes in the tropics which involve the destruction of rainforests are now thought to cause rapid reductions in carbon biomass, resulting in increased levels of carbon in the atmosphere and decreased levels in soils (Don et al., 2011).

1.6 Forest Food Webs – Trophic Interactions

So far, we have only really considered the tree component of tropical forests, since this is the *raison d'etre* of most forest management. We have seen that tropical forests are extremely varied, complex and widespread ecosystems in which animals and plants interact. Such interactions, or food webs, are perfectly natural components of ecosystem function and it is impossible to envisage a forest, whether natural or artificial, that does not contain them. A general tropical food web is illustrated in Fig. 1.8 (Pomeroy and Service, 1986); it is important to note that the abundance of each group in the web is not depicted in the diagram, nor is the likely influence on the 'green plant' trophic level (which, in the context of this book, consists mainly of the forest trees). For example, in a tropical forest there undoubtedly will be vastly more species and numbers of individuals of arboreal invertebrates such as insects than other herbivores such as fruit-feeding birds; both their biodiversity interest and economic importance will also be distinctly different. As far as pests are concerned, the fact that a tree has some of its leaves removed by a moth larva or another has tunnels made in its stem by a beetle is a normal and indeed fundamental component of energy flow and nutrient cycling. Notice also the fundamental importance of decomposers in the food web.

As one example, the almost overwhelming complexity of trophic interactions in a Puerto Rican tropical rainforest is discussed in great detail by Reagan and Waide (1996). As these authors point out, this complexity

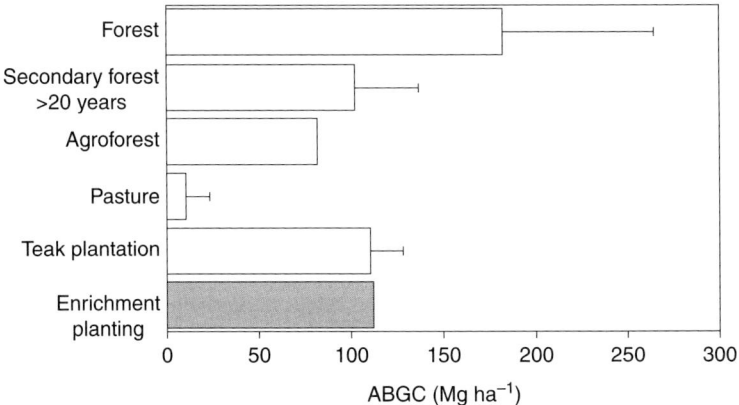

Fig. 1.7. Above-ground (ABG) carbon storage of various types of land in the humid tropics. Carbon storage is based on average rotation periods for each type of land use (i.e. the time over which carbon is stored) (+ error bars). Enrichment planting value estimated from averaged teak monoculture plantation and secondary forest > 40 years old (from Paquette et al., 2009, courtesy Resilience Alliance).

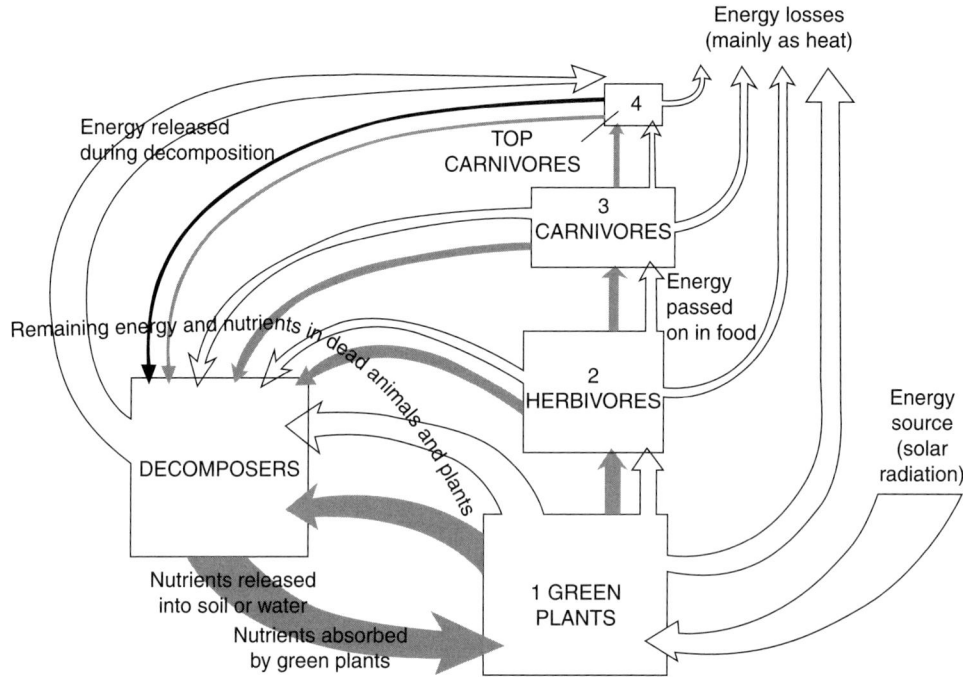

Fig. 1.8. Diagrammatic representation of a general tropical food web. Boxes represent organisms of a particular trophic level (forest insects occur mainly in the herbivore trophic level, though many borers and termites may be included among the decomposers). White arrows indicate the one-way flow of energy; dark arrows indicate the cycling of nutrients; wider arrows indicate greater flows (from Pomeroy and Service, 1986).

is partially a function of the species richness in the forest community, in turn related to the heterogeneity of the forest habitat (see Chapter 2 for details). High species richness may increase ecosystem resilience in a tropical forest following disturbances by increasing the number of alternative pathways for the flow of resources, and indeed it has been suggested that tropical plantations which consist of more than one tree species show higher productivity (Erskine et al., 2006). Generalist species in particular may be able to shift their roles in ecosystem food webs by changing trophic levels according to the changing levels of available resources (Shaner and Macko, 2011). Thus, it is to be expected that when simple plantations consisting of nothing else but one even-aged tree species replace highly diverse natural forests in the tropics, the trophic interactions in these ecosystems must alter as well, with the potential for instability. One result of this instability can be pest outbreaks.

1.7 Forest Pests

Just as humans use taxonomy to classify and 'pigeonhole' animal and plant species, so we do it to ourselves as pest managers. Either through necessity or convenience, forest ecologists end up with labels such as 'entomologist', 'pathologist', 'nematologist' or 'soil scientist'. It is, in fact, an unfortunate consequence of specialization that we persist in considering insects separately from other organisms that play vital roles in tropical forest ecosystems, from both a beneficial or neutral point of view but also as 'pests'. So, in reality, insects and fungi, for example, interact at many complex levels with each other, with the trees and with the general abiotic and biotic conditions around them. A proper understanding of how such interactions bring about tree decline and death is a vital prerequisite for curing the problem, assuming this is possible. More importantly, this knowledge is crucial in preventing these problems in the future. One example will help to illustrate these complexities.

In Queensland, Australia, the decline and death of native trees, in particular *Eucalyptus* spp., has been increasing dramatically since the mid-1960s (Wylie *et al.*, 1993a). This so-called dieback was found to be due to a multiplicity of factors, both natural (e.g. insects, fungi and water relations), and management-related (e.g. clearing of trees, use of fertilizers and improvements to pasture). Figure 1.9 summarizes some of the links in this complex web of interactions (Wylie *et al.*, 1993a). In the case of *Casuarina cunninghamiana*, a species which is common along rivers and streams, the major factor found to be contributing to dieback was severe and repeated defoliation by a leaf-eating insect, *Rhyparida limbatipennis* (Coleoptera: Chrysomelidae) (Plate 9). However, the severity of this insect-related dieback was linked to the extent of tree clearing and the levels of streamwater salinity (Wylie *et al.*, 1993b). The latter factors relate directly or indirectly to tree stress (reduction in vigour), a fundamentally important syndrome in the promotion of susceptibility to insect attack.

In many parts of the tropical world, self-help clinics are being established to which crop producers can turn for help in

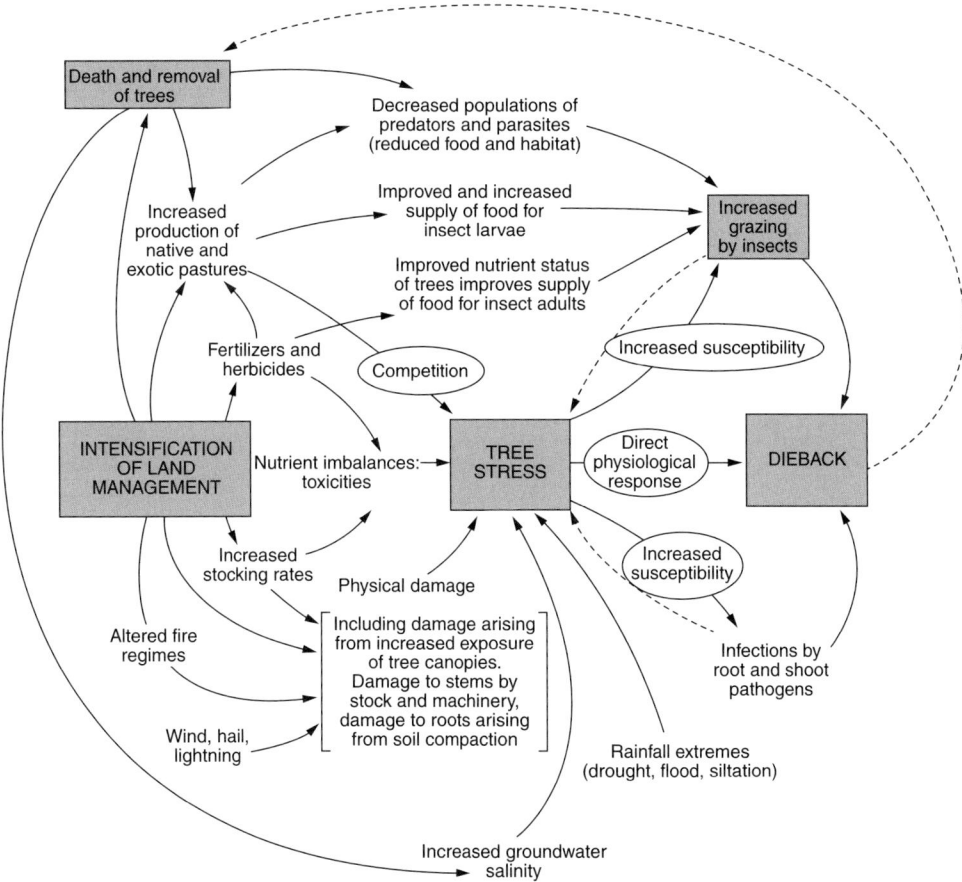

Fig. 1.9. A model of initiation and development of rural trees in southern Queensland (from Wylie *et al.*, 1993a).

diagnosing and remedying (where possible) plant maladies. These plant pest diagnostic services (PPDS) consider all manner of problems (soil type, hydrology, insects, pathogens, etc.) on all manner of crops (bananas and beetroots, peppers and pines) (Smith et al., 2008). Despite this, the current book persists in considering insects as one form of damaging agent to trees but the reader is encouraged to view insect problems as examples of wider pest–tree interactions and to use our examples as concepts that may be equally relevant to fungi, weeds, or even elephants.

1.8 Impact of Forest Insects in the Tropics

Insects have been feeding on plants for over 300m years (Wilf, 2008) and leaving the evidence of their damage in the leaves, shoots and stems of fossilized trees and other plants. The magnitude of this feeding can vary enormously, from a few small holes or patternings to full-scale defoliation and plant death. One simple example will illustrate the magnitude of damage done by insect herbivory; Massad et al. (2011) studied the growth of seedling trees over 4 years in reforestation experiments in Brazil and found that across various treatments, trees achieved a mean height of 240.8 cm, while with 50% herbivory the mean height was reduced to only 101.5 cm. Throughout the book, we shall stress the fundamental importance of 'impact'. It is relatively easy to sally forth into a tropical forest, whether it be a nursery or mature stand, armed with net, knife and specimen tube and return with a large number of creatures found chewing, sucking, biting, boring or merely 'passing through' (the so-called 'tourists'). Indeed, the vast majority of insects encountered in tropical forests are entirely benign. Even termites (Isoptera), which might be thought of as notorious pests of living and dead timber, are, in the main, extremely beneficial to tropical forest systems, being vital components in the breakdown of dead timber and the recycling of scarce nutrients through detritivore food chains. Only a very few species cause economic losses to crops, including trees, and then not all the time. If damage is done, it may take various forms. For example, Zvereva et al. (2010) analysed a large number of published papers to produce syntheses of the types of impacts attributable to sap-feeding insects on woody plants. Growth of various parts of the plants was reduced significantly, as was reproduction and photosynthesis. Clearly, some of these factors are correlated; reduced photosynthesis would be expected to result in reduced tree growth, of both shoots and roots and of the amount of seed produced. All ways round, the forester, as well as the tree, is likely to suffer.

It must be remembered that trees have evolved with insects throughout their evolutionary time and therefore they must be expected to have developed some tolerance and/or defence systems (see Chapter 4). So, even if we find insects causing defoliation of trees (Plate 10) or producing holes in timber, it is still not a reason to swing into action with an eradication programme. Until the value of that damage can be ascertained and then related to the numbers of insects and the cost of control tactics for them, positive action should be avoided if possible. Obtaining sound, dependable data on impacts in tropical forestry is notoriously difficult, though the literature seems full of such information. Table 1.8 provides some examples from which at least two lessons should be learned. First, hardly any scenarios convert the numbers of insects in a pest outbreak, nor the amount of damage done, to financial terms. In an agroforestry situation, of course, where financial profit is not the aim of the project, then the value of the damage is not so important since there is no way of paying for the 'fix'. In big commercial plantations, however, where profit is tied up intimately with repaying interest on loans, increasing gross national products or providing purchasing power, then we need to know how much money we are likely to lose. If the losses exceed the projected cost of control (beyond the economic threshold level), then management of the pest will be cost-effective. The second lesson from Table 1.8 concerns the 'what happens next' situation. Heavy defoliation on a tree may not, in fact, persist for long; density-dependent

Table 1.8. Examples of insect damage to tropical trees.

Insect species	Common name	Insect group	Host tree	Country	Attack mode	Impact	Reference
Aristobia horridula	Longhorn beetle	Coleoptera: Cerambycidae	*Dalbergia sissoo*	Nepal	Stem girdling/boring	43–97% mortality of young trees	Dhakal et al., 2005
Cephisus siccicfolious	Spittle bug	Hemiptera: Cicadellidae	*Eucalyptus urophylla*	Brazil	Sap feeder on leaves and shoots	Up to 99% shoot deformation	Ribeiro et al., 2005
Coccotrypes rhizophorae	Mangrove bark beetle	Coleoptera: Scolytidae	*Rhizophora mangle*	Panama	Bark boring	Up to 72% tree mortality	Sousa et al., 2003
Condylorrhiza vestigialis	Poplar caterpillar	Lepidoptera: Crambidae	*Populus* spp.	South Brazil	Defoliator	50–100% defoliation on 2-year-old trees	Castro et al., 2009
Coptotermes testaceus	Subterranean termite	Isoptera: Termitidae	*Eucalyptus* spp.	Brazil	Root feeder in nursery	10–70% seedlings killed	Wilcken et al., 2002
Dendrolimus spp.	Pine caterpillar	Lepidoptera: Lasiocampidae	*Pinus massoniana*	South China	Defoliation	30–80% defoliation	Wang et al., 2007
Dendroctonus frontalis	Southern pine beetle	Coleoptera: Scolytidae	*Pinus caribaea*	Belize	Bark boring	Up to 90% tree mortality	Snyder et al., 2007
Diorcytria spp.	Pine shoot borer	Lepidoptera: Pyralidae	*Pinus oocarpa*	Indonesia	Shoot/bark boring	70% mortality of young trees	Intari and Ruswandy, 1986
Doratifera stenosa	Nettle caterpillar	Lepidoptera: Limacodidae	*Rhizophora stylosa*	Australia	Defoliation	30–40% leaf loss annually	Duke, 2002
Heteropsylla cubana	Leucaena psyllid	Hemiptera: Psyllidae	*Leucaena leucocephala*	Australia	Sap feeder	Dry matter production reduced by 55% of norm	Palmer et al., 1989
Heteropsylla cubana	Leucaena psyllid	Hemiptera: Psyllidae	*Leucaena leucocephala*	Australia	Sap feeder	Reduction in fodder by 52%; reduction in timber by 79%	Bray and Woodroffe, 1991
Heteropsylla cubana	Leucaena psyllid	Hemiptera: Psyllidae	*Leucaena pallida*	Australia	Sap feeder	Up to 70% yield loss for susceptible clones	Mullen and Shelton, 2003
Hoplocerambyx spinicornis	Sal heartwood borer	Coleoptera: Cerambycidae	*Shorea robusta*	India	Heartwood boring	6–24% damage	Pant et al., 2002

Continued

Table 1.8. Continued.

Insect species	Common name	Insect group	Host tree	Country	Attack mode	Impact	Reference
Hoplocerambyx spinicornis	Sal heartwood borer	Coleoptera: Cerambycidae	*Shorea robusta*	India	Heartwood boring	2.6 million trees killed over 6 years	Joshi *et al.*, 2006
Hyblaea puera	Teak defoliator moth	Lepidoptera: Hyblaeidae	*Tectona grandis*	South-west India	Defoliation	44% loss of volume increment; 50% loss of height increment	Nair *et al.*, 1985
Hyblaea puera	Teak defoliator moth	Lepidoptera: Hyblaeidae	*Tectona grandis*	Andaman Islands	Defoliation	Up to 100% leaves eaten	Veenakumari and Mohanraj, 1996
Hypsipyla grandella	Mahogany shoot borer	Lepidoptera: Pyralidae	*Swietenia humilis*	Honduras	Shoot boring	44% trees attacked	Goulet *et al.*, 2005
Hypsipyla grandella	Mahogany shoot borer	Lepidoptera: Pyralidae	*Cedrela odorata*	Costa Rica	Shoot boring	77% trees attacked after 84 weeks	Newton *et al.*, 1998
Hypsipyla robusta	Mahogany shoot borer	Lepidoptera: Pyralidae	*Khaya anthotheca*	Ghana	Shoot boring	Up to 88% of trees attacked after 2 years	Ofori *et al.*, 2007
Hypsipyla robusta	Mahogany shoot borer	Lepidoptera: Pyralidae	*Toona ciliata*	Thailand	Shoot boring	20–60% trees attacked depending on height class	Cunningham and Floyd, 2006
Leptocybe invasa	Gall wasp	Hymenoptera: Eulophidae	*Eucalyptus grandis*	Vietnam	Gall former on shoots	65–92% damage	Thu *et al.*, 2009
Macrotermes/ Microtermes/ Odontotermes spp.	Subterranean termites	Isoptera: Termitidae	*Eucalyptus* spp.	Ethiopia/ Africa/ India	Stem and root severance	Up to 100% of transplants killed	Cowie *et al.*, 1989
Nephopterix syntaractis	Mangrove caterpillar	Lepidoptera: Pyralidae	*Avicennia marina*	Hong Kong	Defoliation	Over 75% defoliation	Anderson and Lee, 1995
Oiketicus kirbyi	Bagworm	Lepidoptera: Psychidae	*Rhizophora mangle*	Ecuador	Defoliation	80% of foliage reomoved over 1200 ha	Gara *et al.*, 1990
Phytolyma lata	Gall psyllid	Hemiptera: Psyllidae	*Milicia excelsa*	Ghana	Gall forming	Over 40% mortality according to provenance	Ofori and Cobbinah, 2007
Platypus spp.	Ambrosia beetle	Coleoptera: Platypodidae	*Acacia crassicarpa*	Sabah	Stem boring with associated black stain	Up to 50 holes per tree	Thapa *et al.*, 1992

Pteroma plagiophelps	Bagworm	Lepidoptera: Psychidae	Falcataria molyccana	South-west India	Defoliation	Death of 25% of trees and 17% severely damaged after 2.5 years repeat defoliation	Nair and Mathew, 1992
Pulvinaria urbicola	Softscale insect	Hemiptera: Coccidae	Pisonia grandis	Coral Sea	Sap feeding on leaves and stems	16ha trees destroyed	Greenslade, 2008
Schizonycha ruficollis	White grub	Coleoptera: Scarabaeidae	Tectona grandis	India	Root feeding	14–42% nursery seedlings destroyed	Kulkarni et al., 2007
Xyleutes ceramica	Bee-hole borer	Lepidoptera: Cossidae	Tectona grandis	Sabah	Stem boring	10–65% damage	Gotoh et al., 2003
			Gmelina arborea	Sabah	Stem boring	7–12% damage	
Xylosandrus compactus	Bark beetle	Coleoptera: Scolytidae	Khaya spp.	India	Bark boring in seedlings and young saplings	60–70% infestation	Meshram et al., 1993
Several unnamed			Albizia lebbek	India	Seed predation	50–70% of seeds damaged in pods	Harsh and Joshi, 1993
Various species	White grubs	Coleoptera: Scarabaeidae	Acacia mearnsii	South Africa	Root feeding in nursery	2–30% mortality of seedlings	Govender, 2007

Table 1.9. Insect pests recorded from intensively managed teak plantations in various parts of India (modified from Varma et al., 2007).

Pest species (order: family)	Nature of damage	Pest status
Hyblaea puera (Lepidoptera: Hyblaeidae)	Defoliation	Major; already known
Eutectona machaeralis (Lepidoptera: Pyralidae)	Leaf feeding	Major; already known
Sahyadrassus malabaricus (Lepidoptera: Hepialidae)	Stem boring	Major; already known
Mealy bugs (Hemiptera/Homoptera)	Sap feeding on leaves	Minor
Planococcus sp. (Homoptera: Pseudococcidae)	Sap feeding on leaves	Minor
Zeuzera coffeae (Lepidoptera: Cossidae)	Stem boring	Minor, but emerging problem
Helicoverpa armigera (Lepidoptera: Noctuidae)	Leaf feeding/terminal shoot damage	First record; major
Dihammus sp. (Coleoptera: Cerambycidae)	Stem boring	Minor, but emerging problem
Aleurodicus sp. (Homoptera: Aleurodidae)	Sap feeding on leaves	Minor

systems such as food shortage or parasitoid and disease intervention may result quickly in pest declines with no long-term harm done to the crop. Finally, we must not forget that in many cases, several pests and diseases, may assail a tree or plantation at once, adding to the producer's woes. One example is from India, shown in Table 1.9, where Varma et al. (2007) describe various insect pests from teak, *Tectona grandis*, at various locations. The additive impact of these pests is likely to be significantly more severe than each one on its own.

1.9 Prevention Rather than Cure

Not only should we consider insects in the same vein as other potential forest pests such as fungi but also we should consider potential pest problems in the same vein as all other relevant components of a forest ecosystem from which we are trying to derive some sort of product. Figure 6.1 from Savill et al. (1997) shows how a general framework of forest management should include a detailed consideration of pests and diseases at the same stage and with the same degree of seriousness as forest managers and economists consider tree provenance, soil type, seed source and market forces. We will show throughout this book how difficult it can be to solve a pest problem in forestry once it has occurred, especially in large-scale plantations where finances and technology are strictly limited. Far better to manage tropical forests in such a way as to reduce the likelihood of pest damage while maintaining the production aims of the project. Integrating all relevant aspects of pest management with acceptable silvicultural practices should be the prime directive of entomologists and pathologists. Just how far we are from the ideals of seamless integration of tropical forest pest management with commercial forestry is depicted in Table 1.10, where the requirements for pest limitation are laid out under three related topics or perceptions (Speight et al., 2008). These are: (i) the need for basic research into carefully selected and prioritized problems; (ii) the requirement for financial and moral support not only to carry out the research but also to disseminate the results in a fashion which is of significant and practical use to entomologists and foresters on the international scene; and (iii) the acceptance by the industry that such research and the lessons learned from it are of crucial value in all types of tropical forestry, from the simplest tree-growing forest in a village to the largest industrial plantation.

Table 1.10. Components required for efficient tropical forest pest management (modified from Speight et al., 2008).

Research	Support	Industry
Basic ecology, taxonomy and impact of insects (and pathogens)	Databases of pest biology and impact; literature retrieval	Recognition of the equal importance of entomology and pathology in tropical forestry, relative to economics and silviculture
Provenance trials and resistance selection of trees, promotion of indigenous species in high-yield silviculture	Extension and advisory services readily available to commercial and subsistence growers, using well-trained local expertise	Recognition of the importance of prevention rather than cure in pest management, with a willingness to alter forest practices accordingly
Economically viable, appropriate technology systems for pest management, with easy collaboration between research workers internationally	Incorporation of entomology and pathology in international aid and development schemes, enhancing funding for R & D and the provision of support systems	Consultations with entomologists and pathologists at the planning stages of new and/or expanding forest projects

1.10 The Plan of the Book

We hope that this introductory chapter has provided a background to tropical forestry and illustrated the enormous variety of form and function. There are clearly fundamental problems with the ways in which tropical forestry is carried out and the resulting ecological interactions with insects and other potential pests which occur within the forests. The manipulation of tropical forests for efficient production of timber and many other products is bound to continue apace, alongside which it will be crucial to consider forest pests, and insects in particular.

The rest of the book will present details of interactions between insects and forests, both natural and managed, which will be used to plan and implement the best strategies for forestry and pest management combined. Discussions will follow on the types of pests encountered at each stage of forest operations, from pre-planting to post-harvest. At every stage, we want to emphasize that the ultimate goal is integrated pest management, or IPM.

2
Tropical Forests and Insect Biodiversity

2.1 Introduction

It may not be immediately apparent why a book about tropical insect pests should contain a discussion about ecology and biodiversity. After all, it might be strenuously argued that biodiversity and its conservation is exactly the opposite to pest management in the same habitat; in one situation we try to preserve insects and in the other we strive to wipe them out! As discussed in Chapter 1, a great deal of forestry in the tropics is concerned with the establishment and management of monocultures of exotic species of tree, which by their very nature would be assumed to promote very low biodiversity and hence be of no interest to conservationists. The justification for including these topics rests with two related premises.

Firstly, we must emphasize that the basis for modern pest management lies with prevention of outbreaks whenever and however possible. In order to meet this aim, we need to achieve some form of ecological stability in our crops, such that eruptions in insect numbers are rare and if they occur, are soon returned to a normal and tolerable level. There is much debate about the links between diversity and stability (see below), but it must be fundamental to our ability to explain pest problems (and hence avoid them, if possible) that we have some understanding of how our insects interact with the forests in which they live and how the ecosystems which we have created for our own benefits differ from the co-evolved norm. Secondly, we have to accept that tropical forestry has forced upon it, like it or not, a responsibility to reduce as much as is reasonable the damage done to the world's biota. Forest conservation has to be considered alongside forest exploitation wherever possible; the same tract of land may be important for timber, watershed protection and the preservation of rare animals or plants, all at the same time. As Koh and Sodhi (2010) state, with particular reference to South-east Asia, 'all major stakeholders (must) work together to achieve the ultimate goal of reconciling biodiversity conservation and human well-being'. In Sabah (north-east Borneo), for instance, licensed logging of primary and secondary rainforest still continues for economic gain. Some of these sites are replanted with fast-growing exotic tree species such as *Acacia* spp. and *Eucalyptus* spp., which provide essential short-term lumber, while others are enriched by new plantings of native species such as various dipterocarp species. It is hoped the latter eventually will regenerate rainforest. On top of this exploitation we have the very reasonable desire to conserve the myriad of plants and animals found in Borneo rainforests, coupled with the increasing commercial prospects of ecotourism wherein people will pay considerable sums to observe these curiosities. Clearly, not all of these aims are compatible and priorities for action or, if there is space, zones of activity must be established. Here, it is important that we possess an understanding of the ways in which intensive forestry, for example, influences biodiversity and conservation. This is not just so that, as applied ecologists, we can predict the changes caused by our actions but, very importantly, so that we can answer the often ill-informed but very vociferous and influential critics of tropical forest exploitation. Let us not forget that in the majority of tropical countries, forestry (including agroforestry) is carried out for some form of economic gain (see Chapter 1), either as a direct generator of revenue or as a provider of raw materials for immediate utilization. Like it or not, biological diversity in these situations must promote its own cause via its contribution to human welfare (Myers, 1996), not because of a 'feel-good factor' for remote and idealistic first-worlders but pragmatically for local politicians and members of the public directly influenced in economic terms.

2.2 Biodiversity

Biodiversity (short for biological diversity) encompasses all types of natural variation from molecular and genetic levels to species and even subspecies (Hambler, 2004). The species component includes all plants, animals and microorganisms but within this component we can envisage various subdivisions such as populations, communities and food webs. These days we recognize three types of diversity, alpha, beta and gamma diversity, which take into account various aspects of species occurrence within and between communities. Alpha diversity refers to species within a habitat (such as a forest or a pond), beta diversity to the differences between habitats and gamma diversity to species variations on a geographical scale. Loosely put, alpha and beta diversity can be either added or multiplied together to produce gamma (Jost, 2007).

There has been an unfortunate tendency to use biodiversity merely as shorthand for the number of species of organism in a habitat, which has led to the proliferation of lists of species in museums collected from one tropical forest or another. The term 'diversity' can mean a variety of things and indeed species richness (simply the number of species) is used frequently to describe 'diversity' (Kessler et al., 2011). However, Fig. 2.1 illustrates two theoretical habitats, each of which contains the same five species and a total number of individuals summing to 250. The white community has certain species which are very common and others which are very rare, while in the grey community each species is equally common. Species diversity indices are unitless numbers resulting from mathematical formulae which combine the number of species with the abundances of each. There are various species diversity indices in use, such as the Shannon–Wiener, Simpson's, Margalef's and the Williams α index. In all cases, the species diversity of different animal or plant communities can only be compared properly if they are based on the same sample sizes. This can be in terms of area of habitat explored, time devoted to sampling, or the numbers of individuals included in the samples. Clearly, no species diversity index is perfect and we have to compromise between ecological (or statistical) rigour and ease of data collection coupled with widespread comparability. In Fig. 2.1, the

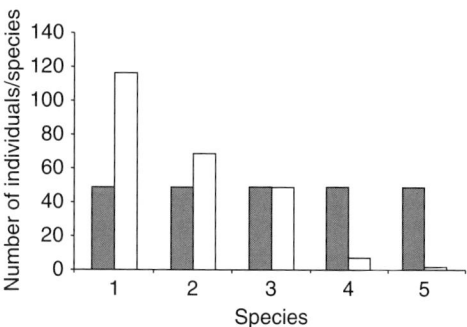

Fig. 2.1. Theoretical species composition ('alpha diversity') of two similar habitats. Shannon–Weiner diversity indices: white community H' = 1.17; black community H' = 1.61.

Shannon–Wiener diversity index, H, is higher for the community with a more equal distribution of species abundance. Thus, species diversity in its strict sense has a considerable amount to say, not only about the numbers of species in a certain place but also about their relative abundances, rarity value and potential population (or demographic) stability. An extra measure may also be useful. Equitability describes the evenness of species abundance distribution in a community. It has a maximum possible value of 1, where every species is equally abundant, as in the case of the grey community in Fig. 2.1.

Figure 2.2 shows three hypothetical model terrestrial ecosystems, developed for experimental manipulations, wherein increasing biodiversity is depicted not only as a larger number of species (the number of circles in the diagram) but also the number of ecological links (envisaged as competition or predation, for example) between these species (Naeem et al., 1994). Such manipulations of relatively simple ecosystems in experimental situations have suggested that different levels of biodiversity of the sort represented in Fig. 2.2 are associated with different levels of ecosystem functions. For example, Naeem et al. (1994) found that plant productivity, CO_2 fixation and plant canopy architectural complexity were all greatest in ecosystems with the highest biodiversity. All of these parameters may influence that component of the system in which we are most interested, i.e. herbivorous insects.

2.3 Variations in Insect Biodiversity in Tropical Forests

It is convenient to use insects as primary indicators of tropical forest biodiversity. The total number of species of all living things in the world today is unarguably enormous and, equally unarguably, impossible to determine. Notwithstanding an unknown but probably huge number of species of nematodes, insects comprise a very substantial proportion of this total, certainly well over 50% of total living organisms and as much as 90% of higher animals (Speight et al., 2008) (see also Chapter 3). As we have seen in Chapter 1, however, tropical forests themselves are very variable and the 'conventional

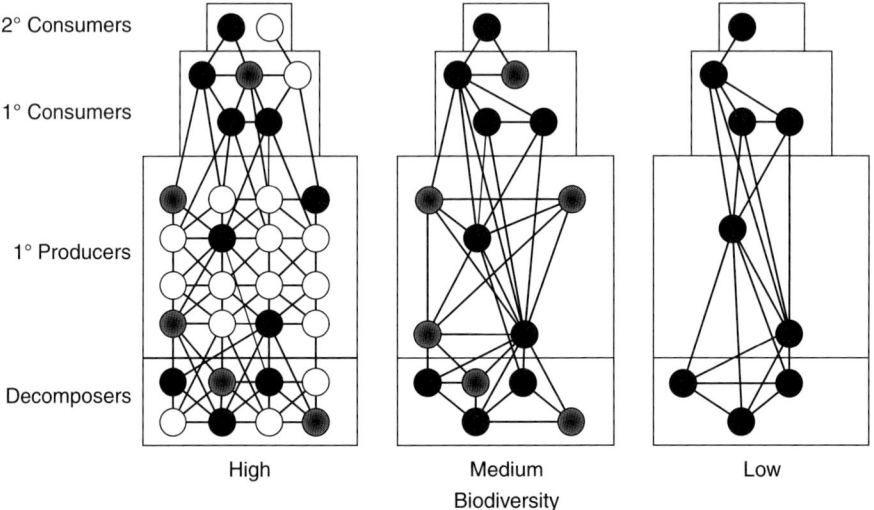

Fig. 2.2. Community diagrams of three types of model terrestrial ecosystem white circles = species which only occur in high-diversity systems; grey circles = species present in all but low-diversity sytems; black circles = species comman to all systems, irrespective of diversity. (from Naeem et al., 1994, courtesy Nature Publishing Group).

wisdom' these days that tells us that these forests in general are the most biodiverse on earth is not necessarily true. The number of species of animals such as insects in a particular habitat is a function of various parameters, which include the size of the habitat, its heterogeneity, its age and its degree of disturbance.

2.3.1 Heterogeneity

One hectare of undisturbed Amazonian forest contains about 175 species of trees larger than 10 cm diameter at breast height (DBH) (Burnham, 1994). Further south, in an Atlantic forest in Brazil, Ribeiro et al. (2009), in a study area of a mere 0.75 ha, counted 1076 trees with DBHs equal to or greater than 4.8 cm, belonging to 132 species and 39 plant families. In addition, many more species of epiphyte, liana and vine festoon the trees, and pioneer species of plant, woody shrubs and herbs emerge from the forest floor. All of this plant diversity, and the habitat heterogeneity it creates, provide numerous niches for animals of all sorts including insects, to exploit. Quite how many insect species there are in tropical forests will never be known for sure and some predictions are wilder than others. Stork (2010) summarizes the current state of knowledge. He suggests that the numbers of species of arthropod in the world are likely to be around 6–8 m (though there are some fairly huge confidence intervals around these data). Notwithstanding the facts that not all arthropods are terrestrial (and of course not all are insects either), Stork goes on to say that arthropods may comprise between 85 and 95% of all terrestrial macro (as opposed to micro) organisms and that most of these arthropods are found in tropical forests. This leaves us with a still unanswered question about the numbers of insect species in tropical forests, but it must lie somewhere in the region of several million.

Not only does each tree species provide different habitats from another species but also there are many niches available for exploitation within one individual tree. The term 'plant architecture' has been coined to describe the complexity of structures found in trees and other plants (Strong et al., 1984); many habitats can be envisaged within one tree and Table 2.1 summarizes some of them and their locations, showing that large numbers of species can be accommodated. Evolution proceeds to avoid competition between species, such that no two species which have exactly the same ecological requirements (or niche) can coexist in the same locality (so-called Gauses's axiom or competitive exclusion principle). We must also remember the food web concept, as mentioned in the previous chapter; herbivores themselves will have their own predators and parasites, who in turn have enemies too, so that, in theory, the total number of species that comprise a food web on a tree can be quite enormous. In summary then, Basset (2001) presents his predictions of arthropod species richness in a hypothetical rainforest canopy (Fig. 2.3). The circles represent the likely relative richness of each group. The majority of species are herbivores, either defoliators such as the larvae of moths and butterflies (Lepidoptera) and the larvae and adults of leaf beetles (Coleoptera: Chrysomelidae), or sap feeding such as the

Table 2.1. A summary of potential habitats for insects in a generalized tree and activities within these habitats.

Region of tree	Habitat	Insect activity
Canopy	Leaves	Defoliation, sap feeding, leaf mining
	Shoots	Boring, sap feeding, chewing
	Flowers, fruit and seeds	Pollen and nectar feeding, boring, chewing
	Twigs and branches	Sap feeding, chewing, boring
Main stems	Bark	Boring, chewing
	Wood	Boring
Roots		Chewing, boring, sap feeding

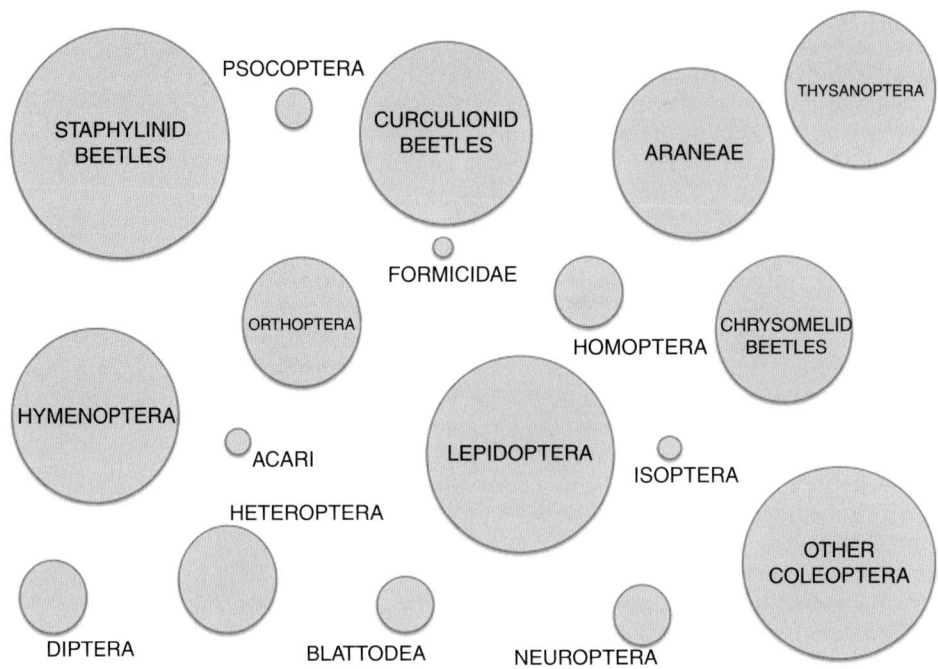

Fig. 2.3. Arthropod species composition in the canopy of a fictitious rainforest. The sizes of the circles are proportional to the species richness of the taxa (from Basset, 2001, with permission Cambridge University Press).

true bugs (Hemiptera: Heteroptera). Predators such as spiders (Aranea) are less species rich, while parasitic wasps (Hymentoptera: Parasitica) are relatively numerous.

Insects may not even have to feed on the trees at all but inhabit areas of tropical forests that exist by virtue of the environmental conditions created by forests, such as soil and litter. In Fig. 2.4, Eggleton and Bignell (1995) represent a complex series of functional groups of termites (Isoptera), which can be found coexisting in a West African tropical rainforest. Some of the groups of termites do have associations with trees directly (such as the wood feeder), others only indirectly. Incidentally, at least at this stage in the book, it is the former category of termites wherein may lie some serious forest pests.

We must not forget that, as Table 2.1 suggests, many tropical insects associated with trees may not occur in the canopy but may bore into stems, bark, or even roots. They may also use trees as food supplies and habitats, not only while the tree is alive but also after its death and, in the latter case, play vital roles in decomposition and nutrient recycling. A large proportion of the invertebrate species in moribund trees and fallen logs are termites (Isoptera) and ants (Hymenoptera: Formicidae) (Torres, 1994) and it is clear that well-rotted logs support the highest biodiversity of animals. This is predominantly a function of the heterogeneous structure of the rotting wood habitat. It is assumed that 'ecospace', or 'environmental mosaic' complexity, increases as decomposition proceeds.

So, it can be seen that forest trees are able to offer a great number of 'potential' niches to insect species; the next question must involve whether or not each tree species offers different niches, and this in turn begs the question as to whether insect herbivores are very host specific. Since, strictly speaking, an ecological niche is a function of its occupant, then it is important to know how generalist insect species might be in a tropical forest. Some

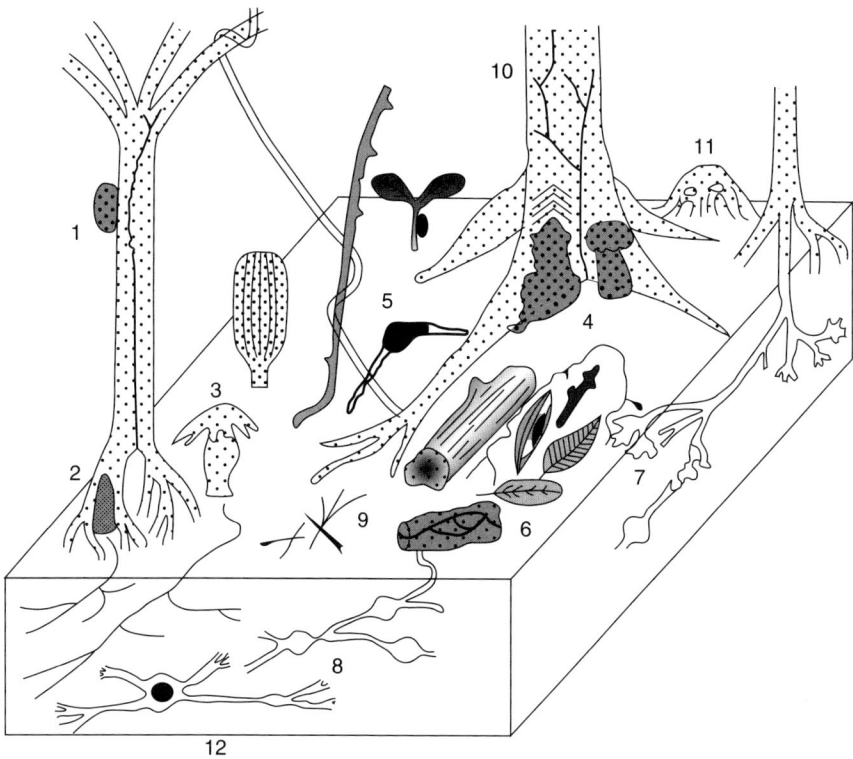

Fig. 2.4. Schematic diagram of the principal microhabitats and probable functional groupings of termites in the humid zone of southern Cameron. 1 = arboreal nesters, wide foragers in canopy and on ground, typically active wood feeders; 2 = termites constructing epigeal mounds on stilt roots, typically soil feeders; 3 = termites constructing free-standing epigeal mounds, typically soil feeders that forage widely in soil profile; 4 = termites in large epigeal mounds often on buttress roots, large colonies, soil feeders, foraging widely; 5 = purse nests made with soil, attached to low vegetation, typically soil feeders or soil/root interface; 6 = termites associated with decaying wood or other organic matter feeding on wood/soil interface; 7 = termites associated with root hairs; 8 = termites constructing subterranean networks over considerable areas, foraging on soil surface; 9 = termites feeding on very fine twigs or dead plant stems; 10 = wood-feeding termites foraging in the high canopy but nesting underground or in heartwood; 11 = large but well-spaced hard 'carton' nests of wood-feeding termites; 12 = entirely subterranean termites foraging in the soil profile (from Eggleton and Bignell, 1995, courtesy Elsevier).

estimates of the total numbers of insect species in the world have assumed that most (if not all) herbivorous insect species are host plant specific, but in reality this is not likely to be the case for all species. Some may, in fact, be extremely generalist in their host preferences. So, if we consider one or two examples of tropical insect pests, it can be seen that their host ranges are rather wide. The mahogany shoot borer, *Hypsipyla robusta* (Lepidoptera: Pyralidae), in Latin and Central America attacks trees in many genera in the large family of the Meliacaea. These include the economically important *Cedrela odorata*, *Swietenia macrophylla* and *Toona ciliate*, planted as an exotic (Plates 57, 58 and 59). The yellow butterfly, *Eurema blanda* (Lepidoptera: Pieridae), is enormously widespread, occurring from West Africa to Southeast Asia, where its larvae attack a large variety of leguminous trees and shrubs, including *Albizia*, *Falcataria*, *Acacia* and *Pithecellobium* (known as Manila tamarind). In fact, for many species, though we know to our cost what crop species they will attack, we still have little or no idea what plant species they are able to feed on in the wild. In other words, host range data are likely to be a highly conservative but vital consideration when planning silviculture and pest management (see Chapters 5 and 6).

In general terms, field surveys tend to show that the dependency of individual insect species on one or only a few host tree species is relatively rare (Moran et al., 1994), though in their southern African study these authors found that the most specialist herbivore species occurred on the taxonomically isolated tree species. In fact, the less taxonomically related the tree species, the fewer similar herbivorous insects that are shared (Fig. 2.5; Novotny et al., 2006). Figure 2.5 shows that the same relationship occurs in temperate as well as tropical situations: the higher the plant 'diversity', the higher the insect 'diversity'. Put simply, it is quite obvious that if an insect is a *Pinus* specialist, then it is highly unlikely to be able to survive readily on *Eucalyptus* or any other non-related family of trees. This has had great benefits for certain tropical forest operations in the past; in southern Africa, for example, pine species such as *Pinus kesiya*, *P. patula* and *P. oocarpa* have been grown very successfully indeed, with no serious pests at all. *Pinus* spp. are totally exotic to the African continent and only when equally exotic pests such as pine woolly aphid, *Pineus boerneri* (Hemiptera: Adelgidae), were introduced accidentally from Australia in the 1960s did significant economic damage ensue (Mills, 1990; see Chapter 5).

None the less, host trees can influence the biodiversity of tropical insects, and some groups of insects which by their very nature tend to be rather host specialist do show links, especially to the abundance and geographical distribution of their hosts. The pine woolly aphid above is one such example, and cerambycid beetles another. Longicorns (or longhorns) attack mainly debilitated or moribund trees, their larvae feeding between the bark and sapwood, producing characteristic gallery systems, prior to 'duck-diving' into the timber to pupate (see Chapter 5). Such insects can be particu-

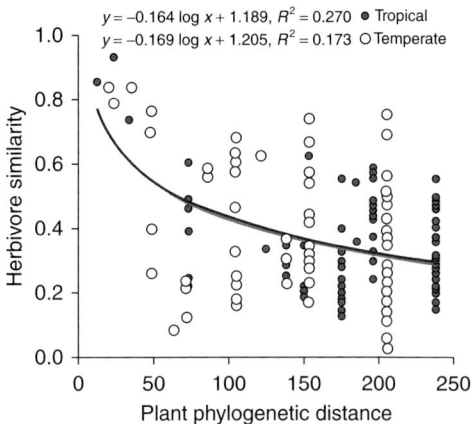

Fig. 2.5. Similarity of folivorous communities between pairs of host species versus the phylogenetic distance between the hosts in a temperate and a tropical forest. The negative correlation between community similarity and phylogenetic distance is significant in both data sets (from Novotny et al., 2006, with permission the American Association for the Advancement of Science).

larly sensitive to forest stand structure. On semi-tropical Japanese islands, for example, Sugiura et al. (2009) found that both richness and abundance of cerambycid beetles were related to the intensity of canopy cover, with gaps tending to promote more beetles (Fig. 2.6). Most cerambycid species are fairly host specific. *Phoracantha semipunctata*, the eucalyptus longicorn (Plates 39 and 40), is one of the most notorious forest pests in the family, now being established and causing economic damage in almost all regions of the subtropical and tropical world. This species is restricted almost entirely to *Eucalyptus* spp., due predominantly to the highly selective oviposition behaviour of the adult female beetle (Hanks et al., 1995b). Some other cerambycids might be a little more polyphagous (Hawkeswood, 1992) but their species richness in the New World region under study was related closely to the diversity and distribution of fairly specific host trees.

Finally in this section, it is worth pointing out that there is some evidence which suggests that as tropical forests become more diverse (in terms of tree species richness), so

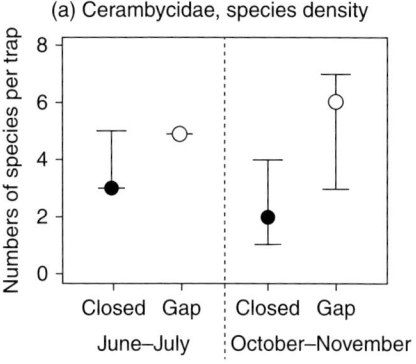
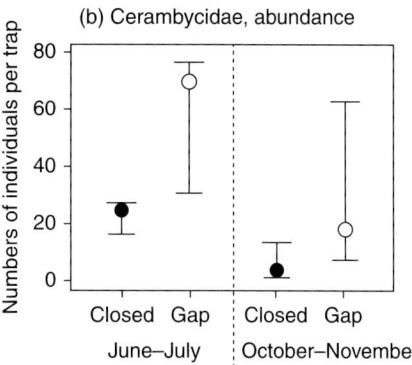

Fig. 2.6. Differences in species richness and abundance of cerambycids between closed-canopy sites and gaps created by killing *Bischofia javanica* trees. Circles = median value; bars indicate range of values (from Sugiura *et al.*, 2009, with permission Springer).

the overall levels of herbivory also increase. Schuldt *et al.* (2010) studied the amount of leaf damage caused by insect herbivores in plots containing varying numbers of tree species in southern China (Fig. 2.7). They concluded that the clear increase in herbivory levels was associated with increases in the numbers of generalist herbivores in species-rich forests. If this principle can be applied widely, the implication is that smaller forests, or fragments of forests, might be expected to show less insect herbivory than larger or continuous forests (Ruiz-Guerra *et al.*, 2010).

2.3.2 Abiotic variations

All else being equal, the species diversity of plants, and forests in particular, increases towards the equator (Novotny *et al.*, 2006; Speight *et al.*, 2008). Warm temperate forests have less biodiversity than subtropical forests, which in turn are less diverse than tropical forests. One explanation for this involves the greater levels of solar energy, higher temperatures and more rainfall in tropical areas, with allied increases in turnover, cycling and productivity (see also Chapter 3). Seasonal rainfall patterns in an area are clearly able to influence the abundance of insects dramatically. Many forest pests exhibit peak populations in wet or dry seasons, so that, for example, the teak defoliator, *Hyblaea puera*

Fig. 2.7. Mean percentage of leaf damage per plot due to insect herbivory in relation to species richness of trees and shrubs in 27 study plots in Zhejiang Province, south-east China (from Schuldt *et al.*, 2010, courtesy Wiley-Blackwell).

(Lepidoptera: Hyblaeidae) (Plate 19), is most common in July and August, during the rainy season, in southern India (Chandrasekhar *et al.*, 2005) and the cypress aphid, *Cinara* sp. (Hemiptera: Lachnidae), peaks in September, during the dry season, in Malawi (Chilima and Meke, 1995). However, seasonal links to variation in biodiversity are harder to detect. One example from Sarawak, Malaysia, describes how the species diversity of soil macrofauna (insects in the main) declines as primary dipterocarp forests are replaced by plantations of *Acacia mangium* (Tsukamoto and Sabang, 2005). One cause of this decline in biodiversity is that soil moisture is reduced in the grassland ecosystems and arthropod species which are intolerant to drought are absent, though able to thrive in the forests.

Within one area and forest type, biodiversity of insects and other arthropods also varies with altitude. Tropical insects are very sensitive to temperature (Chen *et al.*, 2009) and mountain tops tend to be considerably colder than lowlands. Geometrid moth communities were sampled at different altitudes on Mount Kinabalu in Sabah and data collected in 1965 were compared with that of 2007 (Fig. 2.8). The authors estimated that the average altitudes of over 100 species of moth increased by a mean of 67 m over 42 years as a result of climate change warming the region.

2.3.3 Height in the canopy

Tropical forests are, of course, typified by extremely tall mature trees and it is therefore not surprising to find that the distributions of insect activities and communities can vary considerably in terms of vertical stratification within tree canopies (Basset *et al.*, 2003). Take, for example, forest moths in the families Arctiidae (tiger moths) and Geometridae (loopers). Brehm (2007) used light traps to sample moth populations in the understorey and canopy of primary rainforest in Costa Rica and found that assemblages of both families differed markedly between the canopy and understorey (Fig. 2.9). In addition, it was discovered that the species richness and diversity of arctiids was significantly lower in the understorey compared with the canopy, while geometrids

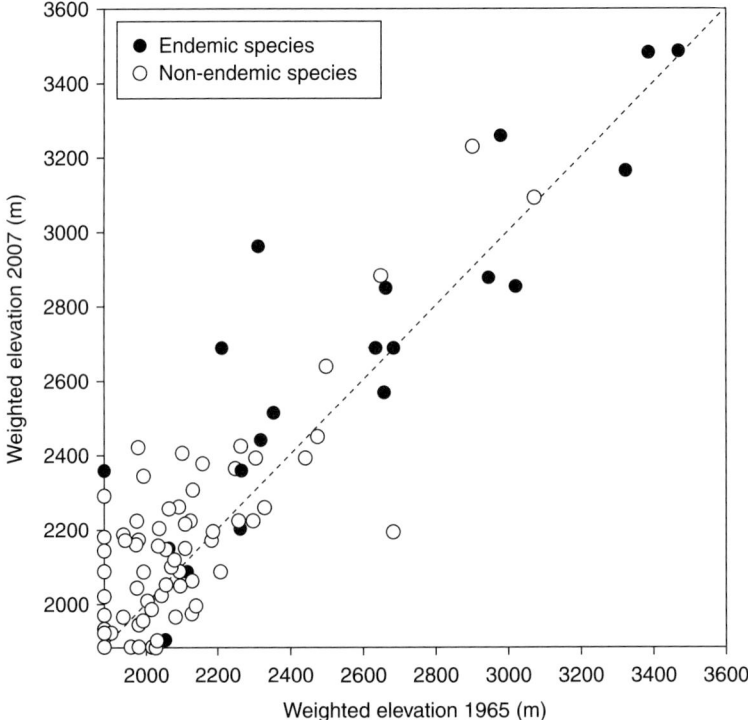

Fig. 2.8. Elevational changes of moths on Mount Kinabalu between 1965 and 2007. Weighted elevations are a means of estimating the average elevation of each species for both sampling years; data points above the 'no-change' diagonal line have moved uphill in 42 years (from Chen *et al.*, 2009, with permission the National Academy of Sciences).

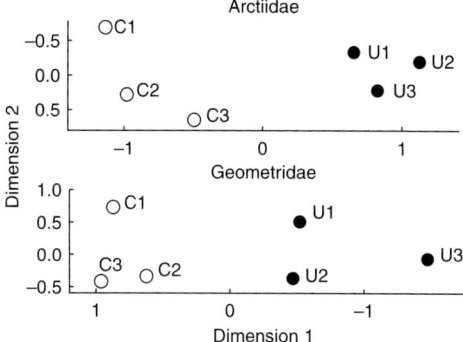

Fig. 2.9. Non-metric, two-dimensional scaling of assemblages of Arctiidae and Geometridae from three canopy (C) and three understorey (U) sites in a Costa Rican rainforest. Note that the y-axis is inverted in the arctiid graph for clarity (from Brehm, 2007, courtesy Elsevier).

showed the opposite relationship. Moving to Sabah, Schulze and Fiedler (2003) carried out a similar experiment, this time studying moths in the family Pyralidae. As can be seen from Fig. 2.10, they also found distinct assemblages occurring in the canopy versus the understorey, with an added dimension of seasonality thrown in. Trophic interactions can also vary vertically in a forest and as Paniagua et al. (2009) found, gall makers in the dipteran family Cecidomyiidae, and their associated parasitic hymenoptera, varied according to height in the canopy. This work, carried out in Panama, showed that parasitoid species richness was higher in the understorey than in the canopy, but gall maker species richness was higher in the canopy. As Fig. 2.11 shows, gall maker–parasitoid trophic webs are thus distinctly different between the two sites in the forest vertical stratification.

2.3.4 Forest type

So far in this account of tropical forest biodiversity, we have concentrated mainly on the stereotypical tropical forest, the rainforest. However, there are many other types of forest in the tropics (see Chapter 1) whose biodiversity is also of great interest. This interest lies both with the conservation viewpoint and also because some of these forests, natural and artificial, have great commercial significance. Because of the various abiotic and regional variations, it is best to compare biodiversities between forest types in localized areas, to remove much of the otherwise confusing variation. A good example concerns the distribution of dung beetle (Coleoptera: Scarabaeidae) species in Queensland, Australia (Hill, 1996). Two forest types abut one another in this region; open-forest, dominated by eucalypts, and rainforest are both natural ecosystems, which may be separated by as little as 10 m. The authors measured the abundance and richness of dung beetles caught in flight intercept traps along transects from open-forest into rainforest. Species richness varied significantly across sites, being lowest in the open-forest and increasing sharply at the forest boundary (called the ecotone) and into the rainforest (Hill, 1996). Many dung beetle species are habitat specific and, in this case, we see that two types of tropical forest differ markedly in the insect biodiversity that they support. Worthy of extra note is the fact that there is a suggestion that the highest abundance of beetles occurs in the open-forest where there are least numbers of species. Some species of scarabaeid can, in fact, be forest pests (see Chapter 5) and high numbers accompanied by low species richness may be of great importance in pest dynamics and resulting crop damage. Still in tropical Australia, Foggo et al. (2001) sampled arthropods (mainly collembola and insects) in rainforests as well as immediately adjacent natural sclerophyll (eucalypt) forests. Figure 2.12 shows that animal communities are clearly different, according to detrended correspondence analysis (DCA), between the two forest types. Not only that, it was found that edge effects (see also later in this chapter) could be detected in both separate habitat assemblages.

On many occasions, tropical forests may have been altered by human activities, with varying degrees of intervention (see Chapter 1 and later in this chapter). Natural primary forests may be logged selectively to form secondary forests, enriched by the planting of native tree species, manipulated

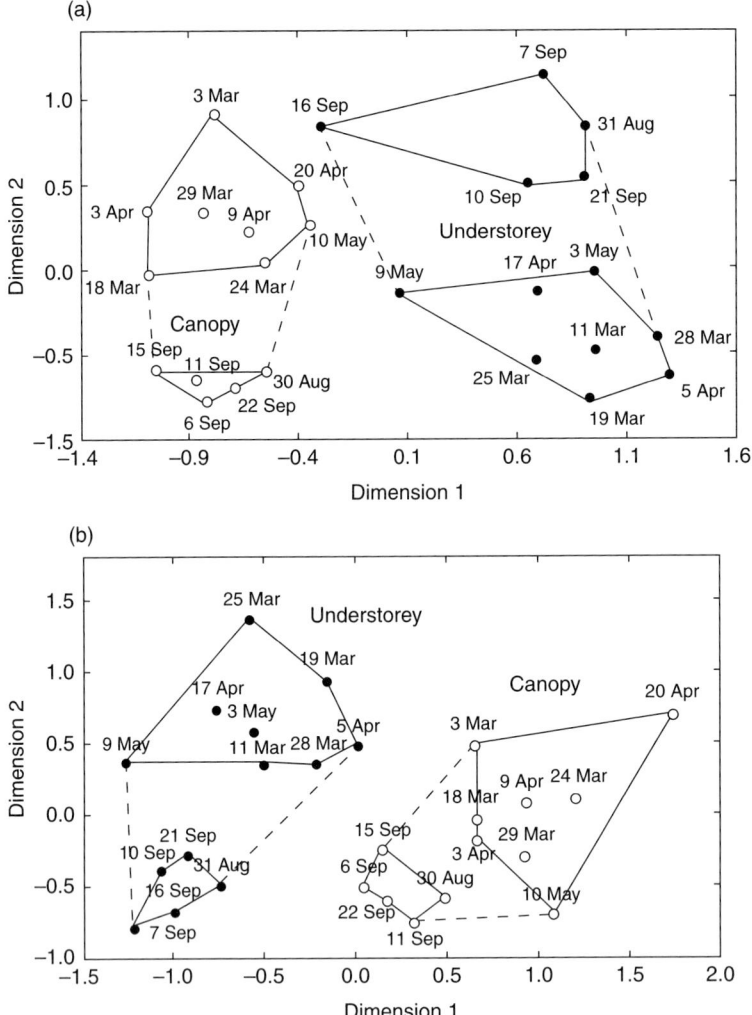

Fig. 2.10. Non-linear, two-dimensional scaling of nightly pyralid moth samples in understorey and canopy samples in Sabah. Samples from the same period of the year are connected by thick solid lines, while sampling periods (March–April; August–September) are connected by thin broken lines (from Schulze and Fiedler, 2003, courtesy the Society for Tropical Forestry).

by the establishment of tree crops within them, such as rattans or fruit, or indeed cleared completely and replaced with exotic plantations of trees (see Chapter 1). All these operations might be expected to alter the biodiversity and ecological stability of the forests and have particular influence on the insects that live in them. These changes may not always occur in predictable ways. Animals such as insects living in these forests must necessarily be affected by these interventions, with consequent effects on their roles in forest ecosystems and food webs. Two examples will introduce these concepts.

Firstly, Maeto *et al.* (2009) investigated the effects of reforestation of degraded grasslands in Kalimantan on parasitoid wasps in the family Braconidae. As Fig. 2.13 shows, parasitoid species richness is much higher

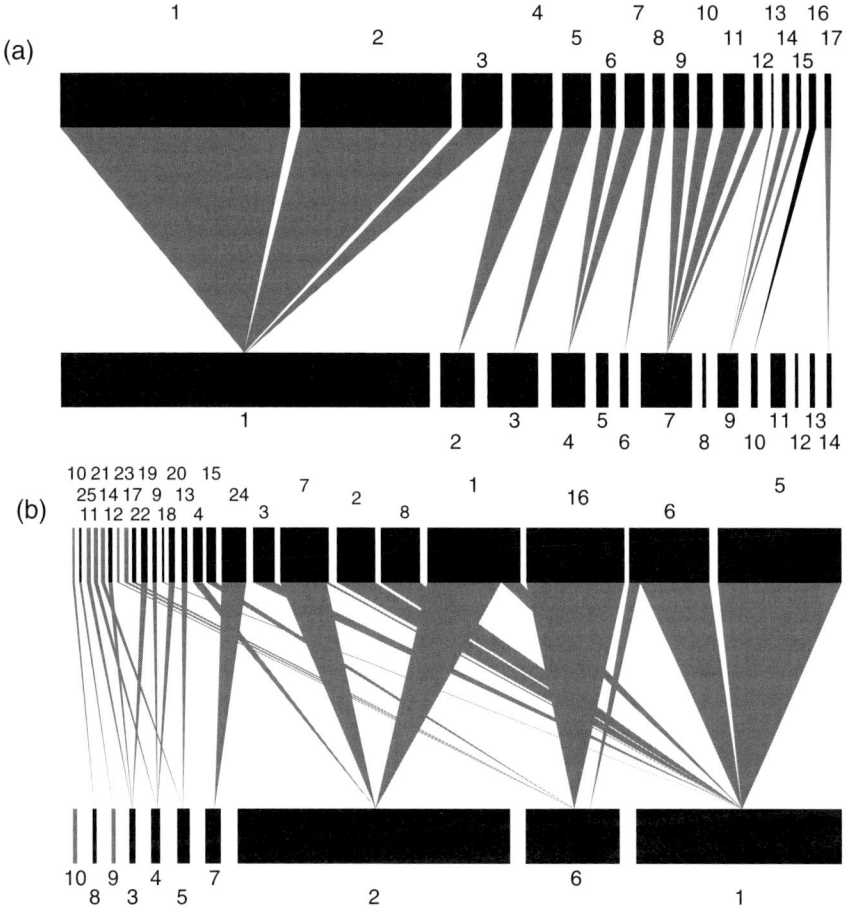

Fig. 2.11. Quantitative gall maker–parasitoid trophic webs for (a) canopy and (b) understorey sites in a Panamanian forest. Lower bars show the density of gall maker host insects and upper bars the density of parasitoids. Numbers relate to individual species of insect (from Paniagua et al., 2009, courtesy Wiley-Blackwell).

in old secondary (but natural) forests than in any of the other habitats sampled. Secondly, Louzada et al. (2010) compared shared species of dung beetles, fruit-feeding butterflies and scavenger flies between primary forest, secondary forest and eucalyptus plantations in the Brazilian Amazon (Fig. 2.14). The differences in insect species between primary forest and secondary forest are considerable and are enormous between primary forest and eucalypts. We shall return to the subject of plantation forests later in this chapter.

2.3.5 Patch size, fragmentation and isolation

Any manipulation of tropical forests is likely to leave remnants or fragments of the original forest that are necessarily smaller than the continuous area of habitat from which they are derived. Furthermore, these patches will be isolated from one another by all manner of dissimilar habitats, including logging roads and farmland. It seems that in nearly all cases studied, tropical rainforest fragmentation leads to a local loss of species, whether they be vertebrates, invertebrates or even plants (Turner, 1996). There may also be a reduction in the general levels of herbivory in small fragments relative to large ones (Faveri et al., 2008). One of the most basic and controversial issues in biodiversity conservation and ecology are the details of the effects of fragmentation and isolation on habitats, and we still have a lot to learn about what happens to the insect species in a rainforest fragment (Didham et al., 1996).

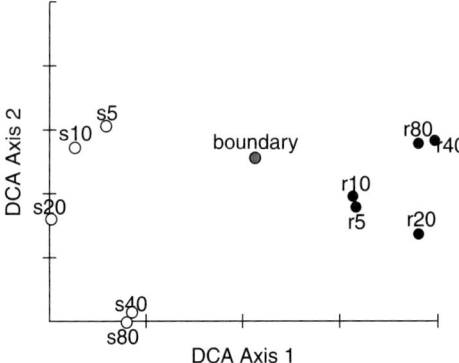

Fig. 2.12. Detrended correspondence analysis of mean faunal abundances along a sclerophyll–rainforest transect in Queensland. Sampling stations are labelled with a letter (s = sclerophyll, r = rainforest) and a distance in metres from the boundary between the two forest types (from Foggo et al., 2001, with permission Springer).

Island biogeography theory and species–area relationships are different forms of the same concept (Whittaker, 1998), which states that: (i) bigger habitat 'islands' (such as a patch of forest) will contain more species than smaller ones of a similar nature, age, degree of isolation, etc.; and (ii) species extinction rates are higher in smaller 'islands'. The species–area equation which links patch or island size to the number of species to be found in or on it is asymptotic, i.e. s (number of species) = a (area) to the power z (a constant less than 1). One example is again from cerambycid beetles in the Bahamas, mentioned previously. In Fig. 2.15, Browne et al. (1993) show that there is a statistically significant increase in numbers of species of longicorn found on an island as the size of that island increases. In this case, the relationship only became meaningful when four other islands had their species counts removed from the analysis due to over- or under-estimations, attributed by the authors to unrepresentative collecting efforts; this shows how tricky it can be to obtain reliable data of this nature. The value of z has interesting implications, too. Cagnolo et al. (2009) studied habitat fragmentation and species loss on various woodland trophic levels in central Argentina and concluded that the slopes of the species–area relationships (z) varied with body size, abundance and

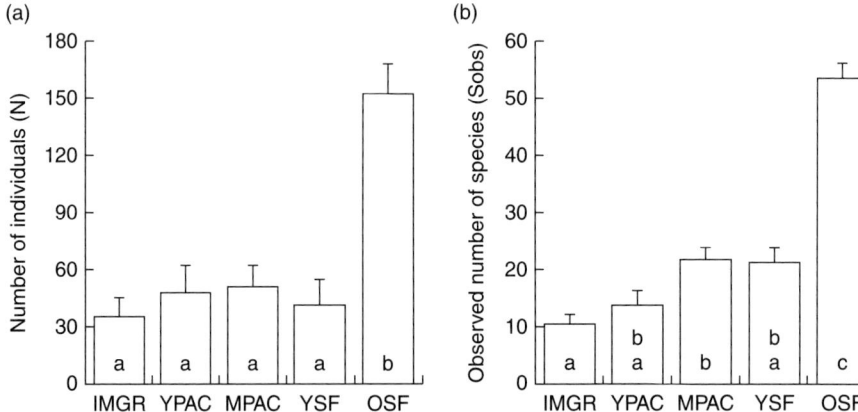

Fig. 2.13. Mean and SD of (a) the abundance and (b) observed species richness of braconid parasitoids in different habitats in East Kalimantan. IMGR = *Imperata*-dominated grasslands; YPAC = young *Acacia* plantations; MPAC = mature *Acacia* plantations; YSF = young secondary forest; OSF = old secondary forest (from Maeto et al., 2009, with permission Springer).

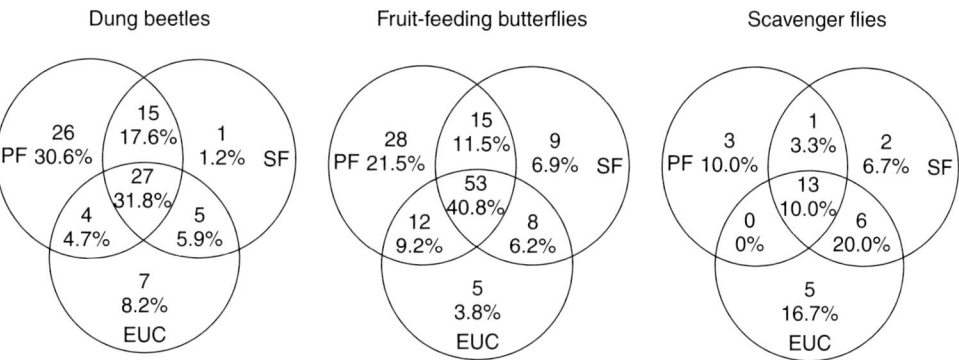

Fig 2.14. Venn diagrams showing the distribution of species exclusive to or shared by three forest types in the Brazilian Amazon. PF = primary forest; SF = secondary forest; EUC = eucalyptus plantation (from Louzada et al., 2010, courtesy Elsevier).

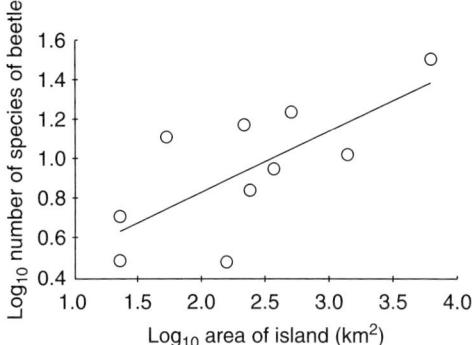

Fig. 2.15. Species–area relationship for longicorn beetles (Coleoptera: Cerambycidae) in the Bahamas (from Browne et al., 1993).

degree of specialization (Fig. 2.16). As might be expected, area-related species loss in forest fragments was most intense for specialist species in high trophic levels, such as parasitic hymenoptera.

Another problem associated with reducing the size of a forest patch concerns the number of individuals, that is, abundance, of insects in the patch. There must be a reduction in resources such as space and food when a habitat is shrunk, so that the carrying capacity or the maximum number of individuals within a species which can be supported must also decline. Coupled with these problems comes that of population viability, whereby smaller populations are more prone to genetic drift and inbreeding or, worse still, to extinctions during environmental perturbations. Working on populations of the army ant, *Eciton burchelli* (Hymenoptera: Formicidae), from Panama, Partridge et al. (1996) concluded that the size of the equilibrium population of this species was related to the habitat size and that, as habitat size increased, then the time it was expected to take before the population became extinct also increased, but exponentially.

It must not be assumed that high species richness is impossible in small forest patches; indeed in a 1 km^2 area of lowland dipterocarp rainforest in Brunei, Orr and Hauser (1996) estimated a staggering 464 species of butterfly, nearly half of the entire fauna of Borneo. It has to be said that in this study, this patch of rainforest was not isolated, in fact it formed part of a continuous tract of forest; the effect on real fragmentation is less encouraging. Decomposer organisms such as termites and dung beetles exert a major influence on nutrient supplies in a forest ecosystem (Fig. 2.17; Didham et al., 1996) and hence the knock-on effects of forest fragmentation on such insects may have fundamental consequences for the functioning of these systems. Smaller fragments have markedly lower population densities and numbers of species. This species number is, in turn, related to the amount of dung decomposed in a given time and though, from a functional point of view, a decrease in population density may be more important than

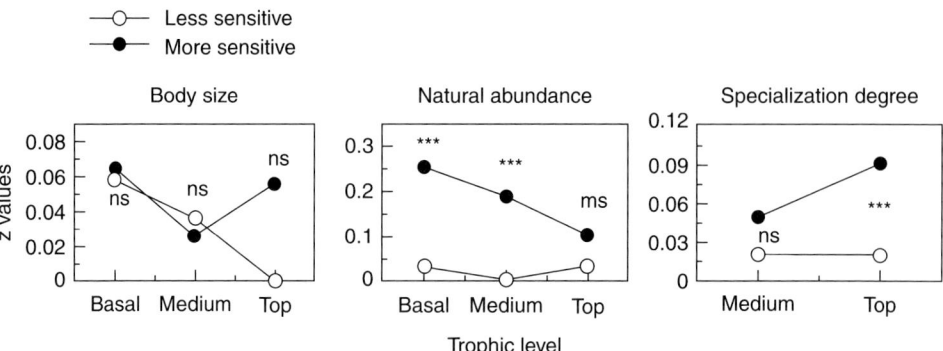

Fig. 2.16. Relationship between species–area slopes (z values), trophic level (basal = plants; medium = leaf mining insects; top = parasitoids) and predicted sensitivity to habitat fragmentation (from Cagnolo et al., 2009, courtesy Wiley-Blackwell).

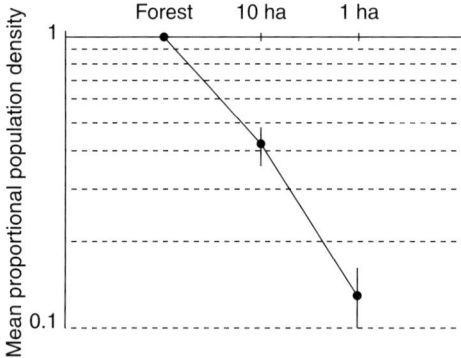

Fig. 2.17. Population density of dung beetles in different-sized patches of forest in Central Amazonia (from Didham et al., 1996, courtesy Elsevier).

that of species richness, both elements of general biodiversity suffer greatly. Plant feeders such as butterflies show a similar relationship (Benedick et al., 2006) (Fig. 2.18). In this example from Sabah, it can be seen that species diversity of fruit-feeding butterflies certainly increases with the size of the forest patch, but not linearly. As forest patch area increases, the relationship with butterfly species diversity becomes asymptotic, for the simple reason that there must only be a maximum number of butterfly species to be found, no matter how large the forest.

Not all the effects of habitat fragmentation are for the worse, at least from the insect's point of view. The degree of dissimilarity between habitats within the fragment and those surrounding it (known as the matrix) is all-important. If animals are able to use the matrix either as a corridor from one fragment to the next or as alternative places to live, then the forest fragment will function differently. Herbivores, including potential pests, may be more abundant in small patches of their host plant than in continuous cover. Isolation, by its very name, suggests that immigration and emigration into and out of a patch of habitat is relatively difficult, especially if the surrounding habitats are very different to that in the patch. This gives rise to the concept of metapopulations, where local populations of a species may have different likelihoods of colonization and extinction. The term 'corridor' has been used to describe a habitat that connects one patch with another, thus reducing isolation and enabling species to move from one patch to the next. Figure 2.19 presents a conceptual framework for how matrix–corridor interactions may function in a mixture of natural and plantation forests (Brockerhoff et al., 2008). Corridors allow the movement of species from one forest fragment to the next and reduce geographical and genetic isolation. Hence, if insects, either pests or beneficials, are able to move from one tropical forest patch to another, fragmentation and isolation may not be such a problem.

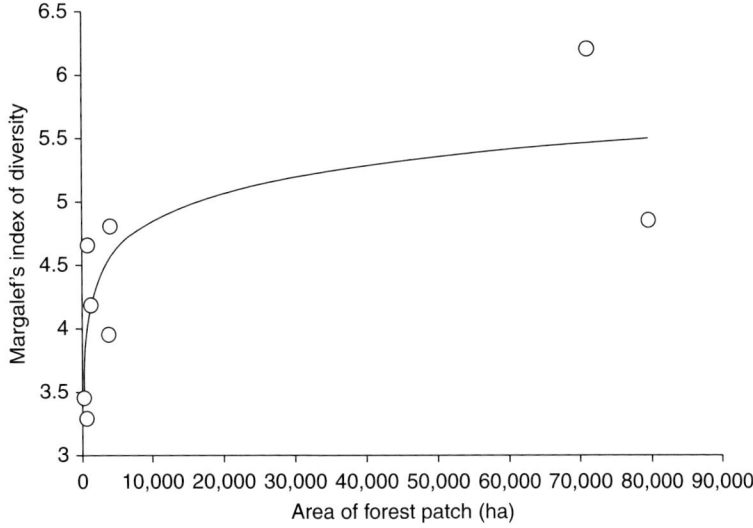

Fig. 2.18. Margalef's index of species richness (diversity) for fruit-feeding butterflies in rainforest patches in Sabah, in relation to area of patch (data from Benedick et al., 2006).

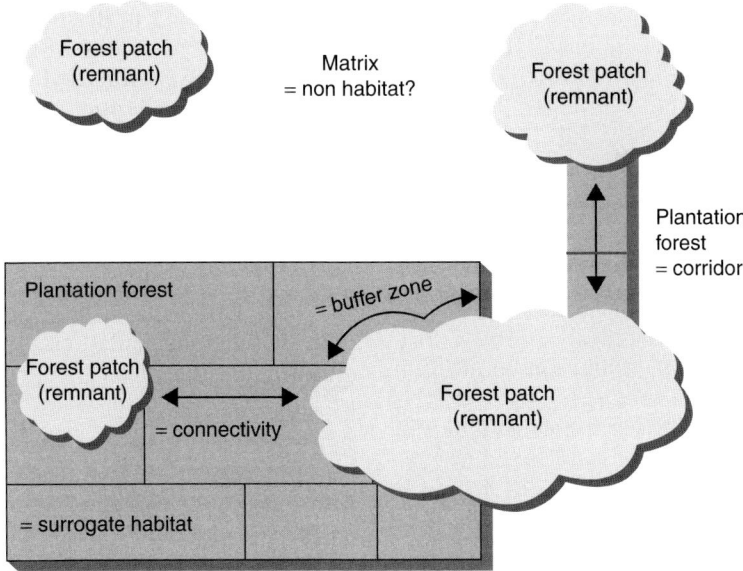

Fig. 2.19. A corridor–patch–matrix model depicting a highly fragmented landscape (from Brockerhoff et al., 2008, with permission Springer).

2.3.6 Edge effects

Another problem with fragments of habitat is that of edge effects. Essentially, an edge effect is a recognizable gradient of an abiotic effect such as temperature or humidity, or a biotic one such as the abundance of plants and animals, away from a recognizable habitat edge (different crop, road, etc.) into the patch of habitat. Abiotic edge effects can be complex;

in a tropical forest in central Amazonia, Camargo and Kapos (1995) found the highest values of soil moisture just inside a forest edge that had been created by cutting, with zones of reduced moisture actually at the edge and again 40–80 m into the forest. Distributions of vegetation might be expected to respond to these edge effects, too. The explanation for Camargo and Kapos's observations was thought to be the fact that the rapid regrowth of natural vegetation at the cut edge of the forest 'sealed' the edge with rapidly transpiring foliage, which depleted soil moisture in their immediate locale but protected the understorey just inside the forest edge from desiccation. Plants themselves can show edge effects, changing their species composition and structure as the distance from the edge of the forest increases. Malcolm (1994) showed how the thickness of foliage of low-lying understorey vegetation decreased up to about 35 m into a forest in Brazil, while overstorey vegetation exhibited the reverse effect, increasing in thickness up to about 60m into the forest. However, in another forest in Brazil, Ribeiro et al. (2009) were unable to find significant differences in tree density, diameter and height between forest interior and fragment edges; so, the situation is clearly complex.

Some studies have been carried out on edge effects and forest canopy arthropods in temperate regions (e.g. Ozanne et al., 1997), but there have been few quantitative studies of edge effects in tropical rainforest fragments (Foggo et al., 2000). On occasion, it has been noticed that species diversity is greatest at the edges of forests and least in the middle, at least for some insect groups. Butterflies tend to be edge or gap species in the main, where the light is brighter, the temperatures warmer and, arguably, there is more space in which to move around. In northern Guatemala, for example, Austin et al. (1996) recorded 535 species of butterfly from forest sites; the largest number of species occurred in forest edge habitats and the smallest number in continuous forested habitats. Butterfly communities in forest fragments in Ghana were found to show edge effects in two out of three sites studied (Bossart and Opuni-Frimpong, 2009), with density point estimates declining away from an edge into a forest fragment. There is, in fact, an argument that suggests that thermophilic insects such as butterflies are not typical forest insects at all, since they prefer open habitats in forest or woodland sites (Hambler and Speight, 2004). If nothing else, however, they are recognized easily and usually received favourably by amateurs and, as such, may act as flagstone species which help the preservation of less subjectively attractive but more important animals and plants in forests.

Edge effect gradients certainly exist. Burkey (1993) showed that the predation on seeds (groundnuts in fact) laid out along transects from edges into patches of tropical forest in Mexico showed increased consumption for at least 200 m into the forest (Fig. 2.20). Unfortunately, this study did not identify fully the predators at work, though both ants and birds were implicated. Leaf-cutter ants in the genus *Atta* are very serious defoliators of trees and other crops in parts of South and Central America (see later), and Urbas et al. (2007) studied their distributions and impacts on trees in relation to forest edges in the Atlantic forest of north-east Brazil (Fig. 2.21). It appeared that leaf-cutter ant herbivory was increased significantly at the edges of forest fragments, an observation, if widely applicable, with clear implications for forest pest management.

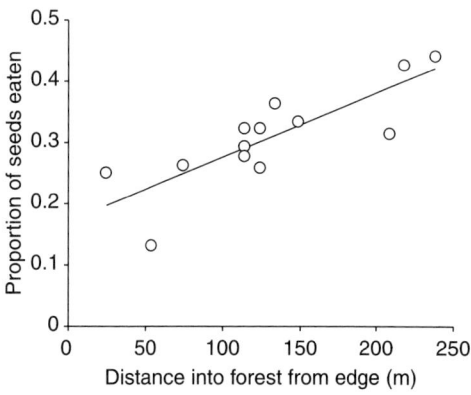

Fig. 2.20. Predation of seeds at various distances into a tropical forest in Mexico (from Burkey, 1993, courtesy Elsevier).

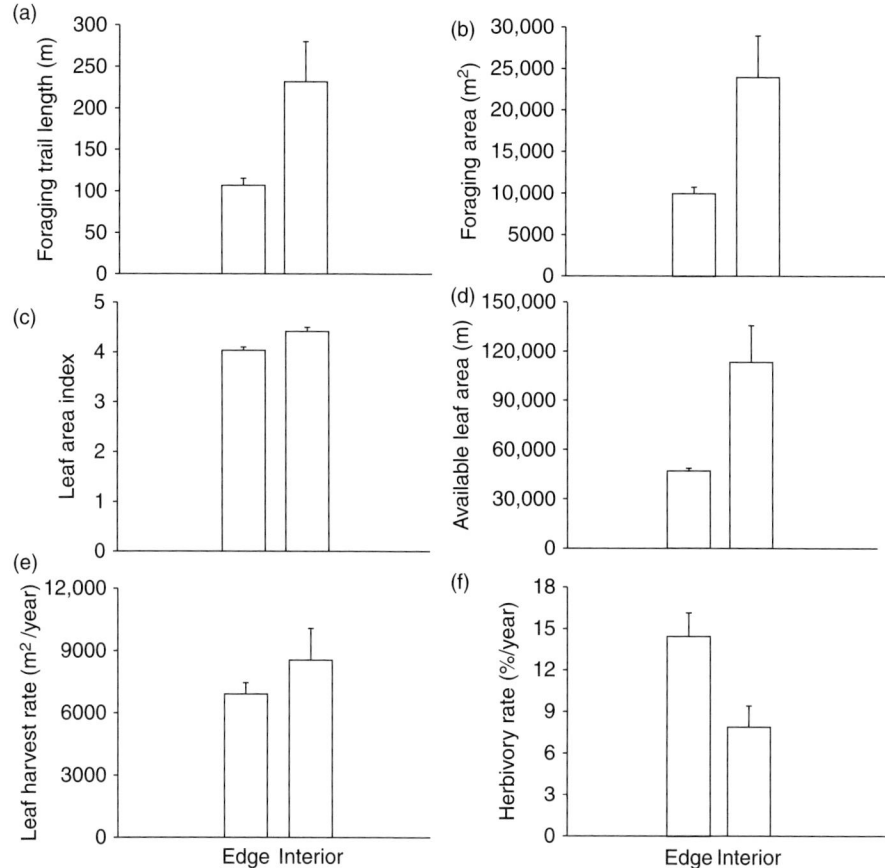

Fig. 2.21. Estimated means (+ SD) of (a) foraging trail, (b) annual foraging area, (c) leaf area index, (d) available leaf area, (e) annual leaf harvest rate and (f) annual herbivory rate of leaf-cutter ant, *Atta cephalotes*, colonies at the forest edge and interior of a remnant of Atlantic forest in north-east Brazil (from Urbas et al., 2007, courtesy Wiley-Blackwell).

2.3.7 Disturbances and logging

Any disturbance to an ecosystem might be expected to influence the animals and plants which live in it. Some types of disturbance are natural and indeed essential for the continuation of the system, so that the creation of open patches in a rainforest by the deaths of trees from natural causes such as wind or old age are vital dynamic events in the life of the forest. Hurricanes, for example, can have catastrophic effects on tropical forests, virtually wiping out all standing trees and other types of vegetation associated with them, such as climbers (Plate 11). Willig et al. (2011) looked at the effects of Hurricane Hugo, a category 4 storm, on the stick insect, *Lamponius portoricensis* (Phasmida), in Puerto Rico (Fig. 2.22). This insect feeds on plants in the genus *Piper*, pepper vines, which require mature forest trees on which to climb. As the graphs show, populations of the insect show no signs of recovery for some years after the hurricane. Even fire can be a perfectly normal phenomenon in some forest systems, such as fire-climax sclerophyll forests, but the large-scale cutting down and removal of large, mature trees cannot be considered as a system with which tropical plants and animals have co-evolved. Selective logging where only certain sizes and species of

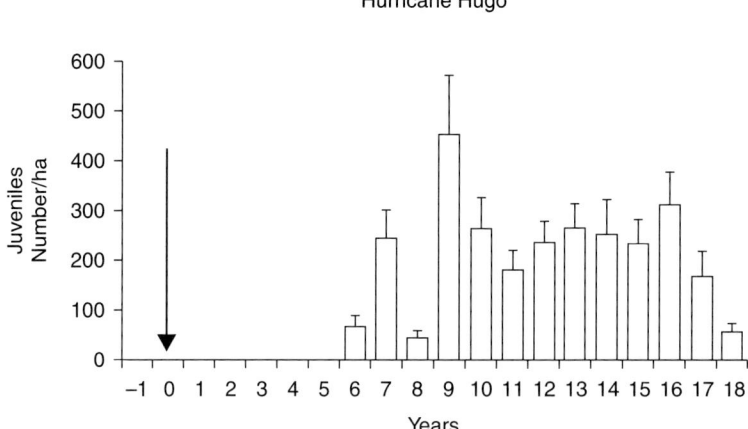

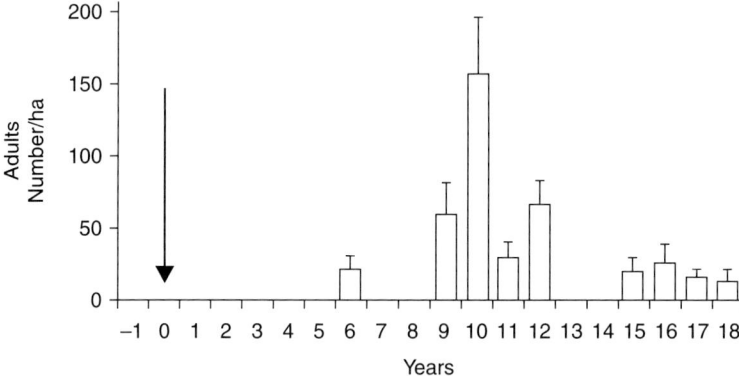

Fig. 2.22. Influence of Hurricane Hugo on the mean density (± SE) of the stick insect, *Lamponius portoricensis*, in Puerto Rico. Arrows indicate time of hurricane; time in years (from Willig *et al.*, 2011, with permission Springer).

tree are removed can be highly disturbing to the inhabitants of a primary forest, even when modern techniques of reduced-impact logging are employed, while clear-felling, where every tree is removed, is obviously catastrophic.

Disturbance in an ecosystem like a forest may affect the biodiversity of animals in different ways, according to their position in a food chain. In Fig. 2.23, it can be seen that, in theory at least, increased disturbance produces high species diversity in the primary producers, intermediate diversity in herbivores and least diversity in carnivores (Huston, 1994). If we consider a hypothetical example from a tropical forest, this means that after logging, and either replanting or allowing natural regeneration, we should expect the highest number of species of plant, more species of insects feeding on the plants, but only up to a point, and relatively few predators or parasites of these insects. In simple terms, the whole balance of the community has been changed, which might also have important implications for pest dynamics. Very often, theory is less easy to replicate in practice, though it is clear that different groups of tropical forest insects vary in their

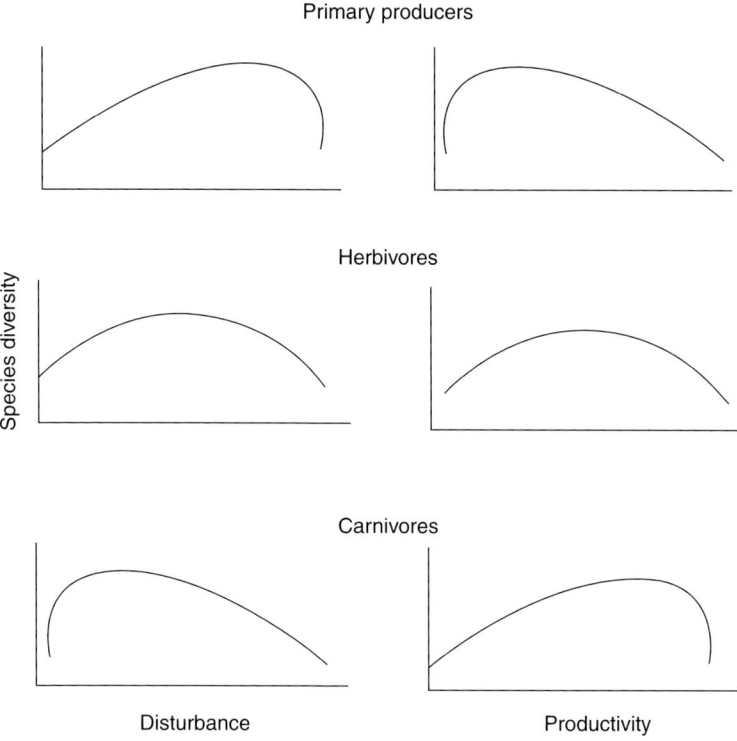

Fig. 2.23. Theoretical patterns of species diversity for three trophic levels in a community according to disturbance and productivity (from Huston, 1994, with permission Cambridge University Press).

response to forest disturbance (Fig. 2.24). Here, Lawton *et al.* (1998) undertook a mammoth study in Cameroon to investigate the effects of varying degrees of forest disturbance on different animal communities and found that butterfly species richness tended to decrease as forest disturbance increased, flying beetles seemed to increase slightly, whereas leaf litter ants peaked in secondary growth forest containing some plantations. All in all, however, the situation was anything but clear.

Butterflies are used frequently as indicators of habitat change. Safian *et al.* (2011) used mashed banana baited net traps to capture butterflies in various types of habitat in southern Ghana and then analysed the similarities between each habitat type using Jaccard's coefficient of similarity (Fig. 2.25). As expected, primary rainforest sites were most similar in terms of butterfly communities, while clear-felled open areas were the most dissimilar. Young planted forests, secondary forests and even forests containing some oil palm formed a midway group of habitats. Undoubtedly, the casual eye would decide that secondary forests were more luxuriant and species rich in terms of plant life than primary ones in the same locality. Animals from mammals to insects are certainly easier to spot than in a primary habitat (for one thing, of course, they are much nearer the ground), but controversy still rages as to the real quantitative effects of disturbance on tropical forests and much research is proceeding. Two more butterfly examples illustrate the effect of forest disturbance. Figure 2.26 shows the results of a principal components analysis (PCA) on butterfly communities in central Kalimantan (Cleary *et al.*, 2009). The cluster of ordination points relating to primary rainforest were clearly distinct from the two logged forest areas that showed more overlap, even though one was logged over

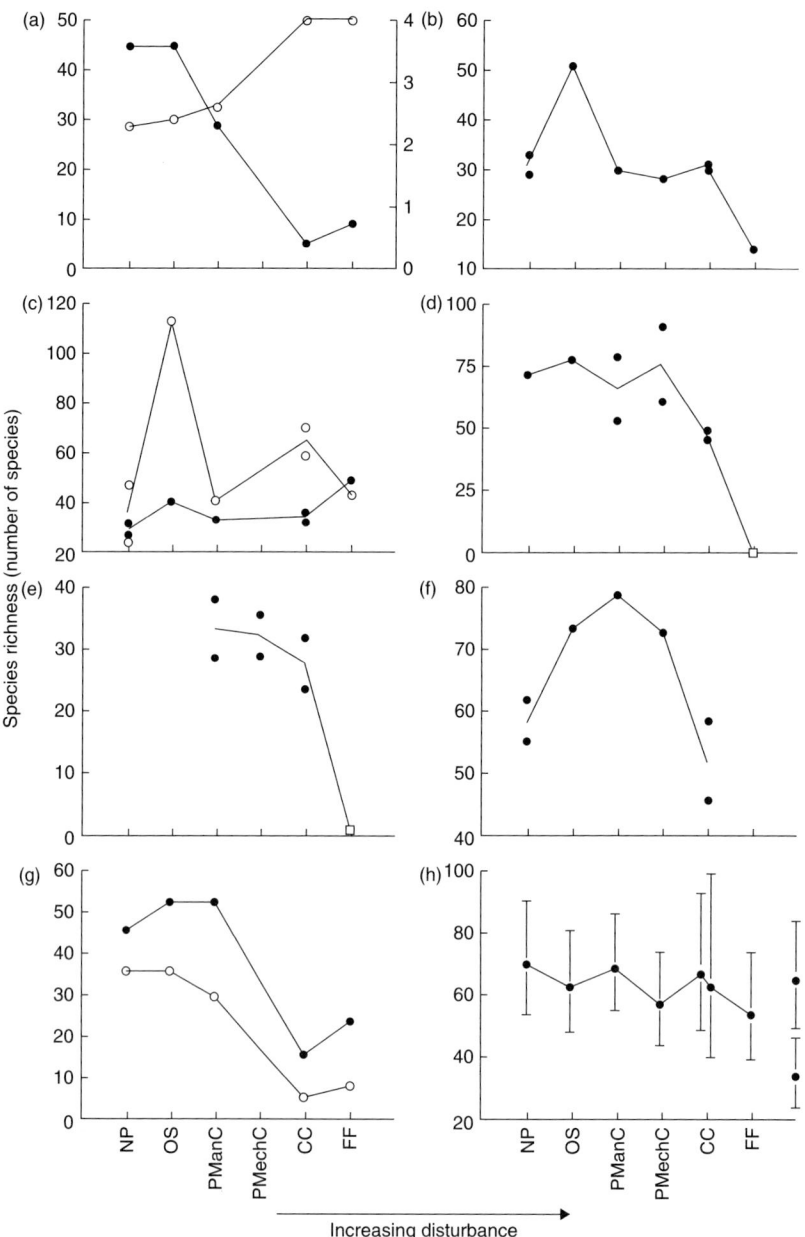

Fig. 2.24. Species richness of various animal groups along a gradient of habitat modification (disturbance) in Cameroon. NP = near primary forest; OS = old growth secondary forest; habitat PManC = old growth secondary forest with some plantation and manual clearance; habitat PMechC = same as previous but with some bulldozer clearance; habitat CC = complete clearance with young plantation; FF = manually cleared farm fallow. (a) Birds (with mean habitat scores on right axis); (b) butterflies; (c) flying beetles (malaise traps = filled circles; flight-intercept traps = open circles); (d) canopy beetles; (e) canopy ants (open squares in (d) and (e) assume richness falls to zero in the absence of canopy); (f) leaf-litter ants; (g) termites; (h) soil nematodes (from Lawton et al., 1998, courtesy Nature Publishing Group).

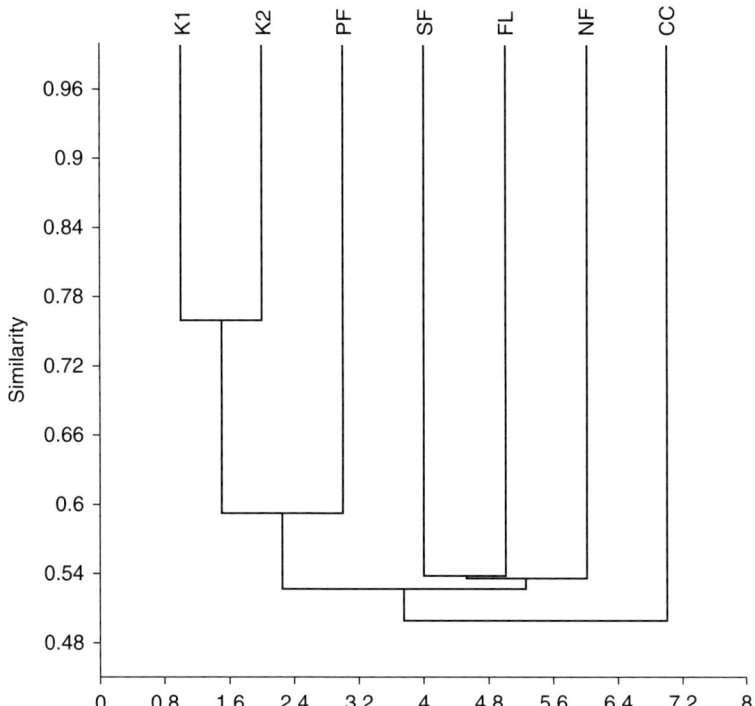

Fig. 2.25. Dendrogram based on cluster analysis of the similarity of butterfly communities in different habitats in Ghana. K1 and K2 = wet primary evergreen forest with only slight disturbance; PF = old primary rainforest with many mature trees and dense shrub layer; SF = secondary forest with a few mature trees and dense understorey vegetation; FL = semi-open area with oil palm and some young trees; NF = young planted forest with even-aged native trees; CC = clear-felled open areas (from Safian et al., 2011, with permission Springer).

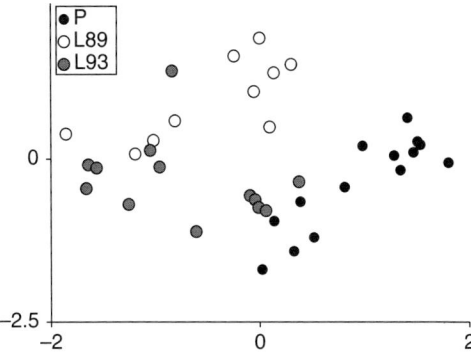

Fig. 2.26. Principal components analysis (PCA) of butterfly community structure in central Kalimantan. P = primary forest; L89 = forest logged in 1989/90; L93 = forest logged in 1993/94. Sampling was carried out in 1998 (from Cleary et al., 2009, courtesy Elsevier).

10 years before the other. Ants are a multi-faceted group of insects, carrying out many ecosystem services in tropical forests, from herbivory to carnivory to scavenging. Studies in eastern Amazonia compared the abundances of 47 ant genera before and after reduced-impact logging (RIL) (Fig. 2.27) (Azevedo-Ramos et al., 2006). Though RIL is supposed to cause significantly less disturbance to the remaining forest when big trees are felled, it can be seen that many ant species are reduced post-logging. Interestingly, though, several genera appeared after logging that were not present before disturbance. Ecosystem services provided by forest insects have also been shown to vary according to the extent of logging, as shown by Fig. 2.28. Here, Slade et al. (2011) studied the

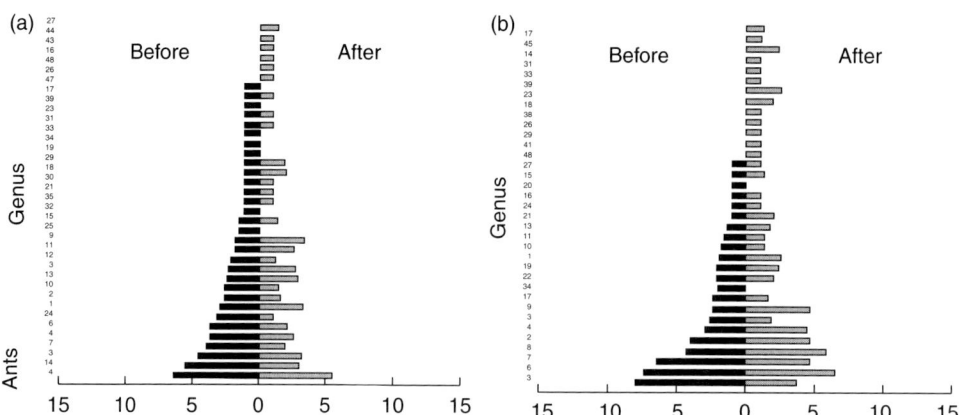

Fig. 2.27. Differences between unlogged and logged forests in the mean abundance of ants at two study sites (a and b) in eastern Amazonia (from Azevedo-Ramos et al., 2006, courtesy Elsevier).

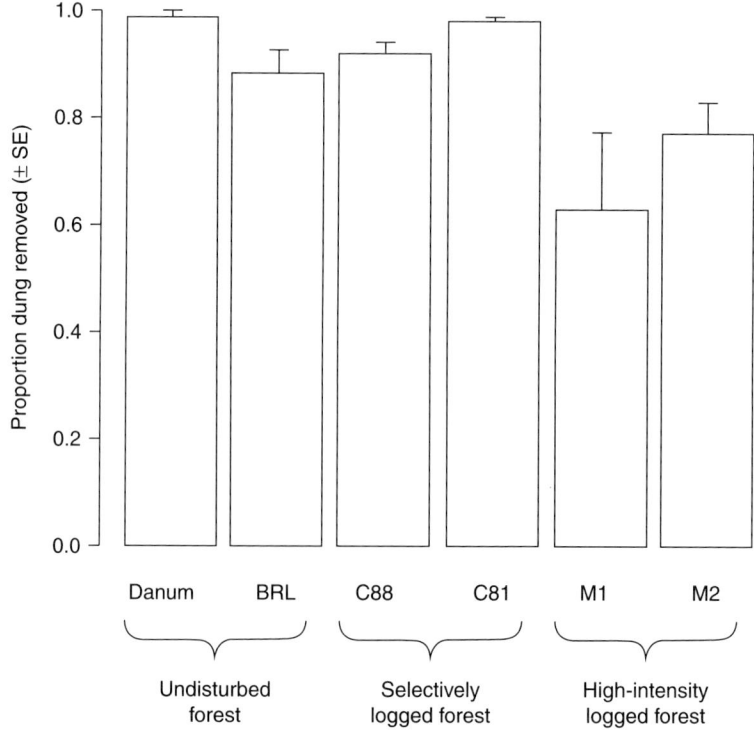

Fig. 2.28. Dung removal by dung beetles over a 24 h period in three types of rainforest in Sabah (from Slade et al., 2011, courtesy Elsevier).

removal of dung by dung beetles (Coleoptera: Scarabaeidae) in Sabah and it was clear that while selectively logged forest did not differ significantly from undisturbed rainforest in terms of the proportion of dung removed in 24 h experiments, heavily logged forests did show a statistically significant reduction in this vital ecosystem service.

2.3.8 Replanting with natives or exotics

Old growth natural forests contain many tree species, a varied age structure of trees from very young to very old, large and small woody debris from twigs to logs and stumps and a wide habitat heterogeneity (Maleque et al., 2009). It is not all that surprising, therefore, that conventional 'wisdom' suggests that plantations of tropical trees will be of very low biodiversity interest, if not completely sterile. However, studies into the insect species assemblages of tropical plantations have shown that many exotic plantations are not biodiversity deserts at all; on the contrary, they do support a rich and varied fauna.

So, it is not unrealistic to use tropical secondary forests in comparisons with much more altered or managed habitats. One study conducted in Sabah investigated the insect diversity of plantations of several exotic tree species (A. mangium, Eucalyptus deglupta, P. caribaea, Falcataria (=Pararaserianthes) moluccana and Gmelina arborea) and compared it with that of natural, albeit secondary, forest in the same locality (Speight et al., 2003). Moths in the superfamily Geometroidea were caught in light traps as indicators of biodiversity and Fig. 2.29 shows that, in terms of the Williams alpha index of diversity, some of the exotics were remarkably diverse. There was, for example, no significant difference between the species diversity of moths caught in natural forest and in eucalypt stands, while even in pine plantations the diversity reached well over 50% of that in the natural forest. It is important to notice that the arthropods on the natural regenerated understorey in the eucalypt stands were remarkably similar to those in the exotic canopy itself; most of the biodiversity of these tropical plantations probably stems from this natural understorey, so that careful silviculture can maintain high levels of biodiversity by promoting indigenous vegetation in forest stands. This may even help in the reduction of forest pest problems. Hawes et al. (2009) also used moths to study the effects of turning natural forests into eucalyptus plantations in northern Brazilian Amazonia (Fig. 2.30). After having standardized the light trap data to take into account the varying light penetration properties of the three habitats sampled, it was clear that, in this case, secondary forest had the highest abundance of moths, whereas eucalyptus plantations had the least. This may, of course, be a good thing from a pest management point of view, since there may be few defoliating moth larvae to attack the plantation trees.

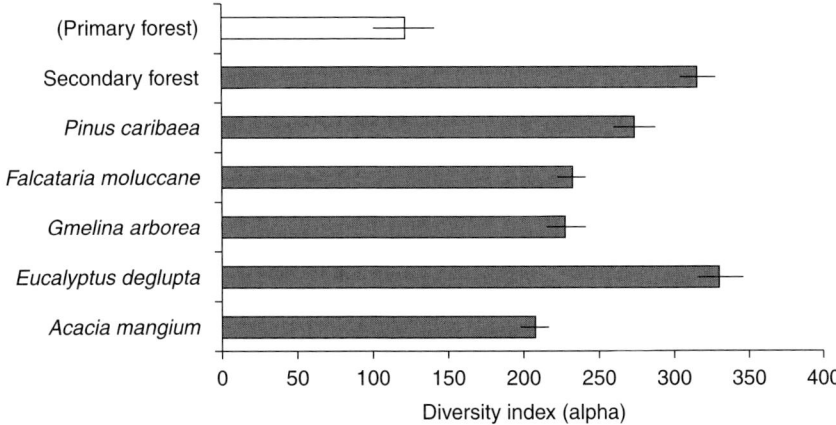

Fig. 2.29. Williams alpha diversity indices (± 95% confidence limits) for moths caught in light traps in different types of forest in Sabah. Primary forest is in brackets because it is located some miles from the others (from Speight et al., 2003).

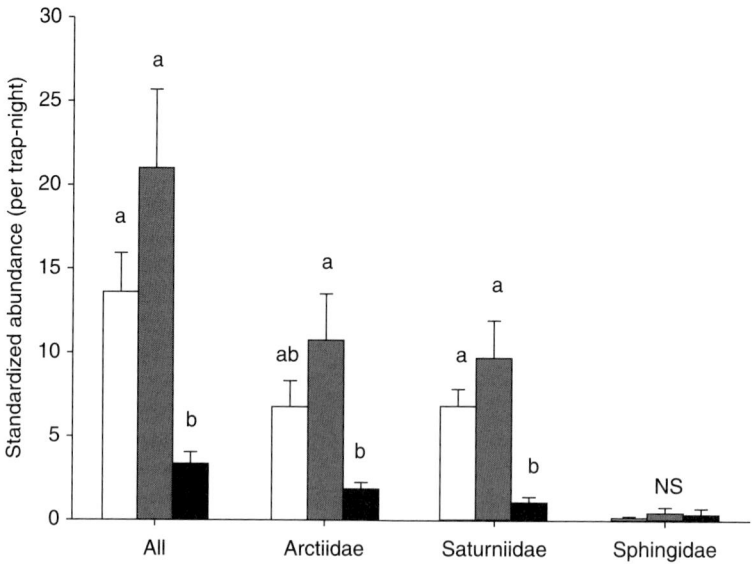

Fig. 2.30. Mean abundance (± SE) of three families of moths and all moths combined, captured in light traps in the Brazilian Amazon. White bars = primary forest; grey bars = secondary forest; black bars = *Eucalyptus* plantations; a and b denote significant differences at $P < 0.05$ levels; NS = not significant (from Hawes et al., 2009, with permission Cambridge University Press).

However, it only takes one species to cause severe defoliation. In a third example, Chung et al. (2000) used flight intercept traps and knockdown mist blowing to sample beetles in primary forest, logged (secondary) forest, plantations of *A. mangium* and oil palm. Canonical correspondence analysis (CCA) (Fig. 2.31) showed that the beetle assemblages in primary forest compared with oil palm were very different, whereas those of logged forest and acacia, though distinct from the first two, were similar to each other.

Certainly, then, converting natural tropical forest to plantations of exotic tree species seems to change the diversity of insect communities living in the forests, which may be a problem for conservationists, but whether or not this will necessarily lead to increased risks from insect pests is another matter, which we will discuss later in the book. For now, Brockerhoff et al. (2008) summarize the situation concerning plantation forests and diversity (Fig. 2.32). Figure 2.32 suggests that there is a continuum of forest types that runs from intensive production, through production and conservation, to pure conservation at the other extreme. As these authors point out, some plantation forests serve multiple purposes of production, protection and conservation and it may be that pest management or prevention systems are more likely to be successful somewhere in the middle of this continuum.

2.4 Biodiversity and Ecosystem Stability – Its Role in Pest Ecology

There may be a tendency for tropical forests to suffer more pest problems as they become more disturbed or perturbed away from natural situations. It is somewhat of a dogma these days to state that more diverse ecosystems are also more stable (McCann, 2000); to put it another way, species-rich communities tend not to exhibit large fluctuations in the abundances of one or more of their constituent species. Proponents of this dogma thus stress the need to promote biodiversity in crop systems to avoid the

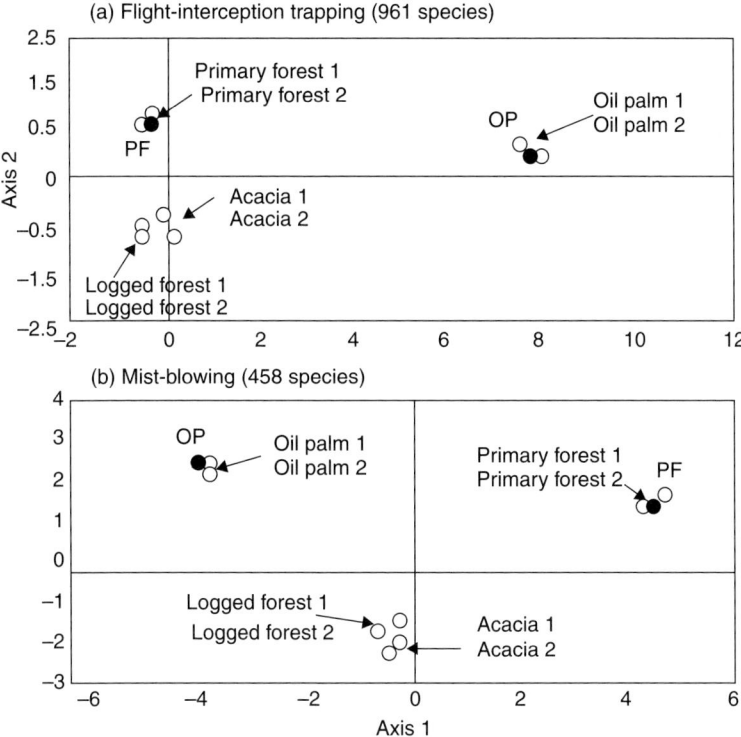

Fig. 2.31. Canonical correspondence analysis of beetle samples collected in four habitat types in Sabah (from Chung et al., 2000, with permission Cambridge University Press).

development of pest outbreaks; this is the basis of ecological pest management. There is nothing wrong with this philosophy in principle, though it may be too simplistic and unreliable in some cases, but one question remains – are biodiverse systems always stable?

Wolda (1992) studied the stability of a large variety of insect populations in relatively undisturbed tropical forests in Panama, emphasizing the variations in abundance of species through time. He found that while some species remained virtually constant in numbers over the 14-year study period, others fluctuated violently. For example, among the group of commonest insect species (those represented by over 1000 individuals), 9% showed variations of over 10% of their mean abundance per year; highest variations exceeded 20%. It has to be said that these figures are not in the same league as some of the veritable plagues of insects seen in certain forest pest outbreaks from time to time, but the work indicates that we should beware of assuming automatically that diverse habitats will never suffer from pest problems. As we shall see in Chapter 8, the promotion of intense monocultures may enhance the probability of pest problems but it is not always the case that trees grown in monoculture have more pests than those in mixtures (Loch and Floyd, 2001).

As we said at the beginning of this chapter, we are concerned with the utilization of tropical forest for human needs of various sorts, and we are not pretending that *no* species are affected by rainforest management, fragmentation or clearance. Undoubtedly, the most sensitive habitat specialists will be the first to become extinct, at least locally. However, since some tropical forest exploitation has to take place for sound humanitarian or economic reasons (or both), then it is important to know what

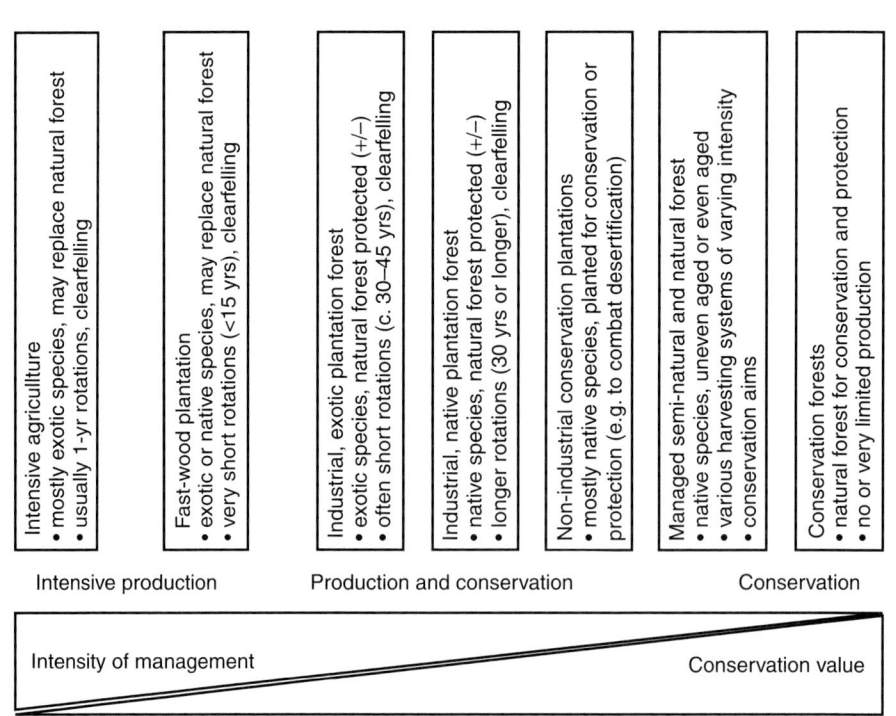

Fig. 2.32. Conceptual model of the relative conservation value of planted forests relative to conservation forests and agricultural land (from Brockerhoff et al., 2008, with permission Springer).

effects we might inflict on natural biodiversity. Brockerhoff *et al.* (2008) have presented some measures which might be taken to optimize biodiversity conservation while still exploiting tropical forest stands for optimal wood production. We would emphasize the vital importance of production silviculture at an industrial scale as well as a local one. Timber in all its forms is a vital commodity in the tropics and we must produce it as cleanly and efficiently as possible in order to protect what biodiversity remains in natural areas. Effective pest management is one key tactic in the pursuit of this aim.

3

Abiotic and Biotic Effects

3.1 Introduction

All animals and plants have evolved in association with their environment, and hence the demands of abiotic (physical factors such as weather and climate) and biotic factors (the influence of other organisms via, for example, predation, parasitism or competition) have had to be met through natural selection and adaptation. The ability of a species or population to increase in numbers, to spread to pastures new or to exploit its food and produce offspring are all influenced by one or both of abiotic and biotic factors, and so we need to understand how these influences function when trying to explain how insect numbers vary in time and space. This is especially important when investigating pest outbreaks.

Most insect populations do not stay at a steady density for long. Some may oscillate gently up and down through time, while others may exhibit enormous peaks, followed by catastrophic crashes. Forest insects are well known to show such cycles (Liu *et al.*, 2007), frequently defoliating their host trees during peak population densities. It has to be said, however, that most quantified examples of oscillations in the numbers of forest insects come from temperate regions. Figure 3.1 shows population trends over about 18 months for the various life stages in the mahogany shoot borer, *Hypsipyla grandella*, from Costa Rica (Taveras *et al.*, 2004). Figure 3.2 looks at the numbers of pest Lepidoptera defoliating *Eucalyptus* plantations in Minas Gerais, Brazil, over 5 years (Zanuncio *et al.*, 2001), while Fig. 3.3 provides data on the impact of just one species, the teak defoliator, *Hyblaea puera*, over 4 years in Chittagong, Bangladesh (Baksha and Crawley, 1998). In all three examples, it can be seen that pest numbers are higher at certain times of year than at others. There is rather clear evidence of peaks and troughs in population densities, but only over a few months or years. These examples are likely to be caused by seasonal changes in weather patterns and/or host plant phenology, rather than long-term cyclical patterns expected of temperate forest insects.

This chapter considers abiotic and biotic factors in detail as separate entities for convenience. Note, however, that most, if not all, of these factors in the environment of a tropical forest insect may be interrelated. The onset of the rainy season, for example, may influence the amount of food available to a leaf-feeding lepidopteran larva, such that competition for food becomes less intense or an increase in air temperature may enable predatory beetles or parasitic wasps to become more efficient at finding their prey or hosts. Wherever possible, we present

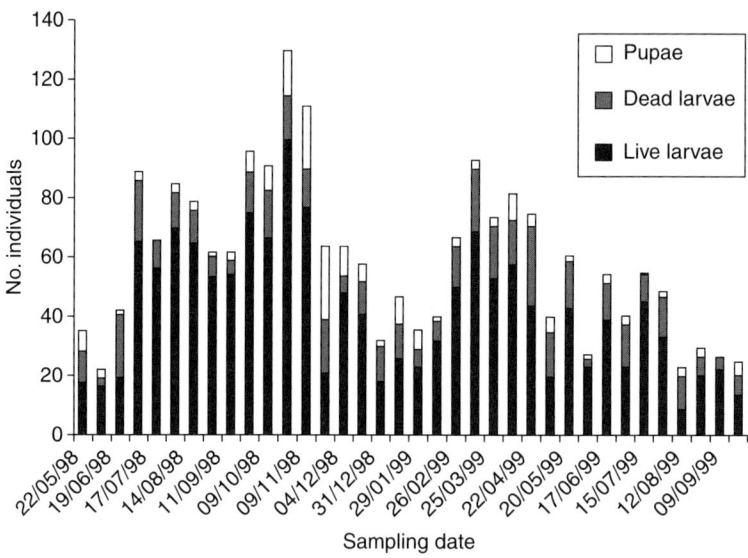

Fig. 3.1. Total numbers of *Hypsipyla grandella* individuals on mahogany trees at Turrialba, Costa Rica (from Taveras *et al.*, 2004, courtesy Wiley-Blackwell).

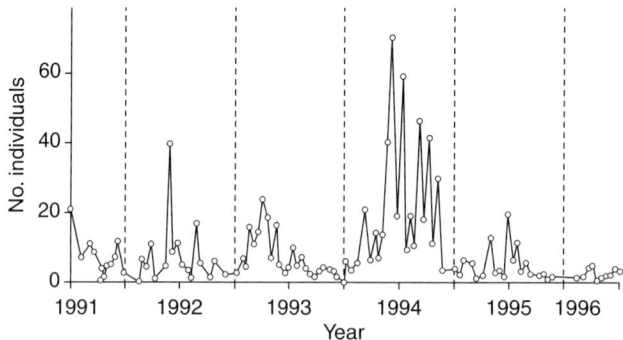

Fig. 3.2. Numbers of individuals of major outbreak species of defoliating Lepidoptera adults collected in light traps per month over a 5-year period in *Eucalyptus grandis* plantations in Brazil (from Zanuncio *et al.*, 2001, courtesy Wiley-Blackwell).

examples from tropical or subtropical forest insects; occasionally, however, important lessons may be derived by examining temperate examples, too.

3.2 Abiotic Factors

Climate and weather (the climate localized in time and space) can have extremely important effects on insects, at both the individual and population levels. Hodkinson (2009) reviewed the ecology of sap-feeding psyllids (Hemiptera–Homoptera: Psylloidea) and after considering all the available literature and analysing life-history parameters for some 342 species of insect, he concluded that 'the main drivers for life-history adaptations are environmental temperature and water availability'. In fact, most tropical climates are rather severe, especially in terms of temperature and humidity, and insects have had to adapt to these difficult conditions (Colwell *et al.*, 2008). During this evolutionary

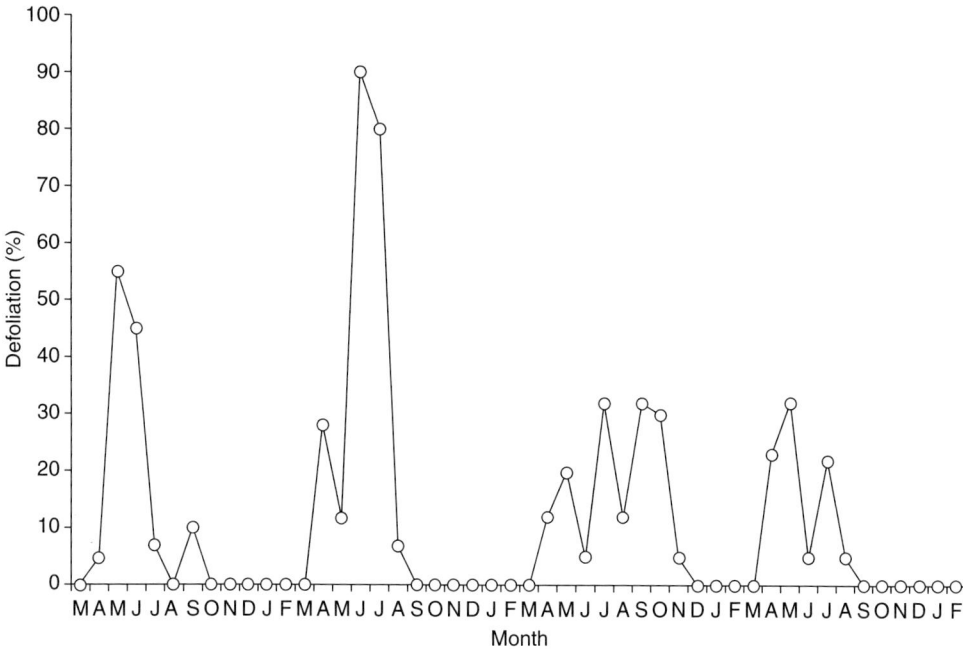

Fig. 3.3. Seasonal incidence of defoliation by larvae of *Hyblaea puera* in an experimental plot of teak in Bangladesh (from Baksha and Crawley, 1998, courtesy Wiley-Blackwell).

process, many species have become 'climate specialists' (Hoffman *et al.*, 2010), wherein they are restricted to a narrow range of altitudes and latitudes. Normally, we are concerned with the changes in weather patterns over no more than a few generations of the insect species and, in terms of pest population dynamics, it is the relatively immediate effects of climatic conditions that concern us most. Again, remember that many factors, though considered separately here, frequently combine, so that, for instance, temperature, rainfall and wind unite to determine relative humidity in a particular locality. Additionally, climate change is likely to influence all such relationships. For example, it is predicted that as global warming continues, insect species are likely to move higher up mountains to remain in their preferred temperature range (Hill *et al.*, 2011).

3.2.1 Temperature

Insects are essentially poikilothermic, in that they are unable to manipulate the temperature of their tissues independently of their surroundings. They are also ectothermic in the main, their major source of heat being from outside their bodies. In general terms, therefore, any changes in the temperature of the immediate environment will have a fundamental effect on the internal temperature of the insect, and hence on its metabolic rate. Metabolic rate determines many processes in an insect's life, including development and growth rate, activity, ability to disperse and reproductive potential, and, as might be expected, there are many examples of direct correlations between temperature and all these factors in both insect pests and their natural enemies. Tropical insects tend only to function effectively and efficiently between extremes of low and high temperatures; outside these ranges, their success can be compromised severely. Take, for example, the sap-feeding psyllid, *Heteropsylla cubana* (Hemiptera: Psyllidae). This notorious pest of multi-purpose trees in the genus *Leucaena* all around the tropical world only exhibits outbreaks above 10°C and below 33°C (Mullen *et al.*, 2003).

Of course, this range of temperatures enables the pest to cover a large part of the tropics, from sea level to quite high altitudes.

Within this thermal range, temperature influences crucial features of an insect's life cycle, such as development rate and survival. Geiger and Gutierrez (2000) studied the thermal ecology of *H. cubana* in the Chiang Mai province of northern Thailand. The development rate increases fairly linearly from around 10°C to nearly 30°C (in keeping with Mullen's data). Survivorship plummets, however, beyond 20°C or so, showing that the quoted temperature range is an absolute maximum, at the higher end at least. From the data, it can be deduced that the optimal temperature for rapid development and high survivorship for *H. cubana* is a narrow 20–25°C.

The slower the development rate, the longer the development time and hence the more protracted the life stage in question. The mahogany shoot borer, *H. grandella*, responds to increasing temperature with a roughly linear increase in development rate and a reciprocal but exponentially decaying development time for eggs, larvae and indeed pupae (Fig. 3.4) (Taveras *et al.*, 2004). As long as the high temperature does not cause mortality, a rapid development rate is of significant benefit, since it reduces the time it takes for a pest to go from a just-laid egg to a newly reproducing adult and thus lowering the probability of being killed by some environmental hazard. Therefore, the generation time for the current example ranged from 104 days at 15°C to 30 days at 30°C. Thus, at the high temperatures, reproductive rate and hence the number of generations per year (voltinism – see below) is optimal for the pest (and, of course, very suboptimal for the trees). Once the pupal stage is reached, temperature also influences the emergence rate of young adult moths, as Fig. 3.5 shows (Taveras *et al.*, 2004). Thus, at high temperatures, all the new adults emerge over just a few days in order to coincide with the best chance of finding a mate, while under cooler conditions, adult emergence takes place over more than 3 weeks. To make matters worse, Taveras *et al.* (2004) found that female shoot borers only copulated at 25°C, showing another example of a climate specialist.

If the seasons remain relatively constant, as they tend to do in parts of the humid tropics, generation times of around 30–35 days can be the norm for forest insects from sap feeders to defoliators, which clearly will result in around ten pest generations per year. Thus, in Bali, Indonesia, scale insects (Hemiptera: Coccoidea) feeding on mangrove saplings have nine or ten generations per year, all of roughly equal duration (Ozaki *et al.*, 1999). In these circumstances, pest population densities can build up very rapidly indeed, causing extensive damage to the host trees. Where climatic conditions merge from temperate to subtropical to tropical, voltinism increases as would be predicted. *Paropsis atomaria* (Coleoptera: Chrysomelidae) is a beetle whose larvae and adults defoliate eucalypt plantations in Australia. This pest occurs in the Australian Capital Territory (ACT), where the climate is temperate, and in central Queensland, where the climate is at least subtropical. In the former region, *Paropsis* has but two generations per year, while in the latter up to four were observed (Nahrung *et al.*, 2008).

All of the above observations are linked to temperature-dependent development rates typically exhibited by insects. Temperature affects the development of insects in a cumulative fashion, so that the effective temperature summation, or day-degree, as it is called, is an important descriptor of how quickly an insect can develop and the next generation begin to reproduce. In simple terms, if the total day-degree summation required for development from egg to adult is, say, 100 day-degrees, then this may be accomplished by 10 days at 10°C, or 5 days at 20°C, and so on. In the case of the olive scale, *Parlatoria oleae* (Hemiptera: Diaspididae), 100% egg laying in Israel requires an accumulation of over 200 day-degrees from the starting date of 15 February in any one year, while complete egg hatch requires around 400 day-degrees (Pinhassi *et al.*, 1996). Though this insect is predominantly a pest of olive trees, it is a widespread forest problem, too, feeding on a very wide range of perennial, woody plants and deciduous trees. In another example, Peres Filho and Berti Filho (2003) investigated the development of egg, larval and

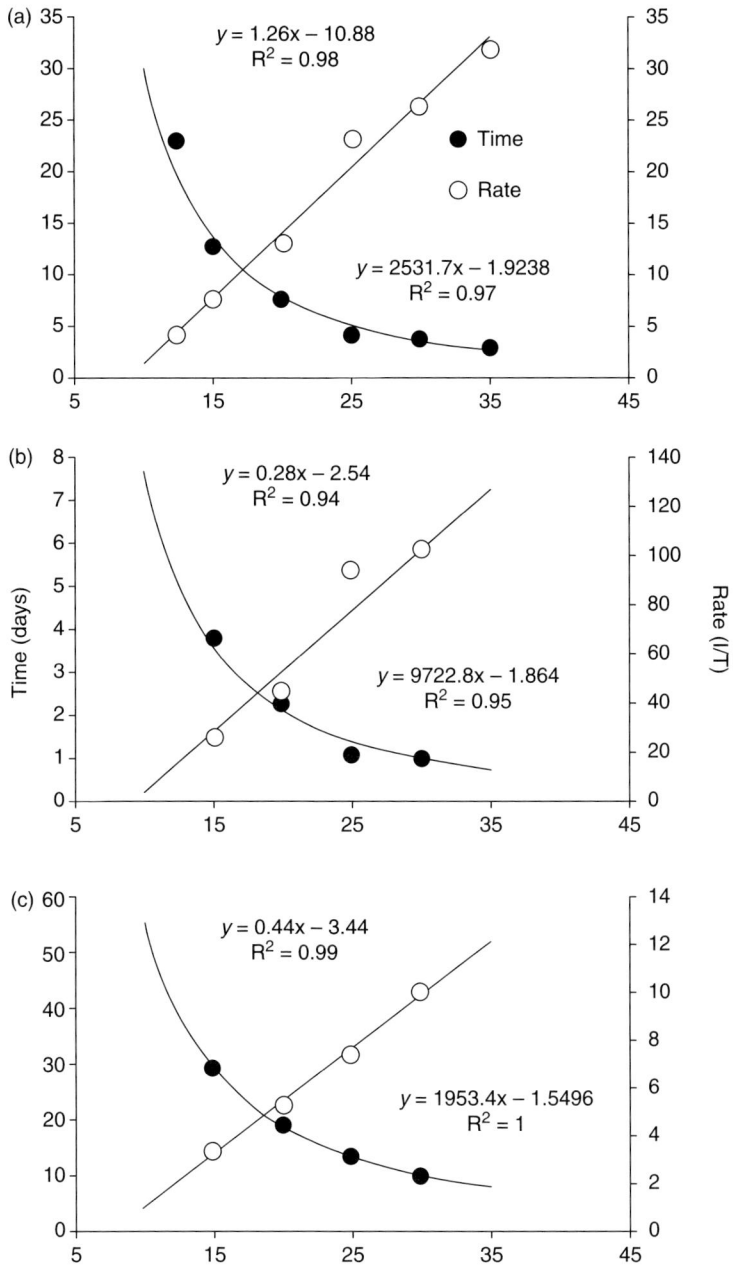

Fig. 3.4. Development time in days (T) and the development rate (1/T) for (a) eggs, (b) larvae and (c) pupae of *Hypsipyla grandella* (from Taveras et al., 2004, courtesy Wiley-Blackwell).

pupal stages of the eucalypt defoliator, *Thyrinteina arnobia* (Lepidoptera: Geometridae), in southern Brazil and found that the egg stage required 156 day-degrees, the female larval stage 496 day-degrees, the female prepupal stage 22 day-degrees and the female pupal stage 133 day-degrees. Thus, a female moth required a total of around 651 day-degrees to develop from a newly hatched larva to a newly emerged adult. This would mean something like 33 days at an average of 20°C.

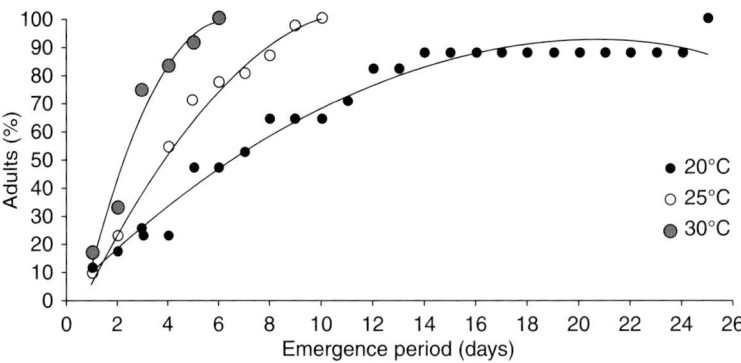

Fig. 3.5. Cumulative emergence of *Hypsipyla grandella* adults from the first day at which emergence occurred at each constant temperature (from Taveras *et al.*, 2004, courtesy Wiley-Blackwell).

Not all of the tropics is hot and some forest plantations are grown at quite high altitudes, with the result that some forest insect pests may be subjected to cold conditions even within the tropics. Low temperatures can stop insect development altogether. Below a certain critical temperature, no development occurs at all. This critical, or threshold, temperature varies from insect to insect but it is usually around 7–10°C (Ye, 1994). Below this temperature, insects merely 'tick over', being unable to grow or reproduce, while remaining alive. In fact, death does not usually ensue until conditions get much colder, and even then, the precise lethal temperature can depend on other climatic factors as well as simple temperature. Adult monarch butterflies, *Danaus plexippus* (Lepidoptera: Nymphalidae), are subjected to subzero temperatures while they spend the winter in high-altitude sites in Mexico. Normally, monarchs exhibit some degree of supercooling, whereby they can withstand below-freezing temperatures to some extent (Anderson and Brower, 1996). However, their capacity to resist the fatal effects of freezing is influenced by wetting and exposure to a clear night sky. Butterflies with wetted bodies die at significantly higher temperatures compared with dry ones, so that those protected from rain under a forest canopy are more likely to survive cold weather. In addition, the use of the canopy as a 'blanket' (Anderson and Brower, 1996) reduces further reductions in temperature.

The latter factor, involving diurnal variations in temperature (and humidity), can be very significant in tropical situations where dense natural forest is replaced by more open and uniform plantations, nurseries or even fields and pasture. In the Atlantic lowlands of Costa Rica, for example, where temperatures in an open pasture in full sun could reach a staggering 51°C, army ants, *Eciton burchelli* (Hymenoptera: Formicidae), tend to remain in adjacent forest fragments where conditions are significantly cooler (Meisel, 2006). Clearly, this is an extreme example, but if an insect pest or its enemies are able to live outside the natural forest, all of their temperature-influenced faculties will be enhanced, possibly to the greater detriment of the new and economically valuable habitat.

3.2.2 Altitude

Various geographical factors are, of course, correlated with temperature. The basic concept of tropical versus temperate insects is clearly temperature related. On a smaller, regional basis, altitudinal variations in climate and vegetation are likely to be linked more to temperature than anything else. It is not surprising to find that temperatures decline as altitude increases in any fixed locality, with predictable effects on insects. Generally, most insect herbivores are expected to be less abundant as altitude increases, though this prediction does not always come about in nature. So, for example, Beck et al. (2011) found that moths in the families Arctiidae and Pyralidae did show reductions in abundances (measured by light traps) in the Ecuadorian Andes (Fig. 3.6), whereas the Geometridae seemed to show a peak in abundance at intermediate altitudes. Under natural conditions, the composition of tropical forest will also, of course, change with altitude. Humid lowland rainforests may give way to cloud forest, followed by dry rainforest, and eventually true forests may disappear altogether as high altitudes are reached. Within a plant species, organic chemistry is known to differ with altitude. Assays were carried out on the foliage of yunnan pine, *Pinus yunnanensis*, from subtropical south-western China (Hengxiao et al., 1999). As altitude increases, the dry weight organic nitrogen content of pine needles decreases, while the levels of defensive (or at least secondary – see Chapter 4) β-pinene increases. It is hardly surprising, therefore, to find that the distributions of forest insects also vary with height above sea level.

Moths have been used frequently to investigate changes in tropical forest structure and function (see Chapter 2), and indeed they have been found to show changes in species diversity with elevation (altitude) (Axmacher et al., 2009). Beck and Kitching (2009) studied one moth family, the hawkmoths (Lepidoptera: Sphingidae), in various parts of South-east Asia and estimated species richness (number of species) from a variety of data sources. Figure 3.7 shows that in all cases, from Malaysia, Indonesia and the Philippines, insect richness declined with altitude. In some cases, such as in Sumatra and Sulawesi, the decline was fairly gradual, whereas in others, such as Irian Jaya and Luzon, richness remained fairly constant until 2000 m or so, whereupon it plummeted. Clearly, this is an association with larval food plants to some extent at least, as well as with temperature and rainfall (see later), and specialist herbivorous insects are likely to be determined by the distributions of their host plants, which themselves will be limited by elevation. Thus, in the Peruvian Amazon the distributions of communities of hawkmoths are linked to the floristic composition of forests at different elevations, with montane cloud forest, for example, showing the highest number of endemic moth species (Ignatov et al., 2011).

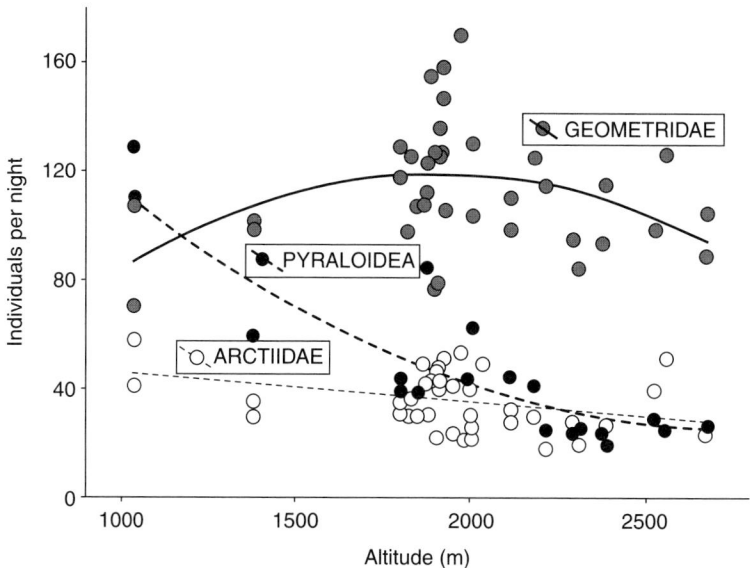

Fig. 3.6. Altitudinal patterns of mean moth abundances per night (corrected for moonlight and temperature effects) in Ecuador (from Beck et al., 2011, courtesy Wiley-Blackwell).

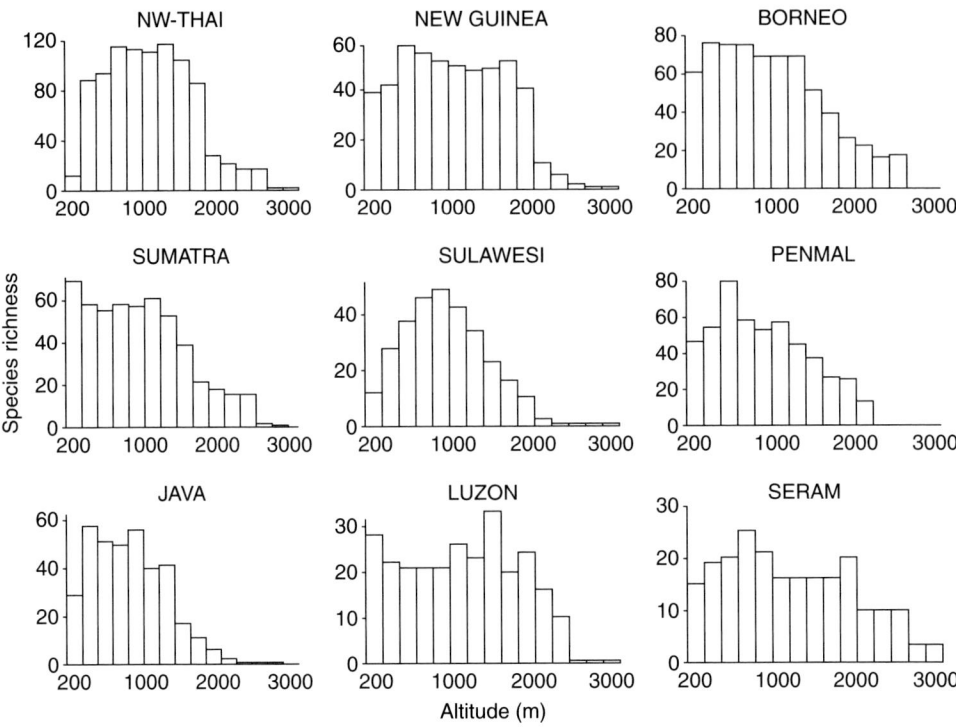

Fig. 3.7. Interpolated species richness of Sphingidae in altitudinal bands of 200 m elevation in various parts of South-east Asia (from Beck and Kitching, 2009, courtesy Wiley-Blackwell).

3.2.3 Sunlight

Too much sunlight is usually unfavourable to insects and trees but the balance between benefits to the tree and detriments to the pest are variable, such that general predictions are difficult to make. Bright sunshine is linked closely to high temperatures and low relative humidities, both of which, if extreme, may be detrimental to insect growth and survival. On the other hand, many examples suggest, though often not conclusively, that forest conditions where some degree of shade is provided, especially in the middle of the tropical day, may diminish pest problems. It can be seen from Fig. 3.8 that the percentage of *Milicia excelsa* leaves infested with galls formed by sap-feeding psyllids (Hemiptera: Psyllidae) rises exponentially with the increasing intensity of full sunlight to which the host trees are exposed (Nichols et al., 1998). In a similar vein, Opuni-Frimpong et al. (2008a) studied the effects of shade on the growth of African mahoganies in the genus *Khaya*, with particular reference to attacks from the mahogany shoot borer, *H. robusta* (Lepidoptera: Pyralidae) (Fig. 3.9). Leaf area index (LAI) is a measure of the degree of canopy closure in the various shade treatments, a surrogate for the amount of light getting to the experimental trees. As can be seen, the deep shade treatment in particular exhibits much less sunlight than the medium shade and the open treatments, and though tree survival after 48 months for both species declines somewhat in the deep shade compared with the open sites, the per cent attack by the pest is reduced to virtually zero in deep shade. Clearly, planting trees susceptible to insect attack in deep shade might be an important pest prevention tactic (see Chapter 8), except that, in this case at least, tree growth rates are probably reduced too much by the deep shade to make the system commercially viable.

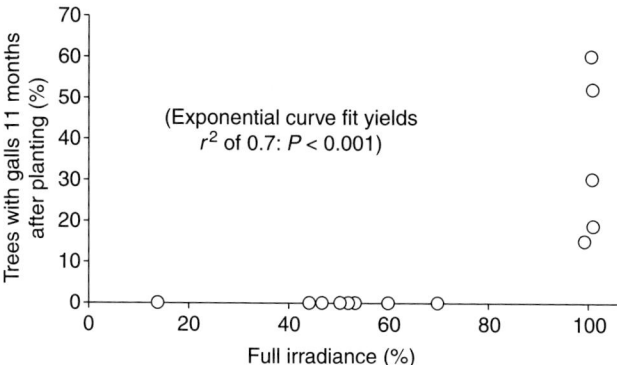

Fig. 3.8. Relationship between percentage of full sunlight (irradiance) in artificial forest gaps and percentage of *Milicia excelsa* seedlings infested with psyllid galls in Ghana (from Nichols et al., 1998, courtesy Elsevier).

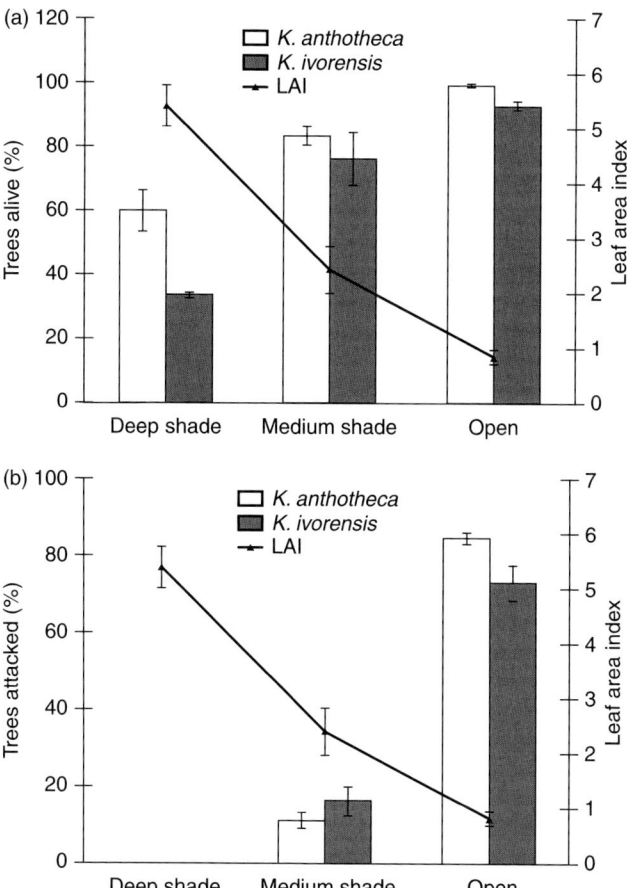

Fig. 3.9. Effect of three levels of overstorey canopy shade on (a) per cent survival of seedlings of *Khaya anthotheca* and *K. ivorensis* at 48 months old in Ghana and (b) per cent of trees attacked by *Hypsipyla robusta* larvae. Line indicates mean (± SE) leaf area index (LAI) (from Opuni-Frimpong et al., 2008a, courtesy Elsevier).

3.2.4 Day length (photoperiod)

In temperate regions of the world, seasons are characterized not only by large temperature variations between summer and winter but also by very significant changes in the hours of daylight, the so-called photoperiod. These variations can have very important influences on both plant and animal physiology and ecology (Speight et al., 2008). In contrast, tropical countries have much less variation in light and dark hours throughout the year, but in some cases an influence on insect populations has been detected. One example of this concerns the leaf beetle communities feeding on *Araucaria* forest in Parana state, Brazil (Linzmeier and Ribeiro-Costa, 2008). Both larvae and adults of leaf beetles (Coleoptera: Chrysomelidae) can be serious defoliators of tropical trees and this study found that the abundance of the beetles varied considerably over a 2-year period (Fig. 3.10). Statistical analyses of biotic and abiotic data showed that, of the environmental variables tested, day length (hours of sunlight) explained more of the variations in beetle numbers than either rainfall or temperature. Beetles were much more abundant during periods of relatively long days and short nights. The authors suggested that this observation was likely to be related to seasonal changes in the host plant (see Chapter 4).

3.2.5 Rain

Van Bael et al. (2004) suggest that since temperature remains relatively constant in one place in the tropics, rainfall may be more influential on insect populations, especially where it varies seasonally. Some tropical forests remain fairly wet all through the year, while others may be seasonally dry. Resultant variations in plant phenology can have significant knock-on effects for insect herbivores residing in these different types of forest (Connahs et al., 2011). Rain can influence tropical insects directly in various

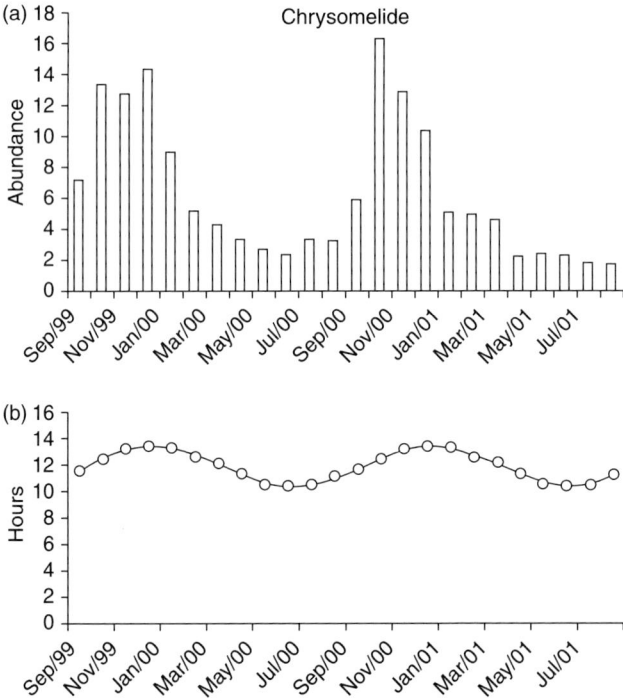

Fig. 3.10. Fluctuations in (a) the abundance of chrysomelid beetles in *Araucaria* forests of Brazil in relation to (b) day length measured as hours of solar light (from Linzmeier and Ribeiro-Costa, 2008, courtesy Sociedade Brasileira de Entomologia).

ways. It can damage them physically if too heavy, it can enhance the likelihood of their contracting diseases by increasing microclimatic humidity, it can prevent them foraging if the microclimate is too dry, it can reduce the temperature around them by evaporative cooling or via the simple fact that clouds hide the sun. In addition, the presence and absence of rain has enormous influence on the host trees, which can run the entire spectrum from flooding and waterlogging to drought. Since most tree species will perform only optimally under a relatively narrow rainfall regime, the ability of herbivorous insects to use them as food will depend very much on their vigour, or lack of it (see Chapter 4).

In reality, it is often difficult to distinguish direct effects of rainfall on insects from indirect ones, i.e. those mediated via the host tree, but in general it appears that damage to trees by insect herbivores is higher in wet conditions (Brenes-Arguedas *et al.*, 2009). Predictably, then, the majority of tropical herbivorous insects are most abundant and/or feed most voraciously from the start of the rainy season and, in a large number of cases, this is when pest damage becomes most severe. Chewers such as weevils (Coleoptera: Curculionidae) (Wolda *et al.*, 1998), sap feeders such as plant bugs (Hemiptera: Auchenorrhyncha) (Novotny and Basset, 1998) and borers such as mahogany shoot borer, *Hypsipyla* spp. (Lepidoptera: Pyralidae) (Floyd and Hauxwell, 2000), all reach maximum densities and impart most impact on trees during the wet or rainy season. Figure 3.11 shows an example from

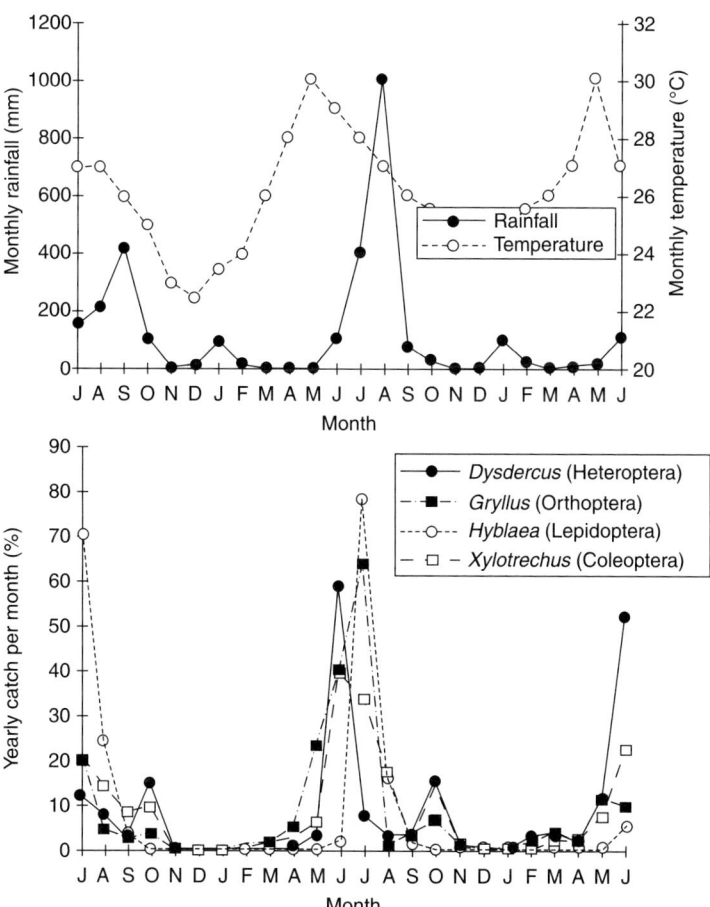

Fig. 3.11. Rainfall, temperature and seasonal captures of four insect pest species in light traps in Madhya Pradesh, India (from Khan *et al.*, 1988).

India where the abundances of various forest insects are presented along with rainfall data for the region. Khan et al. (1988) show how light-trap captures of four insect species with very different lifestyles vary with season, in seemingly close synchrony with rainfall patterns. *Gryllus* sp. (Orthoptera: Gryllidae), for example, is a field cricket best known as a general nursery pest, capable of large-scale destruction of tree seedlings, while *Xylotrechus* sp. (Coleoptera: Cerambycidae) is a cambial and wood borer in the larval stage of more mature trees. *Dysdercus* (Hemiptera: Pyrrhocoridae) and *Hyblaea* sp. (Lepidoptera: Hyblaeaidae) are sap feeders and defoliators, respectively. Note, of course, that light traps will only measure the relative activity of flying adult insects; certainly in the case of *Xylotrechus* and *Hyblaea* only their larvae do direct damage to trees. Indeed, if we look at total arthropod abundance on trees themselves, as in Fig. 3.12, it becomes difficult to separate the effects of rainfall, temperature and tree phenology (Didham and Springate, 2003). None the less, the implications for pest dynamics in tropical forestry are fairly clear and it is easy to show that herbivory is often highly seasonal and correlated with rainfall, as Fig. 3.13 demonstrates. In this example from Uganda, leaf consumption on the tree *Neoboutonia macrocalyx* is related positively to the

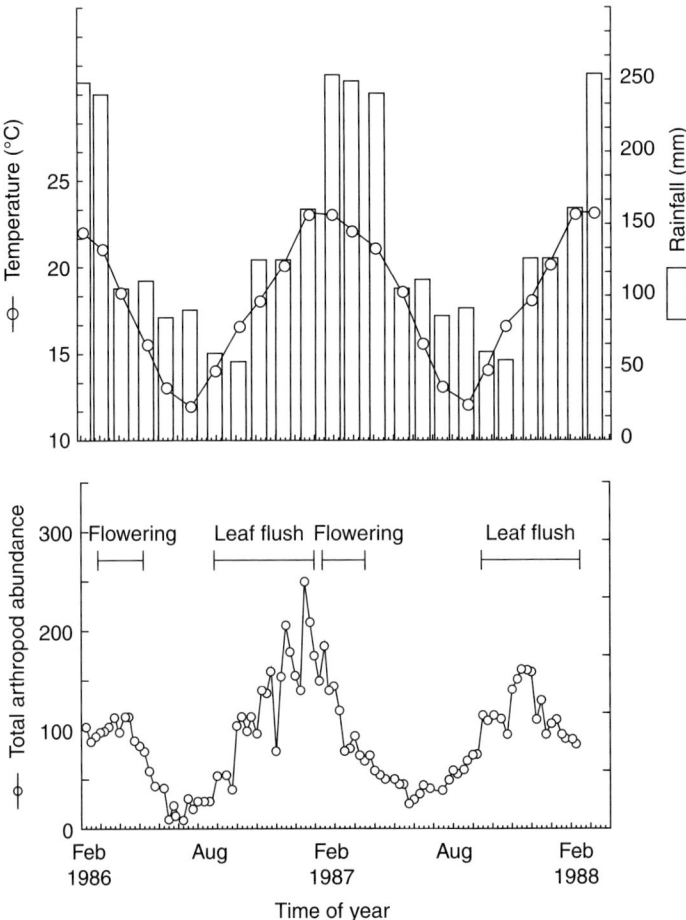

Fig. 3.12. The relationship between canopy arthropod abundance in the crowns of *Argyrodendron* trees in Queensland and variations in biotic and abiotic seasonality (from Didham and Springate, 2003, with permission Cambridge University Press).

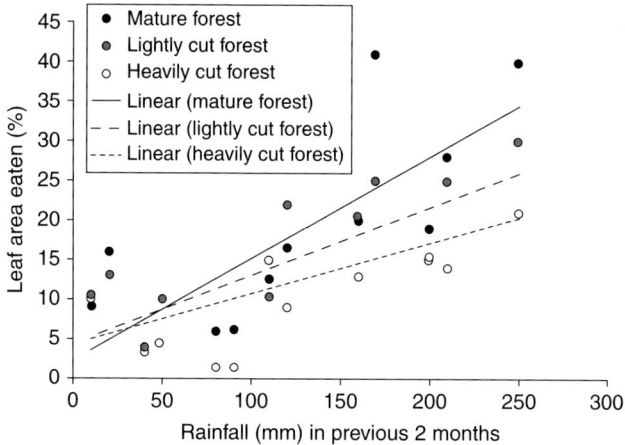

Fig. 3.13. Relationship between rainfall in previous 2 months and the area of *Neoboutonia macrocalyx* leaves eaten by insect herbivores in Uganda. All regressions significant (from Kasenene and Roininen, 1999, courtesy Wiley-Blackwell).

rainfall in the previous 2 months (Kasenene and Roininen, 1999). Increased rainfall stimulates the plant to produce new, flush foliage which is highly nutritious and palatable (see also Chapter 4).

Drought is, of course, a situation caused by lack of rainfall. There are certainly many examples where lack of rainfall can be correlated positively with pest attack. Bark and wood borers are notorious for their associations with dry sites and drought-stressed trees (Hulcr *et al.*, 2008) (see also Chapter 4). Abbott (1993) presents long-term data on the incidence of the pine bark borer, *Ips grandicollis* (Coleoptera: Scolytinae), in plantations of *P. radiata* in Western Australia (Fig. 3.14) *I. grandicollis* was introduced into South Australia in 1943 and, in a separate introduction, appeared in Western Australia in 1952. It migrated northwards and entered subtropical Queensland in 1982 and finally crossed over into the true tropics 12 years later. The species was detected in the more southerly regions from the 1950s onwards but appeared only in outbreak conditions where plantations were suffering significant losses, from the 1970s and particularly into the 1990s. As Fig. 3.14 shows, a decline in annual rainfall over the years is too coincidental with the increasing frequency of beetle outbreaks to be unrelated.

Droughts can also appear to cause big increases in herbivory, which seems at first glance counter to the association between rain and herbivores explained above. In Fig. 3.15, Itioka and Yamauti (2004) show how the numbers of lepidopteran larvae, the biomass of lepidopteran adults and the damage they do to canopy vegetation increase dramatically after a drought event in a rainforest in Sarawak. In reality, the drought (part of the 1998 El Niño Southern Oscillation – ENSO – event; see later in this chapter) ended with heavy rains in April 1998, which resulted in copious and simultaneous leaf flushing of many rainforest tree species. The drought itself was not responsible for the increase in herbivores and herbivory, it merely acted as a stimulant to luxuriant foliage production.

3.2.6 Relative humidity

Relative humidity (RH) is a microclimatic variable that derives from a combination of temperature and moisture (rainfall) and thus might reasonably be expected to influence forest insects in related ways. Insects are prone to desiccation and thus, in many cases, high humidity levels enable them to venture forth, breed, grow and disperse. Take, for example, white grubs. White grubs are the soil-dwelling

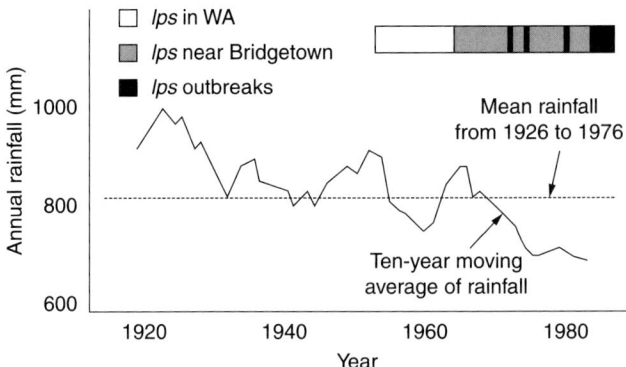

Fig. 3.14. Outbreaks of *Ips grandicollis* in 15,000 ha of *Pinus radiata* in Western Australia in relation to declining rainfall patterns (from Abbott, 1993).

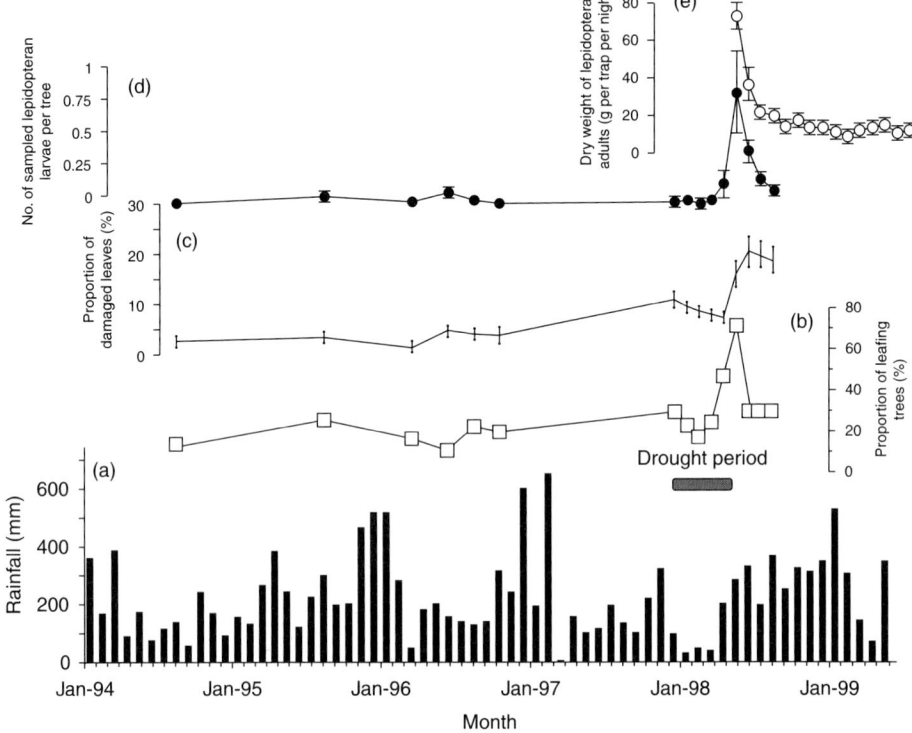

Fig. 3.15. (a) Monthly rainfall – black bars; (b) percentage of leafing trees – open squares; (c) percentage of damaged leaves – vertical bars = SE; (d) mean number of lepidopteran larvae per tree (± SE) – closed circle; and (e) mean dry weight (± SE) of adult lepidoptera caught per night in light traps in rainforest in Sarawak. Drought period (January, February and March 1998) indicated by horizontal grey bar (from Itioka and Yamauti, 2004, with permission Cambridge University Press).

larvae of scarab beetles (Coleoptera: Scarabaeidae) and can be serious pests of nurseries, feeding on the roots of tree seedlings (see Chapter 5). In central India, adults of scarab species in the genus *Holotrichia* emerge from pupae as the RH rises rapidly after pre-monsoon rain, quickly mate and lay eggs in nursery beds of teak seedlings (Fig. 3.16) (Kulkarni *et al.*, 2009). The graph shows a clear link between peak rainfall and maximal

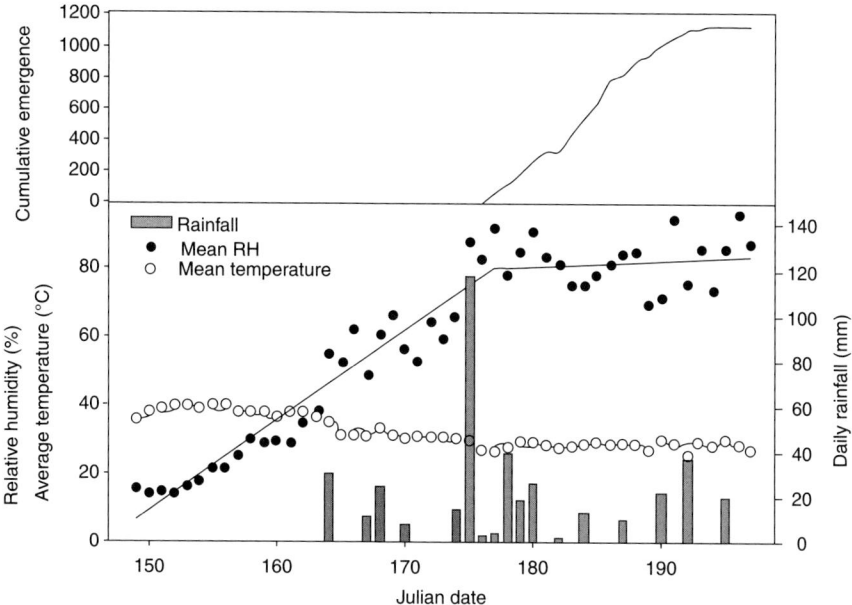

Fig. 3.16. Cumulative emergence of adults of *Holotrichia* spp. in central India, associated with rainfall, temperature and relative humidity in 2003 (from Kulkarni *et al.*, 2009, courtesy Wiley-Blackwell).

humidity, whereas temperature seems less associated. Planting teak in nurseries during the pre-monsoon period may thus exacerbate white grub problems.

The activity of naturally occurring pathogens, fungi in this case, may be influenced by RH. The larvae of *Pteroma pendula* (Lepidoptera: Psychidae) are bagworms which defoliate *Acacia mangium* in parts of India. Fungal infections are correlated positively with rainfall and about 90% of infected bagworms in one study were found in the middle or bottom of tree canopies, where the RH was much higher than the relatively dry upper canopy (Sajap and Siburat, 1992). In contrast, where insects associate with pathogens such as fungi via symbiotic relationships, such as with bark and ambrosia beetles (Coleoptera: Scolytinae), greater humidity in the forests promotes more beetle success and abundance (Hulcr *et al.*, 2008). It has even been suggested that manipulation of the RH in a forest may be useful in diminishing the incidence or activity of insect pests. Certainly, various types of termite are thought to be very susceptible to desiccation (Cornelius and Osbrink, 2010) and cannot forage far in conditions of low humidity. In Indonesia, for example, Teugh-Hardi (1995) suggests that subterranean termites in the genus *Coptotermes* (Isoptera: Termitidae) are influenced greatly by RH, and hence if the first thinning of *A. mangium* stands is carried out after canopy closure, followed by pruning and the elimination of undergrowth, the air humidity in the forest will be decreased and provide unfavourable conditions for the termite's growth. This may work for one pest, but of course it might adversely influence the mortality of the bagworms mentioned earlier by drying out the middle and lower canopies and hence reducing the incidence of fungal infection.

Finally, it appears that high humidity can interfere with chemical communication in some insects. As mentioned earlier in this chapter, the bark beetle, *I. grandicollis* (Coleoptera: Scolytinae), is a serious introduced pest of pines in many parts of Australia from Tasmania to Queensland. Flight experiments carried out by Bassett *et al.* (2011) have shown that very high humidities adversely affect the responses of adult male *Ips* to pheromone baits (Fig. 3.17). This factor may be a

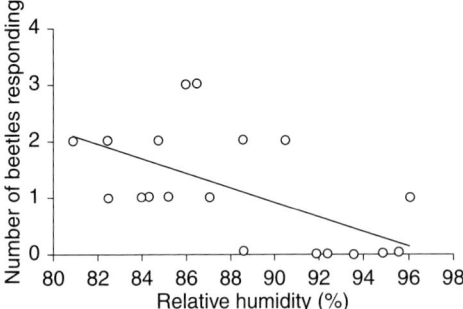

Fig. 3.17. Number of *Ips grandicollis* adults arriving at sex-pheromone baited filter paper in relation to humidity (from Bassett et al., 2011, courtesy Wiley-Blackwell).

'double-edged sword' in that the spread of the pest may be reduced in very humid regions but it may also reduce the usefulness of pheromones in pest monitoring and control programmes.

3.2.7 Wind

The movement of air currents can occur on a large scale, over long distances and with great strength, and can be of great importance in the migration and dispersal of forest insect pests. Alternatively, though, small air movements can be very important in insect communication, especially when adults are attempting to find an individual of the opposite sex for mating using chemical cues.

In 1988, K.S.S. Nair of the Kerala Forest Research Institute (KFRI) in south India published an account of a dramatic event which occurred in a teak plantation in Kerala State (Nair, 1988). To quote: 'A group of at least two million *Hyblaea puera* moths [teak defoliator – see Plate 19] descended on about 20,000 newly flushed teak trees in a 30-ha patch within the plantation. Together they deposited between 50 to 100 eggs per leaf on most of the tender leaves in the top canopy, and by morning, all the moths had disappeared. Within a fortnight, the trees were stripped clean by the feeding caterpillars, and the falling frass sounded like mild rainfall on the dry leaves. Other nearby plantations remained untouched. No one knows where the moths came from or where they went.' Clearly, aerial migrations of pests can occur on an enormous scale and wind is the primary factor determining how far they can travel and in what direction.

Chandrasekhar et al. (2005) studied the dispersal and migration of *H. puera* using molecular markers and randomly amplified gene-encoding primers (RAGEP) to identify and track various populations of the moth. They concluded that outbreaks tended to be caused by long-range movements of the insect rather than local, endemic dispersal. The patterns of long-range dispersal of *H. puera* are illustrated in Fig. 3.18. Here, the probable migratory paths of the adult moths are shown moving up from southern India, in April or May, and across from east to west to infest the central regions of Madhya Pradesh (Bhowmick and Vaishampayan, 1986). These migrations are fuelled by monsoon winds and the distribution of outbreaks in time and space, at least on a regional scale, are a function of the wind strength and direction. Teak plantations usually are supplied with the defoliator by more local movements of adult moths from natural forests where the pest is normally endemic but long lasting, again directed by wind currents. Stimuli to leave high-density patches of pests may result from intraspecific competition (Baksha and Crawley, 1998). One of the major problems with the management of pests such as *Hyblaea* that are carried by the wind in this fashion is the unpredictable nature of air currents, which makes pest forecasting a tricky and unreliable business.

Often, when winds blow over relatively long distances, aerial insects become concentrated at particular heights in the air column (Wood et al., 2010) and come together as a result of weather patterns in areas known as convergences (Drake and Farrow, 1989). In these situations, very high pest densities can be reached rapidly, since very large numbers of reproductive adults breed together in one place, or large densities of dispersing larvae, carried on silk threads, can be deposited on small areas of habitat. Such pests include armyworms, *Spodoptera* spp. (Lepidoptera: Noctuidae), in parts of

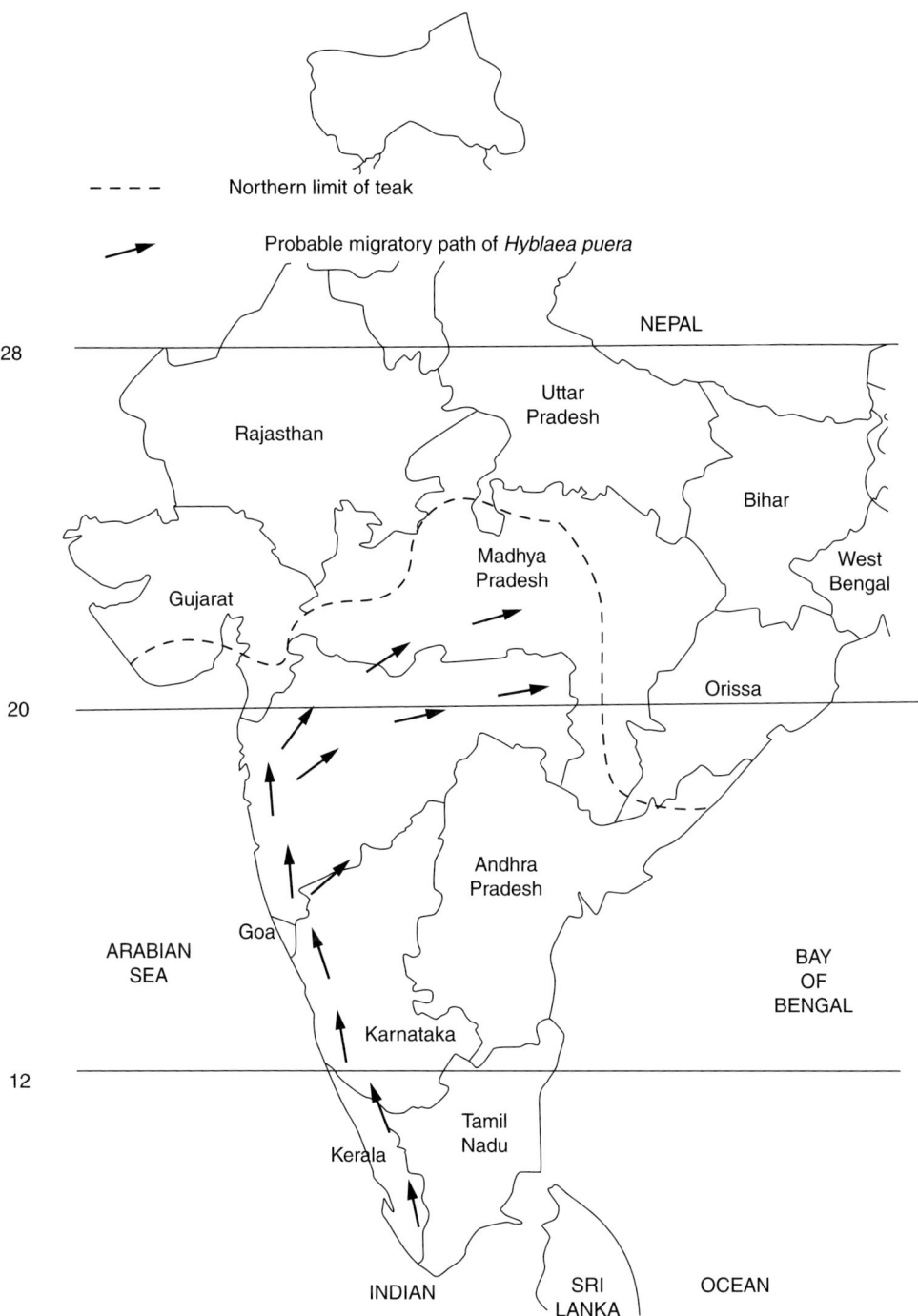

Fig. 3.18. Teak growing and the migration pattern of *Hyblaea puera* on monsoon winds in India (from Bhowmick and Vaishampayan, 1986).

Central and South Africa, spruce budworms, *Choristoneura* spp. (Lepidoptera: Tortricidae), in North America, and locusts, *Schistocerca* or *Locusta* spp. (Orthoptera: Acrididae), in various parts of the Near East, North Africa and South America. An extra benefit for the pests is that these wind convergences are often areas typified by wet weather, so that in semi-arid situations the large numbers of converging insects are provided with newly sprouting vegetation to eat.

Relatively small distances can also be covered by forest pests on the wind. In Ecuador, for example, the bagworm, *Oiketicus kirbyi* (Lepidoptera: Psychidae), occurs on occasion in defoliating outbreaks in mangrove forests dominated by trees such as *Rhizophora mangle* (Rhainds et al., 2009). These outbreaks are found to spread from initial foci in the direction of the prevailing winds. The agents of dispersal in this case are young to mid-stage larvae 'ballooning' on silk threads.

Even if forest insects remain in one place, wind, or at least the gentle movement of air currents, can be a vital component of communication between individuals, in particular those which use chemicals, or pheromones, to attract mates and/or to locate suitable host plants (see also Chapter 4). A small molecular mass volatile scent released by a small point source such as an adult female moth or a small bark beetle (Coleoptera: Scolytinae/Platypodidae) of either sex will not penetrate at all far in still air by diffusion alone. Light winds through a forest stand produce instead a pheromone 'plume', which provides insects of the same species but usually opposite sex with both a direction in which to fly and a concentration gradient up which to progress. The management of certain forest pests with synthetic pheromones (see Chapter 8) would not work without wind movement.

A final effect of wind on forest insects is the most extreme, indeed potentially catastrophic. Tropical storms, hurricanes and cyclones tend to destroy living forests, converting them into so much dead material lying on the ground (Plate 11), just waiting for potential forest pests to use them as breeding sites. Opening of the canopy then enables luxuriant secondary growth to spring up. Severe Tropical Cyclone Harry hit the Atherton Tablelands in north-east Queensland in March 2006, causing severe damage to large areas of rainforest (Grimbacher and Stork, 2009). Beetle sampling before and after the cyclone revealed significant changes in forest structure and associated beetle faunas, including the potentially pestiferous bark and ambrosia beetles (Coleoptera: Scolytinae) and the leaf beetles (Coleoptera: Chrysomelidae) (Fig. 3.19). The most significant changes involved the increase in numbers of scolytines (note the log scale on the *y*-axis of graph (c)). If this had been a plantation forest, the potential for serious damage to surviving trees would have been high via mass outbreaks of beetles breeding in the cyclone-damaged dead wood.

3.2.8 Climate change

It must be clear by now that, in most cases, climatic and weather factors combine to dictate the behaviour and dynamics of insect populations. For example, Kumar et al. (2010) used principal components analysis (PCA) to study the combined effects of temperature humidity and rainfall on a beetle defoliator of *Gmelina arborea* trees in India, *Craspedonta leayana* (Coleoptera: Chrysomelidae). They found that rainfall plus minimum and maximum temperature were among the most influential factors on pest population dynamics. Because of these complex multivariable interactions, it is difficult to predict accurately the effects of climate change – one variable may go one way, while others do the reverse, as it were. Elevated temperatures, higher levels of atmospheric carbon dioxide, and of course sea level rises, may be unavoidable features of the next few decades, but it is difficult at the moment to be sure about how these problems will affect the numbers and the impact of tropical forest insects. From what little is known, it is suggested that the most dramatic effect on, for example, insects in moist tropical forests will be an increase in the rates of herbivory linked, to some extent

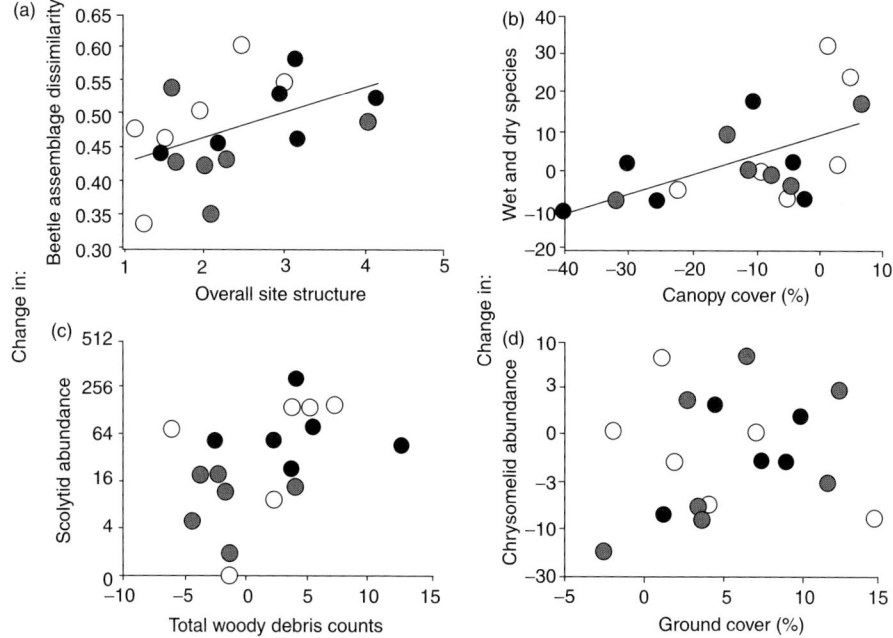

Fig. 3.19. Relationships between pre-Cyclone Harry and post-Cyclone Harry relative values of 18 rainforest sites in north-east Queensland. (a) Overall site structure and beetle species composition; (b) canopy cover and the abundance of wet- and dry-adapted beetle species; (c) wood debris counts and the abundance of scolytine beetles; and (d) cover of vegetation < 2 m high and the abundance of chrysomelid beetles. Different symbols denote types of remnant forest (from Grimbacher and Stork, 2009, with permission Springer).

at least, to changes in plant productivity as a result of higher atmospheric CO_2 levels (Rao et al., 2009). If extremes of drought become more widespread or prolonged because of water shortages and high temperatures, increased tree stress is likely to occur (Martinez-Ramos et al., 2009), especially in plantations on unsuitable sites (see Chapter 4), which may worsen pest damage. We know also that indirect measurements can be linked to changes in forest insect behaviour, so, for example, Srygley et al. (2010) found that sea surface temperature (SST) anomalies correlated with El Niño events were related to numbers of adult butterflies, *Aphrissa statira* (Lepidoptera: Pieridae), migrating in Panama (Fig. 3.20), though these relationships need not imply direct cause and effect. Increased soil salinity as a result of tidal inundations in low-lying areas similarly may exacerbate risks from insect pests. As yet, however, no solid predictions can be made for tropical forest insects.

3.3 Biotic Factors

Biotic factors which may influence the lives and population densities of tropical forest pests derive from living organisms which are themselves capable of growth, reproduction and dispersal, at least in the majority of cases. Such organisms may be members of the same species, or other species belonging to the same trophic level, and are hence competitors for a resource such as food or space. In this respect, the ability of the host tree to provide insects with large quantities of high-quality food is of paramount significance. This topic is discussed in detail in Chapter 4.

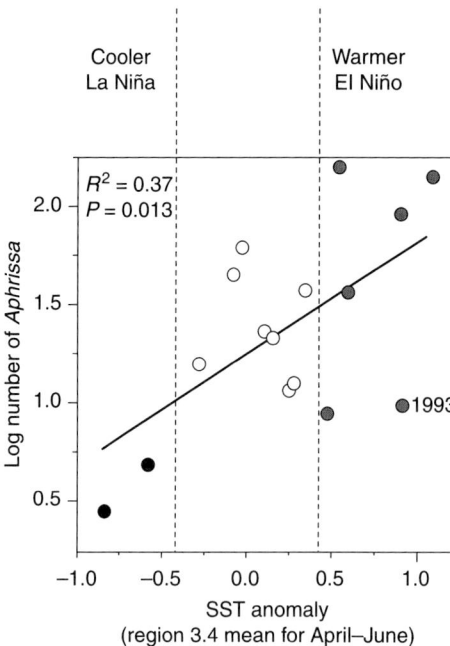

Fig. 3.20. Log-transformed migratory rate of *Aphrissa* butterflies in Panama in relation to the average sea surface temperature (SST) anomalies. Grey circles = El Niño years; Black circles = La Niña years; White circles = normal years (from Srygley et al., 2010, courtesy Wiley-Blackwell).

Alternatively, these organisms may belong to a higher trophic level and are known as natural enemies. These can be split into three subgroups for convenience, the predators, the parasitoids and the pathogens. Predators kill and eat their prey and include a wide range of free-living animals, from vertebrates such as birds and mammals to invertebrates such as other insects. Parasitoids are a much more specialized group of insects, where the whole larval stage of the natural enemy is spent feeding either externally or internally on the body of the host insect. The host (our pest) therefore dies during one completed generation of the parasitoid. This is distinct from the more normal lifestyle of a parasite, where the enemy can undergo many generations in close association with the host without killing it, as is the case of fleas or tapeworms. Finally, pathogens are defined best by listing the major groups rather than by attempting to produce a coverall description (see also Chapter 8).

In most cases, for competitors or natural enemies to have a significant impact on the dynamics of a pest species or population, they must respond to fluctuations in the density of the pests in such ways as to regulate the numbers of the latter.

3.3.1 Population regulation

Strictly speaking, population regulation is the maintenance of the numbers of a pest species at a relatively constant level. In Fig. 3.21, two hypothetical populations are presented, both with the same average density (equilibrium) but one with much higher peaks and lower troughs through time compared with the other. By definition, the so-called 'boom and bust' population with large oscillations in population density is less well regulated than the population with small oscillations.

The mechanism via which regulation functions, and from which population cycles occur such as those depicted in Fig. 3.21, is known as density dependence, defined as a proportional increase in the effect of a factor such as competition or predation, as the density of the pest population also increases (Fig. 3.22). A factor capable of density dependence may be of great value in preventing pest outbreaks or in reducing epidemic numbers of pests to less damaging levels. The efficiency of this density-dependent response at varying pest densities is of fundamental significance in the life of a pest, and in particular to its outbreak dynamics. It is important to distinguish between hypotheses that concern the role of natural enemies (known as 'top-down' regulation) and hypotheses that concern the role of host plant quantity and quality via competition (known as 'bottom-up' regulation) (see also Speight et al., 2008). Additionally, populations may not respond immediately to changes in density and top-down or bottom-up factors, and a time lag may occur between cause and effect. This is termed 'delayed density dependence' (Beckerman et al., 2002). Finally, there is still debate as to whether top-down regulation is predictably more important in population dynamics than bottom-up in certain habitats or ecosystems (Hunter, 2001). So, we might suggest that insect pests in tropical forests are always regulated by their food supply (our forest trees) via intraspecific competition (see next section) and if this were the case, we might also suggest that top-down regulation (for instance, via the predatory or parasitic effects of natural enemies) might be ineffective. Life is, of course, never that simple.

3.3.2 Competition

Competition among herbivorous insects is mainly for food, though space may also become a limiting factor. Chapter 4 discusses the importance of the host plant in the ecology of tropical forest insects in detail and so only the basics of competitive interactions will be presented here.

We can recognize two distinct types of competition, one between members of different species (interspecific competition) and the other between members of the same species (intraspeciflc competition). Interspecific competition (also known as 'interference' competition) is the less readily detectable of the two and is most often represented by its result in terms of niche separation between two insect species. One example involves leaf-cutter ants (Hymenoptera: Formicidae) from Costa Rica. Two species, *Acromyrmex coranatus* and *Atta cephalotes* (Plates 16 and 17), both have small foraging workers (3.4 ± 1.4 and 3.3 ± 1.0 mg, respectively) (Wetterer, 1995) which cut leaves from small herbaceous plants in the forest and return them to the nest to provide a substrate for fungal growth. Clearly, because these two species feed on the same food resource and are the same size, there would be a high likelihood of interspecific competition. It is suggested that this is avoided by little or no geographical overlap between the two species in the country.

Intraspecific competition is much more of an everyday sort of event in the life of a

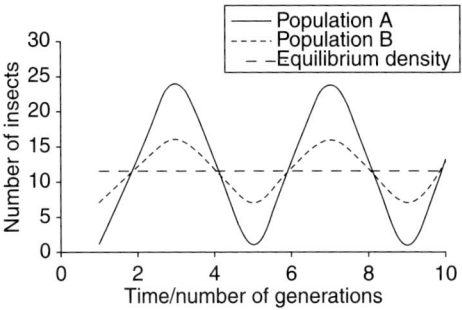

Fig. 3.21. Variations in density of two hypothetical insect populations.

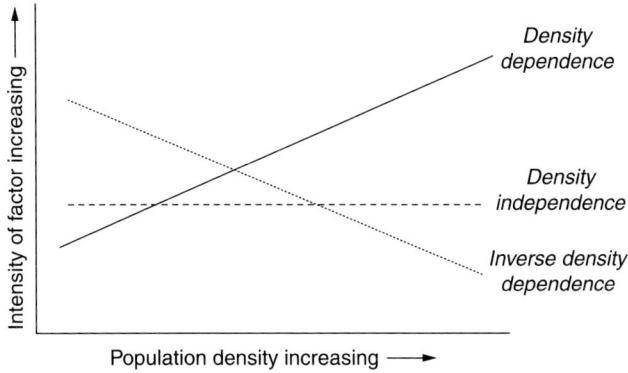

Fig. 3.22. Diagrammatic representation of three types of density dependence. 'Intensity' on the y-axis implies a proportional change. Note that for population regulation to operate, the gradient of the line must be positive.

tropical forest insect. Food supply, whether it be sap, seeds or leaves, is a finite resource which can only support a maximum number of insect individuals, the carrying capacity of which will be a function of quality of food as well as quantity. Thus, if a fixed amount of leaf material is of low nutritional status, then it will support fewer individuals of herbivores than it would if the quality were higher. Therefore, as insect density increases, during, for example, a pest outbreak, there will be little food to go around and all potential consumers are likely to suffer via reduced growth rates, lower fecundities or starvation. This is termed 'resource limitation' (see also Chapter 4).

Such is the case with bagworm moths (Lepidoptera: Psychidae). A high density of larvae on one host plant is known to reduce the 'performance' of each individual insect (Rhainds et al., 2009). This performance may be manifested in terms of larval growth rate, mortality, in the final size of each pupa or adult female, which in turn is linked directly to female fecundity or the rate of dispersal from a high-density population (Strevens and Bonsall, 2011). Scale insects are sap-feeding pests of many forest trees and agricultural and horticultural crops common in both temperate and tropical regions of the world. Sooty beech scale, *Ultracoelostoma assimile* (Hemiptera: Margarodidae), attacks *Nothofagus fusca*, a tree species native to New Zealand, by feeding on the sap beneath the bark (Wardhaugh and Didham, 2005). The fecundity of beech scale declines as their density increases, and this density-dependent response appears not just in a small patch of bark but over various spatial scales up to and including the whole tree (Fig. 3.23). The authors point out that the intraspecific competition between feeding scale insects increases with spatial scale, suggesting that these insects affect the sap content of a wide area of the host tree, not just in the local area of trunk where they are feeding. In a different example, this time from China, outbreaks of the bark beetle, *I. cembrae* (Coleoptera: Scolytinae), were investigated by Zhang et al. (1992). Here, the number of egg niches and adult offspring produced per maternal gallery declined exponentially

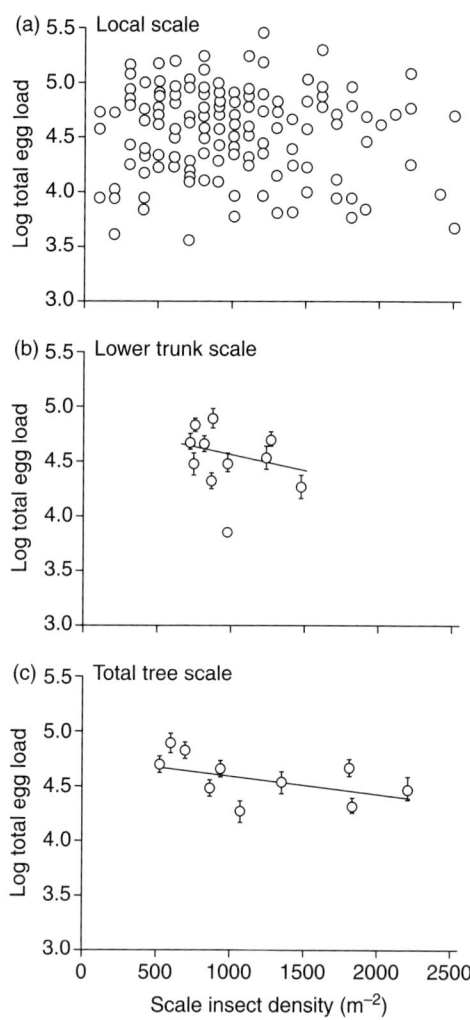

Fig. 3.23. Relationship between log10 total egg load and (a) local (10 × 10 cm^2) density; (b) lower trunk density (± SE); and (c) total tree density (± SE) (from Wardhaugh and Didham, 2005, courtesy Wiley-Blackwell).

with attack (pest) density and the egg-to-adult mortality increased exponentially as the attack density increased, all else being equal. The exponential relationships were attributed to intraspecific competition between female adult parents for breeding space and between larvae for limited food resources.

In general terms, then, it is easy to imagine that outbreaks of forest insects are typified by reductions in the food resource, leaves, sap or bark, as the case may be. This reduction, or in fact complete disappearance on occasion, results in very significant reductions in pest numbers in subsequent generations, either by reducing survival or fecundity or by initiating migration to pastures new. This bottom-up resource limitation has to be borne in mind when reviewing the likely efficiency of top-down natural enemy regulation.

3.3.3 Natural enemies

As mentioned above, natural enemies that may have potential in biological control in tropical forestry fall into three basic categories. Predators, from birds and mammals to spiders and other insects, search for their prey and eat it. Parasitoids are a special type of parasite; in fact, they appear to be a halfway house between true parasites and predators because they consume their host, in this case the pest insect, and complete the entire generation of themselves on or in one pest item. They are comprised mainly of Hymenoptera (parasitic wasps) and Diptera (parasitic flies). Finally, we have the pathogens, or disease-causing organisms, which in tropical forestry at least consist mainly of viruses, protozoa, bacteria, fungi and nematodes. The likely efficiency of each of these groups has to be considered in the light of the pest's population dynamics, as mentioned above. It is common to observe fluctuations or oscillations in natural enemy density in some way linked to similar fluctuations in that of the pest, as exemplified in Fig. 3.24. *Dalbergia sissoo* is a major timber tree in parts of India such as Punjab, but it can be heavily defoliated by various moth larvae, including *Plecoptera reflexa* (Lepidoptera: Noctuidae). As the figure shows, pest densities fluctuate, probably seasonal (see earlier in this chapter), and these fluctuations appear to be mimicked by the numbers of the pest larvae parasitized by various hymenopteran parasitoids (Prasad *et al.*, 2002). There is a tendency for peak parasitism levels to occur either at the peak of the pest numbers, or in fact slightly later. What is not clear is whether the pest drives the natural enemy dynamics or vice versa.

Perhaps the most fundamental problem concerning natural enemies such as predators and parasitoids is that they may be less efficient at dealing with high numbers of prey or host items in a given time. Figure 3.25 illustrates a basic functional response shown by a parasitoid of the red gum psyllid, *Glycaspis brimblecombei* (Hemiptera: Psyllidae) (Plates 31 and 32) (Ferreira Filho *et al.*, 2008). The graph shows that although mortality to the pest increases for a while, the response asymptotes, whereupon no more hosts can be dealt with no matter how high their population density reaches. The major problem is lumped together under the heading of handling time, which is a combination of various functions that the parasitoid or predator has to perform in dealing with its enemy. So, for a predator, it has to seek out its prey wherever it may be located (search efficiency), it then has to subdue or kill it (attack rate), eat it and then digest it. A parasitoid, on the other hand, while still having to find its host, merely has to lay an egg in or on the pest once it has been subdued. In the final analysis, we have to discover whether or not a functional response and its equivalent numerical response (where more enemies invade denser patches of pests) is in fact a density-dependent process. In theory, as can be seen from Fig. 3.26 (Prasad *et al.*, 2002), when the relationship between defoliators of *D. sissoo* and their percentage parasitism is plotted against one another, only part of the response (i.e. the section at low to medium pest density) performs in a density-dependent fashion. The crucial point is that beyond a certain pest density (around ten in this example), the response becomes density independent, and therefore non-regulatory. The key to successful biological control (see Chapter 8) is therefore either to ensure that pest densities never exceed the level where density-dependent regulation breaks down or, if they do, to use other mechanisms to reduce the density again to a level where biological control can take over.

One real-world example which exemplifies these problems involves the notorious

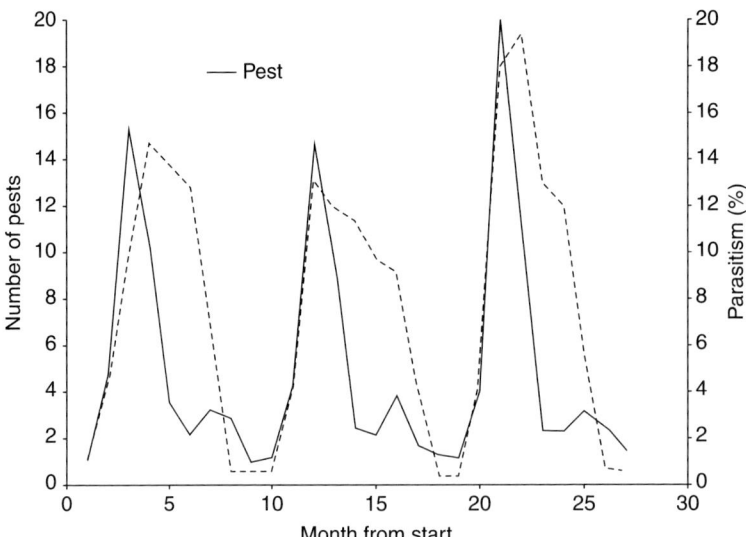

Fig. 3.24. Variations in numbers of noctuid moth larvae, *Plecoptera reflexa*, on *Dalbergia sissoo* trees in India, with percentage parasitism by hymenopteran parasitoids over 28 months (data from Prasad et al., 2002, courtesy Indian Forester).

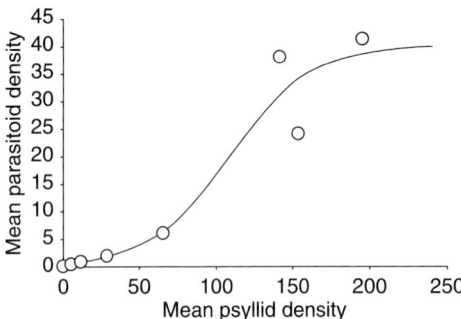

Fig 3.25. Numbers of red gum psyllid, *Glycaspis brimblecombei*, in relation to numbers of the psyllid-specific parasitoid, *Psylleaphagus bliteus*, on *Eucalyptus camaldulensis* in Brazil (data from Ferreira Filho et al., 2008).

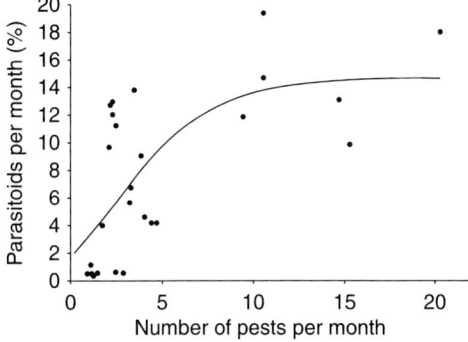

Fig. 3.26. Numbers of noctuid moth larvae, *Plecoptera reflexa*, on *Dalbergia sissoo* trees in India, plotted against percentage parasitism by hymenopteran parasitoids (data from Prasad et al., 2002).

leucaena psyllid, *H. cubana* (Hemiptera: Psyllidae) (Plate 33), an extremely serious pest of the multi-purpose tree, *Leucaena leucocephala*, in many parts of the tropical world (see Chapter 5). Much work has been carried out on the potential of natural enemies to regulate this pest. In the Philippines, some 15 indigenous predators were identified, including spiders (Aranaea), beetles (Coleoptera) and bugs (Hemiptera) (Barrion et al., 1987). *Curinus coeruleus* (Coleoptera: Coccinellidae) is a ladybird predator of the psyllid with which attempts have been made to reduce pest problems. The ability of *C. coeruleus* to eat *H. cubana* nymphs is illustrated by the functional response curves

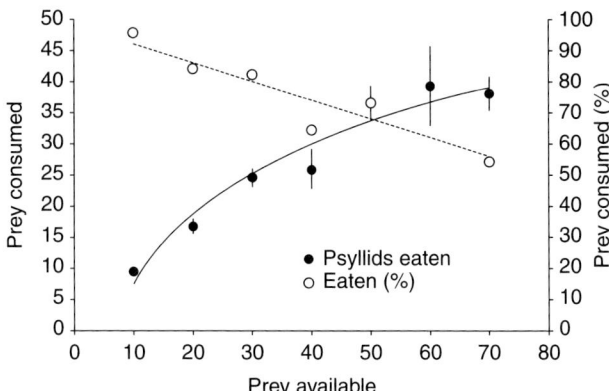

Fig. 3.27. Functional response curve for the ladybird *Curinus coeruleus* on late instar nymphs of leucaena psyllid, *Heteropsylla cubana* (data from da Silva *et al.*, 1992).

in Fig. 3.27. This functional response begins to asymptote at around a prey density of 50–60, with a maximum consumption of 40 or so nymphs every 24 h (da Silva *et al.*, 1992). Thus, maximum mortalities appear to lie in the range of 50–70% but, as the figure shows, percentage mortality plots decline as pest density increases, indicating a lack of a density-dependent relationship required for regulation in biological control. The role of the host plant is also important. Species with wider pinnules show less prey consumption at any given prey density than those with narrow pinnules. Significantly, in this experiment the wide-pinnuled species was *L. leucocephala*, the most widely planted and the least resistant (see Chapter 6). The narrow-pinnuled species, *L. diversifolia*, is much less susceptible to the psyllid. It appears that psyllid nymphs on wide pinnules are able to evade predators by moving to the opposite side. In fact, the ladybirds only find their prey by bumping into them in a random fashion, which is a more successful strategy on narrower pinnules since the psyllids have fewer places to go to avoid being found. However, not all natural enemies are inefficient in this way.

On many occasions, tropical forest insects show little or no evidence of significant natural predation or parasitism. The koa moth, *Scotorythra paludicola* (Lepidoptera: Geometridae), is a species endemic to *A. koa* forests in Hawaii (Haines *et al.*, 2009) which suffers outbreaks occasionally but with no sign of native parasitoids. In contrast, pathogen-caused diseases are commonplace. Indeed, there is much less controversy about the potential of pathogens such as viruses in forest pest epidemics. These agents are true 'plague' organisms, which spread through a population of susceptible hosts most efficiently at very high host densities. Their abilities to regulate are not limited to less than epidemic situations. To give but two examples here from tropical or subtropical countries, the major mortality factor for the defoliating bagworm, *Malacosoma incurvum* (Lepidoptera: Lasiocampidae), was found to be a nuclear polyhedrosis virus (NPV) in Mexico (Filip and Dirzo, 1985), while up to 100% of larvae of the Asian gypsy moth, *Lymantria dispar* (Lepidoptera: Lymantriidae), were killed by another NPV in Korea (Pemberton *et al.*, 1993). Entomopathogenic fungi are predicted to have particular utility on tropical crops in years to come (Li *et al.*, 2010), while bacteria such as *Bacillus thuringiensis* are used extensively as pest control agents in many parts of the tropics (see Chapter 8).

In summary, the actions of predators and parasitoids would seem to be limited to low or medium host densities, while pathogens alone have potential for the suppression of certain high-density pests. The potential importance of food supply must also be considered, which, as we have already mentioned, is likely to be of fundamental significance via intraspecific competition and resource limitation (see also Chapter 4). It is also worth reiterating at this point that all abiotic factors such as climate are unable to act in a density-dependent way and are hence incapable of being regulatory, though they may well influence the survival, growth rates and general efficiencies of pests and enemies alike.

4

Insect–Host Tree Interactions

4.1 Introduction

If insects did not use trees as a source of food, the topic of forest pest management would not exist. Similarly, if trees did not evolve mechanisms in attempts to prevent themselves from being eaten, then insects might become so abundant as to be unstoppable. The quantity and quality of food which trees provide for insects vary enormously according to many factors such as the age of the tree, site, genetics and so on. Evolution via natural selection has come up with relationships between trees and insects that eat them whereby, in natural communities, neither participant suffers permanent depletion or excessive damage. It is therefore very important to understand how growing trees away from their natural situations, regions or ecologies may alter the food and/or defences produced by the trees and the abilities of insects to use them. We may then be able to manipulate our silvicultural practices so that trees no longer provide high-quality food for insects; this is the basis of vigour promotion and tree resistance. Bigger (taller) trees support both a greater abundance of herbivorous insects and a higher species richness (da Costa *et al.*, 2011) by virtue of their complex 'architecture'. However, it is not simply the amount of living space or the number of niches available on a particular tree that influences insect associations but also the quantity and quality of the physical and chemical nature of the host.

4.2 Host Nutrients

Basically, animals may be considered as nitrogen-based life forms; a great deal of our tissues is made up of proteins, and polypeptides and amino acids circulate in our bloodstreams in relatively high concentrations. In addition, our major excretory products, whether urea, uric acid or ammonia, are all nitrogen based. In essence, we are demanding and wasteful consumers of organic nitrogen. In contrast, a plant such as a tree is composed primarily of cellulose and its relatives; they are carbon-based life forms. If plants really can be said to excrete anything, then it is carbon dioxide, again a compound based on carbon. If viewed simplistically, therefore, it is clear that plants are extremely unsuitable as food for animals. Insects and humans alike should feed on other animals in order to gain the right quantities of the most essential nutrients with the least waste and energy expenditure. Any animal that does feed on a plant has to accept that this is a suboptimal diet and should evolve strategies to make the most of a poor resource, in particular the crucial lack of plant organic nitrogen.

Figure 4.1 presents an approximate summary of the composition of various plant tissues in terms of dry weight (DW) total nitrogen (Mattson, 1980). A figure for animals (including insects) is also presented for comparison. Notice also from the figure that some parts of a tree are much worse than others when it comes to sources of nitrogen. Seeds and fruit are relatively rich in this limiting resource, whereas wood and especially sap are very deficient indeed.

An added complication involves the type of nitrogenous compound provided by plants as food for insects. Plants can contain many different nitrogenous compounds, from proteins to amino acids, and the presence, absence or balance of essential compounds may be more important to a particular insect than the gross levels. It is not very surprising, therefore, to find that insects are able to respond to changes in host plant organic nitrogen levels and to maximize their growth rates or reproductive potentials accordingly. Figure 4.2 illustrates the results of one investigation into the relationships between plant nitrogen and insect performance. Coley et al. (2006) studied the growth rates of 85 species of Lepidoptera in Panama and related this to the levels of nitrogen on host plant leaves (Fig. 4.2). Specialist species (those with a small host plant range) grew faster than generalists (those with a broad host range), linked to their higher efficiency at dealing with the particular morphology and chemistry of their specific hosts with which they would have co-evolved (see later). In both types of lepidopteran, growth was faster on young rather than old leaves, which may be linked to the lack of physical or chemical defences (see below) and higher nutrient levels in young foliage.

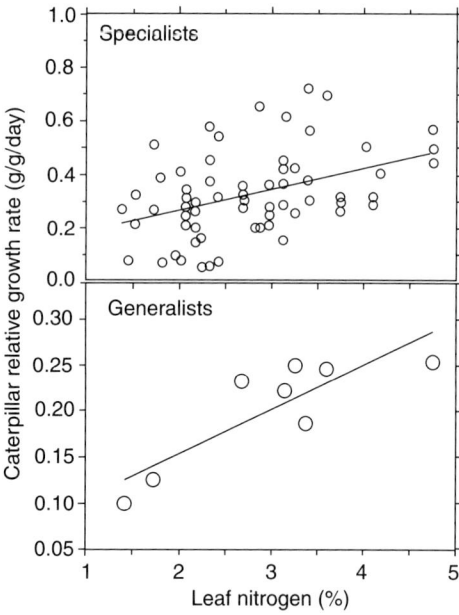

Fig. 4.2. Effects of leaf nitrogen content on relative growth rates (g/g/day) of lepidopteran larvae (from Coley et al., 2006, courtesy Wiley-Blackwell).

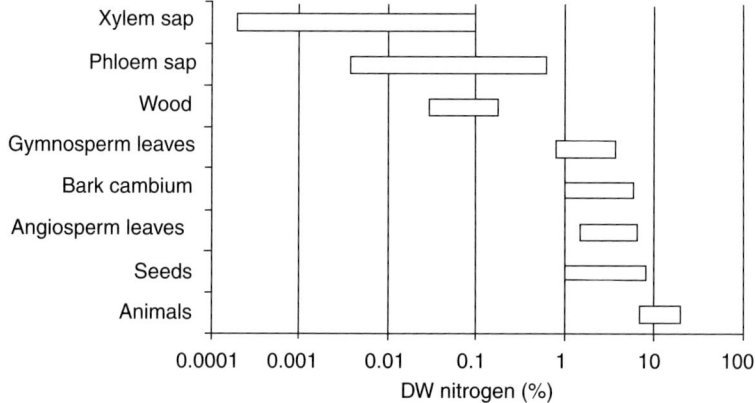

Fig. 4.1. Approximate nitrogen content (per cent dry weight) of various plant tissues compared with that of animals (from Mattson, 1980, redrawn with permission Annual Reviews Inc).

As might be expected, therefore, new, young leaves are preferred as food by insects rather than older leaves, and Fig. 4.3 shows more details of why this may be so. *Ouratea* is a genus of South American evergreen tree, used for timber and medicinal oils. Kursar and Coley (2003) looked at the variation in the nitrogen content of leaves of this tree, again in Panama, as they matured. Young, expanding leaves clearly have higher levels of nitrogen than older ones and this leaf nitrogen then declines markedly after full expansion is obtained. Note that not only are young leaves a better source of nutrients for insect herbivores but also that they are significantly less tough, and thus much more palatable and easier to eat.

Host tree nitrogen levels can often be influenced by certain forest management practices. Watering and fertilizing of trees are, of course, merely anthropogenic variations on natural events (rain and nutrient cycling), but their experimental manipulation provides useful insights into how these basic external factors can alter the nutritional status of trees for insect herbivores and also how forest management, starting in the nursery and continuing after planting, may influence pest impact. In 1991, Thomas and Hodkinson studied the effects of different watering regimes on birch (*Betula pendula*) and defoliating lepidopteran larvae which fed on the trees (Thomas and Hodkinson, 1991). In general, the trees responded to a shortage of water by increasing the concentrations of soluble (organic) nitrogen in their leaves (Fig. 4.4), though the opposite effect, that of waterlogging, did not have so much of an influence. As might now be expected, the digestive efficiency and growth rate of the insects feeding on the trees reflected the increased soluble nitrogen content of the water-deficient trees. This is one example of many to which we will return in Section 4.4, when the links between tree stress and insect attack are discussed in detail.

Insects seem able to distinguish the levels of nutrient chemicals in their host plants, and respond accordingly. Figure 4.5 shows how leaf-cutter ants, *Atta laevigata* (Hymenoptera: Formicidae), in Brazil selectively harvest foliage from plants with significantly higher levels of nitrogen, phosphorus and potassium than the random average in the forest (Mundim et al., 2009). Soil improvement using fertilizers is an artificial way of altering the levels of nitrogen and other nutrient chemicals in plants, and hence the performance of herbivorous insects. It must be remembered, of course, that plants are unable to take up organic compounds from soil and water into their root systems and beyond, and the relationships between inorganic soil nutrients such as nitrogen, phosphorus or potassium (NPK) and the levels of organic, soluble, nutrient nitrogen in plant tissues are not necessarily clear-cut or predictable. However, a simple case involves a species of *Eucalyptus* in New South Wales, Australia. Young

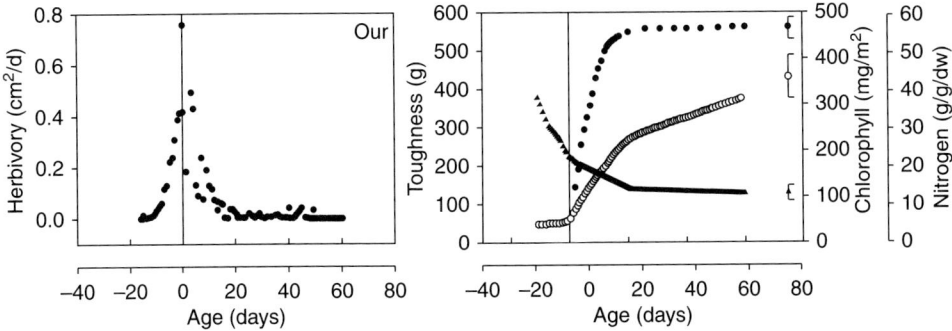

Fig. 4.3. Rate of leaf damage, leaf toughness (solid circles), nitrogen (solid triangles) and chlorophyll content (open circles) for *Ouratea* spp. in Panama. Age zero is the day on which leaves reach full size (from Kursar and Coley, 2003, courtesy Elsevier).

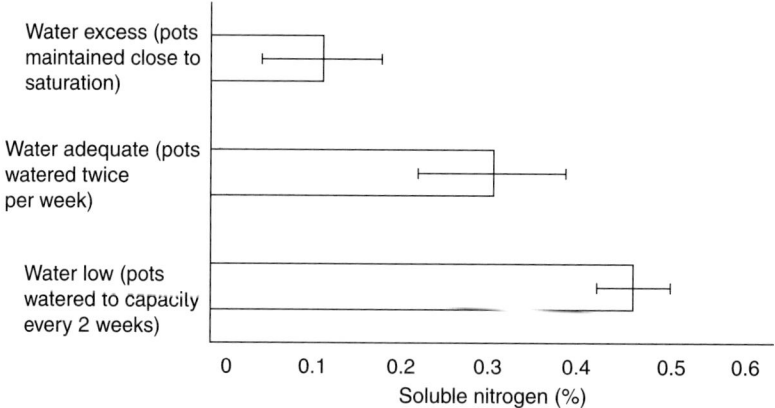

Fig. 4.4. Influence of watering regime on soluble nitrogen levels (± SE) in leaves of birch (from Thomas and Hodkinson, 1991, courtesy Wiley-Blackwell).

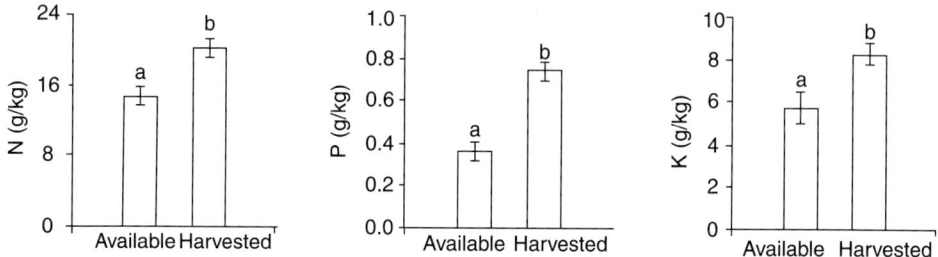

Fig. 4.5. Leaf content of three nutrients (nitrogen, phosphorus and potassium) comparing general nutrient levels in random leaves, with those leaves chosen to be harvested by leaf-cutter ants, *Atta laevigata* (from Mundim *et al.*, 2009, courtesy Wiley-Blackwell).

eucalypts are fed upon by a whole range of insects, including leaf beetles (Coleoptera: Chrysomelidae), scarab beetles (Coleoptera: Scarabaeidae), sawfly (Hymenoptera: Symphyta) larvae and scale insects (Hemiptera: Coccidae) (Fox and Morrow, 1992; Elliott *et al.*, 1998). Beginning in 1979, 2 m tall trees were treated with various types of slow-release fertilizers each year and then leaves were collected from experimental trees and analysed for nutrient content. Both organic nitrogen and phosphorus increased after the soil in which the trees were growing was treated with NPK (Fox and Morrow, 1992), thus providing the numerous defoliating insects with higher levels of organic nutrients. We might predict, therefore, that the type of soil in which trees are growing should have an important influence on levels of insect herbivory via altering host plant nutrients (Sevillano *et al.*, 2010). Such associations are complex and may also involve tree defences and vigour, topics we shall address later in this chapter.

4.3 Host Defences

Trees do not take the depredations of insect herbivores 'lying down'. Throughout the co-evolution of insects and trees, mechanisms have appeared to prevent, or at least deter, the exploitation of tree material by insects. These defence systems may involve physical factors, chemical factors or combinations of both, a classic example being the bark of conifers. Franceschi *et al.* (2005) suggest four phases in the defence systems of external plant surfaces such as bark which act sequentially when the earlier system fails to protect the tree. These phases are (i) a defence that repels or inhibits successful attack by insects or other pests, (ii) the killing or compartmentalizing of the pest, (iii) the sealing and repairing of damage caused by successful pests and (iv) the acquisition of a defence system should future attacks occur. At all these stages, a mixture of mechanisms can be involved and, for convenience only, we present a brief review of each type separately.

4.3.1 Physical defences

We have mentioned already how leaves become tougher as they grow. Leaf toughness is a simple physical defence against chewing which relies on the fact that insect defoliators, especially small ones, are unable to feed on or digest tough plant material. Grasshoppers (Orthoptera: Acrididae) are universal chewers of many forest trees and can be particularly troublesome in tropical nurseries, for example. Locusts such as *Chortoicetes terminifera*, for example, show reduced growth rates and longer development times when fed on tough versus tender leaves (Clissold *et al.*, 2009), while in Paraguay, grasshoppers in the genus *Baeacris* (Plate 126) (Orthoptera: Acrididae) have caused considerable losses to very young (3-month-old) *Eucalyptus grandis* transplants, chewing away the young red/green bark, girdling the trees and causing them to fall and die (Plate 125). Slightly older transplants which have been held back longer in the nursery and have tougher, more mature grey bark appear to be untouched by the small grasshoppers (see Chapter 10). Tough leaves can be thought of as possessing physical strength in terms of resistance to tearing and shearing (Peeters *et al.*, 2007), and herein lies the problem faced by chewing insects who have to overcome this resistance mechanically. Smaller or less powerful insects may find the task impossible.

Leaf or bark toughness does not appear to deter many sap feeders; some actually use stomata anyway to gain entry for their piercing mouthparts into the vessels, and once the leaf surface has been penetrated, other defence systems need to be employed by the plant. However, other physical defences may be employed by plants to defeat sap feeders and defoliators alike and some herbivorous insects, especially the smaller individuals, may be deterred from feeding by hairs or trichomes on leaf or stem surfaces. Many trees support hairy or spiny leaves and though some hairs may have evolved to help with moisture retention and spines to prevent large mammals from eating leaves, insects also may be affected. A lot of evidence suggests that leaf trichomes are very effective at reducing pest densities, and/or the damage done by them, in agricultural crops such as potatoes (Kaplan *et al.*, 2009), though there are fewer such examples from tropical trees. However, Nahrung *et al.* (2009) studied herbivory levels on the hardwood tree *Corymbia citriodora* in Queensland caused by the leaf beetle *Paropsis atomaria* (Coleoptera: Chrysomelidae) and compared these levels with two other *Corymbia* genotypes with differing amounts of trichomes on their leaves, as well as varying plant chemistries. In Fig. 4.6, it can be seen that *C. citriodora* (CCV), which is the pest's preferred host, has glabrous (no trichomes) leaves, whereas the hybrid (CT × CCV) and non-host *C. torelliana* (CT) both have hirsute leaves with many trichomes. Another similar example comes from India, where Jacob and Balu (2007) looked at resistance between teak clones to the teak defoliating caterpillar, *Hyblaea puera* (Lepidoptera: Hyblaeidae). Table 4.1 clearly shows that herbivory levels are much higher on the leaves without trichomes and though this is but one of several physical and chemical differences between host and non-host trees, possessing physical structures such as trichomes is likely to contribute to tree defences against insect herbivores.

Certainly in the case of the notorious leucaena psyllid, *Heteropsylla cubana* (Hemiptera: Psyllidae) (see Chapter 5), an almost worldwide pest of the multi-purpose tree *Leucaena leucocephala*, plants with hairy surfaces to their leaves were only found among the resistant varieties of the tree in Indonesia (Suhendi, 1990). Hairs may, in fact, combine physical defences with chemical ones; certain tree species in the genus *Arbutus* from Mexico have glandular hairs on their leaves which are secretory (Becerra and Ezcurra, 1986). Other species

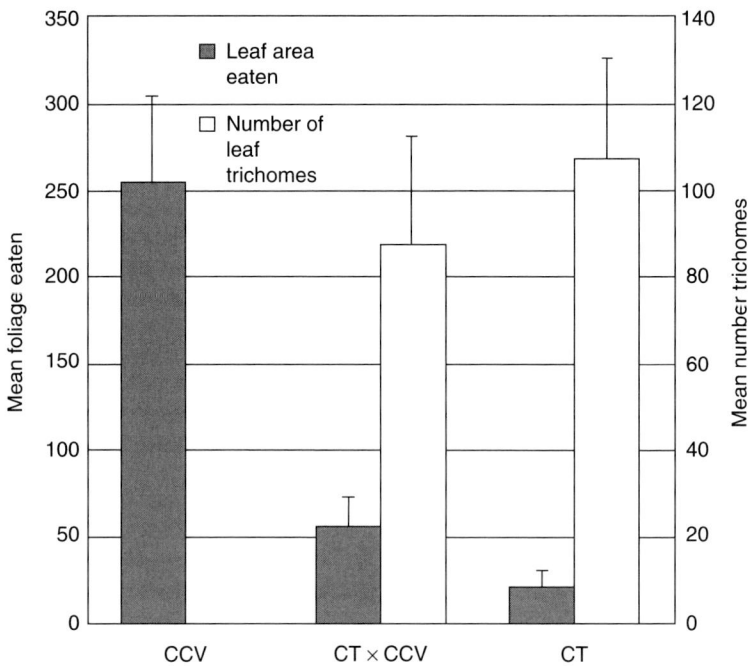

Fig. 4.6. Mean (± SE) amount of foliage eaten by one male/female pair of adult *Paropsis atomaria*, with the mean (± SE) number of leaf trichomes, on each of three *Corymbia* genotypes. CCV = *Corymbia citriodora*; CT × CCV = artificially created hybrid between CCV and CT; CT = *Corymbia torelliana* (data from Nahrung et al., 2009).

Table 4.1. Trichome type and density on teak clone leaves in relation to levels of defoliation by *Hyblaea puera* larvae (from Jacob and Balu, 2007).

Category	Clone	No. trichomes/cm^2 upper surface	No. trichomes/cm^2 lower surface	Comments
Most susceptible	APKKR-4	180 ± 10.4	240 ± 14.2	Very few diffused type of upper cuticular thickenings; no spines; isolated, multicellular long hairs on lower surface
Moderately susceptible	BHA-30	600 ± 29.9	200 ± 8.8	Diffused type of upper cuticular thickenings; no spines; isolated, multicellular long hairs on lower surface
Non-attacked/ resistant	MYSA-3 APKKA-1 APAKB-1 APNBB-1	1075 ± 63.4 825 ± 48.7 1000 ± 48.9 800 ± 47.2	5500 ± 302.5 5460 ± 322.1 5550 ± 327.4 5520 ± 269.9	Numerous convex projections topped with a single pointed hard trichome (Type I), numerous sharp spines (Type II) and blunt spines (Type III) on the upper surface; high density of soft, long and short trichomes on the lower surface

of *Arbutus* have hairs with no secretory properties, while a third group have no hairs at all. Field surveys in Mexico on the damage caused by the butterfly *Eucheira socialis* (Lepidoptera: Pieridae) revealed that damage was greatest on trees with no hairs at all, intermediate on trees with simple hairs and least on those with glandular hairs. The problem of leaf toughness and hairiness can be overcome, though this process involves a somewhat extreme set of adaptations. Most leaf toughness is a feature of leaf epidermises and, as we have seen, has evolved as a simple but often effective defence mechanism to avoid being eaten. The section of leaf sandwiched between the upper and lower epidermises of a normal leaf, the parenchyma, is much less well defended and is much more palatable to insects if they are able to utilize it. Such insects are called leaf miners, and many moths (Lepidoptera), beetles (Coleoptera) and flies (Diptera) have larvae that tunnel in leaf parenchyma (Sinclair and Hughes, 2008).

A final type of physical defence in trees which is frequently accompanied by chemicals is that of sap or resin pressure in main stems and shoots. A healthy, vigorous tree has a continuous transpiration stream flowing from the roots to the canopy, which provides a simple physical defence whereby any hole made in the stem will result in copious flows of sap or resin under an internal positive pressure. Attempts by young larvae of longicorn or jewel beetles (Coleoptera: Cerambycidae/Buprestidae) or small adults of bark or ambrosia beetles (Coleoptera: Scolytinae/Platypodinae) to tunnel against this flow will result in their being repulsed, or even drowned. Indeed, the phenomenon of insects fossilized in amber arises from resin exudations of various conifer species which trap unwary insects and preserve them for millions of years. If the sap or resin exudations also contain toxins such as terpenes, the effect is even more successful. However, any reduction in the pressure of this system will allow insects to attack bark and shoots. Such reductions may arise from defoliation, pathogen attack, arid soil conditions or the simple but effective system of felling the tree. In general terms then, bark beetles (Coleoptera: Scolytidae), and many other borers too, are known as secondary tree pests in that they can only gain entry to the stems of weakened trees through reduced resin pressure or sap flow. In fact, only a handful of the 7500 or so species of bark and ambrosia beetles worldwide are aggressive (i.e. they are capable of overwhelming the defences of healthy trees) (Krokene, 1994). The vital links between tree vigour and insect attack are discussed in detail in Section 4.4 and again in Chapter 6.

4.3.2 Chemical defences

Nutrients are not the only chemicals to be found in plant tissues such as leaves or sap. In addition, there are many other compounds which appear not to be essential for plant growth; these chemicals are known as secondary metabolites, which rather than being beneficial to insects, may deter or even poison them. A vast array of such compounds can be found in plants, from alkaloids such as nicotine and morphine through to the highly toxic cyanogenic glycosides to phenolics such as tannin, the latter group probably being the most important in influencing herbivory (Eichhorn *et al.*, 2007). It is not always clear whether these compounds have evolved directly to defend the plant against herbivores or are merely metabolic by-products that have taken up a later role as toxins or deterrents. Whatever the origins of these metabolites, it is well known that many specialist insects have overcome the chemical barriers erected by them, and not only are the insects now able to feed successfully on the previously defended trees but also in many cases they have turned the plant chemicals to their own advantage, using them as host recognition cues or feeding stimulants. A few insect species even store (sequester) plant poisons in their bodies and advertise the fact that they are now toxic to their own predators using warning, aposomatic colourations.

One famous plant secondary metabolite comes from the neem tree, *Azadirachta indica*, and, not surprisingly, is called azadirachtin. This compound belongs to a group of chemicals called limonoids, for which a wide range of biological activity has been recorded. One claim to fame of azadirachtin is as an insecticide or antifeedant, and a very large number of trials have been carried out against a myriad of insect pests all over the world, though most have little commercial value. It is somewhat ironic, therefore, to realize that neem trees in parts of the Sahelian region of Africa are suffering severe dieback and death as a result of heavy attack by various sap feeders, in particular the scale insect *Aonidiella orientalis* (Hemiptera: Diaspididae) (Boa, 1995).

Some of these chemicals are now used by herbivorous insects as recognition stimuli for their particular and specific host plants; without the presence of this sort of compound in their diets, some insects simply will not feed, despite the fact that the food is perfectly nutritious (as well as any plant ever can be). This process is frequently the basis for host plant specializations exhibited by forest insects the world over. That host plant recognition is determined at least partly by plant chemicals is shown by experiments carried out on the olfactory discrimination of the leucaena psyllid, *H. cubana* (Hemiptera: Psyllidae) (Plate 32). In the Philippines, Lapis and Borden (1993) prepared chemical extracts of leaves of various species of *Leucaena* by grinding up leaf material in the solvent hexane. Female psyllids were then given choices of extracts from different tree species, but always responded most strongly and positively to extracts from *L. leucocephala*, the highly susceptible (and the most widely planted) species. Not all forest pests have very small host tree ranges, however. The autumn gum moth, *Mnesampela privata* (Lepidoptera: Geometridae), feeds on at least 40 species of *Eucalyptus* and one of *Corymbia* (Steinbauer and Matsuki, 2004), though its degree of success measured in growth rates and final pupal weights varies considerably between tree species. These authors extracted oils from a variety of *Eucalyptus* species and plotted their concentration against the final dry weight of gum moth pupae feeding on each tree species (Fig. 4.7). Though only a small subset of the total eucalypt species utilized by the pest, it is fairly clear that as leaf oil concentrations increase, so the weight of pupae decreases. We have seen earlier in the book that pupal weights frequently are related closely and directly to adult fecundity and offspring viability, so leaf chemistry may well have a very important influence on insect success.

As mentioned earlier, leaf miners are a special type of insect herbivore that tunnel

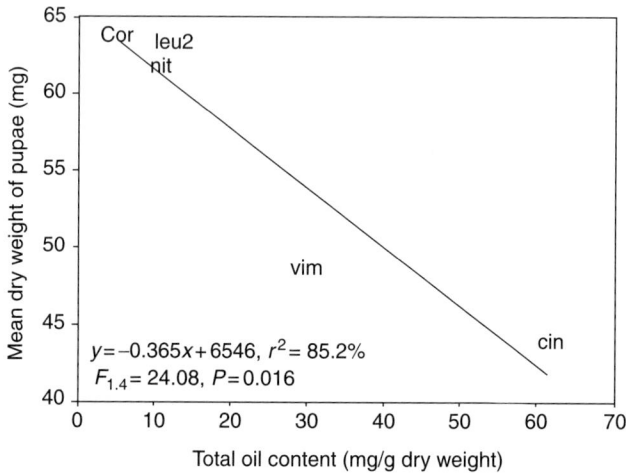

Fig. 4.7. Mean weight of autumn gum moth pupae in relation to total oil content of the leaves of various *Eucalyptus* and *Corymbia* species on which they fed. Cor = *C. eximia*, leu2 = *E. leucoxylon*; vim = *E. viminalis*; nit = *E. nitens*; cin = *E. cinerea* (from Steinbauer and Matsuki, 2004, courtesy Wiley-Blackwell).

between epidermises, where they should be relatively protected from environmental extremes, enemies and both physical and chemical defences located in or on the external surfaces of leaves. So, chemical defences such as phenols may or may not relate to leaf miner activity. Sinclair and Hughes (2008) studied the occurrence of a large number of leaf-mining insects in Australian Myrtaceae tree species and discovered, rather counter-intuitively, that leaf phenol concentration was marginally higher in species attacked by leaf miners (Fig. 4.8). If phenols were acting as direct defences against the miners, then the reverse finding would have been expected, and it was likely that phenols in this case might be effective in reducing leaf attack by other herbivores, especially those feeding externally rather than internally, who might compete with or even destroy leaf miners. Thicker leaves were also shown to be mined less frequently, which might be associated with extra epidermal thickness rather than any extra space for leaf miners to inhabit.

Levels of defences in plant material do not remain constant any more than do nutrients, and defences can change considerably as trees get older (Barton and Koricheva, 2010). Tannins, for example, build up quickly in leaves of eucalypts and oaks as the leaves grow and expand; only the youngest leaves are therefore devoid of protection from this source. In general, it appears that young leaves should be better defended against at least generalist herbivores than old leaves, and saplings should be better defended than mature trees (Junker et al., 2008). However, examples from the field can be more complex than this. In Papua New Guinea, for example, a distinct age effect was noted with insect attack on hoop pine, *Araucaria cunninghamii*, as follows: (i) the branchlet mining scolytine beetle, *Hylurdrectonus araucariae* (Coleoptera: Scolytinae), severely infests those trees aged between 3 and 15 years, but not those in other age classes; (ii) trees in age classes 7–12 years are most susceptible to heavy attack by the weevil *Vanapa oberthur* (Coleoptera: Curculionidae); (iii) the defoliator *Milionia isodoxa* (Lepidoptera: Geometridae) most severely attacks trees aged between 5 and 13 years; and (iv) hoop pine trees older than 21 years appear more resistant to fatal injury by the termite *Coptotermes elisae* (Isoptera: Rhinotermitidae) (Wylie, 1982b). Explanations for this vary; the case where insects attack the foliage may relate to changes in the physiochemical characteristics of the branchlets as the tree matures. In the case of stem borer attack, the tree's physical size and bark characteristics may be important. Older trees can sustain, without damage, a level of insect infestation that would deform or kill younger trees.

4.3.3 Induced defences

A great deal of interest has focused on changes in tree defences as a result of insect attack. This process is known as 'induced

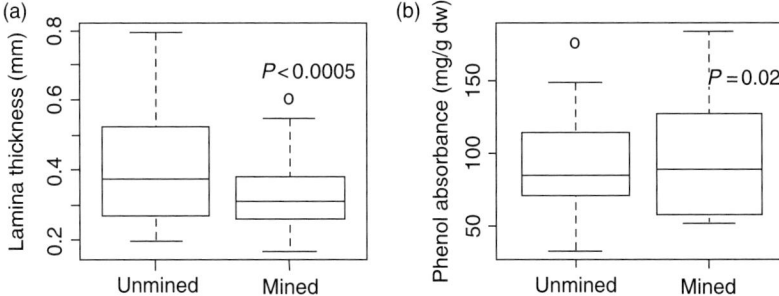

Fig. 4.8. Boxplots showing (a) leaf lamina thickness and (b) phenol concentration in Myrtaceae leaves for mined and unmined groups of trees (from Sinclair and Hughes, 2008, courtesy Wiley-Blackwell).

defence' and has important consequences for pest management in that pest outbreaks may decline on their own, without intervention from foresters or entomologists, as a result of the tree becoming less palatable or more toxic as attacks proceed. These induced defences are likely to depend on the intensity of initial herbivore damage (Utsumi et al., 2009).

One example of an assumed induced defence system can be seen in the case of leaf-cutting ants. The genus *Atta* contains a large number of species of ant which can, on occasion, remove extremely large quantities of plant material from tropical forests, both natural and plantation, in Central America and parts of South America (Plates 16 and 17). Leaf material is cut from leaves in a typical half-moon pattern and returned to the ant colony, where it is used as a mulch for the growth of exosymbiotic fungal gardens by the ants. Howard (1990) looked at *Atta colombica* in a deciduous forest in Panama and discovered that leaf-cutting ants abandoned their host plants long before they had defoliated them completely. As the amount of canopy of the host tree, *Bursera simaruba*, defoliated by the ants increased, its palatability decreased. In this case, it was not clear whether feeding pressure was causing chemical defences to increase in leaves. Indeed, it may be that induced changes in plant chemistry may only rarely be the immediate cause of ant abandonment of plants, but in a forest habitat these changes are thought to add to the existing variation in food quality among the trees.

Changes of this nature in plant chemistry as a result of insect feeding have been demonstrated in tropical trees in Mexico by Pascual-Alvarado et al. (2008). They studied the changes in leaf chemistry in a variety of dry forest tree species after attacks by gall-forming insects (Fig. 4.9.). In all but one case, galling resulted in significant increases in total leaf phenolics. The potential consequences for tropical forest insect pests of these variations in tree nutrients and defences need to be considered when we try to explain the reasons for outbreaks.

In reality, tree nutrients combine with tree defences as a double-action attempt in co-evolutionary terms to reduce damage by

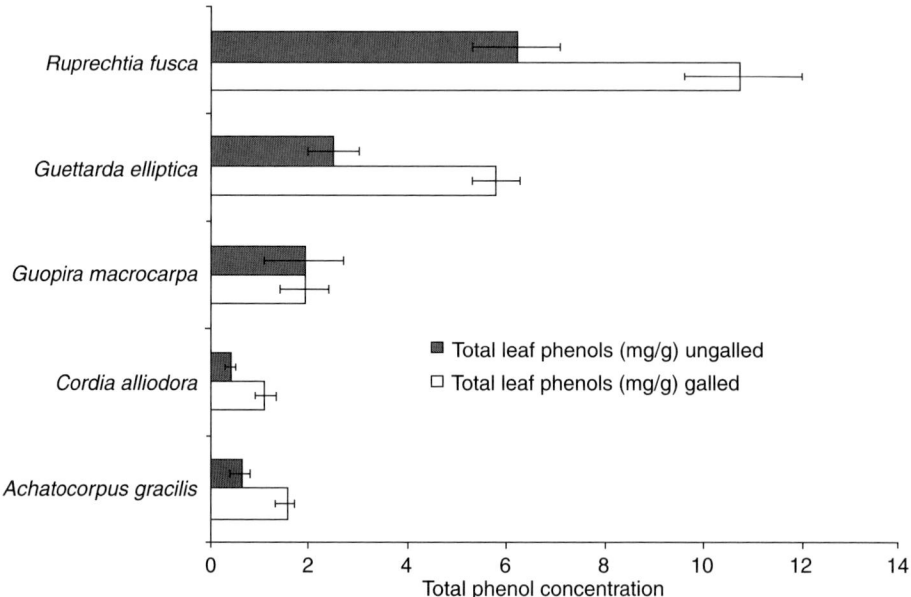

Fig. 4.9. Mean (± SE) total phenol concentrations in galled and ungalled leaves of five host plant species. All but *G. macrocarpa* are significantly different at $P < 0.001$ level (data from Pascual-Alvaredo et al., 2008).

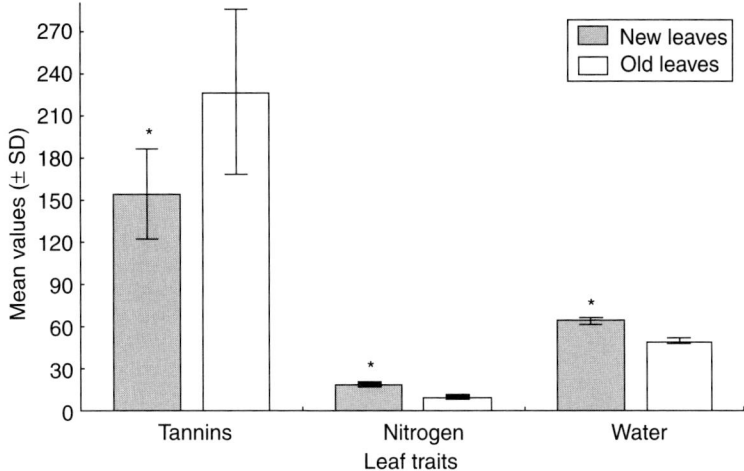

Fig. 4.10. Mean values of tannins (mg/g dry weight), nitrogen (g/kg dry weight) and water content (%) between new and old non-attacked leaves of *Erythroxylum tortuosum* in Brazil (from Ishino et al., 2011, courtesy Wiley-Blackwell).

herbivores, especially to particularly vulnerable plant parts such as flowers and young leaves. Leaves tend to have the highest nitrogen content, the most water and the lowest defences such as tannins (Fig. 4.10; Ishino et al., 2011), as well as being the softest and most easily eaten by insect mouthparts.

4.4 Host Vigour and Stress

It is quite clear that any silvicultural activity which in some way promotes the levels of organic nitrogen and/or demotes the levels of chemical or physical defences in tree tissues is likely to benefit insect herbivores. One of the most fundamental silvicultural properties that can and often does operate in this way is the general concept of the promotion of vigour in trees and the reduction of host plant stress. Table 4.2 shows the effects of attacks by the longhorn beetle *Oncideres rhodosticta* (Coleoptera: Cerambycidae) on mesquite trees *Prosopis glandulosa* var. *torreyana*, in Mexico (Martínez et al., 2009). Both wet and dry sites were investigated and the figure presents some of the differences between these sites 10 months after branches were infested by the pest. Though various parameters were not statistically different (though with one or two intriguing trends), the number of live larvae was more than double in trees on dry sites, with only around 50% of the mortality rate observed in wet sites. Leaf feeders can also be influenced by tree stress. The weevil *Stereonychus fraxini* (Coleoptera: Curculionidae) defoliates ash trees and in an experiment, Foggo et al. (1994) disturbed the roots of trial trees by severing them with a spade. Leaf toughness declined significantly and the area of leaf eaten by weevils increased, also significantly, in the root-damaged trees (Fig. 4.11). There would appear to be a relationship between water levels and the success of insect pests attacking trees. This is the drought stress hypothesis, which suggests simply that, for some insects at least, plants subject to low water levels, drought-ridden in fact, are more likely to be attacked and the pests, once established, will perform better. Water-stressed trees appear to have higher levels of organic nitrogen of various forms in their tissues, caused by heightened hydrolysis of proteins (Seagraves et al., 2011), and it is this increase in plant nitrogen that favours insect herbivores (see above). In addition, chronic water stress is able to reduce defence systems such as resin production, as we have discussed earlier.

Table 4.2. Characteristics of branches of *Prosopis glandulosa* and the pattern of occupation and survival of larvae of the beetle *Oncideres rhodosticta* 10 months after attacks were initiated. Means ± SE (from Martínez *et al.*, 2009).

Variable	Wet site	Dry site	p-value
Length of damaged branch (cm)	49 ± 3.2	43.5 ± 1.12	0.08
Diameter of damaged branch (mm)	8.8 ± 0.35	9.6 ± 0.6	0.08
Distance of first larval gallery from girdle (cm)	4.67 ± 0.9	5.71 ± 0.75	0.4
Distance between larval galleries (cm)	3.5 ± 0.4	4.7 ± 0.6	0.2
Larval gallery size (mm)	10.2 ± 1.23	12.4 ± 3.2	0.6
Larvae length (mm)	4.2 ± 0.9	4.5 ± 0.8	0.8
Total larvae per branch	6.5 ± 0.5	6.4 ± 0.8	0.8
Live larvae per branch	1.0 ± 0.1	2.4 ± 0.1	0.001
Dead larvae per branch	6.0 ± 0.05	3.2 ± 0.1	0.001
Proportion of dead larvae per branch	0.94 ± 0.02	0.71 ± 0.1	0.002

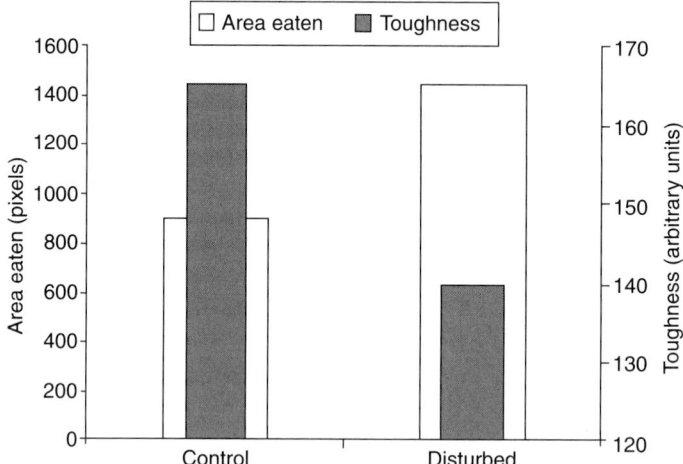

Fig. 4.11. Leaf toughness and amount of leaf consumed by adult *Stereonychus fraxini* according to root disturbance (from Foggo *et al.*, 1994).

Root pruning is only one way of stressing a tree, but it does mimic a very widespread problem in tropical forestry, that of shortage of water and/or nutrients. The lack of tree vigour as a result of water stress has been implicated in numerous stories of tree dieback and death in both tropical and temperate regions. One of the first protagonists of the links between tree disorder, attacks by defoliating and bark- or wood-feeding insects (and, incidentally, root-killing fungi) and climatically derived water stress was White (1986), who invented the stress index (SI). Wylie *et al.* (1993a) used these stress indices in their investigation into the causes of the dieback of *Eucalyptus* spp. in subtropical Australia. The indices took into account both overly wet and overly dry periods in a year. Figure 4.12 depicts a running average of their SI data for one area in Queensland, Australia. Positive values of SI indicate increasingly stressful conditions for trees, and it can be easily seen that water stress is becoming more frequent and commonplace as the century progresses. The peak SI in the 1970s coincided with noticeable and increasing rates of *Eucalyptus* dieback in the region.

The links therefore seem straightforward: increasing water stress, either by waterlogging or drought, reduces tree vigour and

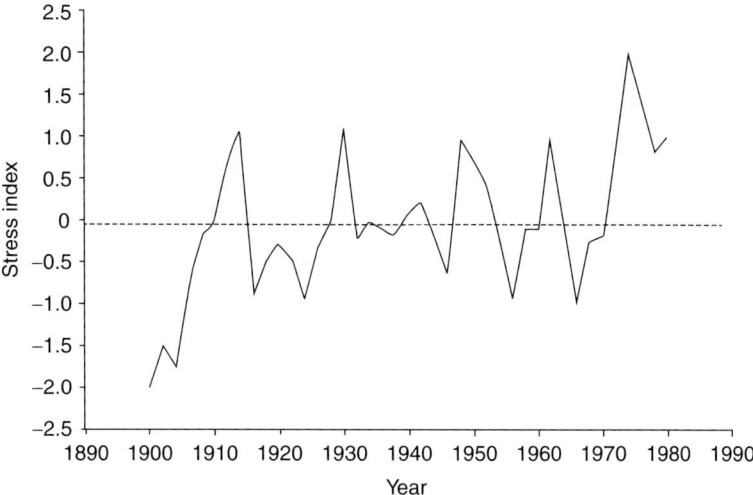

Fig. 4.12. Climatic stress indices for Gundiah, Queensland, Australia. A positive or increasing value indicates trees likely to suffer water stress (from Wylie et al., 1993a).

increases susceptibility to pests and diseases. The moral is therefore equally clear – tropical trees must be grown in a fashion that promotes their vigour as much as possible – tree stress must be avoided at all costs.

This formula is undoubtedly sensible, and in most cases would reduce risk from pest attack. However, not everyone agrees with the simple stress–pest hypothesis, and indeed there are quite a few examples where healthy trees seem to be attacked more heavily than less vigorous ones. None the less, we feel that Fig. 4.13 presents a framework in which to fit most of these rather variable observations. Koricheva et al. (1998) suggest that the relationship between tree stress and herbivore performance (which can be equated with pest attack in most cases) varies depending on the type of insects. Chewing insects (defoliators) respond relatively rarely to increasing stress or declining vigour in trees; their performance declines as serious tree debilitation ensues, probably as a result of serious reductions in foliar quality. Sap feeders, on the other hand, do increase their performance on stressed trees, up to a point, so that many observations of pest problems from aphids, psyllids and coccids in tropical forests can be related to low vigour in the trees. Finally, cambium feeders such as bark and wood borers (the latter spending at least a little time in the cambium in the early stages) are, in the main, unable to colonize healthy trees. They are the true secondary pests and have to wait for tree stress to increase before they can begin to exploit the host. Even they decline eventually; tree bark and wood becomes so debilitated or dry that only a few insect groups can still use it as food. One example of the links between drought-caused tree stress comes from the semi-deserts of Jordan, where a large number of Aleppo pine trees (*Pinus halepensis*) have been debilitated severely for several years by lack of rain and have now been invaded by bark beetles in the genus *Pityogenes* (Plates 12 and 13). By the time this stage has been reached, it could be argued that the tree has left the realms of forestry anyway and become a wood product.

4.5 Co-evolution of Insects and Tropical Forests

Trees are not the easiest of food sources for insects and so, over millions of years, herbivores have evolved a great variety of

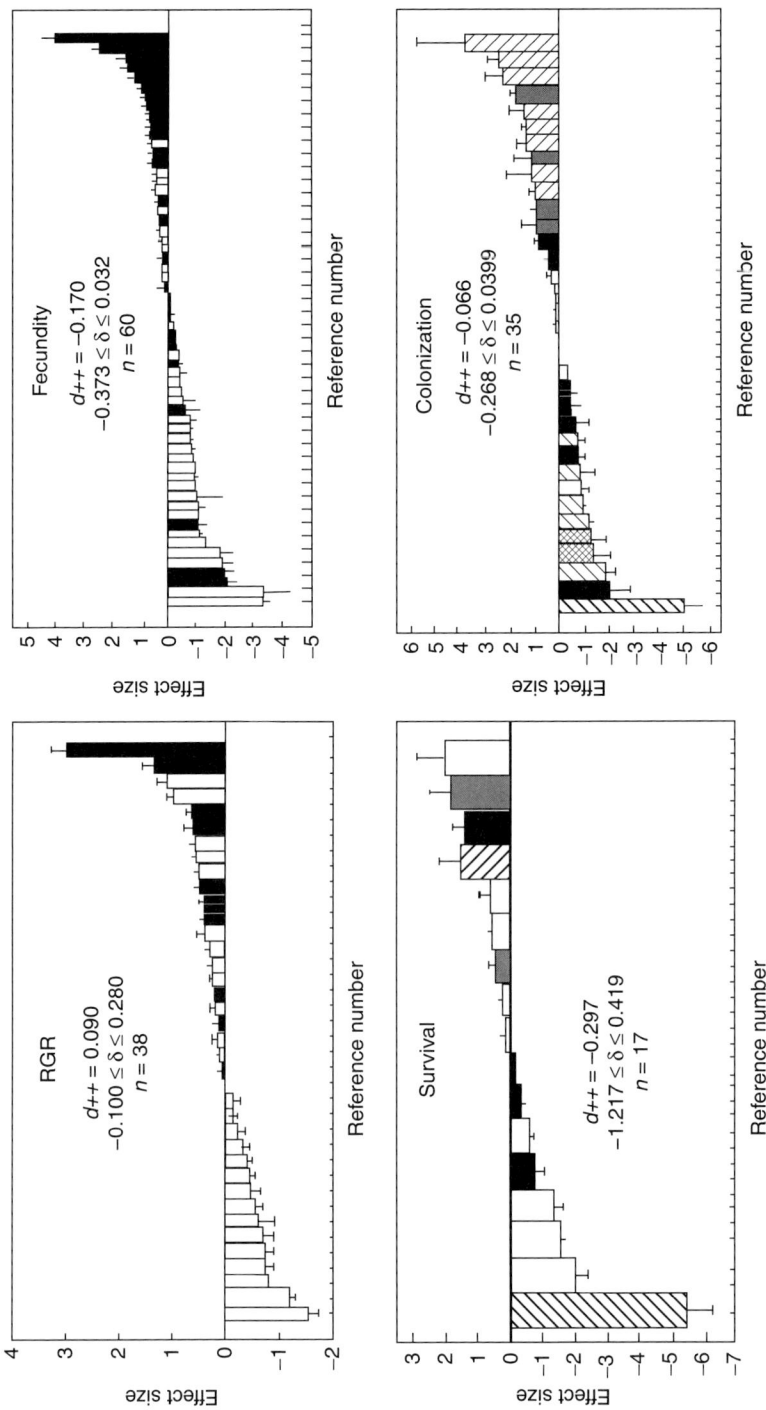

Fig. 4.13. Responses of various insect herbivore guilds to plant stress as reported in the published literature (± 95% confidence limits). Stress in the host plant is indicated by increasing positive numbers on the *y*-axis. RGR = relative growth rate; white bars = chewers; black bars = suckers; cross-hatched bars = miners; right diagonal bars = gallers; left diagonal bars = borers (from Koricheva et al., 1998, with permission Annual Reviews Inc).

mechanisms to optimize their abilities to extract the most they can from their host trees. To this end, many insects have become extremely sophisticated phytochemists, detecting subtle changes in nutrient and defence properties of host plant tissues and responding accordingly. Plant soluble organic nitrogen is the most limiting major resource and any increases in its quantity and quality are likely to be seized by insects and turned to their advantage. It may be useful on occasion to view tropical forestry as a means of growing insect food and to alter our tactics and methods to do this less efficiently.

Co-evolution can be thought of as the evolution of two separate species in relation to one another, such that changes in one species influence and direct changes in the other, and vice versa. One rather hypothetical example might be that of early insects evolving wings to avoid predation by primitive hole-dwelling spiders, the latter in turn then evolving the use of webs to catch their flying prey. In insect and plant associations, the term 'arms race' has been applied wherein insect herbivores attempt to feed on plants and/or breed on or within their tissues, while the plants develop mechanisms to prevent this happening. Another hypothetical example might involve a caterpillar feeding on the leaves of a tree, in response to which the tree produces chemicals or spines which reduce or even prevent this feeding. The next stage is that the insect then develops an immunity or tolerance to the plant's defences, and even uses them to recognize its preferred food. The overall result of co-evolution is therefore diversification in chemical and physical defences in plants and host plant specialization in insects (Menken, 1996). However, not all insect–plant relationships involve aggression and defence. The development in large numbers of angiosperms of pollination by insects is another example of co-evolution, but this time of mainly symbiotic or mutualistic relationships. The modern-day complexities of insect–plant relationships are fundamental to herbivore ecology, and hence the ability or otherwise of an insect to become a forest pest, and form the basis of plant resistance in combating outbreaks (see Chapter 6).

4.5.1 The effect on the insects

As discussed in Chapter 2, some plant-feeding insects are host specialists, whereas others are more generalist. It is assumed that food specialists have competitive and selective advantages over food generalists, since they can utilize optimally the equally specialized host plant to which they are closely adapted. Hence, we find the eucalyptus longicorn beetle, *Phoracantha semipunctata* (Coleoptera: Cerambycidae), which will not attack acacias, the teak defoliator, *Hyblaea puera* (Lepidoptera: Hyblaeidae), which will not eat the leaves of mahogany, or the pine woolly aphid, *Pineus pini* (Homoptera: Adelgidae), which will not feed on cypress, even though all three pairs of tree genera are available in the same locality. It has to be said, though, that in all three examples, this host plant specificity is based at the genus or even family level, but though *P. pini* will feed on all species of *Pinus*, it performs much better on some species than others. Even within a species, certain genotypes of the host are more suitable, as indeed are individuals of the same genotype growing in different sites. These various sorts of specialization are of great significance for pest dynamics and outbreak management.

The result is that closely related insects may, in fact, be adapted to feed on different host plants as a method of reducing competition between members of the same genus. The host plants may also be related fairly closely, even if the insects are widely separated geographically. Futuyama and Mitter (1997), for example, looked at the host plant associations of leaf beetles (Coleoptera: Chrysomelidae) in North America and compared these with chrysomelids in other regions of the world. They found that the great majority of related beetles fed on related host-plant species too, and that genetic variation in the beetles was seen most often in response to plants related closely to natural hosts.

Some insect herbivores may remain or even evolve to be host generalists. Being a specialist is a sensible evolutionary strategy while the host plant remains common and exploitable, but this evolutionary commitment is less successful when environments change and circumstances demand adaptability (Menken, 1996). This adaptability is put to the test particularly when insect herbivores are presented with novel types of host plant, such as exotic tree species in a new country or large areas of resistant clonal trees in a monoculture. It is likely that insect herbivores with more adaptability (i.e. relatively less host plant specialization) will be the first to seize these new opportunities. So, for example, let us return to leaf-cutter ants such as *A. sexdens*. Leaf cutters are restricted entirely to South and Central America, where they are opportunistic feeders in the main, harvesting a large variety of host tree species. When they experience a new host species, thanks to a plantation of exotic (non-native) trees, for example, this is clearly a new resource to try to exploit. Neem, *A. indica*, has also been referred to earlier in this chapter as having foliar chemicals which have insecticidal properties. The tree is native to South-east Asia and hence a complete exotic in Brazil. From all the foregoing chapter, it might be expected that leaf-cutter ants would be incapable of dealing with neem's exotic defences and hence planting this species in Brazil should be a pest-free experience, for a while at least, but as de Souza *et al.* (2009) report, *A. sexdens* has rapidly discovered neem plantations and feeds happily on them.

It has to be remembered, of course, that insect–tree co-evolution works via the ability or otherwise of the two participants to grow to adulthood and to reproduce satisfactorily. This basically admirable goal may be perfectly adequate for the futures of both the insect and the plant in their natural settings, but it may not do at all for the forester. Take the case of the mahogany shoot borer *Hypsipyla* spp. (Lepidoptera: Pyralidae). This complex group of species tunnels into the shoots and tips of a large variety of New and Old World Meliaceae, including *Swietenia*, *Cedrela* and *Toona* (Floyd and Hauxwell, 2000) (Plates 57, 58 and 59) (see also Chapters 5, 6 and 10). Only young trees are attacked seriously, though very few actually die. Instead, leading shoots are killed and apical dominance is transferred to a lateral, which then grows upwards. The tree is perfectly able to grow on and eventually produces a healthy but rather deformed tree. The shape of the main trunk matters little, if at all, to the tree; it is everything to the tropical forester. This classic example of insect–tree co-evolution causes some of the most serious and widespread insect pest damage to tropical forests in the world.

5

Tropical Forest Pests: Ecology, Biology and Impact

5.1 Introduction

Almost every part of a tree can serve as food for insects, and some of the more common groups associated with damage to leaves, shoots, flowers, buds, fruit, twigs, branches, stem and roots are shown in Fig. 5.1. Generally, the species of importance for forestry are contained in the Orders Phasmatodea (stick insects), Orthoptera (grasshoppers, crickets), Hemiptera (plant bugs, leafhoppers, scale insects, aphids, mealybugs), Coleoptera (beetles), Isoptera (termites), Hymenoptera (sawflies, ants, wasps, bees), Lepidoptera (butterflies, moths), Diptera (flies) and Thysanoptera (thrips). Feeding habits vary greatly, not only between but also within pest groupings. Some folivores are generalist feeders with a wide range of tree hosts (e.g. the Asian gypsy moth, *Lymantria dispar*, has over 600 recorded hosts, both conifers and hardwoods), while others are restricted to a single genus and sometimes species (the branchlet mining scolytine, *Hylurdrectonus araucariae*, in Papua New Guinea has been recorded only from *Araucaria cunninghamii*). Some insect species use several different tree parts as food sources during their life cycle (e.g. the larvae of some longicorn beetles tunnel in branches and stems and the adults feed on nectar); other insects such as leaf gall formers restrict their activities to a single tree part.

The impact of insect feeding on a tree is largely a function of the value of the affected part or parts for tree productivity or survival.

The value of older or mature leaves differs markedly from that of expanding or young leaves, which generally have higher photosynthetic efficiencies (see Chapter 4), and ring-barking of the stem cambium by an insect is more important than heartwood tunnelling. However, economic impact is not always linked with tree fitness and, in the case of some stem borers, has more to do with the market quality of the wood product. As emphasized in earlier chapters, the vast majority of insects encountered in tropical forests are entirely benign and only very few achieve tree pest status. In this chapter, we examine the biology, ecology and impact of some of the more important pest species from around the tropics, the examples being chosen to reflect a range of feeding types, host trees and forest situations. The assigning of a pest species to a particular feeding guild (Table 5.1) is on the basis of the principal damage it causes, although some species could easily have been listed under more than one guild.

5.2 Defoliation

Defoliators are insects that feed on leaf tissue, their activities resulting in complete or partial

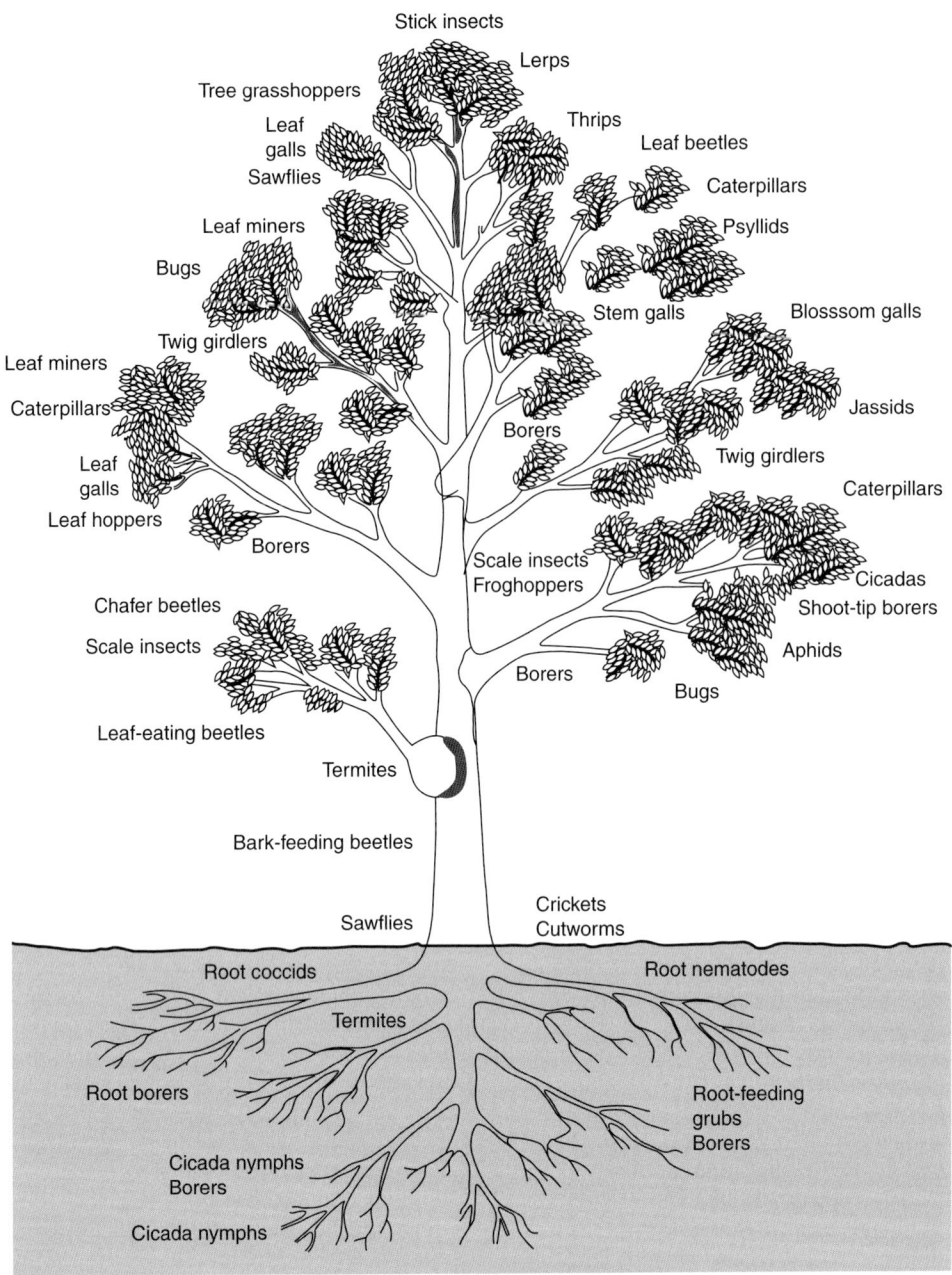

Fig. 5.1. A schematic tree showing the parts affected by various types of pest (from Jones and Elliot, 1986).

destruction of leaves. This feeding group also includes leaf-mining insects, which feed within the leaf just below the upper or lower surface, leaf tiers and leaf rollers, which make shelters from leaves and graze these from within, and leaf skeletonizers, which eat the leaf tissue between the network of leaf veins.

Defoliation by insects reduces the tree's rate of photosynthesis and transpiration. Effects vary considerably, depending on severity,

Table 5.1. List of some of the more important insect pest species in the tropics, categorized according to their feeding habits. Species marked with an asterisk are discussed in detail in this chapter.

Feeding habit and insect order	Insect family	Scientific name	Principal host (genus)	Countries with reported damage
Defoliating				
Leaf chewing				
Coleoptera	Chrysomelidae	*Paropsis* spp.	Eucalyptus, Acacia	Australia
		Paropsisterna spp.	Eucalyptus, Corymbia, Melaleuca	Australia
	Curculionidae	*Gonipterus scutellatus	Eucalyptus	Australia, Africa, South America, California
		Hypomeces squamosus	Many	South and South-east Asia
		Myllocerus spp.	Acacia	India
	Scarabaeidae	*Anomala spp.	Many	Asia, China, India
Hymenoptera	Diprionidae	(Neo)diprion spp.	Pinus	Vietnam, Thailand
	Formicidae	*Atta/Acromyrmex spp.	Many	Southern USA, Central and South America
	Pergidae	*Perga affinis	Eucalyptus	Australia
Lepidoptera	Arctiidae	Eupseudosoma spp.	Eucalyptus	South America
	Geometridae	Buzura suppressaria	Eucalyptus	China
		*Thyrinteina arnobia	Eucalyptus	Brazil, Venezuela
	Hyblaeidae	*Hyblaea puera	Tectona	India and South-, east Asia, South America
	Lasiocampidae	*Dendrolimus punctatus	Pinus	China, Vietnam, Thailand
	Lymantriidae	Lymantria ninayi	Pinus	Papua New Guinea
	Noctuidae	Spodoptera litura	Many	India, South-east Asia, Australia, Oceania
	Notodontidae	Nystalea nyseus	Eucalyptus	Brazil
	Pieridae	Eurema spp.	Acacia, Falcataria, Cassia	Africa, South-east Asia, Australia, Oceania
	Psychidae	*Pteroma plagiophleps	Acacia, Falcataria, Cassia	India, South-east Asia
Phasmatodea	Phasmidae	*Ctenomorphodes tessulatus	Eucalyptus	Australia
Leaf mining				
Hymenoptera	Pergidae	Phylacteophaga eucalypti	Eucalyptus	Australia
Lepidoptera	Gracilliariidae	*Acrocercops spp.	Shorea, Swietenia, Tectona	India, Australia
	Incurvariidae	Perthida glyphopa	Eucalyptus	Australia
Leaf skeletonizing				
Coleoptera	Chrysomelidae	*Craspedonta leayana	Gmelina	India, Bangladesh, Myanmar
Lepidoptera	Noctuidae	Uraba lugens	Eucalyptus	Australia

Continued

Table 5.1. Continued.

Feeding habit and insect order	Insect family	Scientific name	Principal host (genus)	Countries with reported damage
	Pyralidae	*Eutectona machaeralis*	*Tectona*	India, South-east Asia
Leaf tying/rolling				
Lepidoptera	Pyralidae	*Lamprosema lateritialis*	*Afrormosia, Pericopsis*	Nigeria, Ghana, West Africa
	Tortricidae	**Strepsicrates rothia*	*Eucalyptus*	Africa, South and South-east Asia
	Yponomeutidae	*Atteva fabriciella*	*Ailanthus*	India
Sap feeding				
Hemiptera	Adelgidae	**Pineus pini boerneri*	*Pinus*	Australia, Africa, America
	Aphididae	**Cinara cupressi cupressivora*	*Cupressus*	Africa, South America
		Eulachnus rileyi	*Pinus*	Africa, Venezuela, Argentina
	Coccidae	**Ceroplastes* spp.	*Acacia, Melia, Cedrela, Toona, Pinus*	Africa, Australia, India, South-east Asia
		Eriococcus spp.	*Eucalyptus*	Australia
	Coreidae	**Amblypelta cocophaga*	*Eucalyptus, Campnosperma*	Solomon Islands
	Diaspididae	*Aonidiella* spp.	*Azadirachta, Dalbergia, Swietenia, Melia*	Africa, India, South-east Asia
		*Aulacaspis marina	*Rhizophora*	Indonesia
	Margarodidae	*Conifericoccus* spp.	*Agathis*	Australia
	Miridae	**Helopeltis* spp.	*Anacardium, Azadirachta, Eucalyptus*	Africa, India, China, Indonesia
		Lygus spp.	*Pinus*	Southern USA, Mexico
	Pseudococcidae	**Nipaecoccus*	*Casuarina, Dalbergia*	Africa, India
	Psyllidae	*Cardiaspina* spp.	*Eucalyptus*	Australia
		*Glycaspis brimblecombei	*Eucalyptus*	North America, South America, Hawaii, Mauritius
		*Heteropsylla cubana	*Leucaena*	Pantropical
	Thaumastocoridae	**Thaumastocoris peregrinus*	*Eucalyptus*	South Africa, South America, Australia
	Tingidae	**Tingis beesoni*	*Gmelina*	India, Myanmar, Thailand
Bark and wood feeding				
External chewing				
Coleoptera	Cerambycidae	*Penthea pardalis*	*Acacia*	Australia
	Curculionidae	*Plagiophloeus longiclavis*	*Toona*	India, South-east Asia
Lepidoptera	Indarbelidae	**Indarbela quadrinotata*	Many	India, Bangladesh, Pakistan, South-east Asia
Bark boring				
Coleoptera	Anobiidae	*Ernobius mollis*	*Pinus*	Australia

Continued

Table 5.1. Continued.

Feeding habit and insect order	Insect family	Scientific name	Principal host (genus)	Countries with reported damage
	Curculionidae Scolytinae	*Hylastes angustatus*	*Pinus*	Southern Africa
		Ips spp.	*Pinus*	Pantropical
		Tomicus piniperda	*Pinus*	China, Hong Kong
Cambium and surface sapwood boring				
Coleoptera	Buprestidae	*Agrilus* spp.	*Eucalyptus*	Philippines, Papua New Guinea, Africa, Indonesia
		Diadoxus spp.	*Callitris*	Australia
	Cerambycidae	*Phoracantha semipunctata*	*Eucalyptus*	Africa, Australia, America
		Xystrocera festiva	*Falcataria, Acacia*	Indonesia, Malaysia
	Curculionidae	*Aesiotes notabilis*	*Araucaria, Pinus*	Australia
Isoptera	Termitidae	*Macrotermes* spp.	Many	Africa, China, South-east Asia
		Odontotermes spp.	Many	India, Africa, South-east Asia
Sapwood and heartwood boring				
Coleoptera	Bostrychidae	*Apate* spp.	Many	Africa
		Sinoxylon spp.	*Shorea, Acacia, Eucalyptus*	Africa, India, South-east Asia
	Cerambycidae	*Anoplophora* spp.	*Casuarina, Acacia, Eucalyptus*	Taiwan, South-east Asia
		Aristobia horridula	*Dalbergia, Pterocarpus*	India, Nepal, Thailand
		Batocera spp.	*Mangifera, Ficus*	Pantropical
		Celosterna scabrator	*Acacia, Cassia, Prosopis, Tectonis, Eucalyptus*	India, Thailand
		Hoplocerambyx spinicornis	*Shorea*	India
	Curculionidae	*Eurhamphus fasciculatus*	*Araucaria*	Australia
	Curculionidae/ Platypodinae	*Platypus* spp.	Many	Pantropical
	Curculionidae/ Scolytinae	*Xyleborus* spp.	Many	Pantropical
		Xylosandrus crassiusculus	Many	Asia, North and Central America, South Pacific, Australia
Hymenoptera	Siricidae	*Sirex noctilio*	*Pinus*	Australia, South America, South Africa
Isoptera	Mastotermitidae	*Mastotermes darwiniensis*	Many	Australia
	Rhinotermitidae	*Coptotermes* spp.	Many	Pantropical
Lepidoptera	Cossidae	*Chilecomadia valdiviana*	*Eucalyptus*	Chile, Argentina
		Coryphodema tristis	*Eucalyptus*	South Africa
		Xyleutes ceramica	*Gmelina, Tectona*	India, South-east Asia
	Hepialidae	*Endoclita undulifer*	*Gmelina, Eucalyptus*	India

Continued

Table 5.1. Continued.

Feeding habit and insect order	Insect family	Scientific name	Principal host (genus)	Countries with reported damage
Shoot boring				
Coleoptera	Curculionidae/ Scolytinae	*Hylurdrectonus araucariae	Araucaria	Papua New Guinea
Lepidoptera	Pyralidae	*Dioryctria spp.	Pinus	Central America, India, China, South-east Asia
		*Hypsipyla robusta/ grandella	Toona, Swietenia, Cedrela, Khaya, Chukrasia	Pantropical
	Tortricidae	Rhyacionia spp.	Pinus	South America, South-east Asia
Fruit and seed boring				
Coleoptera	Bruchidae	*Bruchidius spp.	Acacia	Africa
Hymenoptera	Torymidae	*Megastigmus spp.	Cupressus, Juniperus, Eucalyptus	Africa, Australia, India, China, Thailand, Mexico
Lepidoptera	Pyralidae	Dioryctria spp.	Many	America, South-east Asia
		Hypsipyla spp.	Many	Australia, India, Central and South America
Gall forming				
Diptera	Agromyzidae	*Fergusonina spp.	Eucalyptus	Australia, Philippines, India
Hemiptera	Eriococcidae	Apiomorpha spp.	Eucalyptus	Australia
	Psyllidae	*Phytolyma spp.	Milicia	Africa
Hymenoptera	Eulophidae	*Leptocybe invasa	Eucalyptus	Africa, Asia, Mediterranean, North and South America
		*Ophelimus maskelli	Eucalyptus	Mediterranean, North Africa
		*Quadrastichus erythrinae	Erythrina	Africa, Asia, South Pacific, Florida
Root feeding				
Coleoptera	Scarabaeidae	Anomala spp.	Cassia, Pinus	India, China
		Holotrichia spp.	Cassia, Shorea, Tectona	Australia, South-east Asia
		*Lepidiota spp.	Araucaria, Acacia, Eucalyptus	Australia, China, South-east Asia
Isoptera	Rhinotermitidae	Coptotermes spp.	Many	India, South-east Asia
	Termitidae	Macrotermes spp.	Gmelina, Eucalyptus	Africa, South-east Asia
		Microtermes spp.	Tectona, Eucalyptus	Africa, South-east Asia
		Odontotermes spp.	Hevea, Pinus	Africa, South-east Asia
Stem and branch cutters				
Coleoptera	Cerambycidae	*Strongylurus decoratus	Araucaria	Australia
Orthoptera	Acrididae	Valanga nigricomis	Hevea, Swietenia, Tectona	India, South-east Asia
	Gryllidae	*Brachytrupes spp.	Eucalyptus, Pinus, Cupressus, Tectona	India, Africa, South-east Asia

age of leaves eaten, position in the canopy, time of year the defoliation occurs, the site and current stresses on the tree (Elliott et al., 1998). For example, small folivore (i.e. leaf-feeding) populations can cause disproportionately large foliage losses by feeding on vegetative buds or unexpanded foliage (Schowalter et al., 1986). Sunlit leaves have higher photosynthetic rates than shaded leaves, regardless of age, and insects feeding on these will affect the host more severely than similar damage to shaded leaves. Trees that are damaged or stressed are less able to cope with the effects of severe defoliation than are vigorous trees. Light (i.e. less than 20%) defoliation normally has no or very little effect on the tree, but moderate (25–50%) and severe (more than 50%) defoliation reduces growth rates, can affect wood properties (shortens fibre length) and can predispose trees to attack by other organisms (Elliott et al., 1998). Defoliation in consecutive years is more damaging than a single severe defoliation, and some species of evergreen conifers can be killed by one complete defoliation if this occurs before bud formation. Site and weather factors interact with defoliation episodes to determine the overall impact on the host tree.

In the sections which follow, examples are given of the various types of insect defoliators, with information on their life history, ecology and pest importance.

5.2.1 Leaf chewing

Paropsisterna spp. and *Paropsis* spp. (Coleoptera: Chrysomelidae)

More than 3000 species of chrysomelid beetle are found in Australia, most of which are never likely to achieve pest status (Waterson and Urquhart, 1995). Damage by these insects in extensively managed native forests generally has been tolerated by forest managers, but with the increasing establishment of fast-growing eucalypt plantations in several states, chrysomelid leaf beetles have become recognized as one of the most serious insect problems associated with these plantations (Nahrung and Allen, 2003). Paropsines cause not only significant losses to eucalypt plantations in Australia, where the beetles and trees are native, but also where the beetles have been introduced accidentally to eucalypt growing regions in countries such as South Africa and New Zealand (Nahrung, 2006).

Some of the more common pests belong to the genera *Paropsisterna* and *Paropsis*, sometimes called tortoise or 'paropsine' beetles, and share generally similar life cycles. As outlined by Elliott et al. (1998), in subtropical and temperate regions the adults overwinter under bark, in crevices or in leaf litter and emerge in early spring to feed on newly expanding foliage, leaving characteristic 'scalloped' damage on the leaf margins. Eggs are laid either singly or in batches on newly expanding leaves or shoots, each species having a particular ovipositional behaviour and site for the eggs. The larvae (Plate 14) of many species consume their egg cases before beginning to feed on the leaf. Species which lay eggs in batches feed colonially and, as the young larvae reach the later instars, they usually move to several feeding sites in the crown of the host plant, feeding singly or in small groups. The larvae of some species feed singly for the entire larval stage. There are usually four larval instars, which are completed in 3–4 weeks, and the larvae then drop to the ground, where they pupate in the litter or soil. Adults emerge in 7–10 days, and in the warmer regions there may be five or more generations per year with adults and/or larvae present on the trees for 8–9 months each year.

In Australia, most studies of the impact of chrysomelid leaf beetles have been carried out in temperate regions. In Tasmania, 1-year-old *Eucalyptus regnans* attacked by *Ps. bimaculata* lost 45.6% and 52.1% of their potential height and basal area increment, respectively, over a 2-year period compared to trees protected by an insecticide treatment (Elliott et al., 1993) (Fig. 5.2). Modelling the impact of defoliation by *Ps. bimaculata* on the growth of *E. regnans* showed that typical defoliation regimes could reduce growth over a 15-year rotation by up to 40% (Candy et al., 1992). In tropical and subtropical eucalypt plantations, *Ps. cloelia* is becoming an increasingly important pest

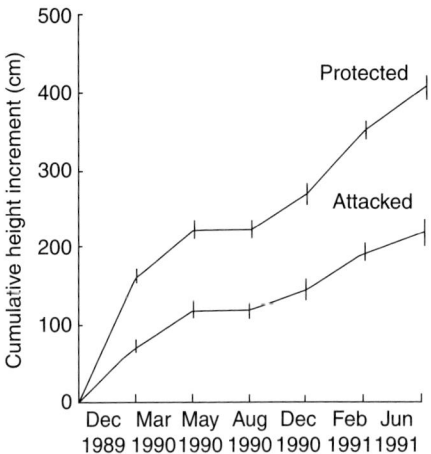

Fig. 5.2. Mean height growth (± SE) of protected and attacked 1-year-old *Eucalyptus regnans*, Florentine Valley, Tasmania (from Elliott et al., 1993).

(Carnegie et al., 2005). In an outbreak of *Ps. cloelia* in *E. grandis* plantations in north coastal New South Wales in 1972–1974, it was estimated that up to 60–70% of the potential annual height increment of the trees could be lost, and more than half the trees died in the following 2 years (Carne and Taylor, 1978). Further north, in Queensland, the relatively recent expansion of hardwood plantations has resulted in the emergence of new pest species and Nahrung (2006) collected at least 17 paropsine species from plantations of *E. cloeziana* and *E. dunnii*, about one-third of which have not been associated previously with eucalypt plantations. The most abundant of these pest species were *P. atomaria*, *P. charybdis* and *Ps. cloelia*.

A wide range of natural enemies of paropsine beetles has been recorded, including parasitic wasps and flies and predatory ladybird beetles and bugs. Cumulative mortalities of *Ps. bimaculata* populations resulting from attack by natural enemies can reach up to 97% by the end of the larval stage (de Little et al., 1990) and up to 90% for *P. atomaria* (Tanton and Khan, 1978b).

Gonipterus scutellatus (Gyllenhal)
(Coleoptera: Curculionidae)

The eucalyptus weevil or eucalyptus snout beetle, *Gonipterus scutellatus*, (Plate 15) is a native of Australia, where it is usually a relatively minor pest in eucalypt forests and plantations, although occasionally it causes economically significant damage. However, it has spread to several other countries, where it has caused severe damage to eucalypt plantations. It was first recorded outside Australia in about 1890 in New Zealand (Withers, 2001) and was discovered in South Africa in 1916 (Mally, 1924). It spread rapidly throughout Africa to Lesotho, Malawi, Kenya, Mauritius, Mozambique, Zimbabwe, Uganda and Malagassy and has been reported from Spain, Italy, Portugal, France, Argentina, Brazil, Chile, Uruguay and California (Richardson and Meakins, 1986; Rosado-Neto, 1993; Cowles and Downer, 1995; Lanfranco and Dungey, 2001; CABI/EPPO, 2010).

Adults lay small yellow eggs on the foliage in black pods, with three to eight eggs in a pod. The eggs hatch in 3–4 weeks and the larvae emerge on the underside of the leaf by chewing through the bottom of the pod and the leaf. All the larvae emerge through one hole and begin feeding immediately on the surface of the leaf, leaving characteristic tracks where the upper surface of the leaf tissue has been removed. There are four larval instars lasting a total of 5–7 weeks. When mature, the larvae drop to the soil and form pupal chambers just below the surface, from which adults emerge after a few weeks. The adults are strong fliers and live for 2–3 months, during which period they lay eggs several times. In Australia, there are usually two generations per year with a life cycle of 10–16 weeks, but there is considerable variation in the countries where the insect now occurs. This depends largely on the incidence of cold weather inducing hibernation and on rainfall inducing flushes of young foliage, which provides the essential diet (Browne, 1968). The larvae feed mostly on the upper crown, where they skim the epidermis of the young leaves, while adults feed all over the tree, devouring the leaves from their edges. In addition, both stages eat the young, soft shoots of some species (Richardson and Meakins, 1986) and continued infestation can produce stag-headed or stunted trees. The impact of defoliation by the weevil on tree height growth is illustrated in Fig. 5.3.

In South Africa, the arrival of *G. scutellatus* led to the cessation of planting of several commercially important eucalypt species, including *E. viminalis*, which had been favoured up until that time. Similarly, in other countries, the beetle caused economic losses in existing plantations and limited the choice of species for future plantings. An Australian wasp parasitoid of *Gonipterus* eggs, *Anaphes nitens*, was introduced into South Africa and several other countries and gave effective control at altitudes up to 1200 m, but was much less effective at higher altitudes where some major plantations of susceptible species occurred (Govender and Wingfield, 2005).

Corymbia Anomala spp. *(Coleoptera: Scarabaeidae)*

The genus *Anomala* is common throughout Asia, and in India alone there are more than 200 species (Beeson, 1941). The adult beetles of many species are defoliators of forest trees, while the soil-dwelling larvae or 'white grubs' are sometimes pests of nursery stock. Typically in India, beetles swarm at dusk in May–July, the earliest activity occurring after the first showers of the monsoon season. Pairing and egg laying take place throughout this period. Eggs are laid singly at night in soil at a depth of 5–8 cm, and one female may lay 30 eggs. The larva tunnels through the soil, eating the fine roots of plants and decaying vegetable matter, and completes its development in about 9 months. Pupation takes place in the shelter of the larval skin and may last a month. The immature beetle remains in the soil until suitable climatic conditions stimulate emergence and flight. The generation is usually annual.

In China, the principal species in tropical regions is *A. cupripes*, which defoliates young plantations of *Corymbia brassiana*, *E. camaldulensis*, *E. citriodora* and *E. exserta* in Guangxi and Gunagdong Provinces. Its pest status has increased considerably with the rapid expansion of the plantation estate in the south (Wylie, 1992). In Hong Kong, swarms of adults sometimes occur in nurseries and plantations during the summer, occasionally causing serious injury to seedlings of *Pinus massoniana* and to the foliage of young stands of *C. citriodora*, *C. grandis* and *E. maculata* (Browne, 1968). In the subtropical Chinese provinces of Fujian and Jiangxi, *A. antiqua* is a pest of *Paulownia*, adults consuming a mean of 882 mm^2 of *Paulownia* leaves a day (Tong and Fang, 1989).

Reported natural enemies of this insect are carabid beetles, ring-legged earwigs and the entomogenous fungus, *Metarhizium anisopliae*. Several *Anomala* species are pests in forest nurseries in Bangladesh (Baksha, 1990).

Atta spp. and *Acromyrmex* spp.
(Hymenoptera: Formicidae)

Leaf-cutting ants are the largest species group of the fungus-growing ants in the tribe Attini and effectively consist of two genera, *Atta* (Plates 16 and 17) and *Acromyrmex* (Ramos *et al.*, 2008). They are sometimes known as parasol ants because of their habit of cutting leaves into more or less circular pieces, which they hold above their heads as they carry them back to the nest. The group is restricted to the New World, *Atta* being

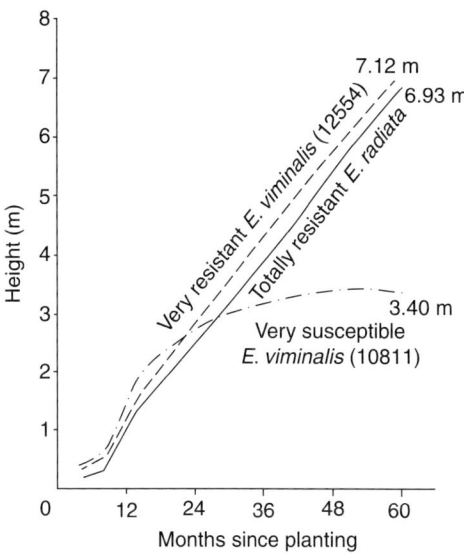

Fig. 5.3. The effect of attack by snout beetle on the height growth of very susceptible and very resistant *Eucalyptus viminalis* (from Richardson and Meakins, 1986).

found in 23 countries and *Acromyrmex* in 25 countries from the USA in the north, through Central America and the West Indies down to Uruguay in South America (Cherrett and Peregrine, 1976). They are among the most prevalent herbivores of the Neotropics, consuming far more vegetation than any other group of animals with comparable taxonomic diversity (Vieira-Neto and Vasconcelos, 2010). The ants are pests of agriculture, horticulture, rangelands and forestry. They have been reported attacking 15 different forest tree crops, including pines, teak and eucalypts, in some 16 countries and their rated importance for forestry in these countries is shown in Table 5.2.

Atta and *Acromyrmex* are social insects which live in large underground nests, where they cultivate fungal gardens on a substrate of harvested leaves, this fungus forming their principal diet. While the fungus provides ants with nutrients and enzymes, the ants, in turn, supply the fungus with a variety of substrates and stimulate symbiotic fungal growth (Marsaro *et al.*, 2004). Following mass mating flights at the start of the rainy season, the founding queen (mated female) will begin to dig the nest. She will usually bring with her spores of the fungus garden from her original nest site for incorporation into this new nest and will then find a suitable place to lay her eggs. She fertilizes the spores with her own waste, to aid in development of food to feed to the larvae. The queen will continue to raise her first brood until these workers are mature enough to assist her with more eggs and the rest of the colony. Colony growth is slow in the beginning, then proceeds rapidly and reaches maturity in about 5 years.

Species of *Atta* characteristically build and maintain physically defined foraging trails, which may extend 250 m or more in length (San Juan, 2005). They also use trail pheromones. When a worker ant finds forage, she will straddle the piece and cut only that portion which is equal to her leg span. She will rotate and shear the leaf with her mandibles and then carry the disc-shaped portion back to the nest, where other workers process the leaf matter and add it to the garden, which is fertilized with faecal droplets. Undigested or decaying parts of the garden are tossed aside or removed from the garden area as debris. As is usual with social insects, tasks are shared among colony members. The smallest workers tend to remain in the nest and serve as gardeners and nursemaids. The medium-sized workers forage, while others maintain the compost heap or deepen and expand the nest. Soldiers fight off other ants, guard the nest and may also control the 'traffic' in a busy working colony to ensure the proper disposal of forage. A nest may cover 600 m^2 and may be more than 6 m deep, containing hundreds of fungus gardens and millions of workers: a colony of *A. sexdens* may contain up to 8 million workers (San Juan, 2005). The nest architecture is complex and adapted to ensure temperature control of the fungus garden.

Table 5.2. Importance of leaf-cutting ants as forestry pests (updated from Cherrett and Peregrine, 1976).

Country (in order of forestry importance, most northerly part)	
USA	***
Mexico	*
Belize	***
Honduras	***
El Salvador	***
Nicaragua	****
Costa Rica	***
Panama	*
Columbia	**
Venezuela	**
Trinidad and Tobago	*****
Surinam	*****
French Guiana	*****
Brazil	*****
Ecuador	***
Peru	***
Bolivia	**
Paraguay	**
Argentina	*****
Uruguay	*****

Notes: ***** Major importance, one of the five worst insect pests in the country; ****considerable importance, one of the 20 worst insect pests; ***moderately important pests causing sporadic economic damage; **minor pests, infrequently producing significant damage; *unimportant, only recorded on one or two occasions as doing any damage.

The impact of leaf-cutter ants on forests is dependent on a range of factors, but the degree of ecosystem disturbance is of great importance. For example, Jaffe and Vilela (1989) found relatively low nest densities of *A. cephalotes* in primary rainforest ecosystems in the Orinoco–Amazon basin, namely 0.045 nests/ha. They attributed this to a need of *A. cephalotes* colonies for a certain degree of sunshine (or light) on their nest heaps. The number of clearings in the forest was proposed to be the limiting factor for colonization of new forest sites by this species. Elsewhere, higher nest densities of *Atta* have been reported in more disturbed forests, with up to 30 nests/ha found in *Pinus* plantations in the south-western savannahs in Venezuela (Jaffe, 1986) and nearly 300 nests/ha in 16-year-old eucalypt plantations in Brazil (Oliveira *et al.*, 1998). Human intervention clearly increases *Atta* nest densities in affected ecosystems. Urbas *et al.* (2007) demonstrated edge effects on foraging and herbivory of *A. cephalodes* in a large remnant of the Atlantic forest in north-east Brazil. Equally-sized *A. cephalodes* colonies located at the forest edge removed about twice as much leaf area from their foraging grounds than interior colonies (14.3 versus 7.8%/colony/year). This greater colony-level impact in the forest edge zone was a consequence of markedly reduced foraging areas (0.9 versus 2.3 ha/colony/year) and moderately lower leaf area index in this habitat, whereas harvest rates were the same. The reduction of the foraging area was attributed to the greater abundance of palatable pioneer plants. The release from natural enemies may also be a contributing factor (Almeida *et al.*, 2008). Some species of *Acromyrmex*, such as *A. landolti balzani*, forage without trails, a behaviour that may be explained by their dependence on ephemeral or homogeneously dispersed resources, which makes main trails unnecessary (Poderoso *et al.*, 2009).

Numerous studies have assessed the rates of leaf fragment input into various leaf-cutting ant nests and some of these are summarized in Lugo *et al.* (1973). Leaf input for *A. cephalotes* ranged from 10.9 g/h for a nest in Guiana to 290 g/h for a nest in Costa Rica. The efficiency of leaf transport to the nest has been estimated at 50–70% (i.e. many leaves are dropped along the way); so herbivory may be much higher than leaf input figures suggest (Lugo *et al.*, 1973; Fowler and Robinson, 1979). Leaf-cutting ants are pests of *Pinus* spp. in several countries; for example, in Venezuela, *A. laevigata* causes economic damage to young *P. caribaea* plantations at densities of 5 nests/ha (Hernandez and Jaffe, 1995), *A. cephalotes* and *A. sexdens* attack *P. caribaea* seedlings in Costa Rica (Ford, 1978) and *A. insularis* attacks several *Pinus* species in Cuba (Tsankov, 1977). In Brazil, *Atta* and *Acromyrmex* attack *A. angustifolia* (Schoenherr, 1991) and are the most severe pests in that country's more than 6 Mha of *Eucalyptus* plantations (Marsaro *et al.*, 2004). Zanetti *et al.* (2000, 2003) found that leaf-cutting by *Atta* spp. in eucalypt plantations reduced wood production, *E. camaldulensis*, *C. citriodora* and *E. tereticornis* being more affected than *E. cloeziana* and *E. urophylla* (Fig. 5.4). *A. sexdens* is a pest of rubber trees (*Hevea* spp.) in nurseries, sometimes affecting more than 70% of trees (Calil and Soares, 1987). In Costa Rica, a species of *Atta* reduced growth of *Acacia mangium* by 50% in trials (Glover and Heuveldop, 1985).

Natural enemies of leaf-cutting ants include nematodes, mites, ant-decapitating phorid flies, birds and armadillos, as well as other ants such as *Nomamyrmex* army ants (Erthal and Tonhasca, 2000; Powell and Clark, 2004; San Juan, 2005; Guillade and Folgarait, 2011).

Perga affinis Kirby (Hymenoptera: Pergidae)

Sawflies (Plate 18) are common pests of woodland and urban trees in many parts of the world and are one of the more dramatic groups of eucalypt-defoliating insects in Australia (Jordan *et al.*, 2002). Some have the potential to become serious pests in eucalypt plantations. The steelblue sawfly, *Perga affinis*, which has several subspecies, occurs widely throughout Australia and feeds principally on *Eucalyptus* spp.

Adults emerge in the autumn from pupation sites in the soil beneath previously infested trees (Fig. 5.5). They do not feed and only live long enough (about a week) to find

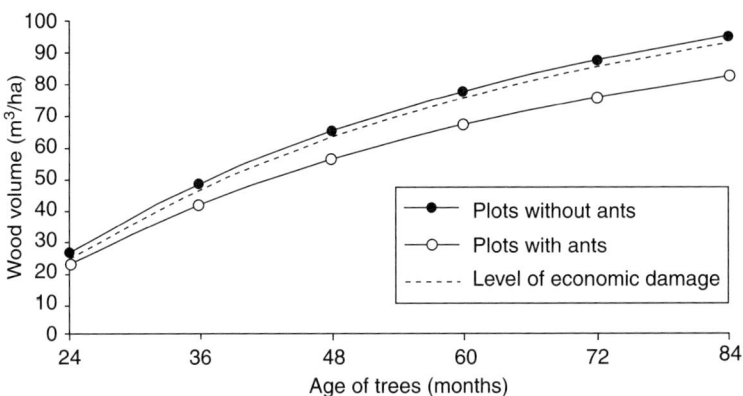

Fig. 5.4. Average wood production for *Eucalyptus* spp. (m³/ha) as function of plantation age (months) considering stands without leaf-cutting ant nests, plots with similar density of nests of the area studied (16.8/ha) and plots with density of nests similar to level of economic damage. Municipality of João Pinheiro, State of Minas Gerais, Brazil from 1991 to 1996 (from Zanetti *et al.*, 2003).

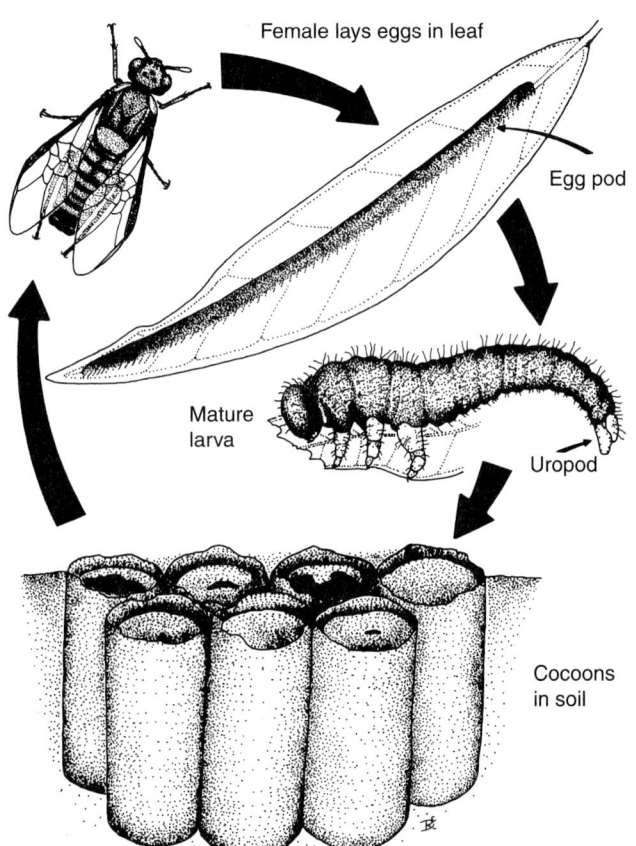

Fig. 5.5. Typical life cycle of *Perga* spp. (from Kent, 1995).

a host tree and lay eggs. Females lay about 65 eggs per batch, inserted by the saw-like ovipositor at regular spacing along the midrib of the leaf, and these hatch in approximately 30 days. Larvae congregate on the leaf surface in a rosette pattern, with their heads facing outwards. They move out and feed on the leaf margins at night, re-forming clusters before dawn (Elliott et al., 1998).

As the larvae grow, usually by about the third instar, they no longer congregate in rosettes on leaves but cluster during the day in large masses surrounding branches and tree boles. When searching for foliage, larvae tend to move upwards and outwards, so the feeding by the older larvae occurs on the terminal shoots first. The tree therefore gets defoliated from the top down. Colonies of larvae from different egg batches can amalgamate and form very large masses. Larvae communicate with each other by vibrational signals, the most likely function of which is to maintain group cohesiveness and coordinate movement (Fletcher, 2007). A heavily infested tree can become completely defoliated, at which time the entire colony moves down the tree and across the ground to a nearby tree. There are six larval instars and when the larvae are fully grown, they leave the trees and burrow into the litter and soil beneath the tree, constructing leathery cocoons which are connected in a compact mass. In the cocoon, the larva moults to a prepupa, which lasts through the summer, when pupation occurs and adults emerge shortly after. The prepupal stage is usually completed during one summer, but some individuals may remain a prepupa for up to 4 years.

Egg mortality results mostly from desiccation and disease, with some eggs being eaten by birds. Larval colonies moving along the ground to other trees can suffer high mortality if nearby trees are more than 10 m away and larvae get caught on the ground during the hottest part of the day. Cocoons of *P. affinis* are parasitized by wasps and flies.

Carne (1969) suggested that changes in the physical environment influencing abundance were common across large areas. Good spring rains soften the soil surface, allowing larvae easy entry for cocoon formation, and also result in good foliage production, which provides favourable ovipositional sites for females in the autumn. Dry winters cause the soil to harden, causing high mortality of larvae attempting to enter the soil to pupate. Hot, dry summers cause desiccation of pupae in the soil (Elliott et al., 1998). Most defoliation by *P. affinis* occurs in late winter and early spring and, at this time of the year, it has least impact on tree growth and health. Trees therefore often survive repeated defoliation in successive years. A single defoliation probably has minimal impact on tree health (Carne, 1969). However, studies by Jordan et al. (2002) of *P. affinis* ssp. *insularis* attack in *E. globulus* plantations in north-western Tasmania demonstrated that sawfly damage caused slow growth and increased mortality of trees. Mild and severe sawfly damage resulted in 16% and 31% reduction in the basal area of surviving trees, respectively, and the effect was consistent across races and families.

Thyrinteina arnobia (Stoll)
(Lepidoptera: Geometridae)

Among the Lepidoptera which defoliate *Eucalyptus* spp. plantations in Brazil, *Thyrinteina arnobia* is regarded as the most important because of its consistent damage and increasing adaptation to this exotic host (Filho et al., 2010). The species occurs widely in Central and South America, from El Salvador down to Uruguay. It has numerous native hosts, including many Myrtaceae, and attacks 13 species of *Eucalyptus*, the principal being *E. grandis* and *E. saligna*.

The female lays her eggs (average 752) on thin branches and these hatch in about 10 days. First instar larvae disperse by direct movement or are carried on the wind by silk threads. There are six larval instars lasting 35–40 days. As is typical of the Geometridae, larvae at rest adopt an erect position, held by the abdominal prolegs, mimicking tree branches. Pupating caterpillars construct a rudimentary cocoon that is fastened by silk lines to eucalypt leaves or to low-lying vegetation. Pupation lasts 4–10 days. Adults emerge from the cocoons at night and are short-lived: about 4 days for males and 7 days for females. Mating and oviposition occur at night.

Caterpillars can be very destructive, especially to new plantations. Initial damage is to the lower part of the crown and attack proceeds upwards to the top of the tree. Outbreaks of this pest are usually not noticed until the larvae are in their fifth instar, when defoliation is rapid (Anjos et al., 1987). Patterns of defoliation vary; in some instances, outbreaks appear to start on the margins of plantations and progress towards the centre, and in other cases they start in the centre and move outwards in a circle, leaving large 'clearings' of defoliated trees. Adults are present throughout the year, but numbers are lowest from October to December. All recorded outbreaks have occurred during the period January–September.

Plantations aged 6 months to 20 years have been attacked. It is suggested by Anjos et al. (1987) that new plantations are attacked only when there are older trees nearby that have a resident population of T. arnobia. Berti Filho (1974) provides estimates of the average leaf area consumed by an individual caterpillar in each instar I–VI (Fig. 5.6). According to calculations by Mendes Filho (1981; cited in Anjos et al., 1987), it would take 663 caterpillars to defoliate completely a 10-year-old E. saligna of 18 cm diameter at breast height (DBH), 4976 caterpillars for a 21 cm DBH tree and 11,610 caterpillars for a 24 cm DBH tree. Early instar caterpillars tend to feed on the leaf surface, but later instars consume the whole leaf. Studies by Oda and Berti Filho (1978) of T. arnobia defoliation in 2.5- to 3.5-year-old plantations of E. saligna showed volume losses 1 year after the event of 25.6 m^3/ha for trees that had been defoliated completely and 8.3 m^3/ha for trees which had been 50% defoliated. In percentage terms, this represents volume losses of 10.4% and 13.2%, respectively. Estimated impacts for some recorded outbreaks are summarized in Table 5.3. In addition to loss of growth increment, defoliation by T. arnobia can also result in tree mortality. There is considerable variation among eucalypt species in their resistance to attack by the pest; E. grandis and E. saligna are highly susceptible, while E. camaldulensis has been rated as highly resistant (Oliveira et al., 1984; Lemos et al., 1999; Filho et al., 2010). Most reports of damage to eucalypt plantations by T. arnobia have been from Brazil, but Marturano and Vergara (1997) recorded that in 1996 this insect caused defoliation on 320 ha of eucalypt plantations aged 2.5 to 3.5 years in eastern Venezuela.

T. arnobia has many natural enemies, including parasitic tachinid and sarcophagid flies, ichneumonid, chalcidid, eulophid and pteromalid wasps and predatory pentatomid bugs, carabid and cicindelid beetles and birds (see, for example, Grosman et al., 2005; Pereira et al., 2008). Mortality rates of more than 80% have been recorded in the immature stages (Batista-Pereira et al., 1995).

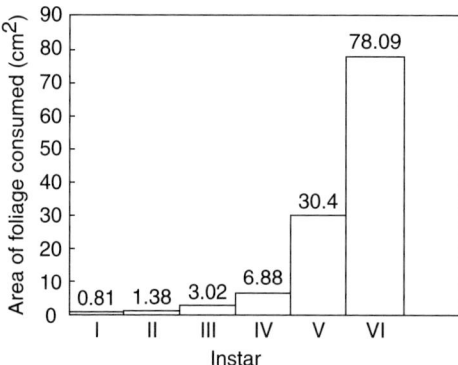

Fig. 5.6. Leaf area consumed by individual larvae of Thyrinteina arnobia in different instars (from Berti Filho, 1974).

Hyblaea puera Cramer
(Lepidoptera: Hyblaeidae)

The teak defoliator *Hyblaea puera* is one of the best known of the lepidopterous defoliators in the tropics, both because of the value of the principal tree on which it feeds and because of its pantropical distribution. Its range extends from the southern USA through the West Indies and Central America into South America as far south as Paraguay, through southern and East Africa, India, China, throughout South-east Asia to Australia and the Pacific. It is now believed that *H. puera* is not a single species but rather a species

Table 5.3. Losses associated with known outbreaks of *Thyrinteina arnobia* in eucalypt plantations in Brazil (from Anjos *et al.*, 1987).

Year	Localities	Area defoliated (ha)	Losses in yield in the next year[a] (stere)[b]
1948	Bauru	600	4,814
1949	Rio Claro	200	1,608
1961	Barra Bonita	200	1,608
1967	Cel. Fabriciano	448	2,401
1973	Ribeirao Preto	4,500	36,180
1973	Itu	827	6,649
1973	Sao Miguel Arcanjo	1,800	14,472
1973	Itupeva	10	81
1973	Itapetininga	200	1,608
1973	Suzano	200	1,608
1974	Ribeirao Preto	600	4,824
1975	Paulinia	200	1,608
1975	Cataguases	200	1,072
1981	Presidente Olegario e Joao Pinheiro	15,000	80,400
Total		24,985	158,933

Notes: [a]Based on defoliation levels of 50% in half the area and 100% in the other half and calculated according to Oda and Berti Filho (1978); [b]stere = a unit of volume equal to 1 m^3.

complex (CABI, 2005). The life history, hosts, population dynamics and impacts of the insect are summarized by Nair (2007). Aside from teak, *Tectona grandis*, it has numerous dicotyledonous hosts, most of which are of minor importance for forestry. It is believed that during non-outbreak periods the insect thrives on hosts other than teak, but there are few data available on the periods of infestation or population levels on these hosts (Nair, 2007).

The moths shelter by day but are active fliers at night and can migrate considerable distances during outbreaks. Eggs are laid singly on leaves, particularly on young, tender foliage, and a female may lay up to 1000 eggs (Beeson, 1941). Early instars feed on the leaf surface, causing skeletonized patches. The third instar cuts out a circular or rectangular flap at the edge of the leaf and folds it over flat, fastening it with silk so as to form a shelter. It skeletonizes older leaves and totally consumes newly formed leaves. Instars IV and V also fold leaves to form shelters and totally consume tissue between the larger side veins to leave bare 'ribs'. Moulting takes place in the leaf folds. The mature fifth instar larva also pupates on the leaf in a triangular fold or, if the crown is stripped, on undergrowth or in soil litter. The life cycle of *H. puera* from egg to the emergence of the adult varies from 14 to 47 days, depending on climatic conditions, and there can be up to 14 complete generations/year. The climate of the locality, particularly rainfall, is one of the main determinants of seasonal abundance (Loganathan and David, 1999). The population density is usually lowest in the season of mature foliage and leaf fall. In southern India, the insect is most abundant in April to early June, while in the north it is abundant in late July to September (Beeson, 1941). Diameter growth of teak generally stops between the beginning of October and mid-November, and defoliation at this time of year has little effect on the current annual increment but may affect future height growth and quality if buds are killed. The frequency of severe defoliation is high in stands aged 11–45 years and is at its maximum in stands 21–30 years old. Nair (1988) estimated that, during outbreaks, a 30 ha teak plantation might have over 450 m larvae.

Nair et al. (1985, 1996) studied the impact of defoliation by H. puera (Plate 19) over the period 1978–1982 in young teak plantations at Kerala, India. They found that the insect caused very significant loss of increment, 44% of the potential growth volume remaining unrealized because of its attack. On a per hectare basis, unprotected trees had a mean annual increment of 3.7 m^3/ha and protected trees 6.7 m^3/ha. Projections suggested that protected plantations could yield the same volume of wood in 26 years as unprotected plantations would do in 60 years.

In Bangladesh, defoliation of teak by H. puera is a regular annual feature, although epidemic outbreaks such as occur in India are rare (Baksha, 1990). The insect is of importance in Sri Lanka (Tilakaratna, 1991), Myanmar (Beeson, 1941), Thailand (Hutacharern, 1990), Malaysia (Abood et al., 2008), China (Chen and Wu, 1984) and Indonesia (Intari, 1978). In the Solomon Islands, it is mainly a pest of nursery stock (Bigger, 1980, 1988). Outbreaks have also been recorded on species of Avicennia, Bruguiera and Rhizophora mangroves in Asia (Palot and Radhakrishnan, 2004), the Caribbean (Saur et al., 1999) and South America (Mehlig and Menezes, 2005). On Brazil's Amazon coast, for example, defoliation of Avicennia germinans by H. puera occurs every 2 years. A 4-year study by Fernandes et al. (2009) showed that leaf loss averaged 13% in the first and third years. They concluded that the conversion of A. germinans leaves into frass by the defoliator favoured nutrient cycling in the mangrove itself and provided a nutrient supply for neighbouring aquatic systems over a short period of time.

Natural enemies of H. puera include parasitic wasps (Chalcididae, Ichneumonidae, Braconidae and Eulophidae) and flies (Tachinidae), and predatory pentatomid bugs, mantids and carabid beetles (Mukhtar et al., 1985; Sudheendrakumar, 1986; Nair, 2007). Birds are also common predators of this insect. In teak plantations in Kerala, India, Zacharias and Mohandas (1990) have recorded 48 species of birds feeding on H. puera, and in Papua New Guinea, the insect is a significant food resource for several bird species in teak plantations (Bell, 1979).

Dendrolimus punctatus Walker
(Lepidoptera: Lasiocampidae)

Pine caterpillars belonging to the genus Dendrolimus (Plate 20) have long been recognized as serious pests of pine forests in China and recorded outbreaks date back to 1530 (Peng, 1989). Each year about 3 Mha of forest are infested, resulting in a growth increment loss of 5 mm^3 (Peng, 1989). D. punctatus is the most widespread and destructive species (Zhang et al., 2003), occurring in 13 provinces, including 3 in the tropical south of the country. It is also an important pest in Vietnam and Taiwan (Billings, 1991). Its main host is P. massoniana, but it has been recorded from several other Pinus species.

Each female moth lays an average of 300–400 eggs, with a possible maximum of 800, in groups on the foliage of the host. Newly hatched caterpillars disperse by 'ballooning' (blown in the wind on silk threads). The larvae feed openly and voraciously on the needles, and there are usually six instars. Mature larvae spin cocoons, in which they pupate, on branches or needles of the host tree or on adjacent vegetation. Adult emergence from the cocoon occurs at dusk and mating and oviposition take place at night. The moths are strong fliers and can migrate up to 20 km.

In the hot and dry regions of Guangxi and Guangdong Provinces in south China, there are three to four generations per year and two to three in warm, humid Hunan (Xue, 1983). In Vietnam, D. punctatus has three to five generations a year, and population densities of up to 700 larvae per branch have been recorded (Bassus, 1974). Severe defoliation may result in loss of growth and resin production, and sometimes in tree mortality. Studies in China by Ge et al. (1988) showed that after almost 100% defoliation of Pinus spp., nearly 25% of trees died and the volume growth of surviving trees was reduced to 31% of normal. Badly affected trees took 3 years to recover. Root rot infection of plantation P. elliottii in southern China has been linked to defoliation by D. punctatus (Liu and Liang, 1993). In general, outbreaks of the pest seem to be fairly erratic, and hence hard to predict (Zhang et al., 2008), though in some localities outbreaks appear every 3–5 years or so (Li, 2007). Billings (1991) recorded that in Vietnam in 1987, 56,500 ha had been affected by the moth larvae, with 33,500 ha suffering from severe defoliation. Young trees may be defoliated completely, and even die as a result. Those between 7 and 15 years old are most commonly attacked.

Biotic agents (parasitoids and entomogenous fungi) have dominated integrated pest management strategies used against this insect in China since the 1970s (Li, 2007) (see also Chapter 10). Xu et al. (2006) list 58 species of parasitic wasps and 20 species of tachinid flies. Trichogramma wasps, particularly T. dendrolimi, which parasitize eggs of D. punctatus, have been widely employed and reportedly are effective (Wu et al., 1988). The number of parasitoids required for suppression varies with the age of the stand, the density of the trees and the density of the egg masses but, in general, 1,050,000/ha are required (Hsiao, 1981). There are many other natural enemies, one of the main in Hunan being the ichneumonid wasp parasitoid *Casinaria nigripes* Gravenhorst, which attacks first to fourth instar larvae. Ma et al. (1989) showed that rates of parasitism by *C. nigripes* in forests were influenced by stand type and growing conditions, with the greatest parasitism in mixed stands, followed by closed pure stands and then sparse stands. Ants are the most important predators in Vietnam and China and can regulate populations of the pest effectively (Bassus, 1974; Hsiao, 1981).

Pteroma plagiophleps Hampson
(Lepidoptera: Psychidae)

Up until the late 1970s, the bagworm, *Pteroma plagiophleps*, was known only as a minor defoliator of tamarind, *Tamarindus indica*, in India and Sri Lanka. In 1977, it caused extensive defoliation to a young plantation of *Falcataria moluccana* at Kerala (Nair et al., 1981) and since that time has become an important tree pest in several countries in South and South-east Asia (Nair, 2007).

P. plagiophleps belongs to a family of moths whose larvae construct individual bag-like shelters made of host material and silk, within which the larva remains concealed and only the head and thorax protrude when feeding (Plate 21). The adult male is a normal winged moth but the female is wingless, with a poorly developed body. The female does not leave the bag but mates from inside, and her eggs also mature and hatch there. Each female usually produces 100–200 offspring, which emerge from the bag and disperse on silken threads. They settle on leaves and construct the protective bag, which they carry about with them, enlarging it as they grow. Pupation takes place in the bags hung from the branches. The generation time of *P. plagiophleps* is about 10–11 weeks (Nair and Mathew, 1992).

While up to five generations a year have been observed in the field, outbreaks leading to heavy defoliation generally occur only once or twice a year, usually in small patches in plantations (Plate 22). Studies by Nair and Mathew (1988, 1992) in a 20 ha plantation of *F. moluccana* showed that repeated defoliation over 2.5 years caused the death of 22% of trees and moderate to severe damage to 17%. Heavy infestations affecting a large number of trees have been recorded for *F. moluccana*, *Delonix regia* and *E. tereticornis*. In the case of *E. tereticornis* at least, there is evidence to suggest that such infestation is related to host stress. In the last couple of decades, the insect gradually has extended its host range and importance as a forest pest (Pillai and Gopi, 1990a; Howlader, 1992), attacking plantations of *A. nilotica* and even mangroves, *Rhizophora mucronata* (Santhakumaran et al., 1995). The latter record indicated the hardiness of *P. plagiophleps*, since it was feeding on saplings which became submerged during high tides and on taller mangroves which regularly received salt spray. In Indonesia also, *P. plagiophleps* is a sporadic pest, with severe defoliation occurring in some endemic patches of *F. moluccana* in Sumatra (Nair, 2007).

Natural enemies appear to play a decisive role in regulating the populations of *P. plagiophleps* larvae, and 18 species of parasitoids, all hymenopterans, are known (Nair, 2007). A 25–38% reduction in populations of this pest has been recorded on some occasions, due mainly to parasitism by ichneumonid and chalcidid wasps (Mathew, 1989).

Ctenomorphodes tessulatus (Gray) (Phasmatodea: Phasmatidae)

The tesselated phasmatid *Ctenomorphodes tessulatus* occurs in coastal areas of Queensland and New South Wales in eastern Australia. It is a pest of native hardwood forests and has a wide range of hosts, mostly in the family Myrtaceae (Elliott *et al.*, 1998).

The adult insect reaches approximately 120 mm in length, the male being winged and the female wingless. Females lay eggs that drop to the forest floor from the canopy and become incorporated into the litter. The eggs hatch in late August and early September and the nymphs ascend nearby trees to feed on foliage. They pass through six instars and adults begin to appear during December. There is usually one generation per year, 2- or 3-year cycles occurring only rarely in most populations. Adults live for up to a year.

All known outbreaks of this insect are from native forests and there are no records from plantations. Outbreaks in southern Queensland over the period 1974–1976 resulted in widespread defoliation and tree death, particularly of *E. tereticornis* (Wylie *et al.*, 1993). Trees bordering road corridors and on pastoral or farm land were most seriously damaged.

Several wasp parasitoids have been reared from eggs of *C. tessulatus*, notably species of *Myrmecomimesis* and *Loboscelidia* (Heather, 1965). There also appears to be high mortality among eggs, caused by pathogens. Nymphs are probably preyed upon by birds, since these have been observed feeding on other phasmatids. In the case of one outbreak in northern New South Wales, forest fires had regularly burned over the area in the years preceding this event. Hadlington and Hoschke (1959) conjectured that fires in the spring and early summer would not affect the nymphs which had entered the tree canopies but would kill the egg parasites, which did not emerge from the eggs until December–February. It is possible that the reduction in egg parasitoid numbers due to early season fires could allow *C. tessulatus* populations to build up. However, in one outbreak area, nymphs and adults disappeared prematurely, indicating that at least one other major regulating factor was operating.

5.2.2 Leaf mining

A large number of insect species 'mine' or feed on tissue between the upper and lower surface of leaves. Most species are in the order Lepidoptera, but some species of Coleoptera, Diptera and Hymenoptera also mine leaves (Elliott *et al.*, 1998). The shape of some mines is characteristic of the species that constructs them and is therefore useful in identification. Because of the restricted size of their habitat, leaf miners are small and their larvae are grub-like and usually flattened dorsoventrally.

Acrocercops spp. (Lepidoptera: Gracillaridae)

The genus *Acrocercops* is a large one, occurring in both tropical and temperate regions around the world. In Australia, where there are more than 100 species, at least 11 species of these leaf miners attack *Acacia*, *Angophora* and *Eucalyptus* (New, 1976; Elliott *et al.*, 1998). The mines are generally of the 'blotch' type, being wide as well as long, and appearing as a blister on the leaf surface. Larvae of most species develop from egg to pupa within a few weeks, with the adults emerging a few weeks later, suggesting that several generations per year are possible.

A. plebeia is a species which attacks several *Acacia* spp. in subtropical Queensland. Adult moths lay eggs on the phyllode surface, usually one per phyllode and near the midrib, giving the larvae maximum choice in mining direction (New, 1976). On hatching, larvae immediately enter the epidermis of the phyllode. There are five larval instars, with a total development time of 14–25 days. The first two instars make a long, narrow and sinuous mine and the remaining instars make a blotch mine (Plate 23). At maturity, the final instar larvae cease feeding and empty the gut. A circular flap is then cut through the phyllode cuticle and the larva emerges to spin a cocoon either on the phyllode surface, on the bark or (more rarely) drops to the ground and pupates in the litter.

In India, there is also a large number of *Acrocercops* spp. which mine the leaves of forest trees (Beeson, 1941). Yadav and Rizvi (1994) record this group as major pests of wasteland plantations of *Syzygium cumini*, and *Acrocercops* sp. near *telestris* is an important pest of cinnamon (Singh *et al.*, 1978). *A. gemoniella* mines the leaves of nursery stock of the milk tree, *Manilkara hexandra*, damaging up to 20% of leaves in some nurseries in central India (Jhala *et al.*, 1988). In the USA, a gracillariid leaf miner, believed to be a species of *Acrocercops*, has shown potential for the biological control of *Melaleuca quinquinerva*, an Australian tree which has become a pest in the Everglades of Florida (Burrows *et al.*, 1996).

5.2.3 Leaf skeletonizing

Skeletonizers strip away leaf tissue, leaving a network of veins, or a leaf 'skeleton'. The very early instars of many species of moth cause this type of damage, but some species are specialist skeletonizers for a considerable part of their larval life.

Craspedonta leayana (Latreille)
(Coleoptera: Chrysomelidae)

This chrysomelid, formerly known as *Calopepla leayana*, is an important defoliator in plantations of *Gmelina arborea*, a fast-growing timber species, in India, Bangladesh and Myanmar (Baksha, 1997; Singh *et al.*, 2006). It has also been recorded from Thailand.

The adults aestivate and hibernate in bark crevices, clumps of grass, soil litter and other sheltered places for about 8 months in a year and become active when the new leaves begin to expand (Browne, 1968). They cut large circular holes in the leaf and also eat the young buds and shoots. Eggs are laid in clusters, either on the underside of the leaf or on the shoots inside a frothy secretion, which hardens to form an 'ootheca'. One female is capable of laying as many as 18 oothecae, each of which may contain up to 100 eggs (Ahmed and Sen-Sarma, 1983, 1990). After about a week, larvae emerge and begin feeding on the lower surface of leaves or on shoots, making discoloured, irregularly shaped feeding patches. There are five larval instars and it is the fourth and fifth instars which cause complete skeletonization, leaving only the midrib and main veins. The larvae of *C. leayana* have an interesting defence mechanism. The excrement, instead of being discarded, is extruded in long, fine, black filaments, often twice the length of the body, which are formed into bunches and attached at the anal end. When disturbed, the larvae of all instars flick these filaments up and down in a defensive action. Pupation occurs on the leaves. There are usually three generations a year, the first two taking about 46 and 43 days, respectively, and the hibernating generation taking about 239 days (Ahmed and Sen-Sarma, 1990).

Defoliation of *G. arborea* is first noticeable at the beginning of the rains, usually in April or May, and may continue until October. A heavy attack causes the leading shoots of young trees to dry up, and the trees remain leafless for about 4 months of the growing season and eventually become bushy. Two or more consecutive complete defoliations will kill trees (Beeson, 1941). There are no other recorded hosts for *C. leayana* (Nair, 2007).

Chalcidid wasps, *Brachymeria* spp., particularly *B. excarinata*, have been recorded as pupal parasites of *C. leayana* (Mohandas, 1986; Singh *et al.*, 2006) and a eulophid wasp, *Tetrastichus* sp., is an egg parasitoid (Baksha, 1997). Factors such as temperature, rainfall and the quality and quantity of food are important contributors to seasonal fluctuations of this insect. In Assam, India, Kumar *et al.* (2010) found that temperature was the principal component influencing populations of *C. leayana*.

5.2.4 Leaf tying/rolling

Larvae in several moth groups construct shelters for protection against predators such as birds, the simplest of these constructions being a webbing-together or rolling of leaves. The larvae feed on foliage contained within the shelter. This habit is common among species of Tortricidae and Pyralidae.

Strepsicrates rothia Meyrick
(Lepidoptera: Tortricidae)

This insect is widely distributed in the tropics of the eastern hemisphere, being recorded in Ghana, Nigeria, Mauritius, India, Pakistan, Sri Lanka, Taiwan and Malaysia. It is associated principally with the foliage of *Eucalyptus*, but also occurs on other dicotyledonous trees such as guava and mango (Beeson, 1941; Browne, 1968).

The larva of Strepsicrates Species rolls a single *Eucalyptus* leaf, forming a shelter in which it feeds and rests (Plate 24). When the leaf turns brown and withers as a result of the feeding, the larva crawls out to a fresh leaf to repeat the process. Pupation takes place in the rolled-up leaf. The life cycle is completed in 3–4 weeks, the egg stage lasting 3–4 days, larval development 10–21 days and pupal stage 5–8 days (Chey, 1996; Wagner *et al.*, 2008). The adult lives for only a few days in captivity.

According to Chey (1996), *Strepsicrates rothia* is the most serious defoliator of *E. deglupta* in Sabah. Damage to the shoots of young trees can be severe, and in nurseries infestation levels of 20% have been recorded in Sabah (Chey, 1996) and of more than 50% in Ghana (Wagner *et al.*, 2008). In Ghana, *E. tereticornis* is the most preferred host species, with *E. alba*, *E. cadamba* and *C. citriodora* preferred in descending order. Wagner *et al.* (2008) noted that the rate of infestation of *S. rothia* in a forest nursery at Yenku was about twofold more on shaded beds than on unshaded beds. Infestation is not confined to seedlings, and Chey (1996) states that trees above 10 years of age in plantations are also attacked. In Sabah, a braconid wasp, *Ascogaster* sp., parasitizes *S. rothia*.

5.3 Sap Feeding

Sap-feeding insects suck liquid or semi-liquid material from succulent parts of the host plant, which can be leaves, stems, roots, fruit, flowers or even seed (Elliott *et al.*, 1998). Most are true bugs (Hemiptera), but thrips (Thysanoptera) can feed in a similar fashion, except that penetration is relatively shallow.

Sap feeders affect tree vitality by extracting sap required for normal functioning of the plant, such as shoot extension and leaf expansion. This results in stunting, distortion or wilting, depending on the size of the pest population, the insect species involved, the location of the feeding site and the season of attack (Elliott *et al.*, 1998). As well, some species, such as certain lerp (protective case)-making psyllids and mirid bugs, can inject a toxic saliva into their hosts, causing necrosis of plant tissue. Sap feeders excrete a clear, sugary liquid (honeydew), which can coat the surfaces of leaves and stems and on which grows sooty mould fungi. Where large populations of insects are present, as with aphids and coccids, the blackening of the plant surface by the mould can cause reduced photosynthetic efficiency and the loss of host vigour.

Feeding by sap-sucking insects provides access for pathogenic fungi into plants, and aphids, mirids and cicadellids are particularly important in the transmission of viruses. Cicadas damage their host physically during oviposition, slitting the bark of stems, while thrips and other sap feeders may cause gall formation.

The known world fauna of aphids (Aphidoidea) consists of 4401 species placed in 493 genera, of which 1758 species in 270 genera spend all or part of their life feeding on trees (Blackman and Eastop, 1994). Aphids are predominantly a northern temperate group, with remarkably few species in the tropics. In contrast, psyllids, which have similar ecology and host relations to aphids, are very common in the tropics. Blackman and Eastop (1994) believe that aphids have failed to diversify in the tropics because of one particular primitive feature of aphid biology, their cyclical parthenogenesis (i.e. a life cycle consisting of one generation of sexual morphs and several generations in which only parthenogenetic females are produced). Aphids moving into the tropics lose the sexual phase of their life cycle, and in so doing, lose the potential to evolve and diversify. Despite this, several aphid species which have become established in the tropics have flourished and are now a problem.

Pineus pini (Macquart), *Pineus boerneri* Annand (Hemiptera: Adelgidae)

Pine woolly adelgids, *Pineus* spp., are native to the temperate zones of the northern hemisphere and feed on conifers. Nymphs and adults suck plant juices from needles, shoots or stems of pine and cause shoot deformity and loss of height growth. Excess plant juice excreted by adelgids as honeydew is a favourable medium for growth of black sooty moulds on foliage and stems (Diekmann *et al.*, 2002). Two species in particular, *P. pini* and *P. boerneri*, have become serious pests where they have been introduced accidentally into several tropical and subtropical countries in the southern hemisphere during afforestation programmes using *Pinus* spp. Both species have a confused taxonomy. *P. pini* (Plate 25), which is native to Europe and introduced into North America, Australia and New Zealand, has sometimes been referred to in the literature as *P. laevis* and quite often mistaken for *P. boerneri*, which is of East Asian origin (Blackman and Eastop, 1994). *P. boerneri* has been recorded under the name of *P. laevis* in Australia, New Zealand and Hawaii, as *P. havrylenkoi* in South America and as *P. pini* in East Africa (Day *et al.*, 2003). The introduction of *P. boerneri* into Africa is believed to have been via pine scions imported into Zimbabwe and Kenya from Australia in 1968 (Odera, 1974), and genetic analysis by Blackman *et al.* (1995) has confirmed an Australian link. Since then, *P. boerneri* has spread to a further six countries in eastern and southern Africa, mostly by the movement of infested nursery stock (Zwolinski, 1989), and also infests *Pinus* plantations in Brazil (Oliveira *et al.*, 2008).

Adelgids are related closely to aphids and occur only on conifers. In Europe and North America, they have complicated life histories which may involve a primary and a secondary host belonging to two different genera (Zondag, 1977). However, in most areas where they have been introduced, they are restricted to *Pinus* spp. In Africa, for example, *P. boerneri* has a host range of over 50 species of exotic *Pinus* (Chilima and Leather, 2001). Pine woolly adelgids have both winged and apterous adults. The wingless females reproduce parthenogenetically, but there is some doubt about the ability of the winged forms to reproduce (Odera, 1974). The young, or 'crawlers', which hatch from the eggs move about on the foliage, find a suitable spot to insert their mouthparts and start sucking sap. Some crawlers are dispersed by wind. The insect moults several times before it reaches adulthood, which usually takes only a few weeks. The generations overlap and several generations are produced in a year.

Various studies have shown that population fluctuations of *P. boerneri* are related to rainfall, but the results are sometimes contradictory. For example, in Kenya, Mailu *et al.* (1980) found that there was a marked decrease in numbers during periods of heavy rainfall and a significant increase during dry weather, whereas in Hawaii, Culliney *et al.* (1988) found the reverse. In Kenya, the heavy rainfall caused high mortality early in the life cycle by washing eggs and crawlers off the host tree. In Hawaii, high population densities during periods of increased rainfall were attributed to the higher nutrient value of trees at such times, which favoured increased survival of the insects. Trees under stress are likely to be infested by *P. boerneri* more heavily than are unstressed trees (Madoffe and Austara, 1993).

In the early stages of infestation, or on lightly infested trees, colonies of the adelgid occur under scales of the bud base, on shoots at the base of a needle fascicle, or on needles inside a fascicle (Zwolinski, 1989). Later, the infestation spreads to the new growth and down to the bark of thicker branches and the stem. The insects produce waxy white threads, which form a dense woolly cover over the colonies and give infested parts of the tree a greyish appearance. Chlorosis of foliage, shoot dieback, malformation and, in severe cases, death of trees can result from heavy infestations of *P. pini* (Plate 26). In Kenya, Odera (1974) reported 20% mortality in some study plots, and Mailu *et al.* (1978) found that severe stunting of needles caused by adelgid feeding could result in loss of half the tree yield. In South Africa, Zwolinski *et al.* (1988) reported a 23% decrease in seed production of *P. pinaster* due to this insect. Studies of the impact of *P. boerneri* on the growth of *P. patula* seedlings in Tanzania showed that infested seedlings had 12.2% less diameter growth and 14.1% less height growth than uninfested seedlings in a 24-week period (Madoffe and Austara, 1990).

P. boerneri has numerous natural enemies and some of these have been used in biological control programmes in several countries (Day *et al.*, 2003). In Hawaii, populations of the adelgid have been regulated below economically significant levels by a fly, *Leucopsis obscum* (Culliney *et al.*, 1988). In Kenya and Zimbabwe, coccinellids *Exochomus* spp. are important predators of *P. boerneri*, although Mailu *et al.* (1980) record only 12% reduction in populations of the pest due to these and other predators.

Cinara cupressivora Watson & Voegtlin,
Cinara cupressi (Buckton)
(Hemiptera: Aphididae)

Cypress aphids belong to the genus *Cinara*, whose members are commonly known as giant conifer aphids (Diekmann *et al.*, 2002). There are about 200 described species of *Cinara*; approximately 150 of these are from North America, 20 from Japan and the oriental region and 30 of European or Mediterranean origin (Blackman and Eastop, 1994). They infest twigs and branches and sometimes roots of conifers, causing dieback and tree mortality (Ciesla, 1991a,b). As is the case for *P. pini*, the taxonomy of the species previously referred to as the cypress aphid, *C. cupressi*, is confused and it now appears to be part of a species complex (Watson *et al.*, 1999). The species causing tree damage in Africa is actually *C. (Cupressobium) cupressivora* (Plate 27). In their native habitat, these aphids generally are not considered major forest pests, but *C. cupressivora* has had severe economic and social impact in eastern and southern Africa following its accidental introduction there in the 1980s, presumably on infested planting stock, and *C. cupressi* is an important pest in South America.

The native range of *C. cupressivora* is most likely the region from eastern Greece to just south of the Caspian Sea (Watson *et al.*, 1999). It is now widely distributed throughout eastern, central and southern Africa, the margins of western Europe, countries bordering the Mediterranean Sea, the Middle East, Yemen, Mauritius and Colombia (Day *et al.*, 2003). It was first discovered in Africa in Malawi in 1986, and subsequently in Tanzania, Kenya, Burundi, Rwanda, Uganda, Zimbabwe and Zaire. It attacks a wide range of exotic and indigenous trees in Africa, including *Callitris*, *Cupressus*, *Cupressocyparis*, *Juniperus*, *Thuja* and *Widdringtonia* (Malawi's national tree), but is particularly severe on the widely planted *Cupressus lusitanica* (Plate 28).

As outlined by Ciesla (1991a), the life cycle of the cypress aphid is complex. During the summer months, only females are present and reproduce parthenogenetically, giving birth to live young. There are two forms of adults, winged and wingless. In temperate climates, as cool weather approaches, both males and females are found and eggs are produced instead of live nymphs. The eggs are laid in rough areas on twigs and foliage, where they overwinter. In warm climates, parthenogenetic reproduction continues throughout the year. Several generations are produced annually and the lifespan of a single generation is about 25 days. Adults and immature insects are found in clusters of up to 80 individuals on the branches of host trees, where they suck the plant sap. Their saliva is toxic to some trees and can cause branch dieback and tree mortality, especially when large numbers of aphids are present. Dieback usually occurs from the inner crown outward and from the lower crown upward. These insects also produce large amounts of honeydew that covers branches and foliage and on which sooty mould grows, interfering with photosynthesis. Damage appears to be more severe in the dry season.

C. lusitanica, a highly favoured plantation and agroforestry species in Africa, is extremely sensitive to feeding by *C. cupressivora*. In addition to the loss of timber, tree mortality caused by the aphid has increased the fire hazard in many rural areas, particularly where cypress hedges have been planted around homes (Ciesla, 1991a). Damage assessments by Symonds *et al.* (1994) in a stand of 65-year-old *C. lusitanica* in Kenya revealed a tree mortality of 12%. In Malawi, the estimate of financial losses on the standing crop by 1990 was over US$3 m (Odera, 1991, cited in Obiri, 1994). Murphy (1996) estimates that in southern and eastern Africa until 1990, the value of cypress trees killed by *C. cupressivora* totalled £27.5 m sterling, and there was a loss in annual growth increment (including that from already dead trees) of £9.1 m sterling/year. In Argentina and Chile, *C. cupressi* affects forests of *Austrocedrus chilensis*, causing defoliation and tree mortality (El Mujtar *et al.*, 2009; Montalva *et al.*, 2010).

Conifer aphids are attacked by a range of predators that include ladybirds, syrphid flies, lacewings and bugs, but they are far less specific than the corresponding predators of adelgids and have a less significant impact on individual prey species (Mills, 1990). Also, in contrast to the adelgids, parasitoids are a very important part of the natural enemy complex, particularly braconid wasps such as *Pauesia* spp. *P. juniperorum* was released as a biocontrol agent against *C. cupressivora* in Kenya and Malawi in 1994 and is now widespread in these countries. Releases of this parasitoid were also made in Uganda in 1995 and more recently in Chile (Day *et al.*, 2003; Montalva *et al.*, 2010). There is a wide range of tolerance among species of Cupressaceae to attack by cypress aphid, which indicates that resistance breeding may offer a viable, long-term solution to the aphid problem (Day *et al.*, 2003).

Ceroplastes spp. (Hemiptera: Coccidae)

The genus *Ceroplastes* occurs widely throughout the tropics and contains many species which feed off forest trees. Few are serious pests, but when they do occur in large infestations they can reduce tree vigour seriously and even cause mortality (O'Dowd *et al.*, 2003).

One of the more important species is *C. rubens*, commonly known as the pink wax scale or red wax scale (Plate 29). It occurs throughout the Australian and Oriental regions and also parts of Africa. Smith (1974) provides a description of its life history and habits. The adult female can lay from 600 to 700 eggs, which are deposited in a cavity beneath the body of the scale. The first instar nymphs or crawlers emerge from beneath the parent scale and spread out over the tree. These generally settle on the midrib of leaves, on needles and on young twigs, and begin feeding through long tubular mouthparts inserted into the plant tissue. There are three nymphal instars and the final moult to the adult takes place after about 10–12 weeks, by which time the soft-bodied scale is protected by a globular covering of pink wax. Adults reach maturity in 4–6 months. In Papua New Guinea, Merrifield and Howcroft (1975) reported severe attack of *P. caribaea* by *C. rubens*, heavily infested trees being characterized by sparse crowns, considerable darkening of foliage by a dense covering of sooty moulds and reduced height increment. There was evidence of clonal differences in the severity of attack. *C. rubens* is also a sporadic pest of *P. taeda* and *P. caribaea* in Queensland, Australia (Elliott *et al.*, 1998).

The Florida wax scale, *C. floridensis*, is a pest of ornamental trees and shrubs in several countries, but occasionally causes damage in forest plantations. An outbreak in *P. caribaea* plantations in tropical Queensland affected 30% of the planted area and was expected to cause reduced growth increment and a decline in tree vigour. The sooty mould interferes with photosynthesis, and sap feeding can cause death of needles and twigs (Elliott *et al.*, 1998). Other *Ceroplastes* of occasional importance are *C. grandis* in Brazil, *C. ceriferus* in the oriental region and *C. destructor* in Africa, Papua New Guinea and Australia (Browne, 1968; Iede and Machado, 1989; Wakgari, 2001).

A range of mortality factors have been reported for these insects which, in the early stages, includes failure to emerge from beneath the mother scale, crowding, desiccation and unsuccessful settling on hosts due to dislodging by wind and rain. Entomogenous fungi and natural enemies such as lacewings, ladybirds and eulophid wasps are also important; in South Africa, for example, more than 21 hymenopteran parasitoids of *C. destructor* have been recorded (Wakgari, 2001). Numerous studies have demonstrated that honeydew-producing scale insects benefit from the presence of ant attendants (Abbott and Green, 2007). In Japan, Itioka and Inoue (1996) showed that ant attendance on *C. rubens* restricted the ovipositional ability of the encyrtid parasitoid *Anicetus beneficus* and reduced its effectiveness as a control agent of *C. rubens*. On Christmas Island, an Australian Territory in the Indian Ocean, *C. ceriferus* and *C. destructor* are among nine species of scale insects occurring in outbreak densities over hundreds of hectares of rainforest, where they cause canopy dieback and tree death (O'Dowd et al., 2003). These outbreaks are being sustained by high densities of the invasive yellow crazy ant, *Anoplolepis gracilipes*. The ants assist the scales by removing honeydew, which prevents asphyxiation, 'nannying' of the mobile crawler stage and protection against natural enemies. In return, the scale insects provide *A. gracilipes* with an abundant source of carbohydrate in the form of honeydew, which fuels their high population densities (Abbott and Green, 2007). The exclusion of ants from plots in affected rainforests on Christmas Island caused a 100% decline in the densities of scale insects in tree canopies in 12 months (Abbott and Green, 2007).

Amblypelta cocophaga China
(Hemiptera: Coreidae)

This insect is a major pest of *E. deglupta* and *Campnosperma brevipetiolata* plantations in the Solomon Islands in the Pacific and a minor pest of several forest tree species, including candlenut, *Canarium indicum* (Bigger, 1988; Ipute 1996). It is also an important pest of agricultural crops such as coconut, cassava and cocoa. Both adults and nymphs feed on young shoots, injecting toxic saliva that results in wilting and dieback of the shoot. The leading shoot is chosen in preference to side branches and when this is killed, the tree responds by putting up several new leaders from below the damaged area. These, in turn, may be attacked and, in extreme cases, the tree becomes very bushy and flat-topped. The new shoots put out at the top of the bush provide abundant food for *A. cocophaga* to flourish, and so the condition is perpetuated (Bigger, 1985).

The genus *Amblypelta* extends from Indonesia to Australia, New Caledonia and Vanuatu, but has the greatest concentration of species in the Solomon Islands, where at least 18 species and subspecies have been recorded (Bigger, 1988).

As described by Bigger (1988), eggs are laid singly, either on a leaf or the stem of the plant. Hatching takes place after 8 days. There are five nymphal instars occupying a total of 33 days. The feeding damage done by the adult insect to *E. deglupta* is often not particularly severe because the insect is mobile and moves from tree to tree, not staying in one place for very long. Nymphs, being flightless, are of necessity less mobile and so their concentrated feeding is more likely to lead to dieback. Damage is most severe in the first 12 months after planting. In trial plots established to examine the effects of clearing of inter-row growth on incidence of *A. cocophaga*, no difference was found between cleared and uncleared plots, and an estimated 37–41% of the potential stand was rendered useless for future timber production due to damage by this pest (Bigger, 1985).

With *C. brevipetiolata*, damage by *A. cocophaga* usually starts when the tree is about 1 m tall and is at an end by the time the tree is 3–4 m tall (about 18 months after planting). Feeding takes place mostly on the lower parts of the leaf midribs and leaf petioles and also on young shoots, resulting in multi-stemming. Cankers often develop at the feeding sites on badly damaged trees and these persist as stem swellings on older trees. Extensive pipe rotting is associated with these swellings. Thomson (2006) mentions that plantations of *Agathis macrophylla* (Pacific kauri) in the Solomon Islands are affected by this pest, but the severity of damage is not stated.

Ants are known to be predators of *A. cocophaga*, particularly *A. longipes*, *Oecophylla smaragdina* and *Wasmannia auropunctata* (Way and Khoo, 1992).

Aulacaspis marina Takagi and Williams
(Hemiptera: Diaspididae)

Mangrove forests are being depleted rapidly and degraded in many countries, and in the past few decades there has been an upsurge in mangrove plantings to restore these forests (Ozaki *et al.*, 1999). The impact of insects on mangroves generally has been considered of minor importance compared to their impact on other types of forests, but it is now clear that this has been strongly underestimated (Cannicci *et al.*, 2008). One emerging pest is the scale insect *Aulacaspis marina*, which has caused significant mortality of *R. mucronata* saplings in Indonesia. Sucking by these insects on the leaves of the host plant causes chlorosis of the leaf, browning and leaf fall. Newly expanded leaves are attacked by crawlers within a generation span of the insect and successive defoliation results in death of the plant after four generations of the pest.

Female scales lay about 140 eggs, which hatch in approximately 7 days. Newly hatched crawlers remain under the scale for a few days before dispersing to the leaves or stems of the infested trees. Densities of more than 200 mature females per leaf have been recorded. Generation time varies from 34 to 42 days and the insect has nine to ten generations a year (Ozaki *et al.*, 1999).

On the Indonesian island of Bali, where 150 ha of abandoned shrimp ponds had been reforested with three mangrove species (*R. apiculata*, *R. mucronata* and *B. gymnorrhiza*), *A. marina* killed 70% of *R. mucronata* saplings in the most heavily infested plantation within 5 months of the initial infestation (Nakamura, 1995, cited in Ozaki *et al.*, 1999). Studies by Ozaki *et al.* (1999) showed that the other two species of mangrove were equally susceptible but were not heavily infested because their saplings, which were not as tall as *R. mucronata* at the site, were periodically completely submerged in seawater. Periodic spraying of seawater has been suggested as an effective way to reduce damage by *A. marina*. At present, serious damage by *A. marina* has only been reported from Bali. However, the species is also distributed in the Philippines and Malaysia, suggesting that it may become more injurious to mangrove plantations in the broader regions as the numbers of similar reforestation programmes increase.

Ants are important predators of *A. marina* in natural mangrove forests, as demonstrated by Ozaki *et al.* (2000) in Bali. On ant-excluded saplings, 90% of artificially introduced female scales survived a 3-day experiment, while only 22% survived on plants foraged by the ants *Monomorium floricola* and *Paratrechina* sp. (Fig. 5.7).

Helopeltis spp. (Hemiptera: Miridae)

Plant bugs of the genus *Helopeltis*, sometimes called mosquito bugs, are serious pests of cultivated plants in the Old World tropics, particularly of major cash crops such as tea, cocoa, cinchona, cashew and pepper (Stonedahl, 1991) (Fig. 5.8). There are 40 known species distributed from West Africa to Papua New Guinea and northern Australia. They have been regarded generally as only minor pests of forest trees, occasional damage being reported to *Swietenia*, *Terminalia*, *Cinnamomum* and *Melia*. However, in more recent years, Wylie *et al.* (1998) have recorded severe damage to young eucalypt and acacia plantations in Indonesia, and there have been similar instances in other countries.

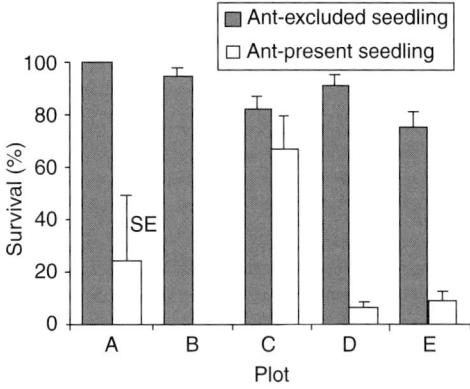

Fig. 5.7. Survival of female *Aulacaspis marina* (mean ± SE) on ant-excluded and ant-present seedlings of *Rhizophora mucronata* introduced into five stands (A–E) in natural forests (from Ozaki *et al.*, 2000).

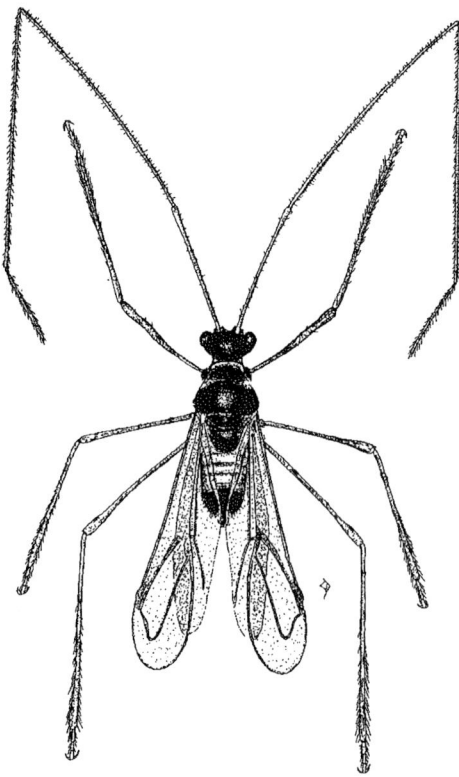

Fig. 5.8. Adult female of *Helopeltis theivora* (from Stonedahl, 1991).

As with many other tropical Miridae, *Helopeltis* spp. exhibit a more or less continuous cycle of generations throughout the year (Stonedahl, 1991). Eggs are embedded in plant tissue singly or in small groups, often on new shoots. Hatching takes place after 6–11 days and there are five nymphal instars taking from 9 to 54 days to complete, depending on species and climatic conditions. Adults may live for up to 30 days. In forest crops, nymphs and adults feed on the leaves and stems of new shoots. Damage first appears as a necrotic area or lesion around the point of entry of the stylets into the plant tissue and progresses to wilt and dieback of the shoot. A single late-instar nymph of *H. theivora* can make as many as 80 feeding lesions during a 24-h period (Das, 1984, cited in Stonedahl, 1991). In heavily infested young eucalypt plantations in Sumatra, feeding by *H. bradyi* and *H. fasciaticollis* resulted in 'bushing' and stunting of the trees (Plate 30). Attack of new growth in the following year left trees with a 'pompom' appearance (Wylie *et al.*, 1998). Rahardjo (1992) reported that the loss due to attack by *Helopeltis* spp. on eucalypt plantations in north Sumatra was 11 m^3/ha. On *A. mangium* in central Sumatra, feeding by *H. theivora* caused distortion of shoots and retardation of growth (Wylie *et al.*, 1998). In south India, widespread and severe shoot dieback of neem trees, *A. indica*, due to feeding by *H. antonii* has become an annual event (Pillai and Gopi, 1990b) and is a limiting factor for the cultivation of neem in some parts (Annamalai *et al.*, 1996). Leaf loss of up to 95% has been recorded, as well as heavy seedling mortalities in forest nurseries. In the Congo, the adaptation of the native *H. schoutendeni* to planted *Eucalyptus* spp. has been a relatively recent development. In 1994, out of 90 host plants inventoried for *Helopeltis*, no *Eucalyptus* species was mentioned (Diabangouya and Gillon, 2001). Since then, severe attacks have been reported for eucalypts, particularly for clones of *E. urophylla*.

Population levels of the various *Helopeltis* spp. in different countries fluctuate throughout the year, but in some cases a build-up in numbers is synchronized with the emergence of new foliage following the cessation of the monsoon rains (Stonedahl, 1991). There are suggestions that the insects do not do well under conditions of heavy rain, high winds or low relative humidity. In West Bengal, Ghosh (1993) found that incidence of mosquito bug was higher where cashew and *E. tereticornis* had been intercropped than where cashew was grown alone and inferred that the eucalypts created an environment favourable for rapid multiplication of the pest.

A wide range of natural enemies of *Helopeltis* spp. has been recorded, including scelionid and mymarid wasp parasitoids of eggs, braconid wasp parasitoids of nymphs and predatory reduviid bugs and ants. In most instances, however, these natural enemies cannot maintain populations of the pests below economic thresholds (Stonedahl, 1991). Several studies in horticultural crop systems have demonstrated the potential for the tree ant, *O. smaragdina*, to reduce damage by *Helopeltis* spp. significantly, but its painful bite makes it unacceptable to plantation workers (Way and Khoo, 1991; Stonedahl *et al.*, 1995).

Nipaecoccus viridis (Newstead) (Hemiptera: Pseudococcidae)

The mealybug *Nipaecoccus viridis*, often referred to erroneously in the literature as *N. vastator*, is an important tropical and subtropical insect pest of many food, forage, fibre and ornamental crops. It has been recorded in 54 countries from Africa and the Middle East, throughout Asia to Australia and the Pacific, and from 96 species of plants (Sharaf and Meyerdirk, 1987; CABI/ EPPO, 2005). Its tree hosts include *A. arabica*, *Albizia lebbek*, *F. moluccana*, *Dalbergia sisso* and *Casuarina equisetifolia*.

This mealybug has been reported reproducing both sexually and parthenogenetically. In the sexual type of reproduction, the adult female lays an average of 667 eggs in an ovisac secreted a few days prior to oviposition (Sharaf and Meyerdirk, 1987). Eggs are laid in batches, gradually causing an increase in the size of the ovisac until it attains a hemispherical shape, from which the common name 'spherical mealybug' is derived. The female dies soon after oviposition, which lasts from 21 to 37 days, and by this time her empty body becomes raised to a vertical position, being anchored to the plant by means of stylets. Eggs hatch in about 10 days. First instar nymphs emerge from the ovisac and search for a suitable spot to begin feeding, preferably near the mother. Males have five instars, while the females have four, development being completed in 19–20 days. The mealybug reproduces continuously throughout the year and there are multiple, overlapping generations.

N. viridis may feed on the host's branches, twigs, shoots, leaves, flower buds, fruit and roots, causing curling and stunting of terminal growth, abortion of flowers, yellowing of leaves and dropping of fruit. In severe infestations, wilting and dieback occur (Sharaf and Meyerdirk, 1987). As well, the insects secrete large amounts of honeydew on which sooty moulds grow, interfering with photosynthesis.

The spherical mealybug is subject to heavy natural mortality induced by unfavourable weather conditions and natural enemies. High humidities have an adverse effect on the hatching of eggs and high temperatures may kill all stages. The insect may feed on different parts of the same host plant in different seasons, partly in response to changing climatic conditions. For example, populations are usually low in winter, but in spring they develop rapidly on the growing buds and on the underside of new leaves. In summer, most of the mealybugs on the exposed parts of the tree perish, and only those in the lower and protected parts remain. In autumn, the surviving insects resume their activities and invade new growth. Over 77 different natural enemies of *N. viridis* have been recorded, comprising 54 hymenopterous parasites, 14 coleopteran predators, seven dipteran predators and two neuropteran predators. Only 13 of these species are considered highly effective against the pest, and several have been used in biocontrol programmes. For example, in Hawaii, the encyrtid wasp *Anagyrus dactylopii*, which was introduced from Hong Kong, has maintained the pest at very low levels. In Guam, natural enemies are also very effective, although Nechols and Seibert (1985) found that the presence of ants decreased the performance of these agents against *N. viridis*.

Glycaspis brimblecombei Moore (Hemiptera: Psyllidae)

The red gum lerp psyllid, *Glycaspis brimblecombei*, is an important pest in several parts of the world, where it feeds on *Eucalyptus* species, especially red gums *E. camaldulensis* and *E. tereticornis*. Sap sucking by adults and nymphs results in early leaf drop, loss of tree vigour and premature death of some highly susceptible species. *G. brimblecombei* is native to Australia and was first detected outside this country in 1998 in California, USA (Brennan *et al.*, 1999). It was found in Baja California, Mexico, in 2000, where it spread rapidly to 21 other states. It was detected subsequently in Florida, Hawaii and Mauritius in 2001, Chile in 2002, Brazil in 2003, Argentina in 2005, Ecuador in 2006, Europe (Spain and Portugal) and Venezuela in 2007 and Peru in 2008 (Halbert et al., 2003; Sookar et al., 2003; Diodato and Venturini, 2007; Santana and Burckhardt, 2007; Burckhardt et al., 2008; Rosales et al., 2008; Valente and Hodkinson, 2009; Huerta et al., 2010).

Female psyllids lay their eggs singly or in scattered groups on succulent leaves and young shoots, and the nymphs and adults feed by sucking plant phloem sap through their straw-like mouthparts. Red gum lerp psyllid nymphs form a cover or 'lerp' which is a small, white, hemispherical cap composed of solidified honeydew and wax (Plate 31). Lerps on leaves can be up to 3 mm in diameter and 2 mm tall and resemble an armoured scale. Nymphs enlarge their lerp as they grow, or they move and form a new covering. The yellow or brownish nymphs resemble a wingless aphid and spend most of their time covered beneath a lerp. Adults live openly on foliage and do not live under lerp covers (Plate 32). All life stages can occur on both new and mature foliage. The life cycle ranges from several weeks to several months, depending on temperature; in Australia, there are two to four generations each year.

High populations of psyllids secrete copious amounts of honeydew on which grows a black sooty mould, affecting leaf photosynthesis and fouling surfaces beneath affected trees. Extensive defoliation weakens trees and can increase tree susceptibility to damage from other insects and diseases, leading to dieback and tree mortality. Halbert et al. (2003) list 22 hosts for G. brimblecombei. In southern California, thousands of mature E. camaldulensis were killed within 2–3 years by uncontrolled populations of red gum lerp psyllid. The removal costs for these dead trees are estimated at millions of dollars (Hoddle, 2010). In Mauritius, Eucalyptus is the main melliferous plant and, if severely damaged, may result in a setback to honey production (Sookar et al., 2003).

Likely pathways for dissemination are plants for planting or cut foliage of Eucalyptus from countries where G. brimblecombei occurs. Long-range dispersal by air transport may be involved, as evidenced in Chile where the red gum lerp psyllid was first detected on E. camaldulensis in 2001 in the neighbourhood of the International Airport of Santiago (Huerta et al., 2010).

Classical biological control of G. brimblecombei by the Australian encyrtid wasp Psyllaephagus bliteus has been implemented in several countries (Daane et al., 2005).

Heteropsylla cubana Crawford (Hemiptera: Psyllidae)

While several psylloid species have spread from their natural range to other parts of the world, few have done so as rapidly or as spectacularly as *Heteropsylla cubana* (Geiger and Gutierrez, 2000; FAO, 2007) (Plate 33). This insect is a serious pest of the widely planted, multi-purpose tree *Leucaena leucocephala*, used as a source of fodder, fuelwood, shade for agricultural crops, reforestation and timber in many countries.

The leucaena psyllid is indigenous to tropical America, where its original range extended from Cuba and Mexico to Argentina (Showler, 1995). It was reported from Florida in late 1983 and the first populations established outside of the Neotropics were found on Hawaii in April 1984. The chronology of its spread is documented by several authors (e.g. Mitchell and Waterhouse, 1986; Muddiman et al., 1992; Napompeth, 1994; Nair, 2007) and is summarized as follows:

- Late 1983 — Florida
- April 1984 — Hawaii
- February 1985 — Western Samoa and Cebu, Philippines
- March 1985 — Mariana Islands
- June 1985 — Cook Islands, Fiji and Niue
- July 1985 — Tonga, Vanuatu and Caroline Islands
- October 1985 — New Caledonia, throughout the Philippines and American Samoa
- December 1985 — Solomon Islands and Taiwan
- March 1986 — Java, Indonesia and Papua New Guinea

- April 1986 — Australia
- May 1986 — Christmas Island
- June 1986 — Sumatra, Bali, Flores, Sulawesi and eastern islands of Indonesia
- September 1986 — Thailand
- 1986 — Malaysia, China, Singapore, Japan, Myanmar, Vietnam, Cambodia, Laos
- 1987 — Sri Lanka, Bangladesh
- 1988 — Andaman and Nicobar Islands, India
- 1989 — Nepal
- 1991 — Mauritius and Reunion
- 1992 — Tanzania, Kenya, Uganda, Burundi, Sierra Leone
- 1993 — Ethiopia, Mozambique
- 1994 — Sudan, Zimbabwe, Zambia, Malawi

The mechanisms by which *H. cubana* dispersed so quickly over such great distances were uncertain. Many psyllids form part of the aerial plankton, and this accounts for some of the range extension (Muddiman *et al.*, 1992). However, it has also been suggested that the spread was facilitated by transportation in or on aircraft. There are several reported instances of introduced psyllid pests being first detected on plants in the vicinity of airports. *H. cubana* lay eggs on young, unopened leucaena leaves, the eggs being attached to the leaf surface by a stalk. After 2–3 days, nymphs hatch and begin feeding. There are five nymphal instars lasting approximately 9 days. Females begin laying eggs within a few days of becoming adults, and a female may lay about 400 eggs in her lifetime. The total life cycle is about 14 days. Both nymphs and adults feed on foliage, causing leaflets to turn yellow, curl and wilt. Massive populations can cause shoot necrosis, defoliation and death of the tree (Showler, 1995). The deposition of honeydew encourages the growth of sooty moulds, which can interfere with the development of adjacent leucaena leaflets.

Leucaena psyllid damage and population trends vary from location to location. Populations generally build up in the wetter months when the trees are growing vigorously, but the effects of their feeding are most severe at the beginning of the dry season when leaf growth slows and psyllid numbers are already high (NFTA, 1990). Populations may fall to very low levels during extended hot, dry periods. In Australia, Bray and Woodroffe (1988) reported a rapid build-up of adults, from almost none to large numbers within 3–4 days after rain. The insects presumably were attracted to new growth on the plants, which was stimulated by the rain. Psyllid damage is often severe when juvenile foliage development is rapid, as on hedges managed for green manure or fodder (NFTA, 1990).

The economic and social impact of *H. cubana* worldwide has been considerable. *L. leucocephala* was widely regarded and promoted as a 'wonder tree' and many countries had extensive plantings of this species. For example, in Indonesia, 1.2 Mha of leucaena were planted in Java as part of the productive taungya agroforestry system for establishing teak plantations (Showler, 1995). In Bali, it shades 12,000 ha of vanilla, and on Irian Jaya, oil palm and cocoa. In the Philippines, Thailand, Australia and Indonesia, the first year of infestation caused an estimated US$525m in damages, with US$316m in Indonesia alone (Geiger *et al.*, 1995). Such damages included not only the direct loss of leucaena as a cash crop for fuelwood, fodder and timber but also decline in crop yields and death of crops where leucaena was used as shade, decline in small farm livestock production and loss of exports of livestock and agricultural produce. Small farmers experienced considerable reduction of income as a result of the psyllid damage; for example, in the Philippines the net monthly income generated from leucaena plantings declined from 1046 pesos in 1984 to 489 pesos in 1987 (Showler, 1995). Another consequence of the psyllid problem in some developing countries has been a loss of confidence in external recommendations, which in turn has affected progress towards larger goals such as reforestation (Geiger et al., 1995).

Classical biological control has been employed against *H. cubana*, the two main agents being the coccinellid predator *Curinus coeruleus* and the encyrtid wasp parasitoid *Psyllaephagus yaseeni*. The predator has been introduced into several countries and has had partial success in Hawaii and Indonesia, but has failed to establish in many seasonal dry areas and is a poor disperser (Geiger and Gutierrez, 2000). The parasitoid has been released in Hawaii, Indonesia and Thailand. It established and dispersed quite readily, having reached Malaysia and the Philippines without human intervention. Geiger *et al.* (1995) report a long-term trend in damage caused by *H. cubana*, this being generally heavy in the first 2 years of infestation, then gradually weakening in duration and severity. This has been attributed to biological control efforts.

Thaumastocoris peregrinus Carpintero and Dellape (Hemiptera: Thaumastocoridae)

Thaumastocoris peregrinus, commonly referred to as the winter bronzing bug or bronze bug, is an emerging pest of *Eucalyptus* in native and non-native regions of the southern hemisphere (Nadel et al., 2009). This sap-sucking insect feeds on eucalypt leaves, causing death of the leaves and leaf drop, resulting in loss of growth increment, dieback and sometimes tree death. In its native Australia, it was virtually unknown until 2001, when severe outbreaks occurred in Sydney's urban forest (Noack and Rose, 2007). It was reported from South Africa in 2003 and in Argentina in 2005. In the initial reports of these outbreaks, *T. peregrinus* was identified incorrectly as *T. australicus* (Carpintero and Dellape, 2006). Since 2003, populations of the pest have grown explosively in South Africa and it has attained an almost ubiquitous distribution over several regions on 26 *Eucalyptus* species (Nadel et al., 2009). *T. peregrinus* reached Zimbabwe in 2007 and Malawi in 2008 and it was recorded as established in Uruguay and Brazil in 2008 (Carpintero and Dellape, 2006; Martinez and Bianchi, 2010; Wilcken et al., 2010).

Thaumastocoris are gregarious insects, with adults and nymphs occurring on the same leaf. Adults live for an average of 16 days and each female will produce about 60 eggs (Noack and Rose, 2007). The eggs are laid in black capsules on the leaves, often in a cluster that can be seen as a large black mark on the leaf (Button, 2007). Eggs hatch in 4–8 days and the total nymphal time is 17–25 days. Typical symptoms of infestation include initial reddening of the canopy leaves, a condition sometimes referred to as 'winter bronzing', although this can occur throughout the year (Plate 34). Subsequently, the foliage changes to a reddish-yellow or yellow-brown colour, and loss of leaves associated with heavy infestations leads to severe canopy thinning and sometimes branch dieback or tree mortality (Nadel et al., 2009).

T. peregrinus has a wide host range, infesting at least 30 *Eucalyptus* species and hybrids. In South Africa, all commercially grown *Eucalyptus* are susceptible to attack, and the pest has been reported from seven eucalypt species in South America (Nadel et al., 2009). In Australia, the insect has become a pest of tropical and subtropical plantations in Queensland and New South Wales, affecting species of *Corymbia* as well as *Eucalyptus*.

Nadal et al. (2009) used DNA bar-coding to investigate the source and patterns of *T. peregrinus* invasions in South Africa and South America and concluded that Sydney was the most likely origin of both these introductions. Extreme long-range dispersal by air travel is thought to be the main mechanism for spread. This hypothesis is supported by information from Brazil, where, in the state of Sao Paulo, the bronze bug was first found in *Eucalyptus* trees adjacent to two international airports in the metropolitan region of Sao Paulo city (Wilcken et al., 2010). The insect may also hitchhike on the clothes of travellers or be dispersed by wind.

Tingis beesoni Drake (Hemiptera: Tingidae)

The lace bug *T. beesoni*, recorded only from India, Thailand and Myanmar, is a serious pest of *G. arborea* in those countries, causing defoliation and dieback in young plantations

(Mathew, 1986; Kamnerdratana, 1987; Harsh et al., 1992; Nair, 2007) (Fig. 5.9).

The life history of the insect is described by Mathur (1979). Eggs are laid over a period of 4–5 days in small batches; they are inserted into the tissue of the tender shoot in a vertical row. They hatch in 2–6 days and the nymphs congregate on the lower surfaces of the foliage and suck sap at the base of the lamina or in the axils of leaves. A feeding cluster may comprise as many as 60 nymphs. There are five nymphal instars taking from 9 to 30 days to complete, depending on the season. Adults are quite active in summer and move about on the under-surface of leaves and new shoots for feeding. They disperse to other plants, but their flight is limited. There are seven generations a year, with considerable overlap. At the onset of cool weather, the adults of the last generation lay their eggs under bark, where they hibernate, hatching the following spring.

Feeding by nymphs and adults causes leaves to become blotched-brown near the base. They wither and fall and the shoots become dry. Eventually, the shoots die back, retarding tree growth. During one outbreak in a 10 ha plantation of 1-year-old trees in India, 67% of the plants were infested, 21% suffering total defoliation and dieback of the terminal shoot (Nair and Mathew, 1988). A study by Harsh et al. (1992) of top dieback and mortality in a G. arborea provenance trial in Madhya Pradesh, India, showed that the problem was caused by T. beesoni in combination with the fungus Hendersonula toruloidea. Meshram and Tiwari (2003) recorded an 80% incidence of top dying in one plantation of G. arborea in Madhya Pradesh and noted that trees aged 2–3 years were more susceptible to this insect. Feeding by the lace bugs in the wet season provided conditions favouring the infection of damaged plants by the canker-causing H. toruloidea. The fungus was capable of invading and killing plants within a year.

5.4 Bark and Wood Feeding

Bark- and wood-feeding forest insects are contained in four main orders: Coleoptera (bark beetles, ambrosia beetles, longicorn beetles, bostrychid beetles, scarab beetles, weevils), Hymenoptera (wood wasps), Isoptera (termites) and Lepidoptera (wood moths). Usually, the larval stages cause most of the damage to the wood and bark. Only in the Coleoptera do the adults of some species (e.g. pinhole borers) tunnel extensively in the wood or feed externally on bark (e.g. some scarab beetles).

The effects of bark- and wood-boring insects on their hosts vary depending on the condition of the tree when attacked, the particular tissue attacked and the activity of associated agencies such as symbiotic fungi, bacteria and predators (Elliott et al., 1998). An obvious effect is structural weakness in stems and branches resulting from extensive tunnelling. This may be exacerbated by

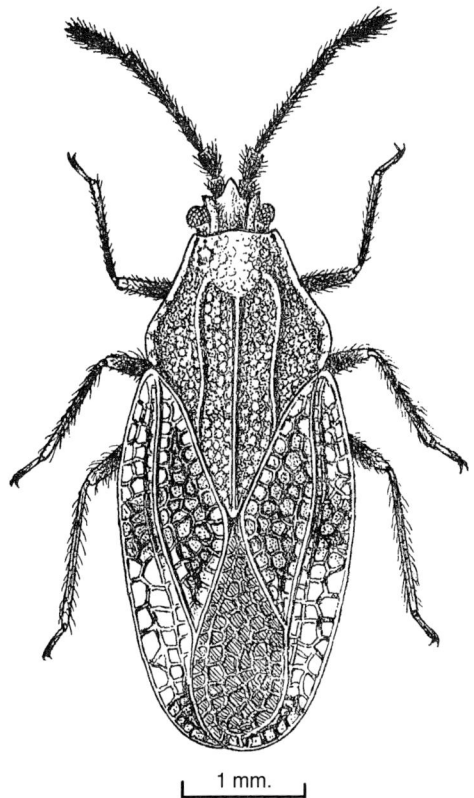

Fig. 5.9. Adult of *Tingis beesoni* (from Mathur, 1979).

the activities of predators excavating for larvae in the tree, as is the case with yellow-tailed black cockatoos (*Calyptorhynchus funereus*) predating larvae of the giant wood moth, *Endoxyla cinerea*, in eucalypts in Australia (Wylie and Peters, 1993). Tunnelling by larvae of longicorn and jewel beetles in stems and branches may interrupt the tree's conductive processes and sometimes results in ring-barking and death of affected parts. Some borers, such as ambrosia beetles, cause damage that is not immediately apparent until timber is cut from the tree. Termites such as species of *Odontotermes* and *Macrotermes* are common and important agents of tree mortality in young plantations throughout the tropics. However, in Australia, termites cause most damage by feeding in the heartwood of trees, causing significant losses in wood volume and downgrading potential sawlog into lower value pulpwood. Several important wood- and bark-feeding insects have very close associations with other organisms such as fungi, with these organisms forming an essential part of the life history of the pest. An example is the sirex wasp and its symbiotic fungus *Amylostereum areolatum*, which not only provides food for the larvae but also contributes to death of the host tree.

The physiological condition of host trees has a major influence on the type of insect borers which attack them and the ability to withstand this attack. For example, drought-stressed trees can be very susceptible to sirex and *Ips* bark beetles, particularly if other factors such as overstocking aggravate these effects. Many borers, such as ambrosia beetles, are attracted to damaged trees.

5.4.1 External chewing

Indarbela quadrinotata Walker
(Lepidoptera: Indarbelidae)

The bark-eating caterpillar, *I. quadrinotata*, which occurs throughout Asia, is a polyphagous pest infesting many fruit trees, street trees and important forest trees such as teak, *T. grandis*, mahogany, *Swietenia* spp., *G. arborea* and species of *Acacia*, *Casuarina*, *Syzygium* and *Terminalia* (Beeson, 1941; Garg and Tomar, 2008; Patel and Patel, 2008). Damage has been reported from India, Pakistan, Bangladesh, Myanmar and Malaysia.

The adult female may lay up to 2000 eggs, which she places in groups of about 15–25 on the bark of the stem or branches of the host. The eggs hatch in 9–11 days. Each larva bores a short tunnel downwards into the wood and shelters there during the day, coming out at night to feed on the outer surface of the bark. Broad irregular patches are excavated in the bark and these areas are roofed in with silk mixed with excrement and fragments of bark. A single larva can damage a considerable area of bark (Plate 35), and in a heavy infestation most of the outer bark of the stem may be destroyed (Beeson, 1941; Browne, 1968). The larval period lasts about 10 months in India and Bangladesh, pupation takes place in the shelter tunnel and moths emerge after 21–31 days. In Myanmar, there are two generations a year.

Feeding by this caterpillar can result in loss of growth increment and, with heavy infestations, girdling and mortality (Sasidharan *et al.*, 2010). The shelter tunnels bored into the sapwood allow the entry of fungi and other organisms and degrade the wood. Zia-ud-Din (1954) also suggests that heavy damage may prevent flowering of some trees. In India, *I. quadrinotata* has been reported as a moderately serious pest of *A. senegal* and *A. tortilis* in the Thar desert (Vir and Parihar, 1993). These tree species are the major component of plantation forestry programmes related to wasteland development, sand dune stabilization, fuelwood and fodder in this arid region. In Tamil Nadu, India, *C. equisetifolia* is a major species used for coastal afforestation and agroforestry and plays a crucial role in the rural economy. Of 40 species of insect associated with *C. equisetifolia* in that state, *I. quadrinotata* is the most economically important pest (Sasidharan and Varma, 2008). Meshram *et al.* (2001) rate *I. quadrinotata* as a major pest of *G. arborea* in India. Baksha (1991) reported a 70% incidence of the pest in a 4-year-old plantation of *F. moluccana* near Chittagong, Bangladesh. *Populus* spp., widely used in farm forestry programmes in Pakistan and India, are also commonly attacked by *I. quadrinotata* (Veer and Chandra, 1984; Gul and Chaudhry, 1992).

Natural enemies of the bark-eating caterpillar include braconid and eulophid wasp parasitoids, ant predators and the entomopathogenic fungus *Beauveria bassiana* (Gul and Chaudhry, 1992).

5.4.2 Bark boring

Hylastes angustatus (Herbst)
(Coleoptera: Curculionidae: Scolytinae)

The pine bark beetle, *Hylastes angustatus*, of European and southern Russian origin, is thought to have been introduced accidentally into South Africa on fresh pine logs (Bevan and Jones, 1971; Govender and Wingfield, 2005). It was found originally in the southern Cape Province, but now occurs wherever *Pinus* spp. are grown commercially in South Africa, as well as in Swaziland and Zambia. Although it is not considered a pest in the winter rainfall areas in the south, it is a serious, though sporadic, pest in summer rainfall regions, causing damage to seedlings of the economically important *P. patula* (Erasmus and Chown, 1994).

As described by Bevan and Jones (1971), there are two distinct phases in the life cycle of *H. angustatus*, a breeding phase and a feeding phase. The former is that period from sexually mature adult, through egg and larva, to young virgin unemerged imago, when the insect is confined to the chosen breeding site. In this phase, the insect is 'secondary', attacking only damaged, dying or dead material such as logs and stumps, and is generally of no direct economic importance. The feeding phase is the period between the emergence of the virgin adult from its breeding site and its eventual arrival at another breeding site when sexually mature. During this time, the beetle feeds on young green bark, this being necessary for proper maturation of the gonads. In this phase, the insect is 'primary', attacking healthy living plants, and its damage is significant.

After feeding, the female enters the bark, excavates a nuptial chamber and is then joined by the male. Following copulation, she constructs a straight gallery about 7 cm long and lays eggs at intervals along both sides of it. Larvae hatch and bore meandering galleries outwards from the mother gallery, and eventually pupate at the end of their tunnel. The duration of the life cycle from egg to egg is about 48 days (Webb, 1974), and there are four to five generations a year in the Transvaal area (Tribe, 1990).

Maturation feeding by *H. angustatus* under bark on roots and root collars of pine seedlings can result in girdling and death of seedlings, which are most vulnerable up to 1 year after planting. Mortalities of up to 52% have been recorded in young *P. patula* plantations in Transvaal (Toit, 1975). The generally accepted loss rate is 15%, above which replacement is necessary. Because *H. angustatus* feeds beneath the bark and mainly below ground, the beetles have already departed by the time the damage is first noticed (Tribe, 1990). The presence of harvesting residues (slash) has been found to impact negatively on the early survival of tree seedlings because this favours build-up of *H. angustatus*, but the practice of slash burning promotes losses due to the fungus *Rhizina undulata*, which requires a heat stimulus for the onset of pathogenicity (Wingfield and Swart, 1994). The discontinuation of slash burning has necessitated the implementation of chemical control for *H. angustatus* in summer rainfall areas of South Africa (Allan *et al.*, 2000).

Various natural enemies of *Hylastes* have been recorded, but these have been considered of little use in commercial plantations.

Ips spp. (Coleoptera: Curculionidae: Scolytinae)

The genus *Ips* contains more than 60 species (Wood, 1982) and is one of the best-known groups of bark beetles, with a worldwide distribution. Several species occur in the tropics and subtropics, and some of these are important pests of *Pinus* spp. They can attack living trees, freshly felled logs and unbarked pine slash (Hanula *et al.*, 2002).

The attack is generally initiated by the adult male, which tunnels into the inner bark/cambial region where the nuptial chamber is carved out. The male produces a pheromone which attracts the females and other males to the log or tree. A male may be joined by up to seven females and mating takes place in the nuptial chamber. Each female bores a tunnel away from this chamber and deposits eggs in niches along the sides. The frass which is produced is pushed back along the tunnel by the female and cleared out through the entrance hole by the male. On hatching, the larvae tunnel in the inner bark. At high densities, the larval tunnels overlap and the phloem is converted into a layer of frass. The larval period may vary from a few weeks to a few months, depending on temperature, and at the end of this period, larvae form a cell at the end of their gallery and pupate. The new adults feed on the inner bark before emerging through exit holes similar to that by which their parents entered.

Ips species carry out 'feeding' and 'breeding' attacks when colonizing the bark of green to semi-green dead pine material or that of apparently healthy trees. In feeding attacks, the inner bark and outer sapwood surface is etched by large numbers of male and female adults in the period prior to reproduction and, as a consequence, the bark peels off. Breeding attacks are made by virgin male and female beetles and/or by fertilized females when entering the inner bark and constructing characteristic gallery systems for breeding (Neumann, 1987).

Different species of *Ips* may show preference for infesting different parts of the host plant. For example, in Jamaica, *I. calligraphus* showed preference for the older, thick-barked regions of the trunk and the larger branches, while *I. grandicollis* preferred the thinner-barked regions of the trunk and smaller branches (Garraway, 1986). Bark thickness also influences certain parasitoids of *Ips*, those species that oviposit through the bark being limited to areas of bark that are no thicker than their ovipositors are long (Riley and Goyer, 1988).

Commonly, *Ips* are secondary pests, colonizing recently cut logs, pine slash and physiologically unhealthy trees. However, sometimes they can assume a primary role, causing tree mortality, particularly when populations are very high. The propensity of *Ips* to attack living trees varies with species; *I. calligraphus* is regarded as one of the most aggressive species (Yates, 1972). In Australia, *I. grandicollis* (Plates 36 and 37) has sometimes caused extensive, although localized, tree death, particularly when trees were stressed by drought or damaged by fire and lightning (Neumann, 1987; Elliott *et al.*, 1998). Following fires in south-east Queensland in 1994, which affected over 8000 ha of *P. elliottii*, *P. taeda* and *P. caribaea* plantations, *I. grandicollis* attacked fire-damaged trees after 6 weeks and was a significant pest in most areas after 10 weeks (Wylie *et al.*, 1999) (Fig. 5.10). Sap staining, caused by fungi (principally *Ophiostoma ips*) carried by the beetle, became significant in attacked stems at the completion of the insect's life cycle (about 4 weeks in summer) (Plate 71). This attack necessitated rapid salvage of the timber and its storage under water spray (Plate 109). The losses caused by *I. grandicollis* and sap stain following these fires were estimated at several million Australian dollars, most of this being in privately owned plantations where salvage was delayed for several months.

Garraway (1986) cited two instances of large-scale attack in pine plantations in Jamaica by *I. calligraphus* and *I. grandicollis*; the first was on fire-ravaged *P. caribaea* in 1979 and the other on a mixed plantation of *Pinus* spp. which had been exposed to flooding and landslides, followed by several months of intense drought in 1980. In the latter case, 22% of the pines were lost over a 6-month period. In the Philippines, Lapis (1985a) reported occasional large-scale mortality of *P. kesiya* in northern Luzon due to *I. calligraphus*, and this species was the principal mortality agent in the deaths of thousands of drought-affected *P. occidentalis* in the Dominican Republic in 1986–1987 (Haack *et al.*, 1990). In Honduras, primary attack of *P. oocarpa* by *I. cribricollis* has been reported and it has been found frequently attacking apparently healthy pines in association with other *Ips* species (Lanier, 1987). In southern USA, three species of *Ips* engraver beetles (*I. avulsus*, *I. grandicollis* and *I. calligraphus*)

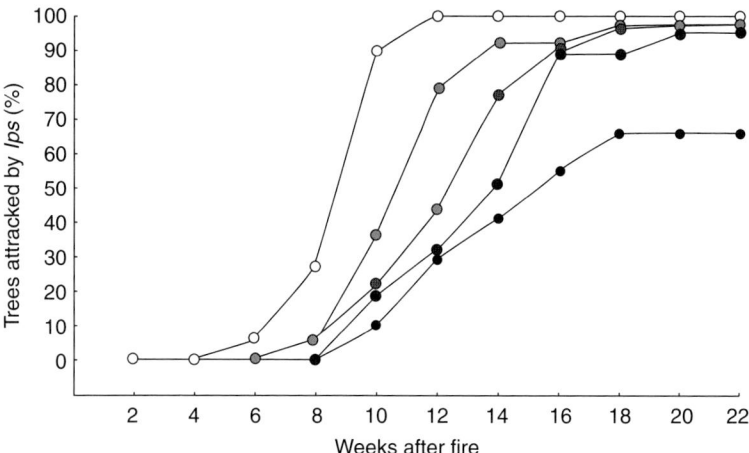

Fig. 5.10. Progress of *Ips grandicollis* attack with time in five plots in fire-damaged pine plantations in Queensland, Australia (from Wylie *et al.*, 1999).

are responsible for documented annual losses of approximately US$6.6m to pine forests (Riley and Goyer, 1988).

Predators and parasitoids are important factors in the regulation of *Ips* spp. bark beetles. In southern USA, Riley and Goyer (1986) found 27 species of insect predators and 10 species of parasitoids associated with *Ips* broods in *P. taeda* and *P. elliottii*, which decreased brood survival by 30.8%. Biological control has been used to reduce populations of *I. grandicollis* in Australia, the torymid wasp parasitoid *Roptrocerus xylophagorum* now being widely established and the braconid wasp *Dendrosoter sulcatus* also being established in the subtropics (Elliott *et al.*, 1998).

5.4.3 Cambium and surface sapwood boring

Agrilus spp. (Coleoptera: Buprestidae)

Agrilus is a cosmopolitan genus containing well over 1000 species. Most of these are of little or no economic importance, but a few species in Africa, Asia and the Pacific are serious pests in forest plantations, causing growth loss, stunting and tree mortality.

The biologies of the main pest species in Papua New Guinea and the Philippines are very similar. Generally, the eggs are laid singly in cracks or crevices in the bark, or under bark flakes, on the lower stem of the host tree. The eggs hatch in about 2 weeks and the larvae burrow directly into the bark down to the wood surface, where they feed on the tissues of inner bark and outer wood (Braza, 1988a). In the course of their feeding and development, the larvae construct frass-filled zigzag tunnels, sometimes more than 2 m long and extending down to the roots. When mature, larvae tunnel into the wood, constructing a chamber, where they pupate. Adults emerge a few weeks later and tunnel to the surface, cutting a 'D'-shaped hole in the bark. The length of the life cycle varies with species of *Agrilus* and the type and condition of the host. In Papua New Guinea, *A. opulentus* completes its life cycle in small-diameter felled trees in 6–7 weeks. In trees with a larger diameter trunk, the life history may take up to three times longer, while in standing, living trees the life history takes at least 9 months to 1 year (Roberts, 1987). Young adults must feed before they can mate and before the females can lay eggs. They usually feed on the new foliage of their preferred host trees.

On thin-barked trees, the zigzag tunnelling of the larvae is clearly visible in the form of raised welts on the bark, giving rise to the names 'zigzag' or 'varicose' borers for these insects (Plate 38). These welts are the result of callus tissue overgrowing old tunnels and are usually not visible where larvae are active.

In Papua New Guinea, the two trees most widely grown in forest plantations in the wet lowlands, *E. deglupta* and *Terminalia brassii*, are attacked by *A. opulentus* and *A. viridissimus*, respectively. Attack is often heaviest on trees which are stressed; for example, *E. deglupta* growing on badly drained soils and *T. brassii* on soils which dry out rapidly (Roberts, 1987). Infested trees show loss of annual increment, and small and suppressed trees are girdled and killed. Estimated growth losses in *E. deglupta* plantations at Madang due to *A. opulentus* totalled US$2.5m over the 10-year rotation (Mercer, 1990). In the Philippines, *E. deglupta* plantations (Papua New Guinea provenance) in Mindanao have been severely attacked by *A. sexsignatus*, with up to 63% mortality. Numerous studies have shown that the Papua New Guinea provenance of *E. deglupta* is most susceptible to attack by *Agrilus* spp., while the native Philippines *E. deglupta* is more resistant (Braza, 1987, 1988a).

In the Sudan, after a long drought during 1979–1984, gum production by *A. senegal* and *A. seyal* decreased and this was attributed in part to attack by jewel beetles, including species of *Agrilus* (Jamal, 1994). In Pakistan, *A. dalbergiae* has caused yellowing and death of unhealthy *D. sissoo* trees in amenity plantings (Sheikh and Aleem, 1983), and in Indonesia, *A. kalshoveni* caused large-scale mortality of scattered trees of all sizes of *Actinophora fragrans* (Kalshoven, 1953; Nair, 2007).

Four hymenopterous parasitoids have been recorded for *A. sexsignatus* in the Philippines, with parasitism rates for eggs and larvae of up to 57% (Braza, 1989; Noyes, 1990). In Papua New Guinea, Mercer (1990) suggests the use of the ant *A. longipes* as a biocontrol agent for *A. opulentus* in *E. deglupta* plantations.

Phoracantha semipunctata (Fabricius)
(Coleoptera: Cerambycidae)

This Australian insect (Plate 39) is now established in many regions of the world (Europe, Africa, the Middle East, North and South America and New Zealand), having been present in South Africa and Argentina since the early 1900s (CABI/EPPO, 2007b). In some places, it is a serious pest of planted eucalypts, attacking and killing young trees (Loyttyniemi, 1991; Hanks *et al.*, 1995a, 2001), but in Australia it is regarded as only a minor pest attacking damaged, severely stressed or newly felled trees.

As described by Elliott *et al.* (1998) for *Phoracantha semipunctata* in Australia, females lay batches of 10–100 eggs in crevices in the bark of dead, dying or stressed trees, in freshly cut logs or in branches down to a diameter of about 150 mm. Dry material is not attacked. The eggs hatch in 10–14 days and larvae tunnel in the inner bark/cambial/outer sapwood region for 4–6 months, making wide galleries, sometimes 1–2 m long, that are tightly packed with frass. When fully fed, the larvae bore deeply into the heartwood and pupate. The pupal period lasts about 10 days. Adults may be found during all months of the year, and generations overlap considerably. Adults can live for more than 90 days (Paine *et al.*, 1995). The length of the life cycle throughout the range of the insect varies with climate and season, and can take from 2 months in the hot tropics up to 1 year in cooler regions.

P. semipunctata occurs widely in Africa from the south through Angola, Malawi, Mozambique, Zambia and Zimbabwe to Morocco and the Canary Islands in the north. In Zambia, where it was first recorded in 1968, severe outbreaks of the insect in eucalypt plantations (mainly *E. grandis*) started at the beginning of the 1970s and tree mortality has reached up to 40% of stocking during a rotation period (Loyttyniemi, 1991). At the beginning of the 1980s, up to 67% of the total area of eucalypt plantations in the country was regarded as severely infested (Selander and Bubula, 1983).

Numerous studies have shown that drought is one of the main factors predisposing trees to attack by *P. semipunctata*. This can be exacerbated by factors such as site and silviculture. For example, in Zambia, Ivory (1977) found that tree mortality associated with attack by this longicorn in eucalypt plantations was highest for trees on soils with a high sand content (Fig. 5.11). Delayed thinning, poor sanitation and damage by termites and fire have also been implicated in attack and mortality (Loyttyniemi, 1991). In Chile, *P. semipunctata* is only considered a hazardous species in the arid and semi-arid regions (Lanfranco and Dungey, 2001). There is considerable variation among eucalypt species in resistance to attack by *P. semipunctata*. In California, Hanks *et al.* (1995b) found that species that were resistant to attack were those that were most tolerant of drought in Australia. Bark moisture content may play a critical role in resistance, the insect being able to colonize trees where this is reduced (Hanks *et al.*, 1991). Water stress has a major influence on the survival and growth of the larvae, as demonstrated by Caldeira *et al.* (2002) in Portugal. *E. globulus* trees subjected to water stress during 2 consecutive years were compared with rainfed and irrigated trees. Larvae of *P. semipunctata* were introduced artificially into the bark of trees of both treatments. Larval mortality was found to be lower and weight gain was higher in water-stressed trees than in rainfed trees, and there was no larval survival in irrigated trees.

Studies of the insect on logs of *E. grandis* in Malawi (Plate 40) showed that most mortality occurred in the larval stage, and the major mortality factor was intraspecific competition. Severe overcrowding led to a reduction in population density from one generation to the next. Overcrowding also led to a reduction in beetle size and to a shift in the sex ratio in favour of males, which being smaller than the females were more likely to complete their development in crowded conditions (Powell, 1982). This effect could reduce the breeding potential of the next generation greatly. Hanks *et al.* (2005) showed that optimal development conditions for *P. semipunctata* larvae, in terms of larval performance and adult body size, were available in large, aged host logs having low densities of larvae.

Numerous predators and parasitoids attack *P. semipunctata* in Australia and several of these have been used in biocontrol programmes in other countries. One of the earliest attempts was the introduction of *Megalyra fasciipennis* into South Africa in 1910. In 1993, parasitism of the cerambycid by this insect frequently reached 50% (Moore, 1993). Three other Australian parasitoids of *P. semipunctata*, an encyrtid and two braconids, are now established in South Africa where an indigenous pteromalid wasp, *Oxysychus genualis*, has also been recorded (Prinsloo, 2004). The egg parasitoid *Avetianella longoi* was released in California in 1993 and Hanks *et al.* (1996) reported high rates of parasitism (up to 91% of all eggs in some instances). This, coupled with the wasp's strong powers of dispersal and efficient location of host eggs, suggests that it may have an important impact on *P. semipunctata* in California, but see Chapter 10.

In southern California, *P. recurva* appears to be replacing *P. semipunctata* rapidly in their shared habitat, for reasons that are not yet clear (Bybee *et al.*, 2004a,b), and there is a similar report of this from Argentina (Di-Iorio, 2004).

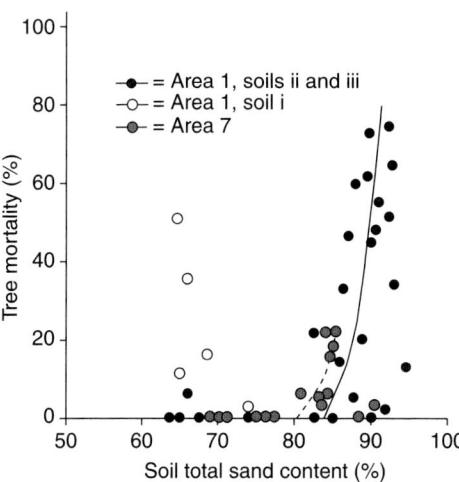

Fig. 5.11. Relationship between attack by *Phoracantha semipunctata* and sand content of soil in Zambia (from Ivory, 1977).

Xystrocera festiva
(Coleoptera: Cerambycidae)

F. moluccana (formerly known as *Paraserianthes falcataria*, *A. falcata*, *A. falcataria* and *A. moluccana*) is a fast-growing leguminous tree, native to Indonesia, Papua New Guinea and the Solomon Islands, which is widely planted in the humid tropics for pulpwood, matchsticks, plywood, lightweight packing materials and community forestry (Nair, 2007). One of the principal pests of this species is the cerambycid borer *Xystrocera festiva*, whose larvae tunnel in the stems of living trees, often causing tree mortality.

Nair (2007) provides a comprehensive pest profile of the insect, drawing on studies by Suharti *et al.* (1994), Hardi *et al.* (1996), Matsumoto and Irianto (1998) and Kasno and Husaeni (2002). Adults are nocturnal and live for only 5–10 days. Eggs are deposited in clusters of over 100, in one or two batches, preferably in crevices on the stem or branch stubs, generally 3–4 m above ground. The average number of eggs laid per female is estimated at 170. Newly hatched larvae bore into the inner bark and as the larvae grow, they feed on the outer sapwood, making irregular downward galleries packed with frass. The larvae remain gregarious, which is unusual in cerambycids. Symptoms of infestation are exudation of a brownish liquid through the bark and the expulsion of a powdery frass. The larval development is completed in about 4 months and each larva bores an oval tunnel upward in the sapwood, in which it pupates. The insect has overlapping generations, with all developmental stages present at any one time.

Infestation by *X. festiva* usually begins when the trees are 2–3 years old and the infestation increases with age. Larval tunnelling can reduce growth rate and timber quality, and heavy infestation can result in ring-barking and death of the tree. The insect is a major pest of *F. moluccana* in Indonesia and Malaysia and also occurs in Myanmar. It is a minor pest of several other tree species, including *Acacia* spp.

A related species, *X. globosa*, is also a pest of *F. moluccana* but, unlike *X. festiva*, larvae are not gregarious and tunnel individually (Matsumoto *et al.*, 2000).

An encyrtid egg parasitoid of *X. festiva*, *Anagyrus* sp., has been released in biological control trials in East Java, which has given promising results.

5.4.4 Sapwood and heartwood boring

Aristobia horridula (Hope)
(Coleoptera: Cerambycidae)

This insect (Plate 41) was first recorded damaging forest trees (*D. paniculata* and *D. volubilis*) in India in the 1930s (Beeson and Bhatia, 1939), but it was not until the last decade or so that it emerged as a serious pest of forest plantations in Thailand, Nepal and India (Hutacharern and Panya, 1996; Dhakal *et al.*, 2005; Nair, 2007). In Thailand, it is regarded as the most important stem borer of *Pterocarpus macrocarpus* and it also causes serious damage to *P. indicus* and *D. cochinchinensis* (Hutacharern and Panya, 1996). Both *P. macrocarpus* and *D. cochinchinensis* are high-value timber species. In India and Nepal, it infests *D. sissoo*, also an important and widely planted tree in South Asia.

Eggs are laid singly in the bark of the host tree, in a crescent-shaped incision made by the adult female, and hatch in 8–10 days. Newly hatched larvae bore extensively in the sapwood and then into the heartwood, where pupation occurs in a chamber plugged by wood slivers (Hutacharern and Panya, 1996). Larval galleries may be up to 70 cm long and are packed with frass (Nair, 2007). In smaller trees, larval tunnelling may extend down to the root (Hutacharern and Panya, 1996). Swelling of the bark, resin exudation and extruded frass are visible symptoms of infestation, and emerging adults cut circular exit holes in the bark. The life cycle is annual. Adult beetles are active during the day and feed on the bark of young branches, sometimes causing girdling and death of small branches (Hutacharern and Panya, 1996). Larval tunnelling in the stems and branches of trees not only degrades the wood but also weakens the trees, making them prone to wind damage, and may cause the death of young trees (Hutacharern, 1995).

The incidence of infestation by *A. horridula* can be high. In Thailand, Hutacharern and Panya (1996) reported infestation levels of 83% of trees in a 16-year-old plantation of *P. macrocarpus*, 33% of trees in an 8-year-old plantation of the same species and 25% in an 8-year-old plantation of

D. cochinchinensis. Also in Thailand, they reported that 100% of trees in a 10-year-old roadside planting of *P. indicus* were infested. In India, incidence of *A. horridula* in *D. sissoo* plantations in West Bengal ranged from 10 to 90%, the older age classes being affected most seriously. It was suggested that the pest had migrated from nearby native forests in which *D. sissoo* was a component into the monoculture plantations (Mishra *et al.*, 1985). Dhakal *et al.* (2005) reported severe damage by *A. horridula* in sissoo seedling seed orchards in Nepal 2 years after planting. Of a total of 6720 trees planted in 1996, 67% were dead in 2002 and only 19 trees showed no symptoms of infestation.

Two other species of *Aristobia* are known to cause damage to forest plantations in Asia. In the Mekong Delta of south-western Vietnam, *A. approximator* girdled and killed about 4000 ha (half the estate) of *E. camaldulensis* and *E. tereticornis* plantations (Wylie and Floyd, 1998). The attack was believed to be stress related and associated with the acid sulfate soils on which the trees were planted (Kawabe and Ito, 2003). In India, *A. octofasciculata* bores in the small branches and stem of saplings of *Santalam album*, causing dieback and mortality (Remadevi and Muthukrishnan, 1998). Tunnelling extends into the heartwood, allowing entry of termites and decay fungi. This combination of agents leads to hollowing-out of the sandalwood heartwood, resulting in an overall loss in volume/weight of almost 20% and a financial loss of US$3000/t (Remadevi and Muthukrishnan, 1998).

Celosterna scabrator Fabricius
(Coleoptera: Cerambycidae)

Celosterna scabrator, commonly known as the Babul borer because of its attacks on acacias such as *A. arabica* and *A. nilotica* in India, is also a pest of other tree species in that country, for example, teak, *Casuarina* spp., *Eucalyptus* spp., *Shorea robusta* and *Prosopis* spp. (Nair, 2007). It also infests *Dipterocarpus alatus* in Thailand (Gotoh, 1994).

The eggs are laid in the bark of young trees with a minimum basal girth of 5 cm and a maximum of 23 cm. According to Beeson (1941), above this dimension the bark is too hard for the insertion of the egg and too dry for the larva to hatch. Females may lay about 40 eggs over a period of several weeks, usually one egg to a stem, and these hatch in 2–3 weeks. The larva bores downwards, hollowing out the main root and keeping its tunnel clean by means of a frass ejection hole just above ground level (Browne, 1968). Tunnels may be up to 60 cm long. The larval period lasts about 9 or 10 months and pupation takes place in the host, lasting about 15–17 days. The beetle emerges by cutting a circular hole in the bark, usually below ground level. The beetles may live 80 days and are also destructive, feeding mainly at night on the bark of young living shoots.

Feeding by the larvae causes cessation of growth and sometimes death of the plant above ground. Incidence of attack of up to 80% has been recorded where trees are growing on unsuitable sites. An endemic pest in scrub and open-thorn forests, it has become of major importance in clear-felled areas replanted with eucalypts. Attack of *Eucalyptus* spp. in plantations in India is common (Ralph, 1985; Sivaramakrishnan, 1986), and Sen-Sarma and Thakur (1983) regard it as a 'key pest' in several states. *C. scabrator* was believed to be responsible for up to 14% mortality of *Prosopis cineraria*, an important agroforestry tree species, in the arid Thar desert in the north of India (Jain, 1996). Further south, in Karnataka, outbreaks have occurred in 1-year-old plantations of *E. tereticornis* and *A. nilotica*, sometimes with high mortalities (Sivaramakrishnan, 1986; Ralph, 1990). Feeding by the adults can result in girdling and breakage of the main stem and branches (Shivayogeshwara *et al.*, 1988). In Thailand, attack rates in 5- to 8-year-old stands of *D. alatus* ranged from 33 to 59%, resulting in loss of growth, lowering of timber quality and some tree death (Gotoh, 1994). Few natural enemies of this insect have been recorded.

Hoplocerambyx spinicornis Newman
(Coleoptera: Cerambycidae)

Known as the sal borer, *Hoplocerambyx spinicornis* is the most notorious forest pest of India because of its periodic outbreaks, during which millions of sal trees (*S. robusta*) are killed (Nair, 2007). Besides India, its distribution includes Afghanistan, Pakistan, Nepal, Bangladesh, Myanmar, Thailand, Indonesia and the Philippines.

A detailed pest profile of this insect is provided by Nair (2007), who summarizes studies dating back to the early 1900s. The female beetle lays her eggs on cuts or holes in the bark of sal trees, normally choosing trees that are freshly dead or highly weakened by various causes. However, during outbreaks even healthy trees are attacked. Each female will lay 100–300 eggs over a lifespan of about 1 month. The newly hatched larvae feed under the bark initially, then in the sapwood and finally bore into the heartwood. A large sal tree may support the development of about 300 beetles, although more than 1000 eggs may be laid on the tree. Coarse dust is thrown out of holes in the bark of infested trees and accumulates at the base of the tree. When larval development is completed, the larva constructs a chamber in the heartwood with an adult exit hole, and turns into a prepupa, then pupa and adult. The adult beetle remains quiescent until it emerges with the onset of rainfall. The length of the life cycle is 1 year.

Extensive galleries in the sapwood made by several larvae cause partial or complete girdling of the tree, leading to its death (Nair, 2007). Outflow of resin from the infested tree traps many young larvae but mass attacks during epidemics kill even vigorous trees. Both the main trunk and crown branches are infested. Nair (2007) presents a chronology of *H. spinicornis* outbreaks spanning the period 1897–2000. One of the most severe was the 1923 outbreak in Madhya Pradesh, which persisted over a 5-year period killing about 7 m sal trees. Another in the same state in 1994–2000 killed more than 3m trees over an area of 500,000 ha. No clear pattern is evident in the timing of the outbreaks. Trapping programmes carried out by the State Forest Department using sal logs yielded a peak of 32.59 m beetles in 1998 before the population declined. The timber of the heavily infested trees is riddled with tunnels and rendered useless, causing enormous economic loss. The circumstances under which outbreaks develop are not understood fully, but they often occur in dense over-mature stands where conditions favour rapid build-up of populations of the insect. Any stress factor which compromises the tree's ability to produce the defensive resin flow may trigger an outbreak.

An elaterid beetle, *Alaus sordidus*, is a predator of *H. spinicornis*, and during an epidemic, up to 10–15% of vacant sal borer pupal chambers were found to be occupied by *A. sordidus*.

Platypus spp. (Coleoptera: Curculionidae)

All *Platypus* spp. are ambrosia beetles, so named because they feed on 'ambrosia fungus', which grows on the walls of their tunnels (Alfaro et al., 2007). They do not feed on the wood itself. *Platypus* species construct galleries in 'green' (unseasoned) wood, either in living trees or in logs and freshly sawn timber. These galleries are usually more extensive in the sapwood, but they can extend deep into the heartwood.

The male beetle generally initiates the attack and is joined by the female. After mating, the female takes over boring of the tunnel, while the male merely removes and ejects the bore dust. Fungal spores are carried into the tunnel by the beetles, either trapped in the hairs on their body surface, in special structures (mycangia), or in their gut. The spores germinate and fungal spores grow on the walls of the tunnel, providing food for the adults and larvae, meeting most if not all of their nutrient requirements (Elliott et al., 1998). The ambrosia fungus discolours the tunnel wall and this dark staining may extend along the grain around the gallery or hole, a condition commonly referred to as 'pencil streak'. Tunnels may extend well into the wood and may be branched or unbranched, depending on the species of borer. Fully mature larvae usually make short side tunnels in which they pupate. New adults emerge by means of the parent tunnels. The length of the life cycle varies with species and with climatic conditions, and can range from 4 weeks to 12 months.

Species of *Platypus* are to be found throughout the world's tropics. Mostly, they are pests of logs and freshly sawn timber, but some attack damaged or unhealthy standing trees, and in a few cases, apparently healthy trees. In Sabah, Malaysia, widespread infestation by an unidentified species of *Platypus* occurred in provenance trials of *A. crassicarpa* aged 1–4 years (Thapa, 1991). Up to 80% of trees were infested, some with almost 300 holes in the lower stem. Despite this, there was no outward symptom of any deleterious effect on the trees. However, associated with the attack, a black stain developed in the sapwood region all along the bole length due to bacterial infection. Testing of the stained wood for suitability in papermaking showed that more bleaching agent than usual was required, but the quality of the paper was in no way affected. Chey (1996) records *P. pseudocupulatus* from the same host in Sabah and *P. solidus* attacking *E. grandis*. Similar attacks have been noted on eucalypts and acacias in several other countries in Southeast Asia (Wylie *et al.*, 1998).

In Fiji, mahogany trees, *Swietenia macrophylla*, are attacked by *P. gerstaeckeri*. Although the insects are unable to complete their life development in living trees, beetles being killed by gum exudation produced by the mahogany, the short galleries that they attempt to form are sufficient to reduce the quality of the timber when the trees are cut down (Roberts, 1977). Trees less than 5 years old are generally not attacked. In plantations, attack generally is related to some forest operation such as thinning, pruning, cleaning, or the removal of sample trees. Site is also important, and the incidence of attack is highest where drainage is bad and soils are poor.

P. hintzi, which is widely distributed in Africa south of the Sahara, may attack healthy trees during periods of temporarily decreased vigour, especially during the dry season (Browne, 1968). In Nigeria, such attacks have occurred on *C. equisetifolia* and *Eucalyptus* spp. and rarely result in successful breeding, but cause degradation of the timber. In South America, *Megaplatypus mutatus*, often referred to in the literature as *P. mutatus* or *P. sulcatus*, is a serious problem in commercial plantations of a number of broadleaf tree species. It is native to tropical and subtropical areas of South America and occurs in Argentina, Bolivia, Brazil, French Guiana, Paraguay, Peru, Uruguay and Venezuela. It attacks only living, standing trees and its hosts include *Acacia, Casuarina, Cedrela* and *Eucalyptus*, but it is particularly damaging to poplars, *Populus deltoides*, in Argentina (Alfaro *et al.*, 2007). Tunnelling by *M. mutatus* degrades the lumber and weakens the tree stems, which often then break during windstorms. Infestation by *M. mutatus* has been recorded in an experimental plantation of brazilwood, *Caesalpinia echinata*, in Brazil. Although the infestation was low level, it is nevertheless of concern because *C. echinata* is at risk of extinction due to exploitation and deforestation, and the wood of this species is highly valued for the manufacture of violin bows (Girardi *et al.*, 2006).

Platypodidae have numerous predators, both of adult beetles outside the nest and of brood within the tunnels.

Xylosandrus crassiusculus (Coleoptera: Curculionidae Scolytinae)

Xylosandrus crassiusculus, commonly referred to as the Asian ambrosia beetle or granulate ambrosia beetle, is a highly polyphagous pest of tree and shrub species, including many economically important horticultural crops and forest trees. Unlike other ambrosia beetles which normally attack only stressed, damaged or 'unthrifty' plants, *X. crassiusculus* is apparently able to attack healthy plants. Infested plants can show wilting, branch dieback, shoot breakage and general decline. Newly planted seedlings are often attacked at the root collar and the resulting girdling can stunt or kill the young tree (EPPO, 2010).

As its name implies, the Asian ambrosia beetle, *X. crassiusculus*, is considered to originate in tropical and subtropical Asia and to have been spread to equatorial Africa hundreds of years ago by early traders. More recently, it has been introduced into the Americas (detected in the USA in the 1970s and in Costa Rica and Panama in the 1990s) and has been reported from several countries in the Pacific, such as New Caledonia,

Papua New Guinea, Samoa, Fiji, Hawaii and Australia.

As described by Atkinson et al. (2005), females bore into twigs, branches or small trunks of susceptible woody plants, where they excavate a system of tunnels in the wood or pith, introduce an ectosymbiotic fungus (*Ambrosiella* sp.) and produce a brood. Like other ambrosia beetles, they feed on the introduced fungus and not on the wood and pith of their hosts. When boring galleries, frass is pushed out in the form of a compact, 'toothpick-like' cylinder, which may reach 3–4 cm in length before it breaks off (Plate 42). Eggs, larvae and pupae (Plate 43) are found together in the tunnel system excavated by the female and there are no individual egg niches, larval tunnels or pupal chambers. The insect breeds in host material from 2 to 30 cm in diameter, although small branches and stems are most commonly attacked. Attacks may occur on apparently healthy, stressed or freshly cut host material. Attacks on living plants usually are near ground level on saplings or at bark wounds on older trees. Females remain with their brood until maturity. Males are rare, reduced in size, flightless and presumably haploid. Females mate with their brother(s) before emerging to attack a new host (Atkinson et al., 2005). In the tropics, breeding is continuous throughout the year, with overlapping generations (EPPO, 2010).

Schedl (1962) listed 124 hosts of *X. crassiusculus*, mostly tropical, in 46 families. The insect has been observed to kill nursery seedlings of mahogany, *S. macrophylla*, in Fiji (ACIAR, 2010), and in Pakistan three important tree species, *T. grandis*, *D. sissoo* and *E. camaldulensis*, are listed as hosts (Khuhro et al., 2005). In Ghana, tree mortality due to this pest has been reported in *Aucoumea klaineana* and *Khaya ivorensis* plantations. In south India, *X. crassiusculus* has been implicated in the death of silver oak (*Grevillea robusta*), one of the most common shade trees grown in almost all the coffee zones (Sreedharan et al., 1991). It has caused death of saplings and seedlings of several hardwood species in the southern USA (Horn and Horn, 2006).

Sirex noctilio (Hymenoptera: Siricidae)

The sirex wood wasp, *Sirex noctilio* (Plate 44), is not considered an important pest in its native range in Eurasia and northern Africa, to the extent that many European texts on forest entomology do not discuss it (Ciesla, 2003). However, it is a significant pest of *Pinus* spp., where it has established in the southern hemisphere threatening approximately 8 Mha of pine plantations (Bedding and Iede, 2005) (Fig. 5.12). The female wasp oviposits eggs into stressed or suppressed trees, along with a phytotoxic mucus and a wood decay fungus (*A. areolatum*) carried by the wasps. Trees drilled by *S. noctilio* soon die, due to a combination of the mucus and the fungus. Carnegie et al. (2006) chronicle the spread of the pest and predict its potential distribution. It was first reported in the southern hemisphere in New Zealand in the early 1900s infesting *P. radiata* plantations and was detected in Tasmania in southern Australia in 1952, from whence it spread northwards in that country over the next 50 years to reach subtropical New South Wales and, very recently, Queensland. It was detected in Uruguay in South America in 1980, in north-eastern Argentina in 1985, southern Brazil in 1988 and Chile in 2001. It was found in the Western Cape of South Africa in 1994 and spread slowly east to KwaZulu-Natal, where it caused extensive damage to *P. patula* plantations. In 2005, an established population was found in the USA (Hoebeke et al., 2005). Carnegie et al. (2006) predicted its spread to southern China, parts of Central America and most of the countries along eastern, mid-western and northern Africa.

As described by Ciesla (2003), the female wasp inserts her ovipositor through the bark of the tree into the sapwood and lays eggs singly. Each female may lay between 20 and 500 eggs, which hatch in about 9 days. During oviposition, a symbiotic fungus, *A. areolatum*, is introduced along with a phytotoxic mucus. The mucus changes the water relations within the tree, creating conditions which are ideal for the growth and spread of the fungus. The fungus rots and dries the wood, providing a suitable

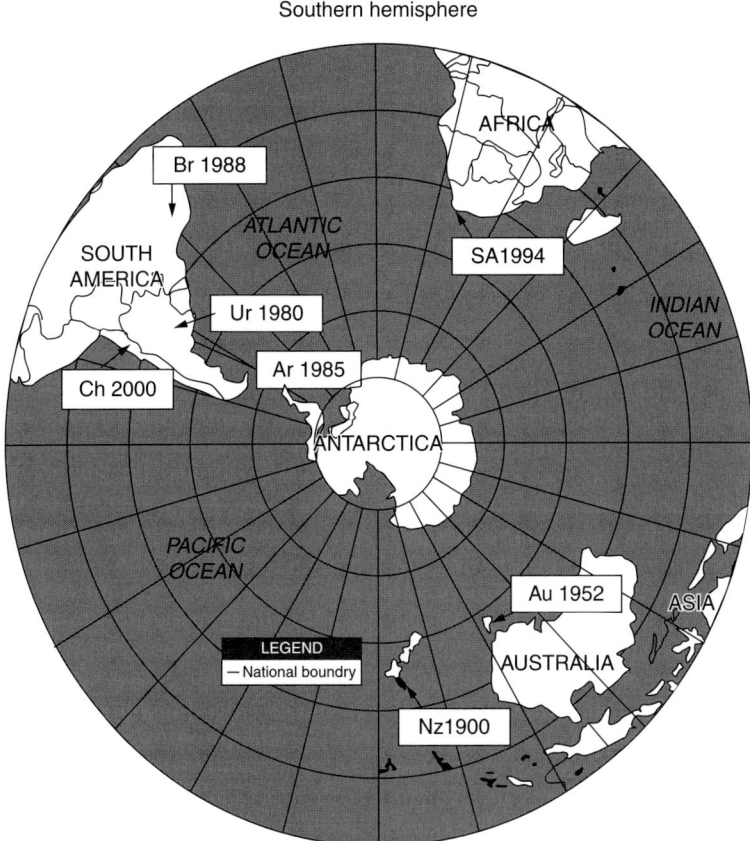

Fig. 5.12. Detection of *Sirex noctilio* in the southern hemisphere. Countries are indicated as: Ar = Argentina; Au = Australia; Br = Brazil; Ch = Chile; NZ = New Zealand; SA = South Africa; Ur = Uruguay. The numbers after the letters indicate the date that *S. noctilio* was first detected in those countries (from Hurley *et al.*, 2007).

environment, nutrients and enzymes for the developing insect larvae, which tunnel in the decayed wood (Plate 45). When feeding is complete, the larvae enter a prepupal stage, then a pupal stage. Pupation lasts 16–21 days and when the female emerges from the pupal skin, she takes up fungal spores and stores them in her abdomen. Adult sirex bore their way out of infested trees and leave a characteristic round exit hole. Males emerge before the females and may outnumber females by 20 to 1. The adults do not feed and live instead on stored fat. Adult lifespan can be 12 days, but a female that has deposited all her eggs may live just 3–4 days. Adults are strong fliers, capable of travelling several kilometres in search of suitable host trees. The dispersal rate in Australia is 30–40 km/year (Carnegie *et al.*, 2006) and in South Africa 48 km/year (Tribe and Cillie, 2004). Typically, suppressed trees are attacked first, but as sirex populations increase, they are capable of attacking and killing more vigorous trees. The first indication of infestation is the appearance of resin droplets and oviposition scars on the bark. Foliage wilts and turns from green to yellow to reddish brown. Perfectly round exit holes appear when the new generation of adult insects emerge. *S. noctilio* can complete a generation in as little as 10 months (Ciesla, 2003).

S. noctilio affects a wide range of Pinus species, P. radiata, P. taeda and P. patula being particularly susceptible (Carnegie et al., 2006). The impact of the pest, where it has established, has varied from minimal to devastating in some areas. In Australia, integrated pest management aims to restrict losses to less than 2% per annum, but there has been the occasional severe outbreak. One example is the outbreak which occurred in the Green Triangle area of southern Australia between 1987 and 1989, when more than 5 m trees with a royalty value of AUS$10–12 m were killed (Haugen et al., 1990). In South America, tree mortality has been over 60% in some stands in Argentina and as high as 70% in some stands in Uruguay (Hurley et al., 2007). In Brazil, 350,000 ha of pine plantations are infested and an estimated US$6.6 m would be lost each year if an integrated pest management programme were not in place (Bedding and Iede, 2005). In South Africa, infestation in some compartments has reached 35% but the overall mean is about 6%, with a total estimated value of damage being R300 m per annum (Hurley et al., 2007).

Integrated pest management for S. noctilio involves a combination of silvicultural measures and biological control. Susceptible plantations are generally 10–25 years old and unthinned stands are more susceptible than thinned stands. Trees under stress (e.g. drought conditions) or injured (e.g. by wind, fire or during logging) appear more likely to be attacked. Therefore, a major preventative measure is to increase stand vigour by thinning (Neumann et al., 1987). The parasitic nematode Beddingia (formerly Deladenus) siricidicola is the primary biocontrol agent for the pest. This nematode is able to breed in vast numbers throughout the tree while feeding on the fungus A. areolatum; then, it enters the wasp larva and begins reproduction when its host pupates. Nematode juveniles sterilize the adult female S. noctilio by entering all her eggs. When nematode-infected wasps emerge and attack other trees, they transmit packets of nematodes instead of fertile eggs. Infection levels can approach 100% and lead to a collapse in the pest population (Haugen et al., 1990). Several species of parasitic wasps have been released for biocontrol, of which the most successful have been Ibalia leucospoides, Megarhyssa nortoni and Rhyssa persuasoria. In combination, these species usually do not kill more than 40% of an S. noctilio population and are therefore not considered sufficient to control the pest on their own.

Hurley et al. (2007) review the success of control programmes in the southern hemisphere and conclude that the results have been variable. In New Zealand, S. noctilio is no longer considered a major threat and an active control programme is not considered necessary. In Australia, infestations are mostly below 1%, although an active control programme remains in place. The pest is still considered a major pest in South America, where biological control has been successful in some areas but less so in others. In South Africa, infestations remain low in the Western Cape but are above 30% in some areas of KwaZulu-Natal and the Eastern Cape, and they are increasing in these provinces. The performance of the nematode in summer rainfall areas of South Africa has been poor, for reasons currently unknown (Hurley et al., 2008).

Coptotermes spp. (Isoptera: Rhinotermitidae)

Coptotermes is a large genus with numerous representatives in the tropics and many species which are injurious to trees. Among the best known are C. formosanus and C. curvignathus (Plate 46) in Asia, C. elisae and C. obiratus in Papua New Guinea, C. acinaciformis in Australia, C. amanii and C. truncatus in Africa and C. niger and C. testaceus in Central and South America.

Like most subterranean termites, they are generally ground dwelling or require contact with the soil or some constant source of water. Nests may be in dead timber, logs and old tree stumps, mounds or in the trunks of living or dead trees. Nest material usually consists of a mixture of termite excrement and earth glued together with salivary secretions. Colonies can be very large, in excess of 1.25 m individuals (Krishna and Weesner, 1970). As with other termites, within each colony are several types or castes which are specialized to perform different tasks. The main three castes are workers (nest and gallery construction, nurturing, food gathering), soldiers (defence) and reproductives (primary king and queen and winged forms which disperse to establish new colonies).

Foraging from the nest takes place via underground tunnels or, sometimes, under covered runways of digested wood and soil particles built above ground. These galleries may be up to 50 m long and, in the case of one colony of *C. formosanus* in the southern USA, cover an area of more than 0.5 ha. Cellulose obtained from plant material is the basic food requirement and termites digest this with the aid of microorganisms in their gut.

Many *Coptotermes* species are destructive pests of timber in service, but some attack living trees as well, causing death of young plantings and hollowing out the centre of older trees (Plates 47 and 48). Some species gain entry through diseased or damaged roots, or through scars or wounds on the lower stem. With others, such as *C. curvignathus*, attack can be primary, independent of decay or wounds, and entry may be made through the roots or above ground. When a tree is attacked above ground level, its stem is encased in a thick crust of earth, the bark is eaten away and the termites penetrate to the heartwood, which is hollowed out and often filled with wood-carton combs (Browne, 1968). Infested trees are frequently wind-thrown.

C. elisae has caused considerable mortality among plantations of *A. cunninghamii* and *A. hunsteinii* in submontane areas of Papua New Guinea (Gray and Buchter, 1969). In some compartments, incidence of infestation was about 7% and nearly all attacked trees died. Groups of up to 20 dying trees were common, each group representing the foraging of one colony from a central nest. A novel method of control was employed in new plantation areas, colonies being destroyed by means of explosives. In established plantations, the primary queens were removed from the main termitarium by excavation. In lowland Papua New Guinea, *C. obiratus* frequently destroys the bark and sapwood of very young teak trees, causing death.

In Africa, *C. amanii* damages plantation trees in several countries, hollowing out pipes in mature trees and killing young trees. It has the habit of forming 'budded nests' by isolation of subsidiary nests, containing supplementary reproductives, from the original colony (Krishna and Weesner, 1970). *C. sjostedti* is a general pest of living trees throughout West Africa (Wagner *et al.*, 2008) and *C. truncatus* is considered to be the most destructive termite in the Seychelles, killing plantation trees such as *Corymbia (Eucalyptus) citriodora*.

The Formosan termite, *C. formosanus*, originally from Taiwan, is a 'tramp' species which has been widely spread by international trade. It now occurs not only in Japan and China but also on various Pacific Islands, including Hawaii, south-eastern USA, Sri Lanka and South Africa. It is able to construct and, if conditions are sufficiently damp, maintain aboveground nests in enclosed spaces without requiring ground connection. It is this characteristic that enables *C. formosanus* to occur in large, viable colonies on board ships or in containers. In Hawaii, it has been recorded from 48 host plants, and in southern China, it caused about 8% mortality in young plantations of *E. excerta* (Wylie and Brown, 1992). Also in southern China, an investigation showed that about 16% of 'ancient and valuable trees' was damaged by termites in Guangzhou and *Coptotermes* spp. were responsible for nearly 96% of that damage (Liu, 1997). In southern states of the USA, Formosan subterranean termites routinely attack standing live trees and can cause extensive damage, resulting in the weakening of the tree to the point of failure (Brown *et al.*, 2007). The rubber termite, *C. curvignathus*, is most notorious as a pest of *Hevea brasiliensis* (Pong, 1974), but attacks a wide range of other trees including *A. mangium* in plantations in Sumatra (Wylie *et al.*, 1998) and in Malaysia (Intachat and Kirton, 1997; Kirton and Cheng, 2007). Studies by Kirton *et al.* (1999) in *A. mangium* plantations in Peninsular Malaysia showed that while *C. curvignathus* was capable of primary attack, its entry into the wood was often facilitated by large pruning wounds, abscission scars resulting from natural pruning and damage by insect bark borers. In Malaysia, infestation of living trees appears to be most frequent on low-lying, moist sites (Browne, 1968). Flooding, however, curtails foraging activity (Sajap, 1999).

In Australia, *C. acinaciformis*, now believed to be a species complex, is responsible for greater economic losses, in the aggregate, than all the other species of Australian termites combined (Krishna and Weesner, 1970). This is due not only to its extensive range and to the severe nature of its attack, but also to its extraordinary success in adapting to urban conditions. It causes severe damage to forest trees in at least six genera, including 25 species of eucalypt, hollowing out large pipes in the stems. Various studies have shown that the majority of trees (66–89%) in the eucalypt savannahs of tropical northern Australia have hollow cores or pipes attributable to termite activity, mainly by *C. acinaciformis*, and the effect of such damage on tree growth and survival has been examined by Werner and Prior (2007). They found that growth and survival of eucalypts increased with tree diameter and decreased with pipe ratio (Fig. 5.13). Contrary to the suggestion that tree hollows are an adaptive trait whereby trees benefit by the release of nutrients, in north Australian eucalypt savannahs the net effect of termite piping on individual tree growth and survival was negative.

C. niger occurs in the Caribbean, Central America and northern South America, where it attacks forest trees such as *G. arborea*, *S. macrophylla* and especially *P. caribaea* (Krishna and Weesner, 1970). It forages freely on the bark of living trees and is thought to be a vector of a nematode-related disease of palms. The heartwood termite, *C. testaceus*, is a pest of *Eucalyptus* spp. in plantations in Brazil, attacking trees 2-years-old or older, destroying their inner portion and leaving the trees hollow (Junqueira et al., 2008).

Ants are the most important natural enemies of this group. When disturbed, soldiers defend themselves by exuding a milk-like latex from a pore, the fontanelle, on the front

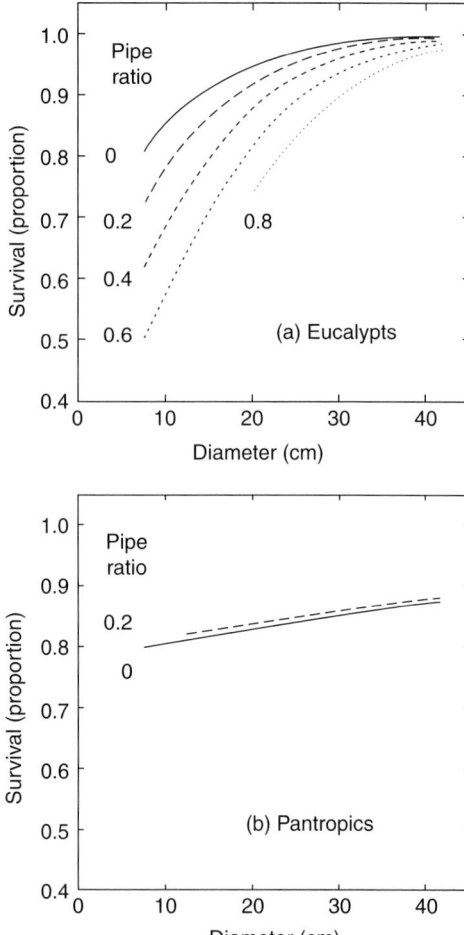

Fig. 5.13. Survival (proportion) of (a) eucalypts and of (b) pantropics in northern Australia savannah woodland following termite attack: modelled values of survival as a function of tree dbh, for a range of pipe ratios. Few pantropics had a pipe ration > 0.2 and there were none > 0.4. No eucalypts < 20 cm dbh had a pipe ratio > 0.6 (from Werner and Prior, 2007).

of the head and this compound is applied to the adversary when the soldier strikes.

Chilecomadia valdiviana (Philippi)
(Lepidoptera: Cossidae)

Chilecomadia valdiviana, variously known as the quince borer, carpenterworm or butterworm, is an emerging pest of *Eucalyptus* spp. plantations in Chile (Lanfranco and

Dungey, 2001). It is native to both Chile and Argentina and typically is associated with *Salix chilensis* and native forest tree species such as *Nothofagus* spp. As well, it infests commercial fruit trees and ornamental species such as *Ulmus glabra* (Angulo and Olivares, 1991). It is widely distributed in Chile, from Atacama in the dry northern tropics to Aisen in the most southerly region (Tierra del Fuego).

The insect infests live trees from 4 cm DBH and larger, with attacks occurring in all portions of the bole. Tree stress is not a prerequisite for attack. As described by Kliejunas et al. (2001), attacks on the host trees begin in the spring. The female lays eggs in groups of 30–50 at branch axils or in natural bark crevices. Each female is capable of laying up to 200 eggs. The newly hatched larvae feed gregariously beneath the bark near the point of oviposition. Their feeding produces a sap flow on the bark that is an ideal substrate for the development of sooty mould fungi. Trees with multiple attacks are easily recognized from a distance by the darker colour of the bole resulting from the sooty mould. Towards the end of summer, the larvae leave the phloem and begin boring deeply into the heartwood. At this stage, they feed individually, boring longitudinal galleries up to 27 cm in length and 1 cm in diameter. They produce large quantities of frass, which is expelled from the gallery and accumulates at the base of the tree. Larvae may grow to 50 mm in length and pupation takes place in the gallery. The adult female may have a wingspan of 48–60 mm and a body length of 30–40 mm; males are slightly smaller. The life cycle can take from 1 to 3 years to complete, depending on both the host species and climatic conditions (Lanfranco and Dungey, 2001). *C. valdiviana* often infests the same tree, suggesting a slow rate of spread (Boreham, 2006).

The insect first came to prominence in 1992 when it was detected infesting *E. nitens* in localized areas in the eighth and ninth regions of Chile, where the majority of the eucalypt plantations are situated. Since then, it has been found associated with *E. gunnii*, *E. camaldulensis* and *E. delegatensis* and there are concerns that the insect could create future serious wood quality problems for the industry.

C. valdiviana does not kill the tree but lowers wood quality through the formation of multiple galleries at all levels along the trunk. Fungi are known to colonize larval galleries, causing both staining and rot. This in turn can weaken the trees and cause stem breakage in strong winds. There are no foliar symptoms of larval presence.

Interestingly, tens of millions of larvae of *C. valdiviana* and another species of *Chilecomadia*, *C. morrie*, are sold as fishing bait and reptile food in the USA and Europe (Thomas, 1995; Iriarte et al., 1997).

Plant protection agencies in the USA and Australia rate *C. valdiviana* as a potential quarantine risk (Tkacz, 2001; Lawson, 2007).

Coryphodema tristis (Drury)
(Lepidoptera: Cossidae)

As with the Chilean carpenterworm, *C. valdiviana*, discussed above, the South African quince borer, *Coryphodema tristis*, is another example of a native insect developing a new and very damaging host association with *Eucalyptus* plantations. *C. tristis* is widely distributed throughout South Africa and is well known as an economically important pest on many fruit trees, including quince, grapevines, apples and sugar pears (Gebeyehu et al., 2005; Boreham, 2006). It also feeds on a wide range of native and exotic trees, including species in the families Ulmaceae, Vitaceae, Rosaceae, Scrophulariaceae, Myoporaceae, Malvaceae and Combretaceae, to which can now be added Myrtaceae.

As described by Gebeyehu et al. (2005), the moth lays its eggs (from 104 to 316 per female) on the bark of the main stem or branches, and soon after hatching the early instar larvae begin to feed on the cambium. As they grow, they tunnel into the sapwood and heartwood, making extensive galleries and pushing frass to the outside of the stems, which makes their presence easy to detect (Plates 49 and 50). Fully-grown larvae range in size from 2 to 4 cm, depending on the size of the stems on which they have fed. The insect takes 2 years to complete its life cycle, approximately 18 months of this being spent in the larval stage. Pupation occurs in the tree and pupae cut holes to the exterior, from which their cases are left protruding after moth emergence. The adults do not feed and live for about 6 days.

Serious damage caused by *C. tristis* on *E. nitens* was first noticed in 2004 in Mpumalanga province, which lies in the high altitude, summer rainfall area of the Highveld of South Africa. There are approximately 25,000 ha of *E. nitens* plantations established in this region. In September 2004, a survey conducted in about 3000 ha of these plantations in 95 compartments showed infestation levels of up to 77% in some compartments, with an area weighted mean infestation of 9.6% for the entire survey area (Boreham, 2006). Infested trees ranged in age from 8 to 13 years. Infestations occurred in both high and low productivity sites and were not restricted to stressed trees. No other *Eucalyptus* species in adjacent compartments was affected.

C. tristis does not cause tree mortality directly, but the open larval galleries provide potential entry points for stain and decay fungi and other pathogens (Boreham, 2006).

Xyleutes ceramica Walker
(Lepidoptera: Cossidae)

The beehole borer, *Xyleutes ceramica*, is best known as a major pest of teak, *T. grandis*, in Myanmar and Thailand but attacks several tree species, particularly *G. arborea*, in other parts of Asia.

According to Beeson (1941), the adult female is short-lived and lays up to 50,000 eggs attached in strings in the crevices of bark. These hatch in 10–20 days and the young larvae are dispersed on silken threads by the wind. A larva can survive without food for up to 6 days after hatching. When it alights on the stem of a suitable host, it seeks a crevice and spins a web, under the protection of which it bores through the bark and into the sapwood, the tunnel then curving upwards and becoming vertical in the heartwood. A tunnel may reach a length of 25 cm and a diameter of 25 mm. The larva (Plate 51) feeds not on the wood but on the callus tissue that develops on the wound around the entrance of the tunnel, or 'beehole'. A feeding chamber is excavated in the bark and outer sapwood and one or more circular holes are drilled to the outside through which frass is ejected. Larval development may take 1–2 years, after which the larva clears a free passage in its beehole, closes the exit hole with a circular disc of silk and debris, and retires to the upper end of the tunnel to pupate behind a loose wad of silk. After 2–3 weeks, the pupa wriggles down the tunnel, cuts through the disc covering the exit hole and protrudes itself halfway out of the hole to facilitate emergence of the moth (Plate 52).

On vigorously growing trees, the exit hole is soon occluded by callus. Within a few years, there is no external sign of the tunnel. Host trees may be attacked by *X. ceramica* throughout their life. When such a tree is felled in a thinning or at the end of the rotation, it contains the accumulated effects of attack by this insect. Attack does not retard growth appreciably or increase mortality, but seriously degrades the timber. An attack of one beehole per tree per year in any period is a very heavy incidence commercially (Beeson, 1941). The vertical distribution of beeholes in the trunk of an average tree varies progressively through its life. In young trees up to an age of 25, there is a definite preponderance of beeholes in the lower portion of the bole, but by age 60 the site of maximum beeholing is towards the top of the bole. This may be related to changes in the nutritive value of the bark as the tree ages.

No epidemics of *X. ceramica* have been recorded and population densities of adults are usually very low. Beeson (1941) stated that a population of 40 moths/ha would be regarded as a high incidence. In Peninsular Malaysia, incidence of borer attack in young plantations of *G. arborea* approached 50% at some localities (Sajap, 1989) and 10% in Indonesia (Suratmo, 1996). The insect has been recorded from *A. mangium* in the Philippines (Braza, 1993). Healthier trees have a higher incidence of attack than suppressed trees and this may be related to the superior quality and quantity of food for larvae in vigorously growing trees.

Ants and birds are important predators of the beehole borer and, according to Mathew (1987), the ichneumonid parasitoid *Nemeritis tectonae* provides effective control of this insect.

5.5 Shoot Boring

Shoot- or tip-boring insects cause the most damage when they attack the apical terminal or leader of the tree, which results in irregular stem growth or multiple branching when secondary terminals take over dominance. Trees which have been subjected to repeated attack by such insects may have a stunted, bushy appearance, or at least malformed or forked boles, and their value for timber production can be reduced greatly or eliminated (Berryman, 1986). Attack on the buds and terminals of very young seedlings will often kill them, or so stunt their growth that they are overtopped by undesirable plant species. Attack on saplings often results in crooked stems, but on pole-sized and mature trees the impact is usually inconsequential because height growth is well advanced and the bole form already established. Damage is usually most severe in plantations and in natural regeneration after heavy cutting, and is often associated with particular site and stand conditions that affect the vigour and exposure of young trees (Berryman, 1986).

Hylurdrectonus araucariae Schedl
(Coleoptera: Scolytidae)

This scolytid occurs only in Papua New Guinea and has just a single host, *A. cunninghamii*, commonly known as hoop pine. Whereas most bark beetles are cambial borers, *Hylurdrectonus araucariae* is unusual in that it is a branchlet miner (Gray and Lamb, 1975). The damage caused by this insect to the country's main plantations of hoop pine at Bulolo and Wau resulted in the abandonment of planting of this species there in the late 1960s.

Hoop pine branchlets consist of sharply pointed leaves or needles arranged along an axis. The nest of *H. araucariae* invariably is initiated by the female (Gray, 1974), which bores into the needles either close to their junction with the axis or directly into the axis. She is then joined by the male, and occasionally an extra male or another pair. Up to four nests may be established in one branchlet. The beetles lay their eggs inside the excavated needles, usually at the base, and these hatch in 5–15 days. The larvae and adults mine into the needles and along the branchlet shaft, consuming most of the tissue except for the epidermis and pith (Plate 53). The length of the excavation is proportional to the age of the nest, and after 100 days may extend for several centimetres. The distal portion of the infested branchlet behind the area of active excavation turns brown, dies and eventually falls off. After 1–3 years of severe attack, most branches have been infested repeatedly and there is nearly complete defoliation of all branchlets on many branches (Gray and Lamb, 1975). The larval stage lasts 10–25 days, the pupal stage 10–15 days and the immature adult possibly 2–10 days to mature and establish a new colony. Nests may contain up to 75 individuals in all stages of development and the maximum number of adults recorded in a nest is 25. The adult may live 60 days or more. The life cycle takes 5–9 weeks and there are 5–10 generations per annum.

Attack by *H. araucariae* on hoop pine is primary, nearly all trees in an outbreak area being infested regardless of condition. However, there is a definite age effect, trees aged 2.5–12 years being the most susceptible (Gray, 1976). The insect is comparatively rare in natural stands of hoop pine but has infested approximately 47.5% of major plantations at Bulolo and 91% at Wau (Gray, 1975). This circumstance was attributed to the fact that most of the plantations were of an age class susceptible to attack and that trees of similar age in natural stands had different physiochemical characteristics of foliage to those of plantation-grown hoop pine. Considerable growth loss and high tree mortality has been recorded in severely infested stands, particularly on poor sites. Most mortality was due to secondary insects, such as the weevil *Vanapa oberthuri*, which attacked the weakened trees. As a result of

H. araucariae damage to hoop pine at Bulolo and Wau, the planting emphasis there switched to *A. hunsteinii*, which is resistant to attack by the beetle.

Apart from spiders, which prey on the adults when they leave the nest, no other natural enemies of *H. araucariae* have been found (Gray and Lamb, 1975).

Dioryctria spp. (Lepidoptera: Pyralidae)

Until comparatively recently, *Dioryctria* spp. (Plate 54) were best known as cone or shoot borers of conifers in Europe and North America. However, with the rapid expansion of plantations of *Pinus* spp. in Asia during the past few decades, mainly destined to provide long fibre pulp, there has been an increase in damage caused by several pests and diseases, including shoot moths in the genus *Dioryctria*. Severe damage to shoots and bark of young *Pinus* spp. has occurred in the Philippines, Thailand, Vietnam, Indonesia, India, Pakistan, Taiwan, southern China and Cuba (Wang *et al.*, 1999; Nair, 2007). As well, species of *Dioryctria* have been reported to attack cones of *Pinus* spp. in the Central American countries of Honduras and Nicaragua (Becker, 1973), and in Florida *D. amatella*, commonly referred to as a pitch moth, is a serious pest of pine seed orchard crops in that state (Meeker, 2008).

As described by Speight and Speechly (1982), eggs are laid around the bases of needles of young shoots. The hatching larvae feed externally at first on the needle bases, spinning small silken tents which become covered in resin and frass. Some mining of the needles may also occur. The larvae may remain on the outside of the shoots for a week or more before boring in, and may reappear on the surface from time to time as they mature. Usually, larvae bore upwards at first towards the shoot tip and later down towards the basal part, the tunnelling extending for up to 30 cm (Plate 55). Pupation occurs in the shoot near the old entrance hole. The length of the life cycle varies with climatic conditions and may take from 2 to 4 months.

Attack results in dieback of leading and lateral shoots with a browning of needles and tips, which break off easily to reveal the hollowed-out shoot. By this time, the larvae usually have already pupated and the adult moths emerged. In moderate infestations, the terminal shoot dies but laterals take over to produce a 'stag-headed' effect, with a resultant loss in form. In more serious cases where the majority of shoots are attacked, the trees become stunted and bush-like (Plate 56). Increment is reduced greatly and the trees become valueless for timber. Larvae of *Dioryctria* spp. sometimes also bore in the bark cambium and can girdle and kill young trees.

In Thailand, especially in lowland areas, several species of *Pinus*, principally *P. kesiya*, are attacked by *D. sylvestrella* and *D. abietella* (Hutacharern, 1978; Speight and Speechly, 1982). In Vietnam, *D. sylvestrella* attacks *P. kesiya* and *P. caribaea*, while in the Philippines heavy attacks on *P. caribaea* by *D. rubella* affect the viability of the plantation programme there (Lapis, 1985b). *D. castanea* is a major pest of *P. kesiya* in India and during one outbreak in a 900 ha plantation in the north-east of the country, all age groups were attacked and every tree affected (Singh *et al.*, 1988). The most important of the *Dioryctria* spp. in Taiwan is *D. pryeri* (Yie *et al.*, 1967), and in Pakistan it is *D. abietella* (Ahmad *et al.*, 1977). Also in Pakistan, Ghani and Cheema (1973) reported severe damage by *D. raoi* to *P. roxburghii* trees in Kashmir, with up to 55% of shoots affected on some trees, resulting in stunting and malformation. Of the tropical pines planted in southern China, *P. caribaea* var. *hondurensis* is the species affected most seriously by *D. alternatus*, whose damage to the apex and main stem is one of the principal contributors to the generally poor stem form of this tree species in the region (Wang *et al.*, 1999). In Yunnan Province in the south-west of China, *D. rubella* causes severe damage to *P. kesiya* var. *langbianensis*. In the rainy season, the damage rate is 35–40% and in the dry season 90–100% (Tong and Kong, 2010). One of the reasons given for such a high rate of infestation is the ability of the larva to attack more than one shoot. Altitude seems important in determining the severity of shoot moth attack; at altitudes over 1000 m, damage is of little significance but may be severe in plantings below this level (Speight, 1983). Other associated factors are uniformly high all-year-round temperatures, soils low in nutrients (especially phosphorus) and the presence of older trees that are already infested.

Numerous wasp parasitoids of the egg, larval and pupal stages of several *Dioryctria* species in the tropics have been recorded (Ghani and Cheema, 1973; Belmont and Habeck, 1983; Thakur, 2000; Nair, 2007).

Hypsipyla robusta (Moore), *Hypsipyla grandella* (Zeller) (Lepidoptera: Pyralidae)

Hypsipyla spp. shoot borers are among the most economically important insect pests in tropical forestry, virtually preventing the cultivation of mahoganies (*Swietenia* spp., *Khaya* spp.), cedars (*Cedrella* spp., *Toona* spp.) and other valuable Meliaceae, primarily of the subfamily Swietenioideae, in their native areas (Griffiths, 2001; Opuni-Frimpong *et al.*, 2008b; Wagner *et al.*, 2008). *H. grandella* is found throughout Central and South America and also occurs on many Caribbean islands and the southern tip of Florida. The species referred to as '*H. robusta*' is widely distributed throughout West and East Africa, India, Indonesia, Australia and South-east Asia (Table 5.4). It has now been shown that *H. robusta* is not one but at least three different species (Horak, 2001; Cunningham *et al.*, 2005), which accounts for some of the differences apparent in the reported biology and behaviour of the species in various parts of its range.

Newton *et al.* (1993) have summarized the biology, ecology and importance of these pests. The two species appear to behave similarly, the total life cycle lasting about 1–2 months, depending on climate and food availability. In *H. grandella*, oviposition occurs during evening or early morning and egg eclosion occurs at night. An individual female lays 1–7 eggs at a time on one or more plants and may repeat oviposition over a period of 6 days, laying 200–300 eggs in all. In comparison, a *H. robusta* female may lay 450 eggs during a 7- to 10-day period (Griffiths, 2001). Eggs are laid singly, or occasionally in clusters of three to four, usually on shoots, stems and leaves, often in concealed positions such as leaf axils, scars and fissures. Eggs hatch in about 3 days and the larvae move towards the new shoots, burrowing into the stem or leaf axil. The larvae (Plate 57) cover their entrance holes with a protective web containing plant particles and frass (Plate 58) and tunnel in the primary stem or branches, feeding on the pith. There are five to six larval stages lasting about 30 days. Larvae pupate in cocoons spun in the stem tunnels, or among the leaf litter and soil around the tree base. Pupation lasts about 10 days.

Larvae are also known to feed on the bark, fruit and flowers of their hosts. In Australia, larvae of *H. robusta* feeding on *Toona ciliata* fruit initially feed externally on the epidermis and later live within, and consume, the seeds and soft tissue of a single fruit before emerging and entering a new fruit (Griffiths, 2001). Neighbouring fruit are joined by a tunnel of silk and frass through which the larvae move. Feeding damage to fruit results in their premature shedding and larvae apparently circumvent this by spinning a mat of webbing across the point of abscission (Griffiths, 1997). Fully fed larvae generally exit the fruit to pupate beneath the bark of the mature trees close to the base, or among surrounding soil and leaf litter, but sometimes pupate in the hollowed-out fruit. In Nigeria, Roberts (1968) reports that while it is usual to find only one larva in each shoot, fruits of *Carapa procera* have yielded up to 26 larvae, and more than one larva is not uncommon in fruits of *Khaya* spp.

The adults are nocturnal, are strong fliers and able to locate their food plants over large distances by means of chemoreception. The exact nature of the attractants emitted by the plant is unknown.

The number of generations a year varies with variations in climatic conditions and availability of new shoots. In areas which are wet all year round, the insects are able to attack continuously as the young trees resprout repeatedly (Newton *et al.*, 1993). In areas with a pronounced dry season, attacks switch from shoots to fruit during the dry period when no unlignified shoots are available.

Table 5.4. *Hypsipyla robusta* damage on species of Meliaceae (subfamily Swietenioideae) grown in various Asian and Pacific countries (from Floyd and Hauxwell, 2000).

Tree species	Bangladesh	Sri Lanka	India	Philippines	Vietnam	Laos	Thailand	Malaysia	Indonesia	Papua New Guinea	Solomon Islands	Australia
Cedrela odorata		✓					✓	✓			x	✓
Cedrela lilloi							✓					✓
Chukrasia tabularis	✓			✓			✓✓	✓				✓✓
Khaya anthotheca								✓				
Khaya grandifolia								✓				
Khaya ivorensis								✓				
Khaya nyasica												
Khaya senegalensis		✓			✓							✓
Swietenia macrophylla	✓	✓	✓	✓	✓	✓✓	✓✓	x	✓	✓✓	✓✓	✓
Swietenia mahagoni	✓	✓	✓				✓✓	✓✓	✓	✓		
Toona ciliata	✓		✓			✓	✓✓		✓	✓		✓✓✓
Toona calantas				x				✓				
Toona sinensis								✓				
Toona sureni					✓		✓	✓✓	✓	✓		
Xylocarpus moluccensis						✓✓						

Notes: Empty cells in the table indicate that there are no records of a tree species in a country. The number of ticks indicates the severity of damage and a cross indicates no damage observed. Countries with a single tick against all species recorded for that country indicates damage has been observed but no indication of relative severity of damage.

Tunnelling by the larvae in the shoots causes shoot mortality, growth reduction, branching and poor tree form (Plate 59). Repeated attacks can result in tree death. Trees may be attacked from the nursery stage through to maturity, but attacks up to the pole stage are most critical from a silvicultural point of view. Newton et al. (1993) cite many examples of damage caused by *Hypsipyla* spp. throughout the tropics, with up to 100% of plantings affected in some cases. Plantation programmes using species of Meliaceae (subfamily Swietenioideae) have been abandoned almost completely in many countries. Despite such pest pressure, there are instances where susceptible species of Meliaceae have been grown with minimal shoot borer damage. This often involved the growing of the desired species in mixtures with other tree species. The reasons for the occasional success are not clear but have been variously suggested to relate to the effects of shade (overhead and lateral), planting density and growth rate, or interactions between these factors. Published viewpoints on this subject are often conflicting.

Many species of parasitoids, principally braconid and ichneumonid wasps, and predators of *Hypsipyla* spp. have been recorded, but these do not provide effective control either in their native country or where they have been introduced to other countries in classical biocontrol programmes (Newton et al., 1993). In Australia and Malaysia, investigations into the use of weaver ants, *O. smaragdina*, as a biological control agent for *H. robusta* show some promise (Lim et al., 2008).

5.6 Fruit and Seed Boring

Several different groups of insects attack the fruit, cones or seeds of forest trees. Some, such as torymid wasps, lay their eggs directly into the seed of young cones and their larvae feed within the seed; others, such as bruchid beetles, lay their eggs on the exterior of the pod or fruit and the larvae tunnel inwards to the seed, which they consume. Insect tunnelling may sometimes result in abortion of cones or fruit before the seeds have developed, or cause seed abortion (Berryman, 1986).

In plantations established for wood production, and even in natural forests, insects that feed on seeds or cones are relatively unimportant, but in seed orchards such damage is more serious and losses can be considerable (Speight and Wainhouse, 1989). In years when the cone crop is small and populations of cone and seed insects are high, the entire crop may be destroyed. Because of the expense in managing seed production areas, even a 5 or 10% loss may be economically intolerable (Berryman, 1986). Problems may also arise when natural seeding is required following logging operations, for example, where only a few select trees are left to provide seed and the insect population is high.

Bruchidius spp. (Coleoptera: Bruchidae)

Species of *Bruchidius*, commonly known as seed weevils, are numerous in the tropics, where they are pests of the seed of various plants, including some important leguminous tree species in the genera *Acacia*, *Albizia*, *Falcataria* and *Prosopis*.

According to Beeson (1941), the habits of all species are very similar. The eggs are laid normally on the skin of the pod or fruit to which they are attached by an adhesive fluid. The larva hatches in a few days and burrows directly into the fruit to the seed on which it feeds. One larva may hollow out several seeds during its development. Pupation occurs in the seed. The length of the life cycle varies considerably according to temperature, humidity and the quality of the food available. In the tropics, development may be completed in a few weeks and there may be numerous overlapping generations during a year. Because of the specificity of legume allelochemical substances, the larvae of a particular bruchid species may feed on the seed of just a single host species (Singh and Bhandari, 1988).

Leguminous trees are important in social forestry, particularly in arid lands where there is a demand for large quantities of sound seed which can be collected locally.

Destruction of the seed crop by bruchids can therefore have serious consequences for rural communities in these areas (Singh and Bhandari, 1988). *A. tortilis* has been suggested to function as a keystone species in the arid and semi-arid regions of Africa and the Middle East but sometimes suffers from extraordinarily high infestation of seed beetles (Noumi *et al.*, 2010a). Tunisia (Noumi *et al.*, 2010b) recorded from 31 to 83% infestation of *A. tortilis* seed by *B. raddianae* and *B. aurivillii*. In Botswana, the degree of seed parasitism of *A. tortilis* by several *Bruchidius* spp., principally *B. albosparsus*, *B. aurivillii* and *B. rubicundus*, ranged from 10 to 82% between trees and years (Ernst *et al.*, 1989). Infestation rates of almost 100% have been recorded for *B. spadiceus* in Tanzania (Lamprey *et al.*, 1974).

One of the reasons for the success of *Acacia* spp. is that they produce substantial quantities of seeds in order to overcome a range of environmental uncertainties, including the high rates of infestation by bruchid beetles (Sabiiti and Wein, 1987). Fire and grazing by large herbivores also contribute to the survival of these plants, not only by stimulating seed germination (fire, animals) and aiding dispersal (animals) but also by reducing populations of the seed beetles. Bruchid larvae are more sensitive to fire than the seed embryo, and this heat treatment leads indirectly to higher germination. Studies by Sabiiti and Wein (1987) in Uganda suggest that as fire intensities increase there will be fewer *Bruchidius* spp. larvae and more viable seed embryos. Lamprey *et al.* (1974) in Tanzania showed that fluids in the gut of animals that grazed and ingested seeds or pods of *A. tortilis* killed a higher proportion of bruchid larvae in the seed than seed embryos, and this contributed to higher rates of germination.

In Australia, *B. sahlbergi* was introduced from Pakistan as one of a complex of biocontrol agents for the prickly shrub *A. nilotica*, a declared noxious plant in arid sheep and cattle grazing regions in the tropical north of the country (Wilson, 1985).

Numerous hymenopteran parasitoids of *Bruchidius* spp. have been reported, but generally their impact on populations of these pests is small (Ernst *et al.*, 1989).

Megastigmus spp.
(Hymenoptera: Torymidae)

Megastigmus is an important genus of seed pests and these wasps are particularly destructive of conifer seed in Europe and North America. Worldwide, there are 51 species of *Megastigmus* which damage the seeds of conifers (Roques *et al.*, 2003) and several species which attack the seeds of eucalypts. The biology and impact of tropical species is less well known, but damage has been reported from southern China, Mexico, Thailand, Africa and Australia.

As described by Drake (1974) for *Megastigmus* on eucalypts in Australia, the insects lay their eggs in the ovules of young seeds during or just after the flowering period. After hatching, they feed off the developing embryos, often leaving the seed coat intact. Each insect consumes only the seed within which it has been deposited and therefore not all the seeds in an infested capsule are necessarily destroyed. The level of attack varies considerably, but between 20 and 50% of eucalypt capsules in localized areas can be destroyed (Drake, 1974).

In southern Mexico, *M. albifrons* damages seed of three species of *Pinus*, affecting up to 27% of seed (Rio Mora and Mayo-Jimenez, 1993). The wasp has an annual generation, but a small part of the population stays in diapause for 2 years, which is probably an adaptation to the irregular periodicity of seed production (Browne, 1968). In Thailand, seeds of *Sesbania grandiflora* are commonly infested with larvae of *Megastigmus*, infestation levels reaching 70% (Helium and Sullivan, 1990). Recently, a new species of *Megastigmus*, *M. zebrinus*, has been described from South Africa, associated with seed capsules of *E. camaldulensis*, an endemic Australian tree. The wasp presumably was introduced along with its host tree. It appears to be a true gall-maker, rather than a parasitoid of gall-forming insects, and now has adapted to the fruit of

S. cordatum, an endemic South African myrtaceous tree (Grissell, 2006).

5.7 Gall Forming

Galls are unusual plant growths which develop as a result of abnormal cell division and/or cell enlargements following infestation of plants by organisms such as insects, mites and fungi (Elliott *et al.*, 1998). They can occur on all plant organs, but are found most commonly on leaves, stems and buds. Insect-induced galls can be caused by species in several Orders, but the most important groups are Hymenoptera, Hemiptera and Diptera. Galls provide food and shelter for the invading insects, but the plant derives no benefit in this relationship, sustaining loss of nutrients, changes in growth architecture and structural weakening. The shape of a gall is often characteristic of a particular insect species, and in some cases there is marked sexual dimorphism, with the males producing a gall of a different shape from that produced by the females.

Galling rarely results in the death of the host tree, but severe dieback of attacked parts can occur (Cobbinah, 1986) and branches may break under the weight of galls (Currie, 1937). Photosynthetic capacity may be reduced where leaves are heavily galled and distorted (Elliott *et al.*, 1998). Some gall-making insects can have a serious impact on seed production. For example, the pteromalid gall wasp, *Trichilogaster acaciaelongifoliae*, is used as a biocontrol agent for *A. longifolia*, which has become a weed in South Africa, and on some sites has reduced seed production by 95–99% (Dennill, 1985).

Fergusonina spp. (Diptera: Agromyzidae)

Fergusonina spp. are pests of myrtaceous trees and are associated in their galls with nematodes in the genus *Fergusobia*. This is the only known mutualistic association between insects and nematodes (Davies and Giblin-Davis, 2004). Six genera in the family Myrtaceae, viz. *Corymbia*, *Eucalyptus*, *Angophora*, *Syzygium*, *Melaleuca* and *Metrosideros*, are known to be hosts of this association, with most records being from *Eucalyptus* and *Melaleuca* (Taylor and Davies, 2008). The galls may appear on stem tips, leaf buds, leaves and flower buds and are a result of the combined effect of the nematodes and fly larvae (Currie, 1937). While several species have been recorded from Asia and Papua New Guinea (Davies and Giblin-Davis (2004), they are most numerous in Australia, where they are commonly known as eucalypt flies.

Detailed studies of these flies in Australia have been carried out by Currie (1937) and are summarized by Elliott *et al.* (1998). In species which infest eucalypt flowers, adult flies emerge from galls in the summer. Following mating, females fly to flowers and lay eggs in the young flower buds. From 1–50 larval nematodes are laid with each egg. Many fly eggs may be laid in the same eucalypt flower bud by a single fly or several different flies. While the fly egg is developing, the immature nematodes feed on the primordia of the stamens, causing rapid proliferation of cells, which form irregular masses of cells inside the now galled bud. On hatching, the fly larvae form cavities between two contiguous cell masses and develop rapidly to the pupal stage by feeding initially on cell sap and later on ruptured plant cells. The nematodes join the fly larvae in these feeding cavities and eventually undergo parthenogenetic reproduction, with the female nematodes laying eggs beside the fly larvae. The nematodes do not harm the fly larvae.

Male nematodes appear in the galls in the autumn and winter. When the female fly larvae are about to pupate, fertilized female nematodes enter these larvae. During the fly pupal stage, the nematodes change from free-living forms to much enlarged parasitic forms. By the time the female fly emerges, the parasitic nematodes are discharging eggs inside the fly body cavity. On hatching, the larval nematodes migrate to the ovaries, penetrate the oviduct and wait there until an egg passes down the oviduct. The nematodes then accompany the eggs into the flower bud where the fly–nematode cycle begins again (Fig. 5.14).

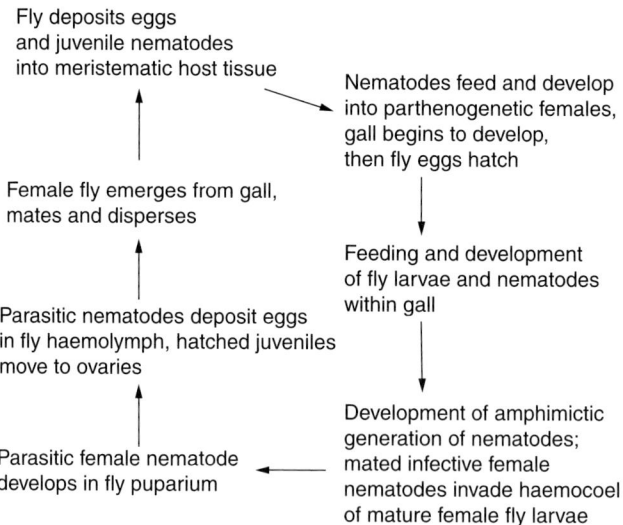

Fig. 5.14. Life cycles of the *Fergusonina–Fergusobia* association (from Davies and Giblin-Davis, 2004).

The frequency of galling of flower buds of some eucalypts by *Fergusonina* varies markedly from year to year and can be low even in the presence of abundant flower buds. Currie (1937) records that galling reduced seed production in some eucalypts and that whole branches might break under the weight of galls. A species of *Fergusonina* is a pest of *E. deglupta* in the Philippines (Braza, 1991) and *F. syzygii* galls flower buds of *S. cumini* in India (Siddiqi *et al.*, 1986). *Fergusonina* is being considered as a potential biocontrol agent for the broad-leaved paperbark tree, *M. quinquenervia*, which was introduced from Australia into Florida in the USA in the early 1900s and has since become a serious weed. It infests over 200,000 ha in the Everglades, causing extensive environmental damage (Goolsby *et al.*, 2001). The high degree of host specificity of the *Fergusonina–Fergusobia* association makes it an ideal candidate for biological control (Goolsby *et al.*, 2000).

Chalcid and braconid wasp parasitoids are the main natural enemies of *Fergusonina* spp. and are thought to exert some measure of control (Currie, 1937). Goolsby *et al.* (2000) reared 11 species of wasps from *Fergusonina* galls on *M. quinquenervia* and recorded a parasitism level of the fly larvae and pupae of greater than 60%. The two most common parasitoid species they reared were *Eurytoma* sp. and *Coelocyba* sp.

Phytolyma spp. (Hemiptera: Psyllidae)

Milicia [*Chlorophora*] is a tropical African genus of forest trees of considerable economic importance because of its natural durability and good working properties (Ofori and Cobbinah, 2007). It occurs naturally in the forest belt of West Africa, extending from Gambia to Nigeria and also in Central and East Africa. Attempts to cultivate *Milicia* spp. on a large scale have been hampered by *Phytolyma* psyllids, particularly *P. lata*, commonly known as the Iroko gallfly, which forms galls on the foliage (Nichols *et al.*, 1999; Wagner *et al.*, 2008).

Eggs are laid in rows, or scattered singly, on the buds, shoots or leaves of the host and hatch in about 5–25 days. The larva burrows into the adjacent tissues, breaking down the epidermal cells, which causes fermentation of the parenchyma. A gall is formed within 1 or 2 days, which encloses the nymph completely. The insect feeds within the gall tissue and there are five nymphal instars lasting approximately 2–3 weeks (Wagner *et al.*, 2008). The gall eventually splits open, usually at the point of original infection, to release the adult. Occasionally, the gall hardens without opening and the trapped insect dies. The total life cycle is completed in 22–45 days.

Phytolyma spp. galls disrupt the plant's translocation processes, and when they erupt to release the adults, death of the leading and lateral shoots can result (Atuahene, 1972; Cobbinah and Wagner, 1995). Terminal dieback causes growth reduction and seedling mortality in many cases. Agyeman *et al.* (2009) studied the impact of *P. lata* on seedling growth of *M. excelsa* in Ghana. They found that infested plants had lower height, stem diameter and biomass growth than uninfested plants. Mean yield losses of stem, branches and leaves of infested plants were 68.9%, 48.3% and 64%, respectively, of matched uninfested plants. Infested plants also had smaller, fewer and highly chlorotic leaves. The insects have a preference for young leaves, which not only ensures that good-quality food is available but also that they are able to complete development before leaf fall (Cobbinah, 1986). Physical properties such as cuticle thickness may enhance the preference for young leaves (Wagner *et al.*, 2008).

Phytolyma spp. attack is more injurious in nurseries and vigorous young plantations than in natural forests and 100% failures have been reported in Ghana (Wagner *et al.*, 2008). Studies by Bosu *et al.* (2006) of survival and growth of mixed plantations of *M. excelsa* and *T. superba* 9 years after planting in Ghana indicated that shade from *T. superba* reduced psyllid galls on *M. excelsa*, though crop growth was slow.

A species of *Trichogramma* parasitizes eggs of *P. lata*, while encyrtid and eulophid wasps parasitize the nymphs. Mantids predate first instar crawlers and, to a lesser degree, exposed adults.

Leptocybe invasa (Hymenoptera: Eulophidae)

The blue gum chalcid, *Leptocybe invasa* (Plate 60), is a new pest of *Eucalyptus* that was found in the Middle East in 2000 and has since spread rapidly to most Mediterranean countries, to many countries in Africa and, more recently, to Asia and North and South America. It is native to Australia, where it is unknown as a pest, but poses a serious threat to young *Eucalyptus* plantations, where it has become established in other countries.

The chronology of the spread of blue gum chalcid, as determined from the literature and from information provided by researchers on the pest, is summarized below (Mendel *et al.*, 2004; Doganlar, 2005; Hesami *et al.*, 2005; CABI/EPPO, 2007a; FABI, 2007; Costa *et al.*, 2008; Tang *et al.*, 2008; Wiley and Skelley, 2008; Jhala *et al.*, 2009; Nyeko *et al.*, 2009; Thu *et al.*, 2009; Botto *et al.*, 2010; Javaregowda and Prabhu, 2010). In some of the earlier records from the Mediterranean region, *L. invasa* was named erroneously as *Aprostocetus* sp. (Protasov *et al.*, 2008).

2000: Israel, Iran, Algeria, Italy
2001: India (Karnataka), Morocco, Egypt, Turkey, Jordan, Syria
2002: Uganda, Kenya, Ethiopia, Tamil Nadu (India), Vietnam
2004: Corsica (France), Spain, Greece
2005: Mainland France, Tanzania, Portugal
2006: Gujarat (India), Thailand, Southern England
2007: Zimbabwe, Guangxi (China), Brazil, South Africa
2008: Florida, Lao PDR, Hainan (China)
2009: Argentina
2010: Chile

The rapidity of spread of *L. invasa* around the world (29 countries in a period of 10 years) is akin to that of the leucaena psyllid *H. cubana*, which spread to 45 countries over a similar timespan (see earlier in this chapter). This spread has been attributed to movement of nursery stock (plants for planting) or to the cut flower trade.

L. invasa causes galls on the midribs, petioles and stems of new shoots of eucalypt trees, including coppice and nursery stock. Heavy infestations can lead to deformed leaves and shoots, a growth reduction of the tree, dieback and, in some cases, tree death.

Adult females insert their eggs in the epidermis of young leaves on both sides of the midrib, in the petioles and in the parenchyma of twigs. Mendel et al. (2004) identified five stages of gall development on E. camaldulensis. The first stage begins 1–2 weeks after oviposition, with the first symptoms of cork tissue appearing at the egg insertion spot, accompanied by a change in midrib colour from green to pink. Towards the end of the stage, the galls are spherical and glossy green. In the second stage, the galls develop their typical bump shape and reach their maximum size (about 2.7 mm). In the third stage, the green colour of the gall changes to glossy pink and in the fourth stage, the gall loses its glossiness and changes to light or dark red. In the final stage, emergence holes of the wasp are noticeable (Plate 61). In Israel, the development time from oviposition to emergence was 4.5 months. Adult wasps may live 3–6 days and may feed on the flowers of their host plant. In the Middle East, two to three overlapping generations per year have been observed. A wide range of Eucalyptus species are hosts to this pest, including: E. botryoides, E. bridesiana, E. camaldulensis, E. dunnii, E. globulus, E. grandis, E. gunnii, E. maidenii, E. robusta, E. saligna, E. tereticornis, E. urophylla, E. viminalis and various clones and hybrids.

L. invasa attacks trees of all ages, from nursery stock to mature tree, but the damage is most severe on younger plants. Mendel et al. (2004) reported that in the Bet Shean Valley in Israel, where L. invasa had reached epidemic levels, juvenile shoots were often killed due to egg overloading. In southern states of India, more than 20,000 ha of 2-year-old eucalypt plantations had already been affected by gall formation within 5 years of first discovery of the insect there (Javaregowda and Prabhu, 2010). In Vietnam, where the plantation industry is reliant on only a few clones of E. urophylla and E. camaldulensis, it reportedly has devastated nurseries and young plantations (Thu et al., 2009). The wasp has been similarly problematic in E. camaldulensis plantations and nurseries in Thailand (Dell et al., 2008). Surveys by Nyeko et al. (2009) in Uganda across 154 stands of young (< 3-year-old) Eucalyptus spp. and clonal hybrids showed that infestation by blue gum chalcid was most severe on E. grandis and E. camaldulensis, with most of the trees having galls in more than 50% of total shoots. Evaluation of growth losses caused by L. invasa on an E. grandis × E. urophylla clone in Guangdong, China, showed leaf loss of greater than 90% and average loss of tree height and tree diameter of 29.2% and 30.8%, respectively (Zhao et al., 2008).

Two species of eulophid wasps from Australia, *Quadrastichus mendeli* and *Selitrichodes kryceri*, which are larval parasitoids of L. invasa, have been introduced into Israel as part of a biological control programme for the pest (Kim et al., 2008). Two local species of Megastigmus have been found to be parasitoids of L. invasa in Turkey and Israel (Protasov et al., 2008) and another in Italy (Viggiani et al., 2001), but are not considered to be natural associates of L. invasa (Protasov et al., 2008). The strategies being adopted by most affected countries for dealing with the pest are the introduction of biocontrol agents and the deployment of resistant or tolerant species or clones.

Ophelimus maskelli
(Hymenoptera: Eulophidae)

Ophelimus maskelli is another gall-forming pest of Eucalyptus that originated in Australia, where it is largely unknown, and has been found recently in several other countries, where it has become a problem. While O. maskelli has not yet been recorded from the tropics, it has been included in this chapter because its spread to date parallels closely that of the blue gum chalcid, L. invasa, which was first reported from Europe and the Middle East and now occurs in several tropical and subtropical countries in Africa, Asia and the Americas. Coupled with this is the striking similarity between the host ranges of O. maskelli and L. invasa.

In the Mediterranean and the Middle East, O. maskelli severely injures eucalypts, particularly E. camaldulensis, the most economically important planted hardwood species in the region (Protasov et al., 2007b). As with L. invasa, this gall wasp was not accompanied by its principal natural enemies, which occurred in Australia, and therefore quickly reached epidemic levels. When it was first discovered in Europe, it was reported erroneously as *O. eucalypti*, a species known to be an invasive gall inducer on *Eucalyptus* in New Zealand (Withers, 2001; Protasov *et al.*, 2007b). The first report of the insect from Europe was in Italy in 2000 and it has been recorded subsequently from Greece, Spain, France, England, Israel, Turkey, Portugal and Morocco. Protasov *et al.* (2007b) detail the biology and impact of the pest in the region. *O. maskelli* shows a clear tendency to oviposit in developed immature leaves, generally in the lower canopy, and it prefers to oviposit on an area of the leaf blade near the petiole. Each female lays an average of 109 eggs and each egg induces a gall. Gall diameter ranges from 0.9 mm to 1.2 mm and there can be 11–36 galls/cm^2 of leaf. Under epidemic conditions, the entire upper leaf surface can be covered densely with galls (Plate 62). Heavy leaf galling can result in premature leaf drop soon after the emergence of the wasps. Both mature and young trees can be infested. In Israel, where infested trees are close to humans, mass emergence of the spring population can cause a nuisance by forming 'clouds' of wasps (Protasov *et al.*, 2007b). The insect produces three generations a year in Israel. Of 84 eucalypt species tested, 14 species were found to be suitable hosts: *E. botryoides*, *E. bridgesiana*, *E. camaldulensis*, *E. cinerea*, *E. globulus*, *E. grandis*, *E. gunnii*, *E. nicholii*, *E. pulverulenta*, *E. robusta*, *E. rudis*, *E. saligna*, *E. tereticornis* and *E. viminalis* (Protasov *et al.*, 2007b).

In the Mediterranean region, *O. maskelli* shares its new habitat with *L. invasa*, and both species may infest the same leaf. *O. maskelli* develops only on the leaf blade, whereas *L. invasa* induces galls on the midrib, the petiole and newly developed twigs. This partitioning should minimize interference and competition, although preliminary data collected by Protasov *et al.* (2007b) suggest that *O. maskelli* is a better competitor and displaces *L. invasa*, or impairs its performance.

Two species of mymarid wasp, *Stethynium ophelimi* and *S. breviovipositor*, and a eulophid wasp, *Closterocerus chamaeleon*, are parasitoids of *O. maskelli* (Huber *et al.*, 2006). *C. chamaeleon* has been released in Israel and appears to achieve effective biological control (Protasov *et al.*, 2007a). In the 16 months since its release at Bet Dagan in Israel, *C. chamaeleon* has spread some 1300 km to the city of Izmir in Turkey (Doganlar and Mendel, 2007).

Quadrastichus erythrinae (Hymenoptera: Eulophidae)

Erythrina is a widespread and diverse genus of trees with a pantropical distribution. It contains some 120 species (Mabberly, 2008), many of which have an important commercial, ecological or social value. Some are grown in tree plantations for timber or pulp, some are used as a living fence or as fodder for stock, but the principal use of *Erythrina* is as a shade tree for crops such as coffee and cocoa and a support tree for vines such as pepper or vanilla. Various species known as coral trees are used widely in the tropics and subtropics as street and park trees, and others are important components of native ecosystems (Rubinoff *et al.*, 2010). Damage to *Erythrina* trees due to an invasive wasp, *Q. erythrinae* (Plate 63), was first documented in 2003 in Reunion, Mauritius, Singapore and Taiwan (Kim *et al.*, 2004; Yang *et al.*, 2004). In 2005, the insect was detected in southern China (Huang *et al.*, 2005) and in Hawaii (Heu *et al.*, 2005). Since then, it has been reported from India, Japan, Vietnam, Sri Lanka, Thailand, Malaysia, American Samoa, Western Samoa, Guam, Florida and Fiji (Faizal *et al.*, 2006; Heu *et al.*, 2006; Prathapan, 2006; Wiley and Skelly, 2006; Uechi *et al.*, 2007; Messing *et al.*, 2009). *Q. erythrinae* originates in East Africa (La Salle *et al.*, 2009a) and is predicted to spread more widely in that continent as well as in Asia, Oceania and South America (Li *et al.*, 2006). The predicted potential range includes tropical rainforest, tropical monsoon, subtropical monsoon and tropical savannah climates. In commenting on the rate of spread of *Q. erythrinae*, Rubinoff *et al.* (2010) note that within 2 years of its discovery, the pest was distributed across a tropical swath from Hawaii to India, a distance of more than 12,546 km, much of it across open ocean. Long-distance dispersal is likely to be by global trade and locally by wind or transport on items such as clothing or flowers.

The erythrina gall wasp inserts eggs into young leaf and stem tissue of the tree. The larvae then develop within the plant tissue, forming galls in leaflets and petioles. As the infestation progresses, leaves curl and appear malformed, while the petioles and shoots become swollen (Plate 64). After feeding is complete, larvae pupate within the galls and the adult wasps cut exit holes through the plant gall material to emerge. Adults may live 3–10 days (Heu et al., 2006) and a single female wasp carries on average 322 eggs (Yang et al., 2004). The life cycle, egg to adult, is about 20 days (Heu et al., 2006). The species has overlapping generations and different stages of development can be found at the same time. Trees with large populations of wasps within the leaves and stems have reduced leaf growth and the plant declines in health. Severe gall wasp infestations may cause defoliation and tree death.

Studies by Messing et al. (2009) in five botanical gardens in Hawaii showed that 59 of the 71 species of Erythrina planted there were susceptible to Q. erythrinae. It is in Hawaii that some of the most severe impacts of the pest have been reported. E. sandwicensi, known as the wiliwili tree, is endemic and a 'keystone' species in Hawaii's lowland dry forest, one of the most endangered ecosystems in the world (IUCN, 2010). It is also an important cultural and ethnobotanical resource (Doccola et al., 2009). In the first few years following the arrival of the gall wasp in 2005, mortality of E. sandwicensi due to the pest was estimated at 10% and the survival of the wiliwili tree was questioned, prompting a major seed-banking effort (Doccola et al., 2009). Widespread and nearly complete mortality, believed greater than 95%, was also reported in Hawaii for the Indian coral tree, E. variegata, a popular landscape tree, and on O'ahu alone the cost of removing dead trees exceeded US$1m (Doccola et al., 2009). In Taiwan, where Erythrina also has great cultural importance, five species of Erythrina were affected across the whole island, severe infestations resulting in defoliation and tree death (Yang et al., 2004). The pest was reported from Sri Lanka in 2006 and at that time was confined to the plains, but there were fears that it would spread to the highlands, where Erythrina was the single most important shade tree of tea (Prathapan, 2006). Similar concerns are held in south India, where Erythrina is widely used as a live standard for trailing black pepper and vanilla (Faizal et al., 2006).

A eulophid parasitoid, Eurytoma erythrinae, from Tanzania has been released in Hawaii as a biological agent of Q. erythrinae and reportedly is very effective (La Salle et al., 2009b). Two other eulophid wasps from Africa, Aprostocetus excertus and A. nitens, also have potential as biocontrol agents against erythrina gall wasp (La Salle et al., 2009b; Prinsloo and Kelly, 2009).

5.8 Root Feeding

A wide range of insects feed on the roots of trees, including white grubs, termites, root weevils, larvae of longicorn and buprestid beetles and root aphids. These insects generally are not a problem for established trees with well-developed and extensive root systems but they can cause serious damage in nurseries and young plantations, where the trees have small and fragile roots. The problem is sometimes intensified when soil cultivation or site preparation removes the other vegetation on the site, and root insects are then forced to feed on the roots of the planted stock (Berryman, 1986).

Among groups of root-feeding insects, termites are the most damaging in the tropics, species of Odontotermes, Macrotermes, Microtermes and Coptotermes (Plate 65) being responsible for considerable mortality of newly established plantations throughout Asia and Africa.

Lepidiota spp. (Coleoptera: Scarabaeidae)

The larvae of some species of scarab beetles, such as Lepidiota spp., feed on the root systems of young trees, often ring-barking and severing the stems below ground. Such insects are commonly referred to as 'white grubs' or 'curl grubs' and are important pests in nurseries and young plantations (Plate 66).

Typically, the adult female burrows into the soil to lay her eggs. Larvae live in the soil, feeding on decaying organic matter when young and, later, on plant roots. They migrate through the soil profile, burrowing deeper in very cold or very hot, dry weather and coming close to the surface in wet conditions. They pupate in the soil, often in a smooth-walled cell. New adults may remain in the soil for a period until stimulated to emerge by heavy rain. Some species feed on the leaves of trees and occasionally can cause severe damage. The life cycle of most species lasts 1–2 years.

In tropical and subtropical Queensland in Australia, the larvae of *L. trichosterna* and *L. picticollis* are among the most important nursery pests of hoop pine, *A. cunninghamii*, necessitating routine treatment of nursery beds with chemical pesticides. In young plantations of hoop pine, local losses of up to 60% have been recorded (Elliott *et al.*, 1998). Attack is usually most severe where plantations have been established on grasslands or on land previously devoted to agriculture. In 2-month-old *E. exserta* plantations on soft, sandy soils in southern China, Wylie and Brown (1992) record losses of up to 50% caused by *L. stigma* white grubs, often requiring a replanting of the whole area (Plate 67). This species is also reported as a pest of 1- to 2-year-old trees (*Swietenia* spp., *T. grandis*, *F. moluccana*, *Anthocephala cadamba*) in taungya plantations in East Java (Intari and Natawiria, 1973). The authors note that in such situations, the trees suffer more than the farm crops. In the Philippines, adults of a species of *Lepidiota* have caused almost complete defoliation of young plantations of *A. auriculiformis* and *A. mangium* in Mindanao (Wylie, 1993). *L. mashona* larvae are very injurious to seedlings of *A. mearnsii* in Zimbabwe and have been known to kill large saplings (Browne, 1968).

5.9 Stem and Branch Cutters

Several insect groups damage trees by severing stems or branches completely. In the case of some species of crickets and grasshoppers, such damage occurs mostly on nursery stock or newly planted trees. However, larvae of several species of longicorn beetle make spiral tunnels or 'cuts' across branches and stems up to 10 cm in diameter, causing them to snap off in wind or under their own weight. An example is *Hesthesis cingulata* in Australia, which severs the stems of eucalypt saplings just above ground level (Elliott *et al.*, 1998).

Strongylurus decoratus (McKeown)
(Coleoptera: Cerambycidae)

This insect is common in hoop pine, *A. cunninghamii*, plantations in southern Queensland, Australia. It prunes mainly branches, but occasionally causes more severe damage, particularly to young trees, through pruning the leader.

As described in Elliott *et al.* (1998), eggs are laid singly into branches or leading shoots of hoop pine. The larva (Plate 68) tunnels in the cambial region for a time before boring longitudinally in the centre of the branch or top. Infested material may range from 1 to 10 cm in diameter. The larval stage occupies the tunnel for 39–41 weeks and pupation takes place at the end of the tunnel. Prior to pupation, the larva makes a transverse spiral tunnel or 'cut' across the stem, which creates a point of weakness where breakage can occur due to the weight of the stem or the effects of wind, leaving behind on the tree a stub which contains the insect. Sometimes, more than one larva may infest a branch and shed sections may then contain immature stages of *Strongylurus decoratus*. In constructing the pupation chamber, the mature larva cuts two sets of two exit holes transversely from its tunnel to the exterior, a short distance back from the pruning cut. It blocks both ends of its tunnel between the two sets of holes with plugs of wood slivers, forming a pupation chamber (Plate 69). Approximately 4 weeks after the holes are cut, the insect enters a prepupal stage, which lasts for 7–10 days, and then pupates for a further 4 weeks. The teneral adult may remain in the chamber for 1 or 2 weeks before removing a plug and exiting through one of the holes cut previously. Egg laying commences after 1–2 weeks.

S. decoratus has affected trees ranging in age from 4 years to more than 30 years and in height from 3 to 31 m. In some cases, trees have been defoliated completely, with only branch stubs remaining, but top damage is generally of most concern. In some areas, an incidence of leader attack of between 15 and 41% has been recorded. Some of the most severe attacks have been linked with below-average site quality, usually in upper slope positions, and with tree provenance (Wylie, 1982b).

A complex of natural control agents operates against larval and pupal stages of the insect, including encyrtid and ichneumonid wasp parasitoids and tachinid flies, and a predatory clerid beetle larva. Cockatoos are important predators of pupating larvae and seem to use the exit holes cut by the larva as visual clues to the location of their prey. Incidence of cockatoo predation sometimes reaches 20%.

Brachytrupes spp. (Orthoptera: Gryllidae)

Crickets, together with grasshoppers, cutworms and white grubs, form the chief pests of young seedlings but they are rarely detected in the act of injury. *Brachytrupes* is a genus of ground-dwelling crickets, occurring in Africa and throughout Asia, which contains several species injurious to seedlings in the nursery and sowings in the open.

The main species in Asia is *B. portentosus*, which makes a deep tunnel with an opening at the ground surface that is surrounded by ejected earth (Beeson, 1941). The eggs are laid at the bottom of the tunnel and the newly hatched nymphs remain in the tunnel for a time and then disperse. Young crickets dig themselves new tunnels every few days and these become longer and more ramified as the insects grow older. The adult stage is reached in a few months and the insect then occupies the same tunnel throughout its life. There is usually one generation a year. The cricket feeds on young seedlings and low shoots, cutting them off at night and dragging the pieces into the tunnel for feeding. In India, this species has been recorded damaging several tree species in nurseries. In north Bihar, for example, it killed up to 30% of *E. tereticornis* and *G. arborea* in one season (Ali and Chaturvedi, 1996) and 8–21% of *D. sissoo* seedlings (Sah *et al.*, 2007). Similarly, in the Jabalpur region of India, *B. portentosus* damage to roots and shoots of *D. sissoo* in nurseries resulted in 20–25% mortality of plants (Kalia and Lal, 1999). Beeson (1941) mentions that it attacks field plantings of *Casuarina* less than 60 cm high, but that taller trees escape. *B. portentosus* is a pest of newly planted trees in southern China, at one site damaging 40% of 3-month-old *E. urophylla* in a 10 ha plantation (Wylie and Brown, 1992). In a community woodlot project in north-eastern Thailand, the cricket was listed by Hutacharem and Sabhasri (1985) as one of the main pests of young *E. camaldulensis*.

In Zimbabwe, *B. membranaceus* is a pest of *Thuya* sp. and *E. grandis*, in some cases affecting more than 50% of plants (Taylor, 1981). This cricket occurs mainly on sandy soils, where it lives in a long, curved burrow up to 1 m in length. It cuts off seedlings, leaving just a stump, and carries them back to the burrow, where they are consumed. The insect also collects plants that have been cut off and left by cutworms and false wireworms.

In both Africa and Asia, indigenous peoples use *Brachytrupes* species as food (Fasoranti and Ajiboye, 1993; Agbidye *et al.*, 2009; Raksakantong *et al.*, 2010).

6
Management Systems I: Planning Stage

6.1 Introduction

Planning is the first, and most essential, step in all projects, large or small, and can be the major determinant of success. This is particularly so for an enterprise such as forestry, where crop rotation time commonly exceeds 15 years, and may be as much as 70 years. Once the forest crop is planted, it is usually prohibitively expensive to harvest it prematurely, should problems arise, and replace it with another of superior species. Evans and Turnbull (2004) caution that careful planning of projects is also necessary to avoid thinking about the economics and silviculture in isolation, for example. Pest and disease considerations most frequently are divorced from other planning processes, and sometimes ignored entirely. Savill and Evans (1986) outline some of the main factors that need to be considered in the planning and establishment of plantations (Fig. 6.1). A possible addition to this model is socio-economics, since no management system, be it silvicultural or entomological, can function in a region or country where the levels of education, literacy, numeracy and/or technology are insufficient to execute the planned tactics and strategies effectively.

The main silvicultural decisions at the planning stage are therefore:

1. What is the purpose of growing the trees?
2. What sites are available for planting and what are their conditions?
3. What tree species are available and suitable?
4. Will these trees grow well and be resistant to pests and diseases?
5. What are the socio-economics of the region?

All five of these have entomological perspectives which will be considered in turn.

6.2 Purpose of Growing Trees

Trees are grown for a great variety of purposes (Lamb, 2011), for timber, pulp, oil, fuel, land rehabilitation, shelter, food, animal fodder, carbon sequestration, or simply for beautification, and indeed, the concept of growing trees solely for logs for industry is completely outdated (Putz, 2008). The importance of insect damage to trees also varies according to the intended use of the trees. For example, a single cossid moth hole in a tree stem intended for use as a utility pole may be sufficient to cause its rejection for this purpose

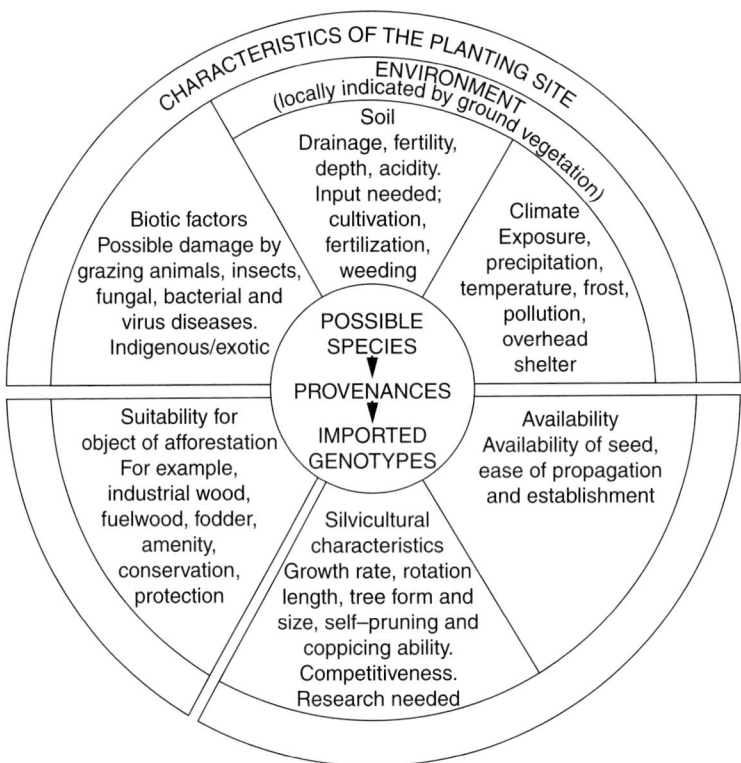

Fig. 6.1. Major considerations in selecting tree species for afforestation programmes (from Savill and Evans, 1986, with permission Oxford University Press).

but is of little consequence for the same species in an amenity planting. Such varying impacts are discussed below for each of the main types of tree plantings.

6.2.1 Industrial forest plantations

Some species of tropical tree are planted over vast areas of the planet; *Gmelina arborea*, for example, probably covers somewhere in the region of 1 Mha globally (Dvorak, 2004) and is used for a wide variety of purposes. Some species have been cultivated for many years under low input systems, such as teak, *Tectona grandis*, in India (Varma *et al.*, 2007), and it is only relatively recently that management practices have become more intense and heavily industrial. The industrial uses of trees such as these run a spectrum of financial returns from high-value products at the one end, such as saw-wood, poles, veneer, plywood, through oils, resins, tannins and fuelwood, to wood chip and fibre for pulp at the other end. Generally, plantations at the upper end of this spectrum are less tolerant of insect damage and the defects they cause than are plantations at the lower end, as is illustrated in the following examples.

Less tolerant of insect damage

The timber of the mahoganies (*Khaya* spp., *Swietenia* spp.) and cedars (*Cedrela* spp., *Toona* spp.) are among the most sought after in the world and command high prices (e.g. *T. ciliata* in Australia has been sold for over AUS$3000 m^3). However, the commercial growing of these species throughout most of the tropics is prevented by the attack of shoot borers, *Hypsipyla* spp. (Lepidoptera:

Pyralidae), whose tunnelling causes shoot mortality, growth reduction, branching, poor tree form and sometimes tree death (see Chapters 5 and 10 and Plates 57–59). Even a single attack by the larva of one of these insects on the leading shoot may be sufficient to alter the architecture of the tree and prevent the formation of a saleable bole. The moths demonstrate incredible host-locating ability and can affect up to 100% of plantings (Cornelius, 2009). Despite many decades of research on *Hypsipyla* in many countries, there is still no really effective control for these pests. Planting of susceptible Meliaceae in areas where *Hypsipyla* spp. occur is therefore high risk. However, because the timber is so valuable, quite high inputs into pest control would be justifiable and affordable, if and when it becomes technically possible (see Chapter 10).

If a tree species seems irretrievably susceptible to pests and diseases, its abandonment as a commercial proposition may be feasible where the areas planted are relatively small. However, with the trend towards the rapid establishment of large areas of a chosen tree species, abandonment is not a realistic option, and there are also instances where replanting has proceeded in the face of known pest problems (Wylie and Arentz, 1993). The imperative is not always solely economic. An example is Mulanje cedar, *Widdringtonia whytei* (*cupressoides*), which is Malawi's national tree. In Africa, it is the most northerly and only tropical representative of the genus and is confined to the Mulanje massif in the south of Malawi (Bayliss *et al.*, 2007). There, the widely scattered stands are mostly derelict, damaged by fire and exploitation, and young healthy stands account for no more than 2% of the total area of this species. The timber is high quality, is resistant to insects and decay and is used for boat building, furniture, panelling and veneers. Its scarcity value is such that theft of trees can be a problem. One threat to this cedar is the cypress aphid, *Cinara cupressi* (Hemiptera: Aphididae), discovered in Malawi in 1985, and despite some resistance in *W. whytei* to the insect, losses can occur (Chapman, 1994) (see Chapter 5 and later in this chapter). Despite the aphid, replanting and afforestation with this species have been recommended, for reasons relating both to the value of the timber and to conservation.

There are numerous examples of the low commercial tolerance of insect-related wood defects in high-value plantations. In Queensland, harvested 30-year-old *Eucalyptus cloeziana* destined for the pole market were found to have been attacked by the giant wood moth, *Endoxyla* sp. (Lepidoptera: Cossidae). Under the state's regulations, the structural integrity of a pole is deemed to have been compromised by insect tunnels in excess of 25 mm diameter. Larvae of this wood moth commonly make tunnels which exceed this size. Just a single large hole is sufficient to reduce the value of the stem by up to 90%. In Myanmar and Thailand, incidence of attack of teak by the beehole borer, *Xyleutes ceramica* (Lepidoptera: Cossidae), of even one beehole per tree per year is considered to be a very heavy incidence commercially (Gotoh *et al.*, 2007), since the timber is seriously degraded by internal burrowing of the larva. Aside from defects, insect attack may affect the commercial viability of a plantation by slowing growth and prolonging a rotation. One of the most notable examples is that of the teak defoliator *Hyblaea puera* (Lepidoptera: Hyblaeidae) in India. Nair (2007) has reported that plantations protected from the pest can yield the same volume of wood in 26 years as unprotected plantations would do in 60 years.

In general, for plantations where a high-value end use is planned, the choice of low-risk tree species and good sites is very important, and pest control may be an affordable option.

More tolerant of insect damage

For plantations producing fibre for pulp, insect defect of timber is generally of little consequence. An exception would be insect attack that resulted in the formation of extensive pockets of dark resin (Plate 70), requiring the use of additional bleaching during the pulp-making process. This can sometimes occur with stem borer attack on certain species of eucalypts, for example, tunnelling by larvae of the longicorn *Phoracantha acanthocera*

(Coleoptera: Cerambycidae) in *Corymbia maculata* and *E. grandis* in Australia (Paine et al., 2011). Sap staining associated with bark and ambrosia beetle attack degrades not only the high-value sawn and veneer timber, but can also affect the quality of the pulp that can be produced from attacked trees. In Queensland, sap staining associated with *Ips grandicollis* (Coleoptera: Scolytinae) attack on some fire-damaged *Pinus* spp. was so severe that the wood could only be used to produce pulp for cardboard and low-value packaging (Plate 71; Wylie et al., 1999). Sap staining can also weaken the binding properties of adhesives on wood fragments in the making of various composite boards. Tree shape is unimportant for fuelwood, but straight stems are desirable in trees destined for wood pulp to aid in rapid debarking. Insect attack which results in malformation of stems, for example, *Dioryctria rubella* (Lepidoptera: Tortricidae) shoot moth attack on *Pinus radiata* in southern China (Bi et al., 2008) and wood moth attack on leaders of plantation eucalypts in Sumatra (Plates 72 and 73), can thus lower the value of the crop. There is a general tendency among forest managers to underestimate the impact of pest attacks which do not result in outright tree mortality, particularly where the crop is destined for pulp. However, there are many examples in the tropics where plantation production targets to meet the requirements of pulp and paper mills have not been met because of growth checks due to insect attack. One such is attack by mosquito bugs, *Helopeltis* spp. (Hemiptera: Miridae), on eucalypt plantations in the island of Sumatra, Indonesia (Wylie et al., 1998). In Malaysia and Indonesia, the rubber termite, *Coptotermes curvignathus* (Isoptera: Rhinotermitidae), causes mortality in young plantations of *Acacia mangium*, but may also tunnel in the main stem without killing the tree. Considerable, and underestimated, losses in wood volume can result from such attack. In Brazil, *G. arborea* pulpwood plantations may suffer significant growth increment loss due to defoliation by leaf-cutting ants in the genus *Atta*. *Gmelina* appears to be one of the most preferred of all tree species, at least in laboratory trials (Peres Filho et al., 2002).

Insect damage to plantations grown for the production of oils or resins, rather than wood or pulp, can also have serious consequences. In the Sudan, gum arabic production is one of the most important sources of foreign currency for the country (Mathews, 2006). The acacias which produce this gum are managed as part of an agro-silvo-pastoral system – they grow naturally or are planted together with crops like sorghum, millet or other oilseeds or vegetables. The trees are also used as forage for domestic animals. Following severe drought between 1979 and 1984, the production of gum in one of the main producing provinces dropped from 9000 t to 750 t due to a combination of insect attack and socio-economic pressures. The long drought resulted in reduced agricultural production and farmers turned to tapping of weakened gum trees, which in turn led to attack by a range of insect pests and consequent tree death. Sandal, *Santalum album*, is a highly valuable tree of which the heartwood yields fragrant sandalwood oil used for perfume and medicines. It is indigenous in southern India, where it is now grown increasingly in plantations. Sandal has a wide range of pests, including treehoppers (Hemiptera: Cicadellidae) (Sundararaj et al., 2006), which transmit a mycoplasma-like organism that is the cause of the devastating spike disease (Shafiullah and Naik, 2003). This is characterized by extreme reduction in the size of leaves and internodes, accompanied by stiffening of leaves. In advanced stages, the whole shoot looks like a 'spike' inflorescence. Diseased trees die within 12–36 months after the appearance of symptoms (Sunil and Balasundaran, 1998b). There are also serious pests such as the bark-eating caterpillar *Indarbela quadrinotata* (Lepidoptera: Indarbelidae), the red borer *Zeuzera coffeae* (Lepidoptera: Cossidae) and the longicorn *Aristobia octofasciculata* (Coleoptera: Cerambycidae). *I. quadrinotata* causes branch dieback, while the other two pests tunnel in stems and branches and can cause sapling mortality. Studies by Remadevi and Muthukrishnan (1998) have shown that insect tunnelling causes a loss of around 200 kg of sandal heartwood per tonne of wood produced by the trees, which at the then price of US$15/kg was a loss of US$3000/t.

Black wattle, *A. mearnsii*, whose bark is used to produce tannin extractives (Govender, 2007), has been grown in southern Africa since 1899. There, the bagworm defoliator *Chaliopsis (Kotachalia) junodi* (Lepidoptera: Psychidae) has had a significant impact on production over several decades. Moderate to severe defoliations have caused losses of bark growth increment of over 3 t/ha bark (Atkinson and Laborde, 1993). Such losses can make even expensive treatments, for example, stem injection with systemic insecticide, cost-effective. Ironically, *A. mearnsii* is an invasive weed problem in other parts of southern Africa (Impson *et al.*, 2008) and research is ongoing to develop biological control programmes against it using herbivorous insects.

6.2.2 Seed orchards and clonal plantings

As outlined by Evans and Turnbull (2004), the aim of seed collection is to obtain large quantities of seed of the best genetic quality as cheaply as possible. Although expensive, seed orchards are the highest quality source. This is a stand of trees where every tree is from selected parents (plus trees) with desirable characteristics and where further improvement by progeny testing and culling may continue over several generations. Seed orchards are a very valuable resource, both for improving the quality of the plantation programmes they service and as a source of revenue through seed sales.

From an entomological viewpoint, the high value of this crop means that expenditure on pest management is easily justifiable. Because of the expense in managing seed production areas, even a 5 or 10% loss may be economically intolerable (Berryman, 1986). Seed orchards are usually of small size and readily accessible, which makes pest monitoring a relatively easy task. Insects may have direct impact on seed production through attack to seeds, fruit and flowers, or indirectly through attack on foliage and other parts. As an example of the latter, in Queensland, Australia, in late 1994, a clonal seed orchard of hoop pine, *Araucaria cunninghamii*, was severely attacked by golden mealybug, *Nipaecoccus aurilanatus* (Hemiptera: Pseudococcidae) (Plate 74). The attack destroyed the cone-bearing zone of 67% of the orthotropic ramets and reduced the plagiotropic ramet pollen production by 54% (Elliott *et al.*, 1998). In a survey of the orchard 3 months after the attack began, a tree mortality of 10.3% was recorded, with a further 32% defoliated almost completely and showing dieback of branches and main stem. The infestation was brought under control by a complex of natural enemies. In Nepal, seed orchard seedling trials of *Dalbergia sissoo* studied the incidence and damage rating of the longicorn beetle, *Aristobia horridula* (Coleoptera: Cerambycidae) (Dhakal *et al.*, 2005), showing that there were highly significant differences in beetle-caused mortalities between seed progenies (see later in this chapter).

The use of vegetative propagation with cuttings to develop clonal plantations is common in the tropics, examples being *Eucalyptus* species in Brazil, *G. arborea* in Malaysia and *P. caribaea* in Queensland. The advantages of clonal forestry include shorter plant production times, control of pedigree, flexibility of deployment, multiplication of valuable crosses and greater crop uniformity. As well, it provides an opportunity to minimize pest and disease damage, for example by the improved use of resistant genes and by allowing the matching of clones to sites to which they are better adapted. However, there are also fears that clonal forestry will reduce genetic variability and increase the risk of pest disasters. Too narrow a genetic base may lead to additional problems in the future and clonal forestry may be at risk of reducing genetic diversity too far (Burdon and Wilcox, 2007).

6.2.3 Village or agroforestry

Agroforestry, the integration of trees into farming systems, has long been practised in village communities throughout the tropics (Plates 75 and 76). However, such plantings are becoming increasingly important as burgeoning human populations and widespread deforestation in the region create ever-greater demand for the services and products that trees can provide. In the Tamil Nadu region of India, for example, farmland is now growing exotic trees such as *Eucalyptus* and *Casuarina*, managed by farmers, to supply an expanding paper pulp industry (Parthiban *et al.*, 2010). *Leucaena* and even bamboo may also be cultivated to put into the pulp mix (Prasad *et al.*, 2009), so agroforestry systems are now producing wood fibre alongside, and indeed intermixed with, food and fodder crops of many kinds. These crops may be agricultural species such as maize, beans and cassava, or they may be cash crops such as coffee or cocoa. Of crucial concern is the potential for pest and disease relationships between the perennial trees and the annual crops (Schroth *et al.*, 2000). For example, can maize mixed with acacias protect them both from attacks? Alternatively, one may provide a pest reservoir for the other. Insects (and pathogens) may impact on these systems, as discussed in the following examples.

Fodder

Leucaena leucocephala, a leguminous tree native to areas of Mexico and Central America, has been widely used through the tropics as a valuable source of high-quality feed for livestock. As outlined in Chapter 5, since 1984 the leucaena psyllid, *Heteropsylla cubana* (Hemiptera: Psyllidae), has spread to most areas where *Leucaena* is grown and has caused extensive damage (see Plate 33). There have been several studies of the effect of the psyllid on various members of the genus (Mullen and Gutteridge, 2002), with particular reference to their relative susceptibility to the pest (see below). An observed long-term trend is that damage is generally heavy in the first 2 years of infestation and then gradually weakens in duration and severity (Geiger *et al.*, 1995). The *Leucaena* story frequently has been quoted as an example of the dangers of the introduction and use of a single-tree species on a massive scale and may prompt changes towards more diversified agroforestry systems in the future. Many other tree species are used as fodder in various countries and several of these have experienced pest problems. For example, in Malawi, a leaf beetle, *Mesoplatys ochroptera* (Coleoptera: Chrysomelidae), causes severe defoliation of *Sesbania* (McHowa and Ngugi, 1994), and in western Burkina Faso, between 50 and 80% of the agroforestry tree *Vitellaria paradoxa* (Shea tree) was found to be infested by the shoot- and fruit-boring larvae of the moth *Salebria* sp. (Lepidoptera: Pyralidae) (Lamien *et al.*, 2008).

Fuel

Fuelwood is immensely important to many tropical countries, and rural economies may be entirely dependent on trees to produce such fuel, either as logs and sticks or converted into charcoal (Plates 77 and 78). For example, just one agroforestry project in the Democratic Republic of the Congo (DRC) produces between 8000 and 12,000 t of charcoal a year using *A. auriculiformis* (Bisiaux *et al.*, 2009). The gross annual revenue for DRC from charcoal is over US$2.5 m, of which the local agroforesters receive around a quarter. In 2007, India was estimated to have between 14 and 24 m trees outside conventional forests, spread over 17 Mha and supplying over 200 Mt of fuelwood annually (Pandey, 2007). Any pest or disease problems in these situations could have very serious consequences.

Lahaul Valley lies just north of the tropic of Cancer, at an altitude of around 3000 m. It is classified as a cold arid desert, not the most suitable place for growing large numbers of healthy trees. However, the people that live there have to use fuelwood derived from various species, such as poplars, walnuts and, in particular, willows (Rawat *et al.*, 2006). *Salix fragilis* provides between 29 and 69% of the total annual fuel used by various villages in the region, but these agroforestry trees are severely attacked by a fungus, *Cytospora chrysosperma*, and two insects, the San Jose scale, *Quadraspidiotus perniciosus* (Hemiptera: Coccidae), and the giant willow aphid, *Tuberolachnus salignus* (Hemiptera: Aphididae). The damage of all three pests plus the ravages of the climate mean that only about 30% of *S. fragilis* survives. The only solution seems to be to abandon this willow species and attempt to grow native Himalayan willows instead, which though less productive, appear to be more tolerant of the conditions and less susceptible to pests and diseases.

Eucalyptus spp. are grown for fuelwood in many countries. In Ethiopia, severe attack by subterranean termites *Macrotermes subhyalinus*, *Pseudacanthotermes militaris* and *Odontotermes* spp. (Isoptera: Termitidae) has occurred on the roots and stems of newly planted trees in every province except in highland areas above 1800 m, and losses sometimes approach 100% (Cowie and Wood, 1989). The authors noted that rural buildings constructed from grass and wood also suffered serious damage, particularly in western Ethiopia. Houses were attacked within 1 year after construction, necessitating at least partial rebuilding within 3–5 years. This clearly results in greater pressure on natural resources of wood, and protection of both houses and fuelwood plantations is therefore essential in order to prevent even further deforestation.

Shade

A very common use of trees in agricultural systems is as shade for crops (Plate 79). There are numerous insect pests whose activities affect tree crown condition and cover, thereby negating the purpose for which such trees are planted and exposing the crops to damage by other agents. Such pests span a range of feeding guilds from leaf eaters and sap suckers to root feeders and stem girdlers. Some individual shade species suffer damage from several important pest groups and an example is *Falcataria moluccana*, native to South-east Asia and one of the fastest growing leguminous trees in the world. It is commonly and extensively defoliated by lepidopterous pests such as the yellow butterfly *Eurema blanda* (Lepidoptera: Pieridae) (Irianto *et al.*, 1997) and the bagworm *Pteroma plagiophleps* (Lepidoptera: Psychidae), the latter defoliating trees once or twice a year in India, causing considerable tree mortality after repeated defoliations (see Chapter 5 and Plates 21 and 22). In Bangladesh, India and Pakistan, where *F. moluccana* is used as shade for tea crops, the bark-eating caterpillar *I. quadrinotata* (Lepidoptera: Cossidae) causes loss of growth increment and sometimes girdling and mortality (Baksha, 1993). In a very similar situation, Suharti *et al.* (2000) reported quite serious damage to *F. moluccana* stems by *I. acustistriata* in Java.

Multi-purpose

As outlined by van den Beldt (1995), the impetus for much of the research and development work on multi-purpose tree species (MPTS) can be traced back to two crises of the 1970s – deforestation and the 'other energy crisis', i.e. fuelwood. MTPS are now widely used in farming systems in the tropics and can perform the roles of both service and production. For example, they may reduce soil erosion, contribute substantial amounts of nitrogen to agricultural systems (nitrogen-fixing trees), offer shelter or make effective boundary fences and, at the same time, provide products such as fruit, lumber, tannins, fodder or fuelwood, either for home consumption or for sale. Tree species such as *L. leucocephala* are becoming used as raw materials for paper pulp, for packaging and for use as fuel in biomass power plants, as in southern India, for example (Prasad *et al.*, 2011). Some of the examples already given of insect damage to trees planted specifically for fodder, or for fuel or shade, also apply to the same tree species when used for multiple purposes. Within a single farming unit in the Philippines, for instance, *L. leucocephala* is used for fuelwood, fodder and leafmeal, the latter becoming a major source of income to small farmers (Showler, 1995). With the arrival of the psyllid *H. cubana* in early 1985, damage to *Leucaena* reached 80% of total leafmeal production and new plantings for leafmeal halted within a year. The infestation resulted in fodder shortages and led to farmers feeding their livestock other plant material such as banana leaves and trunks, maize husks and coconut fronds. However, with the lower feeding value of these substitute materials, most of the animals became weak and susceptible to diseases (Moog, 1992). Farmers were forced to sell their stock early at suboptimal prices. As another example, *Manilkara zapota*, or sapodilla, a tree native to southern Mexico and Central America, is grown in Puerto Rico for fruit, shade and ornamental purposes and its white latex is used in the production of chewing gum. Numerous insect pests have been recorded feeding on fruit, leaves and twigs, but the most important of these are scarab beetle defoliators, *Phyllophaga* spp. (Coleoptera: Scarabaeidae). Trapping in the field showed an average of 920 adults/tree, and a tree mortality of 67% was recorded. These insects are an important limiting factor to the growing of sapodilla in northern Puerto Rico (Medina *et al.*, 1987). Finally, Sileshi *et al.* (2006) report on insect pests of the multipurpose tree *Sesbania sesban*, which can be attacked regularly by various leaf beetles such as *Mesoplatys* sp. and *Exosoma* sp. (Coleoptera: Chrysomelidae) in East Africa.

Taungya

Growing food crops among trees while they are young is a widespread establishment practice, usually carried out for the first 2 or 3 years in the life of the tree crop before canopy closure (Evans and Turnbull, 2004) (Plate 80). Taungya was developed initially as a means of ensuring tree cover following shifting cultivation and has been very successful. It remains an inexpensive system in the tropics, which leads to improved land-use practices but which does not require radical change in lifestyle by the land user. These days, taungyas are being discussed as carbon sequestration mechanisms (see below), as in the case of maize and timber tree mixtures in Chiapas, Mexico (Soto-Pinto *et al.*, 2010). There have been many studies on the interactions between crops and trees and the associated economics of this mix. One example by Mishra and Prasad (1980) examined the growing of oilseed, groundnut and soybean as cash crops in conjunction with *D. sissoo* and *T. grandis* in a newly felled forest area in Ranchi, India. They found that oilseed was not a successful cash crop with either sissoo or teak in the first year of the plantation. Groundnut and soybean were highly remunerative with both tree species but shared a common pest with teak, the hairy caterpillar *Diacrisia obliqua* (Lepidoptera: Arctiidae). The attack on teak was as severe as on groundnut and soybean and resulted in some growth loss and mortality of teak. However, attack by this caterpillar on sissoo was sporadic and minor and had no effect on this species. The recommendation, therefore, was to grow groundnut and soybean in association with *D. sissoo*. In more recent times, Goulet *et al.* (2005) studied the effects of growing Pacific mahogany, *Swietenia humilis*, in a taungya system with maize in Honduras. They found that the young trees were attacked more significantly by mahogany short borer, *Hypsipyla grandella* (Lepidoptera: Pyralidae), in taungya than if they were grown in weedy plots. More maize was produced, but the trees themselves suffered badly.

When food crops are planted with trees, farmers tend and protect both types of crop. Grass invasion is prevented, ground cover is maintained and simple pest control measures may be undertaken. The small scale of most plantings means that farmers are likely to detect problems early and lavish more attention on individual trees. Profit is usually only important after subsistence needs have been met. For fodder and fuelwood crops, gross yield is the most important consideration. The form or shape of the tree is irrelevant, but biomass is crucial. In village systems, there is usually no money or expertise available for sophisticated pest management. However, labour is cheap and plentiful and simple prevention and control techniques can be employed effectively, for example, hand picking and destruction of insects, enhancement of natural enemies (e.g. by providing sources of nectar for flower-feeding insect parasitoids and predators), repellent planting, trap cropping and use of 'folk medicines' (see Chapter 7).

6.2.4 Rehabilitation

Trees have an important role to play in protecting the environment and are planted both for prevention (reducing soil erosion, slowing wind speeds, moderating the force of rain and runoff of water, combating desert encroachment) and for rehabilitation (repairing land degraded by salt scald, water-table salinity, industrial waste or by mining, restoring biodiversity) (Plate 81). Timber or biomass production, form and aesthetics are generally of secondary importance; what matters is that the trees stay alive. Given that many of the sites on which these trees are planted are poor and degraded, and therefore there is marginal to good growth, insect attack is likely and may threaten the success of such plantings.

Casuarina equisetifolia is planted in coastal sand dunes to reclaim the land for agriculture and stop erosion. The bark-eating caterpillar *I. quadrinotata*, mentioned earlier in this chapter, can reach large population densities and cause serious timber damage (Sasidharan and Varma, 2008). Of course, since the trees are not being grown in the sand dunes for commercial timber, the type of damage caused by stem borers may be of no real consequence as long as the trees can continue to grow. This same tree species is used in coastal China in protective shelterbelts, the so-called 'great green wall'. Several insect pests contribute to tree mortality, particularly of young trees, including the cerambycid *Anoplophora chinensis*, the pyralid moth *Euzophera batangensis* (Lepidoptera: Pyralidae), the cossid moth *Z. multistrigata* (Lepidoptera: Cossidae) and the metarbelid moth *Arbela baibarana* (Leidoptera: Metarbelidae) (Huang, 1995). Note that all these pests are mainly secondary, that is, they require some sort of stress or lack of vigour in the host tree (see Chapter 3), so it may well be that the sites in which the casuarinas are planted are suboptimal, even for this hardy tree species.

Yue *et al.* (1994) report on a long-term study on rehabilitation of eroded tropical coastal land in southern China. The procedures followed involved the establishment of pioneer communities (mainly plantations of *P. massionana* and *E. exserta*) and, later, enrichment of these plantations to create a mixed forest. The 30-year project showed that a diverse tropical forest could be re-established on severely degraded sites. However, in the early stages of development of the mixed plantation, several tree species were severely attacked by insects. Most *Arlanthus malabaricus* trees were killed by *Eligma narcissus* (Lepidoptera: Noctuidae), and *Chukrasia tabularis* was badly attacked by the longicorn *A. chinensis* (Coleoptera: Cerambycidae). In one experimental plot of 1 ha, 20 herbivorous insect species were observed. As plant diversity increased, so too did the diversity of other organisms and food webs, with a corresponding decrease in pest outbreaks. In addition to the likelihood that a degraded site will prove challenging for whatever species is planted, the problems of rehabilitation are often compounded by inappropriate species selection due to community attitudes. In some communities, only native tree species would be considered for such plantings, when in fact an exotic may be more suitable. In others, an unsuitable exotic may be used; for example, at one particular mine site in Australia, *P. radiata* was used for rehabilitation for the sole reason that the plants were available cheaply from the local forestry nursery. The plantings failed due to a combination of site stress and insect attack.

6.2.5 Amenity

Amenity plantings differ from the types discussed already in that the cosmetics (form, shape, appearance) are important, while production or conservation values (timber quality, biomass, soil-binding ability) are generally unimportant. Amenity trees are used to enhance the environment of mainly urban areas and may have both aesthetic and practical value. In some countries, individual trees or relatively small stands may have religious significance and be considered sacred by local people (Plate 82) (Khumbongmayum *et al.*, 2006). In Malaysia, for example, the rapid process of urbanization has brought increased public awareness of the aesthetic and environmental value of trees and has resulted in extensive plantings in residential and recreation areas, along roadsides and in open spaces between buildings (Sajap *et al.*, 1996).

One of the most common trees planted is *Pterocarpus indicus* and this has been badly attacked by a leaf miner, *Neolithocolletis pentadesma* (Lepidoptera: Gracillariidae), resulting in defoliation, dieback and general unsightly appearance, to the consternation of the public. Tree stress caused by human activities or 'people pressure' is often a major factor in insect outbreaks on urban trees. In northern Cameroon and the north-eastern states of Nigeria, neem trees, *Azadirachta indica*, in medium to large towns are severely attacked by the oriental scale, *Aonidiella orientalis* (Hemiptera: Diaspididae), but attack is minor in smaller villages and shelterbelts in country areas (Boa, 1995). The scale feeds on foliage, petioles and small branches, causing browning or mottling of leaves, dieback and sometimes tree death. The scale build-up in towns has been linked to depositions of dust on leaves, which may affect the feeding activities of natural enemies. There is a very definite link between the proximity of trees affected most seriously to areas of increased human activity. A good example of how a range of factors associated with human activities may combine to affect the health of urban trees and predispose them to insect attack comes from Queensland, Australia. Bottle trees *Brachychiton rupestre* are used as a feature street tree in Roma in the south-west of the state. The poor condition of many of these trees prompted an investigation recently into the causes (Plate 83). Root damage from several sources was a major contributor. Cars had been allowed to park under the trees, compacting the soil and damaging surface roots. Bitumen had then been laid right up to the base of some of these trees, further restricting infiltration of water. Some trees had received regular watering with bore water which had high sodicity and affected the tilth of the soil. Runoff of oils, herbicides and other pollutants compounded the problem, resulting in tree dieback. Stressed trees were heavily attacked by insect defoliators. Physical damage of the trunks of trees by parking vehicles allowed the entry of weevils, which tunnelled in the pith of the trees, leading to collapse. The result of the investigation was a commitment of resources by the town council and community groups to repair damage and to alter the practices which had led to the problem. This is a feature of urban tree entomology, as distinct from insect attack in most other types of plantings, that money is likely to be available for pest management on individual trees.

6.2.6 Carbon sequestration and biofuel production

The ability of trees in tropical forests to absorb atmospheric CO_2 has long been thought to help in the reduction of greenhouse gas accumulation and hence go some way to negating global warming (Pyle *et al.*, 2008). Forests not only store (sequester) carbon in their tissues, but also maintain carbon stores in soil. Deforestation in many tropical countries clearly removes this carbon absorption service, and when forest removal makes way for agriculture, it may reduce soil organic carbon significantly as well (Tan *et al.*, 2009). Existing forests are being maintained, and new ones are being planted, in many countries as part of bioremediation or carbon offset projects (Glenday, 2008), and various tropical countries are now planting forests in attempts to balance out CO_2 production by fossil fuel power stations in other countries. Enrichment planting (EP) of old or degraded forest sites is one technique being employed (Paquette *et al.*, 2009). Insect herbivory will interfere with the ability of trees to take up CO_2, and indeed, it has been suggested that defoliation by insects may reduce the accumulation of carbon in woody material by around 20% (Schaefer *et al.*, 2010). It is possible that rather than being net sinks of atmospheric carbon, forests experiencing severe pest outbreaks may become net carbon sources (Slaney *et al.*, 2009). It is not just defoliators that can be a problem in this context; Kirton and Cheng (2007) found that several species of termite would commonly kill young dipterocarp trees in enrichment planting experiments in Peninsular Malaysia by ring-barking the roots and lower stems.

Most biofuel (biodiesel and bioethanol in the main) are derived mainly from agricultural crops such as maize or sugarcane, or tree fruit such as oil palm (Koh et al., 2011), but woody biomass can be extracted from forests to convert into biofuels (Landis and Werling, 2010). Woody debris of various kinds can be utilized, such as thinnings or logging residues. Normally, this material would be left *in situ*, but can be collected and processed accordingly. The implications for pest management are mixed; removing material such as stumps and logs reduces the likelihood of insect pests breeding in them and subsequently attacking standing trees (see Chapter 8), but it is possible that removing this material also destroys reservoirs of natural enemies and otherwise desirable forest insects (Walmsley and Godbold, 2010). For now at least, biofuel production from tropical forests is of minor importance.

6.3 Sites Available for Planting and their Conditions

A primary objective of tropical forestry is to enhance productivity (Kumar and Thakur, 2011). However, a common misconception among non-foresters is that trees thrive on adversity and will grow well on sites where other crop species will fail. This is often the expectation of potential participants in farm–forestry ventures, where the most fertile land is reserved for agricultural crops, the next best for grazing and the poorer, less accessible or degraded land for growing trees. This attitude is fuelled somewhat by the success of multi-purpose trees such as the Australian *Eucalyptus* across a range of low-quality sites worldwide. However, trees usually have the same high demands for water and nutrients as do farm crops and will perform best on the better-quality sites. This particularly may be the case for many of the fast-growing but high-yielding tropical tree species such as, for example, *G. arborea* (Espinoza, 2004). The reality in tropical countries is that the sites available are likely to be suboptimal, e.g. grass-covered steep slopes, swampy, rocky, arid or depauperate land. Such sites are likely to engender low tree vigour and may stress trees, predisposing them to insect attack (see discussion in Chapter 4 and the exceptions to the rule). It may also be the case that large-scale plantations of these fast-growing exotic tree species convert indigenous insects into serious pests, who transfer their attentions from unimportant native hosts to the new and tender exotics (Zhao et al., 2007). The common goal of the silviculturalist and the protection specialist is therefore to obtain the best possible matching of tree species with planting site to promote vigour and avoid stress, thereby minimizing pest and disease damage.

6.3.1 Types of problems confronting the planner or silviculturalist and insect examples

Wind-, frost- or hail-prone sites

Hurricanes in the Caribbean (Plate 11) and cyclones in the Far East seem to be evermore frequent, and trees vary considerably in their abilities to withstand high wind damage and recover from it (Duryea et al., 2007). In areas at risk from hurricanes, typhoons or cyclones, it is important to use wind-firm species. In 1988, a hurricane devastated much of the island of Jamaica, damaging 3006 ha or one-quarter of the total pine plantations, mostly *P. caribaea*. Of this, 2540 ha were more than 50% damaged and required salvage logging, clearing and replanting. Damaged trees were colonized rapidly by bark beetles *Ips calligraphus* and *I. cribicollis* (Coleoptera: Scolytinae), which can complete their life cycle within a month. Pheromone-baited trap lines were recommended in areas where excessive quantities of blown-down timber were adjacent to stressed standing trees, the idea being to attract the beetles away from such trees and thus reduce tree mortality. As a consequence of the hurricane, Jamaica embarked on a new seed improvement programme to produce stock that was more wind resistant than that planted previously (Bunce and McLean, 1990). Studies by Haines et al. (1988) in Queensland provided a basis for this programme, having demonstrated the superior straightness and greater wind firmness of the Central

American coastal lowland provenances of *P. caribaea*. Wind firmness is also an important consideration for tree plantations in the northern Philippines, where numerous typhoons cross the islands each year. *Pinus* spp., particularly *P. kesiya*, damaged by these storms are attacked and killed by *I. calligraphus* (Viado, 1979). Frost is generally of little importance in the tropics, except at high altitudes, where it may preclude the use of some species. Trees damaged by frost but still alive may be prone to attack by insect borers such as bark and ambrosia beetles. Hail damage is more common in the subtropics than the tropics. In South Africa, one of the most serious problems in pine plantations is dieback caused by the fungus *Sphaeropsis sapinea* in hail-damaged trees, which is often exacerbated by infestation of the weevil *Pissodes nemorensis* (Coleoptera: Curculionidae) and the bark beetle *Orthotomicus erosus* (Coleoptera: Scolytinae) (Wingfield *et al.*, 1994). *P. radiata* and *P. pinaster* have been particularly susceptible to this type of damage, and *P. patula* and *P. elliottii* are now planted in summer rainfall areas where hail frequently damages trees. In southern Queensland, Australia, hail-damaged cypress pine *Callitris glauca* is prone to attack by the small cypress jewel beetle, *Diadoxus erythrurus* (Coleoptera: Buprestidae), sometimes necessitating early or preventative salvage logging (Elliott *et al.*, 1998).

Dry, arid or desert sites

Of the many factors that determine which tree species can be planted on which site, the amount and distribution of rainfall seems to have the greatest influence. Annual rainfall in the tropics varies from almost nothing in deserts to several thousand millimetres in parts of Papua New Guinea and India. A tree species such as *E. grandis* which grows naturally in wet sclerophyll forests in subtropical eastern Australia could not be expected to do well in the deserts of central Australia and nor would the dryland mulga *A. aneura* prosper on the humid coast. Even when temperatures and annual rainfall are similar, a species such as *E. tychocarpa* from the Northern Territory of Australia does not always thrive in southern Queensland because the distribution of the rainfall is different. In addition, of course, we have the problems of climate change (see also Chapter 3), where extra rainfall, or alternatively drought conditions, prevails to create conditions conducive for new pest outbreaks. One such example would be the increase in outbreaks of the bark beetle *Dendroctonus frontalis* (Coleoptera: Scolytinae) in *P. caribaea* forests in Honduras. Rivera Rojas *et al.* (2010) suggest that climate change causing less seasonal rainfall and higher temperatures may be responsible. Certainly, it seems that prolonged drought provoked an earlier outbreak of the same beetle in Guatemalan pine forests (Haack and Paiz Schwartz, 1997).

Arid or desert regions are usually defined as having less than 200 mm annual rainfall, semi-arid 200–600 mm, dry tropical 600–1000 mm, semi-humid tropical 1000–1800 mm and humid tropical more than 1800 mm (Evans, 1992). Inappropriate land use and the increasing prevalence of droughts in some parts of the world are leading to expanding desertification (Khanmirzaei *et al.*, 2011). When planting trees in an arid region, one could hardly imagine that conditions could get worse. Nevertheless, drought can occur in regions of perpetual aridity. In the Turkana district in north-western Kenya, 500 ha of desert were planted with *Prosopis juliflora* for fuelwood and fodder in 1982. Average annual rainfall is 160 mm but in the drought of 1984–1985 only about 30% of this norm was experienced, and such drought conditions again occurred in 1990–1991. Speight (1996) records that by 1990, 80% of trees were showing dieback (indicated by sparse foliage, dead branches and a significant loss in growth), and by 1991, about 60% of trees had lost all their foliage (Plate 84). Shoots and stems of these trees were attacked by an unidentified longicorn (Plate 85). One of the major constraints on tree planting in dry and semi-arid regions in the tropics is attack by subterranean termites in various genera, including *Coptotermes*, *Macrotermes*, *Microtermes* and *Odontotermes*. In these conditions, newly planted trees represent a source of both food and moisture, and protection using chemical insecticides is often necessary. The problem is particularly severe in India and Africa, where termites can attack young trees within 6 months after planting and mortality can reach high levels.

Shallow soil and stony sites

Rooting depth is extremely important. Shallow soils provide poor conditions for root growth, reducing tree stability and resistance to drought. The small volume of soil may lead to nutrient shortages. Shallow soils may oscillate between being swampy in the wet season and arid in the dry (Evans and Turnbull, 2004). Plates 86 and 87 show attack by the longicorn *A. lucipor* (Coleoptera: Cerambycidae) in a plantation of *A. mangium* in the northern Philippines. These trees were planted on a stony ridge top, a site totally unsuitable for *A. mangium*, and drought stress is believed to have precipitated the attack which affected almost every tree in the stand (F.R. Wylie, unpublished). *A. nilotica* is an all-purpose timber species in India and is commonly attacked by various boring insects such as the stem-boring longicorn *Celosterna scabrator* (Coleoptera: Cerambycidae) (Luna *et al.*, 2006). Attacks and subsequent tree mortality was significantly worse in trees growing on more exposed, less fertile sites with erosion exposing laterite and stones.

Compacted, hardpan sites

Areas of compacted soil are very common following harvesting operations and can cause poor growth. Successive short rotations can magnify the problem. Soil compaction reduces the amount of water infiltration greatly and may result in tree stress and a predisposition to insect attack. When vegetative cover is removed, certain soil types may develop a hard 'ironstone' layer in the soil or at the surface, which is not conducive to good tree growth. Hardpans or claypans deeper in the soil may not be obvious at planting and trees may grow well for a time before encountering this impervious layer. In south-eastern Nepal, severe dieback and patch deaths of teak occurred in parts of 8- to 10-year-old stands (Speight, 1996). Buprestid larvae were present under the bark of these trees, causing stem girdling. Excavation of the roots of dead trees showed the lack of taproots due to a layer of impenetrable, heavy clay close to the surface. Periodic waterlogging occurred on these sites (Plate 88).

Waterlogged, acidic and saline sites

Waterlogging fundamentally reduces the growth and potential survival of most tree species (Smith *et al.*, 2001). It reduces soil oxygen concentrations available in the root zone and this slows growth rates, causes premature leaf shed, inhibits new root formation, restricts root growth to surface soil horizons and thus makes trees vulnerable to wind or drought stress in dry seasons (House *et al.*, 1998) (Plate 89). Soil salinity affects the ability of plants to extract water from the soil, resulting in water stress, a nutrient imbalance, salt toxicity and an associated reduction in growth rates. Changes in the pH of soils can also interfere with nutrient uptake. In south-western Vietnam, plantations of *E. camaldulensis* and *E. tereticornis* have been established on acid sulfate soils which have been mounded to about 80 cm above the original ground level. The trees grew well initially, some reaching a height of 13 m after 4 years (J. Kelly, personal communication). However, soon after, there were reports that about 4000 ha of plantations (half the estate) had been girdled and killed by the stem-boring longicorn *A. approximator* (Coleoptera: Cerambycidae) (Wylie and Floyd, 1998). The attack is almost certainly stress related and associated with extreme acidification. A pH of 2 has been recorded in adjacent canals as a result of runoff from the acid sulfate soils. The root systems of most trees developed poorly in these conditions and there was a high incidence of trees falling over (J. Kelly, personal communication). In Queensland, Australia, Wylie *et al.* (1993) demonstrated a link between insect-related dieback of *C. cunninghamiana* trees growing alongside streams and levels of streamwater salinity. Dieback and death of these trees was the direct result of severe and repeated defoliation by the leaf-eating chrysomelid beetle *Rhyparida limbatipennis* (Coleoptera: Chrysomelidae) (see Plate 9). However, it was worse in areas of increasing water-table salinity associated with extensive tree clearing and intensive land management.

6.3.2 Silvicultural options in the planting of inhospitable sites

The most desirable option for the silviculturalist would be to use another site with less stressful conditions. However, if that is not possible, then one option would be to improve the site available. In arid lands, for example, numerous strategies have been developed to maximize the benefits of what moisture is available, and Evans and Turnbull (2004) detail some of these techniques. Water conservation measures include the collection of surface runoff from a wide area and funnelling it to the planting site, open pit sunken planting and the application of mulches. Periodic flood irrigation of plantation sites during the dry season traditionally is carried out in many countries, and trickle or drip irrigation is sometimes used, as is the insertion of a permeable container filled with water in the root zone of the newly planted tree. Physical barriers such as fences or pallisades can be constructed from local materials to protect young trees against the desiccating and abrasive effects of wind and blown sand, and grasses used to stabilize dunes. For saline sites, subsurface or surface drains help to reduce waterlogging and to allow rainfall to flush salt from the affected area. Deep ripping may improve soil drainage and root penetration on sites with a clay subsoil. Mounding can be beneficial to both tree survival and early tree growth on most saline or waterlogged sites, providing improved drainage and reducing surface salinity levels through associated leaching (House *et al.*, 1998). Cultivation with a rotary hoe along the tree lines may be useful to break up large clods produced by initial mounding and provide a better soil tilth for planting. For soils with a high pH, gypsum can be added, which also improves soil structure and permeability, root penetration and leaching of salts. It is particularly useful on sites with hardpan clay subsoils and cracking clay soils. Finally, for infertile or deficient soils, fertilizers and appropriate trace elements can be added to ensure good plant growth. Alternatives to the use of fertilizers include changing species to a less demanding one, leaving ground cover and litter to protect from erosion and leaching, mulching, animal dung, enrichment of soil via tree leaf fall and planting of nitrogen-enriching species. Picking the right tree species or provenance for the particular site remains one of the key decisions for the silviculturalist in the planning stage. In the past, these decisions were reliant on individual or corporate experience, but with the development of evermore powerful computers, such decisions are assisted by any of a number of databases developed for this purpose in various countries. For example, the Food and Agriculture Organization of the United Nations has produced ECOCROP, which has a database of over 2500 plant species, including many tropical trees. A wide variety of site-specific search terms can be used, including temperature, rainfall, soil pH, latitude and altitude, to produce short lists of suitable tree species. CABI's Forestry Compendium (FC) is a similar package with which tree species can be matched to the available site conditions. Figure 6.2 shows an extract from the Species Selection module in the FC, using an imaginary set of search criteria to select the most suitable (and hopefully therefore pest free) tree species to plant. The location chosen arbitrarily was Mexico, with an altitudinal range of 0–100 m, on saline and infertile soils with end products including light construction. As can be seen, only five tree species fitted all the criteria, three of which had 'disadvantages' against them (triangles). In the case of *Melaleuca quinquenervia*, for example, the FC states that 'this species seeds profusely and can become a weed...'.

Climatic mapping programs such as CLIMEX are available to predict the suitability of trees species and also the likely behaviour

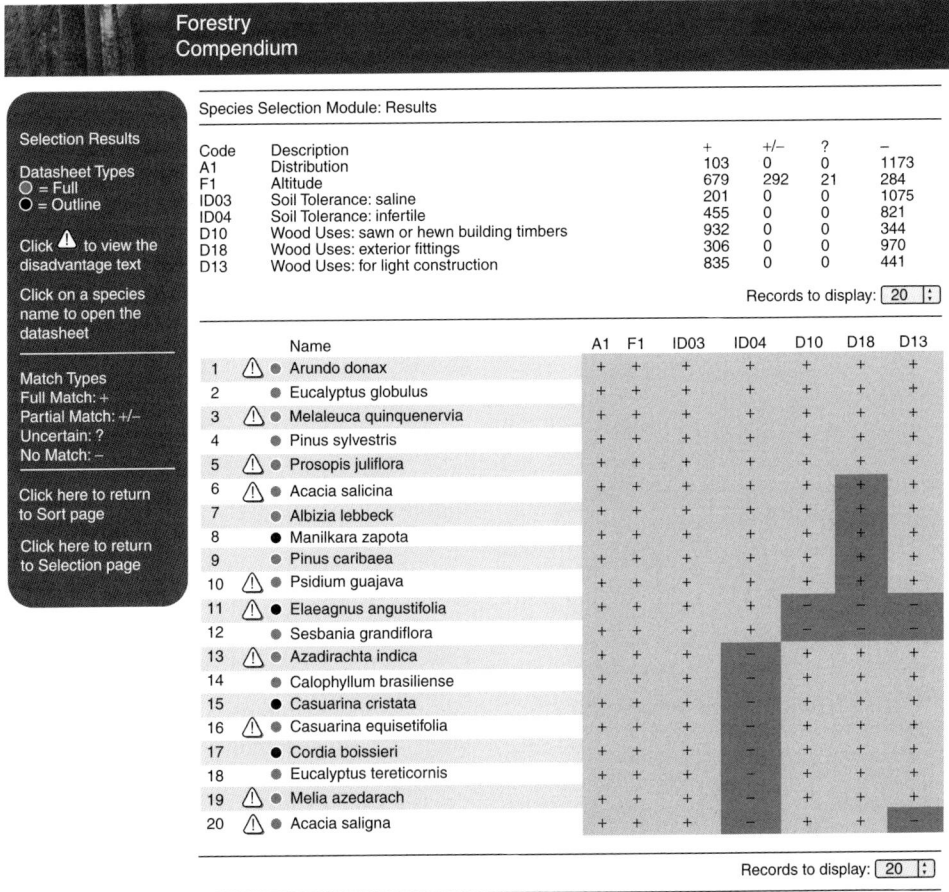

Fig. 6.2. Extract from Species Selection Module of CABI Forestry Compendium to identify the most suitable tree species for a particular (imaginary) site (CABI, 2010).

of insects and diseases under different climatic conditions (see, for example, Pinkard *et al.*, 2010). If more than one tree species is suitable and available for planting on a particular site, then further selection can be made not just on the basis of environmental resistance but also on demonstrable inheritable genetic resistance properties, as discussed in Section 6.4.

6.4 Tree Selection and Plant Resistance

Febles *et al.* (2008) screened 40 leguminous tree species to attacks by leaf-cutter ants, *Atta insularis*, in Cuba. After 2 years, they found that 11 species and genera were not harvested by the ants, thus showing that, for whatever reason, some trees were favoured by insect herbivores and others rejected. We have seen in Chapter 4 how tree-feeding insects have co-evolved with their food plants and have, in the main, become specialized to a relatively small set of related species of trees, which they are able to exploit. The efficiency of this exploitation can vary considerably between suitable tree species, or even within a species, between different genotypes or provenances. Some of the observed variations in feeding success may be environmentally derived and modified, and hence only be applicable to one set of trees in one place and time. Others, however, may be related to the tree's genetics and are therefore potentially inheritable by its offspring. For example, Snyder et al. (2007) found that coastal populations of P. caribaea var hondurensis in Belize were significantly less attacked by the southern pine beetle, D. frontalis, than those inland at higher altitude. Was this due to a difference in the tree's environment or to some inherited trait which was local-site independent? Manipulations of variability in the ability of an insect herbivore to exploit a tree which is based genetically within the plant can be extremely important in pest management. Planners, if they so wish, may choose to plant a tree species or genotype which is less palatable, better defended or simply unrecognized by insects. This is the science and practice of plant resistance.

6.4.1 Environment or genetics?

In the case of P. caribaea in coastal Belize mentioned in the previous section, resistance to bark beetles was found to be associated with differing levels of a chemical called estragole in the tree's oleoresins; this could still, however, be a function of the tree's environment (certain site conditions promoting the production of the resin) and not necessarily inheritable. In a somewhat similar investigation, Andrew et al. (2010) looked at the occurrence of another defence chemical, this time sideroxylonal, in eucalypts in Australia. They discovered that there was a significant genotype × environment interaction at the population level, so that some type of genetic variation existed in chemical defences which might be mediated or moderated by where the trees were grown.

In Chapters 5 and 10, we discuss the almost global problem of the eucalyptus longicorn beetle, P. semipunctata (Coleoptera: Cerambycidae). This pest is a particular problem on trees which are growing on dry, semi-arid or otherwise water-stressed sites (Plates 39 and 40). Species of Eucalyptus planted on a dry, sloping site with discontinued irrigation in southern California (33°N approximately) could be split into two groups according to pest impact. One group of species was found to be seriously damaged by Phoracantha and many trees were actually killed by the pest, while the other suffered very few deaths indeed, even though some of the trees showed signs of larval feeding (Hanks et al., 1995a). The two groups were characterized according to their known reaction to drought conditions and the soils in which they grew naturally in their native home, eastern Australia. These results show a clear variation in the resistance properties of Eucalyptus species to Phoracantha in California, but the important question for a planner is concerned with the basis for this observation. Obviously, growing eucalypts on arid sites should, if possible, be avoided (see Section 6.3), but if such sites are the only ones available, then species such as E. camaldulensis (red river gum) and E. sideroxylon (ironbark) are a much better 'bet' than E. grandis (flooded gum) or E. saligna (Sydney blue gum). In fact, in the California trials, not one individual of the last species survived.

However, it must be concluded that this observed resistance has a genetic base – it is as heritable property. Thus, all individuals of E. camaldulensis would be expected to show resistance to longicorn attack, even on dry sites where other species succumb. An additional problem has to be faced in that, although resistance to Phoracantha may be demonstrated in some eucalypt species or genotypes, their performance as trees may not go hand in hand with pest resistance. Thus, in work on Eucalyptus clones in Morocco, Belghazi et al. (2008) found that certain clones which resisted Phoracantha better than others showed only weak or intermediate timber production.

In many cases, as we might now expect, insect damage combines with environmental factors to determine a tree's aggregate performance, as shown in Table 6.1. In this example, Mullen et al. (2003) evaluated the influence of the feeding of the psyllid H. leucocephala, in combination with various indices of temperature, rainfall and soil type, on *L. leucocephala* collected from 18 different locations around South-east Asia and Australia. All but rainfall was found to have a significant influence on dry matter (DM) production, while all significant relationships were positive, except that of pest pressure. Detecting true genetically based resistance to the pest would be difficult. A rather similar problem has arisen in Costa Rica, where trials have been carried out in attempts to protect various species of mahogany from attacks by the mahogany shoot borer, *H. grandella* (Lepidoptera: Pyralidae). Details of the biology and pest status of the borer are presented in Chapter 5 (see also Chapter 10 and Plates 57–59). Figure 6.3 shows that *Cedrela odorata* is attacked much earlier after establishment than *Swietenia*; *Swietenia*, in fact, appears to be free of damage for the first 50 weeks, whereas *Cedrela* is attacked almost immediately after planting out (Newton *et al.*, 1998). Can this be considered to be evidence of heritable resistance? Looking further, the same authors found that attacks to *Cedrela* were correlated with various measures of soil minerals (Fig. 6.4), such that as the levels of soil calcium and total base minerals increased, the number of attacks per tree declined. Thus, there appears to be an environmental component to the supposed resistance of host trees. In the final analysis, the ability of the pest to cause economic damage has to be considered. Figure 6.3 shows that, in fact, both species of tree end up severely damaged, since only one or two attacks by the borer in young trees can render the crop unmarketable. Resistance in this case is not as yet a commercial proposition.

Table 6.1. Relationships between dry matter (DM) yield and mean monthly environmental variables during two phases of growth of *L. leucocephala* accessions from various locations (* = significant at $P < 0.05$; ** = significant at $P < 0.01$) (from Mullen *et al.*, 2003).

Environmental factor	Relationship to yield (r^2) in establishment phase	Relationship to yield (r^2) in post-establishment phase
Maximum temperature	0.37*	0.30*
Minimum temperature	0.41**	0.75**
Rainfall	0.26	0.19
Fertility index	0.68**	0.71**
Acidity index	0.84**	0.74**
Psyllid pressure	−0.39*	−0.38*

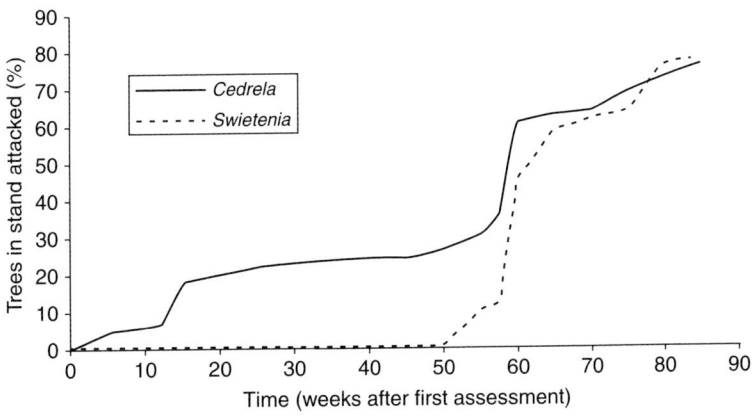

Fig. 6.3. Cumulative number of two mahogany species attacked by *Hypsipyla grandella* in Costa Rica (from Newton *et al.*, 1998, courtesy Elsevier).

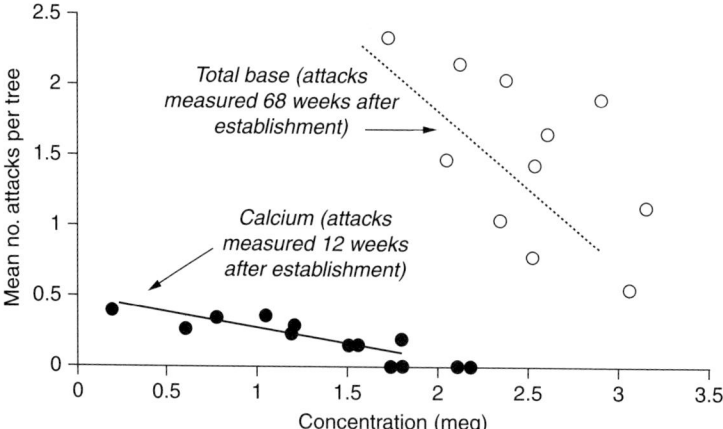

Fig. 6.4. Relationships between attacks by larvae of *Hypsipyla grandella* on *Cedrela odorata* and soil minerals in Costa Rica (from Newton et al., 1998, courtesy Elsevier).

6.4.2 Inheritable tree resistance

As we have seen in previous chapters, trees have evolved various methods to combat insect herbivory, such as physical and chemical defences. The occurrence and efficacy of such defences can vary not just between tree families and genera but even within genera and species. The teak defoliator moth, *H. puera*, is the most serious defoliator of teak in most parts of the world (Chapter 5), but some of the clones of *T. grandis* have much hairier leaves than others. These hairs, or trichomes, are physical, and sometimes chemical, methods by which trees deter defoliation (see Chapter 4) and teak clones with more trichomes seem much more resistant to attacks by larvae of *H. puera* (Jacob and Balu, 2007). The trichomes impede the movement of small moth larva and prevent them actually reaching down to the leaf surface to feed. Clearly, everything else being equal (growth, form, resistance to other pests and diseases, etc.), hairy-leaved clones should be the ones selected for planting.

H. puera is not the only defoliating insect to attack *T. grandis*, and while certain genotypes, provenances or clones might show resistance to this pest, other insects may show a different response. The teak skeletonizer, *Eutectona machaeralis* (Lepidoptera: Pyralidae), is another common defoliator of teak, which, though smaller than *Hyblaea*, can cause very significant damage from time to time. Laboratory feeding trials of *Eutectona* were carried out in Madhya Pradesh (central India) in a similar fashion to those described above for *Hyblaea*, and again, significant differences in consumption of leaves were detected between clones of teak (Meshram et al., 1994). Third instar larvae ate an average of 2.7 cm^2 from most resistant clones, as compared with an average of 8.2 cm^2 from the susceptible clones. Neither set of trials compared both species of teak pest at once, however. Whether least preferred leaves contained antifeedants, toxins or simply low nutrient levels (see Chapter 4), which would have a similar effect on most if not all insect herbivores, is not yet known. An example of the complexities that can arise when variable resistance to insect feeding is considered for more than one insect and tree species is illustrated by the example of the pine woolly aphid. Pine woolly adelgids, *Pineus* spp. (Hemiptera: Adelgidae), are now very widespread pests of pines from Europe to Australia, Africa to South America (see Chapter 5 and Plates 25 and 26). They attack a large number of *Pinus* species but do show some response to tree genotype, and also to site conditions. *P. patula* is widely planted in the subtropics and tropics and is a native of Mexico.

In central and southern Africa, for example, young *P. patula* can be infested quite badly by woolly adelgid, and though the actual impact on the trees has to be elucidated, clear differences in infestation levels have been observed in various countries, including Tanzania (Madoffe and Austara, 1993). In this case, the frequencies of infested trees in plantations were equally high on both good and inferior sites, but population densities were highest on poor sites. This observed suitability of poor sites for high adelgid development is in keeping with the plant stress hypotheses discussed in Chapter 4. There are also genetic components to woolly adelgid resistance in pines, however. Chilima (1995) reported that in Malawi, where the woolly adelgid was discovered in the 1980s, *P. kesiya* was consistently the most infested, while *P. radiata* was one of the least infested. Furthermore, this trait is without doubt genetically based, at least in *P. radiata*, since pest-free parent trees produce pest-free offspring. Quite how powerful this resistance is to woolly adelgid when the trees are planted on poor or degraded sites is another matter, but from this work in Africa, *P. radiata* would seem to be a suitable species for planting in areas where *Pineus boerneri* is common.

The teak defoliator is native to the Indian subcontinent, but a lot of pests of plantation tree species are exotics, *Craspedonta (Calopepla) leayana* (Coleoptera: Chrysomelidae) an important case in point (see also Chapter 5). *G. arborea*, on which the beetle feeds, is a tree species that is very widely planted in the tropical world these days, but it is originally native to South and South-east Asia. Kumar *et al.* (2006) studied the attacks of *C. leayana* on various clones of *G. arborea* from north-east India and West Bengal. They graded resistance or susceptibility to the insect using the amount of leaf area consumed by the pest and found that between 8 and 26% of the *Gmelina* clones appeared to be highly or moderately resistant (Fig. 6.5.). However, the paper does not record what levels of defoliation were considered indicative of resistance, and whether these observations hold true for all planting locations is not reported.

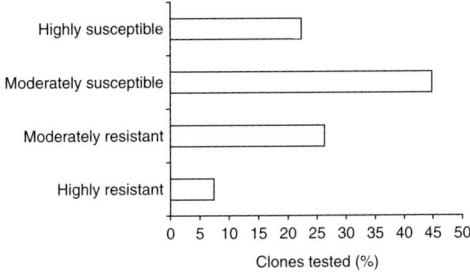

Fig. 6.5. Levels of resistance to *Craspedonta leayana* shown by various clones of *Gmelina arborea* in India (from Kumar *et al.*, 2006, with permission *Journal of Experimental Zoology India*).

We know that tree genotypes and provenances vary considerably in their herbivore relationships, but we must not forget that insects vary too, often much more than trees. A single population of insects in a forest may be very diverse genetically due to factors such as sexual reproduction, rapid generation times, dispersal and migration, as in the case, for example, of mahogany shoot borers in the genus *Hysipyla*. This worldwide genus has but two quoted species (*robusta* in the Old World and *grandella* in the New World), but the genetics of the system show clearly that there are many different genotypes of the moth with multifarious ecologies and behaviours (Alves, 2004). Undoubtedly, even if taxonomy insists that but one species of insect occurs in two different continents, it is safe to assume that the gene frequencies and hence phenotypic characteristics of the populations thousands of miles apart will be very different. In this way, then, it is quite likely that even with the same tree clone on an exactly equivalent site but in two different parts of the world, the supposedly same pest species will perform very differently in the two scenarios.

For this reason, tree resistance testing tends to be most meaningful within relatively small geographical regions, where the insect populations might be assumed to be fairly homogeneous genetically. It is the role of the silviculturalist, geneticist and entomologist together to screen trees for their relative susceptibilities to the most serious insect pests (and diseases) in that particular region, so that carefully considered plans can be laid down before any planting begins. This is particularly the case with the enormous problem of termites (Isoptera), which feed on the roots of young transplanted trees and kill them in the first year or two of life.

6.4.3 Screening for resistance

So, now we know that trees can exhibit genetically inheritable resistance to insect pests (and pathogens, too), it becomes the job of the forest planners to find out which of the available tree species or genotypes are most likely to have dependably low incidences of pests and pathogens at the same time as showing good silvicultural properties. This is where screening comes in. Figure 6.6 shows steps proposed for detecting resistance (or tolerance) to pests (and diseases) (Ingwell and Preisser, 2011). As can be imagined, a whole process such as the one illustrated may take some years to complete and may benefit from the assistance of local foresters and tree users. The forestry literature is full of papers and reports detailing the results of performance trials for tree species and genotypes which monitor all sorts of tree properties, only a few of which involve resistance to pests and diseases. We present just a few examples here.

The cypress aphid, *C. cupressi* (Hemiptera: Aphididae), is an extremely serious exotic pest which causes extensive loss of growth and mortality in plantations of various tree species in the Cupressaceae, especially *Cupressus lusitanica*, widespread in many countries, including Kenya, Tanzania and Malawi (see also Chapter 5). Trials in Kenya have shown that some members of the family of trees are much less susceptible to the aphid than others, everything else being equal. In this case, damage was assessed using 'eyeball' scores for the incidence of aphids on trees, from a score of 1 where no aphids were present, to a maximum of 5 where all branches and twigs were infested and the foliage yellow or brown with death of twigs and branches in the entire crown (Obiri, 1994). However, though a particular tree species may be in some way unsuitable for the pest, its silvicultural characteristics make it completely useless as a

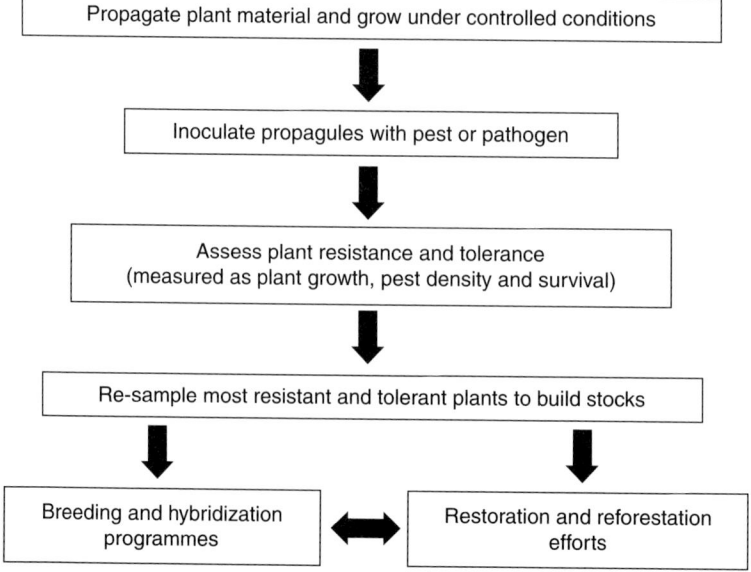

Fig. 6.6. Steps in determining whether or not trees that survive attacks from pests or diseases are resistant to the attacker (from Ingwell and Preisser, 2011, courtesy Wiley-Blackwell).

forest tree; *Thuya orientalis* is a case in point. Its height at 15 years old, 3 m, and its mean annual increment, 0.2 m, show that it grows far too slowly, despite being very resistant to the pest. In general, the preferred species, *C. lusitanica*, is highly susceptible to the aphid, but even then, careful testing of genotypes or families of this species do show marked variations in attack and damage levels (Mugasha *et al.*, 1997; see below).

More closely related tree species have also been screened for insect pest resistance in attempts to choose the best tree species for planting in specific localities. Pham *et al.* (2009) studied the impact of the gall wasp, *Leptocybe invasa* (Hymenoptera: Eulophidae) (see also Chapter 5), on various eucalypt species in nurseries and young plantations in North Vietnam. Both pest and host tree are exotic to this part of South-east Asia, but a big demand for rapidly growing and highly productive timber demands pest- and pathogen-free species. As Table 6.2 shows, the most susceptible hosts in North Vietnam are *E. camaldulensis*, *E. tereticornis* and *E. grandis*, in fact three of the most commonly planted species worldwide. Nevertheless, some variation has been observed between species and provenances, and these results can be taken forward for more breeding and selection trials.

Variations in pest resistance within a genus or even a species are potentially very important. Take the example of the multi-purpose tree, *Leucaena*. This genus of leguminous trees has been planted throughout the tropical world as an agroforestry tree which can be used for shade, fodder and timber. The most popular species is *L. leucocephala*, and in many countries this has been found to be heavily attacked by the leucaena psyllid, *H. cubana* (Hemiptera: Psyllidae) (see Chapter 5 and Plate 33). The damage caused by this sap feeder can

Table 6.2. Impact of *Leptocybe invasa* after 3 months in the nursery, as determined by damage incidence, damage severity and tree survival in the field, as well as the severity of damage by the pest in young field plantations (from Pham *et al.*, 2009).

Tree species	Damage incidence (%)	Damage severity	Tree survival (%)	Field assessment of pest severity
Corymbia henryi	0	Nil	44.4	Nil
C. citriodora ssp. *citriodora*	0	Nil	70.4	Nil
C. tessellaris	0	Nil	55.6	Nil
C. polycarpa	15	Low	88.9	Low
Eucalyptus pilularis	9	Low	48.2	Nil
E. microcorys	10	Low	59.2	Nil
E. cloeziana	0	Nil	44.4	Nil
E. globulus ssp. *maidenii*	25	Low	96.3	Nil
E. smithii	14	Low	7.4	Low
E. moluccana	36	Low	55.6	Nil
E. coolabah	31	Low	74.1	Low
E. pellita	12	Low	85.2	Nil
E. urophylla	19	Low	70.4	Low
E. robusta	26	Low	74.1	Medium
E. saligna	41	Medium	37.0	Medium
E. grandis	92	Severe	44.4	Medium
E. tereticornis ssp. *tereticornis*	81	Severe	77.8	High
E. camaldulensis ssp. *camaldulensis*	80	Severe	44.4	Medium
E. camaldulensis ssp. *obtusata*	76	Severe	96.3	High
E. camaldulensis ssp. *simulata*	88	Severe	85.2	High

be so bad that the tree simply cannot be grown in some places, and there has therefore been urgent need to reduce pest numbers. The search for resistant *Leucaena* species and provenances has been intense and Wheeler (1988) has produced a damage score system for the pest (Table 6.3).

Using this system, Mullen *et al.* (2003) assessed the resistance to the psyllid in 116 accessions of *Leucaena*, from 27 species and subspecies, using the data scale shown in Table 6.3. They found a continuum from highly resistant to highly susceptible (Fig. 6.7.). Genotypes within *L. collinsii*, *L. pallida*, *L. diversifolia* and *L. esculenta* were very resistant, whereas *L. leucocephala*, for example, the first widely planted commercial species

Table 6.3. Ratings scale used to assess psyllid damage to *Leucaena* (from Wheeler, 1988).

Rating	Clinical symptoms
1	No damage observed
2	Slight curling of leaves
3	Tips and leaves curling and yellow
4	Tips and leaves badly curled, yellowish, covered with sap
5	Loss of up to 25% young leaves
6	Loss of 25–50% young leaves
7	Loss of 50–75% young leaves
8	100% loss of leaves and blackening of lower leaves
9	Blackened stems with total leaf loss

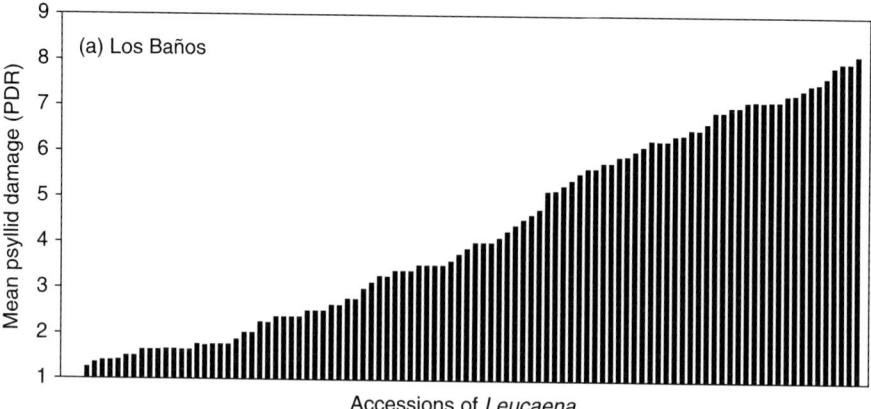

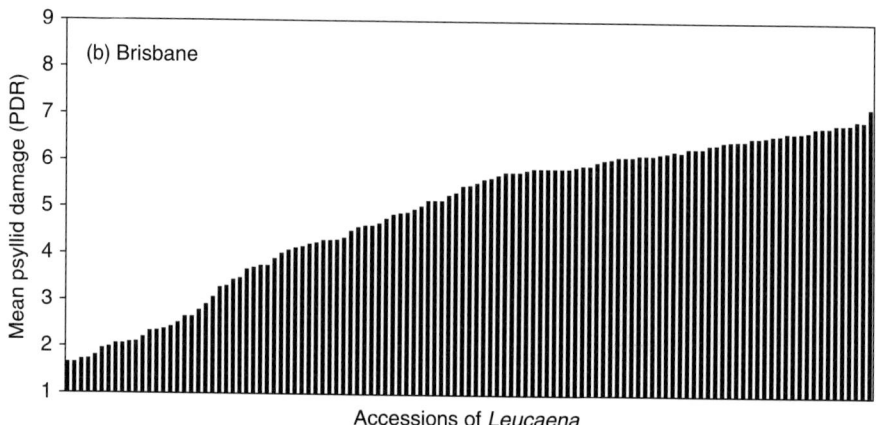

Fig. 6.7. Psyllid resistance to *Leucaena* accessions measured by mean psyllid damage ratings (PDR) during periods of high psyllid pressure in (a) Los Baños, Philippines, and (b) Brisbane, Australia (from Mullen *et al.*, 2003, with permission Springer).

in the world, was highly susceptible. In addition, this screening was carried out in two distinctly different localities, Brisbane in southern Queensland, Australia, and Los Baños in the Philippines, showing some differences in resistance due to varying climatic conditions but with general consistency of resistance. Moving to Tanzania, Edward et al. (2006) used the same data scale as in the previous example and compared psyllid infestations with various silvicultural characteristics (Table 6.4). Some interesting results are presented.

On aggregate, the very best all-round tree species/provenance was *L. diversifolia* (Batch 155510), with *L. diversifolia* Ex. Mexico and *L. diversifolia* Ex. Veracruz as runners up. Two of the worst (not shown in table) were *L. collinsii* Ex. Chiapas and *L. shannonnii* also Ex. Chiapas. It is crucial to realize, therefore, that such selection trials are only really applicable for local or regional conditions and their results should not be expected to apply over large geographical scales. Furthermore, resistance is not a predictable feature of an entire tree species, as can be seen from the table. Some provenances of *L. diversifolia* had very good pest resistance and silvicultural properties, while others within the species were very poor. The basis of resistance in this situation is probably linked to the levels of tannins in the foliage of *Leucaena*, though correlations are weak (Wheeler et al., 1994). The problem is that farm animals such as cattle and goats, for which the fodder is grown in many countries, do better on leaves with smaller tannin concentrations, just like psyllids.

6.4.4 Tree resistance in practice

It now almost goes without saying that tropical forestry, whether large-scale industrial or small-scale agroforestry, should select tree species or genotypes at the planning stage which have been shown to be attacked less by insect pests in the locality, as long as inheritability has been demonstrated and site factors do not override resistance mechanisms. It is also, of course, of prime importance that the trees to be grown are going to be commercially useful, but it does not seem overcomplicated to attempt to choose trees suited to the planting sites that show pest resistance. A simple example of this tactic comes from Morocco, where the careful selection of tree species according to their resistance properties to particular pests and suitability to local climatic conditions is considered the best way to establish successful plantations (Belghazi et al., 2008). In this semi-arid region, *Eucalyptus* plantations

Table 6.4. Ordinal ranking of tree parameters showing differences between *Leucaena* species and provenances planted at Morogoro, Tanzania (from Edward et al., 2006).

Species/ provenance	Survival (%)	Height	DBH	SPH	PSY	FBIO	SBIO	BBIO	WBIO	Mean	Overall rank
L. diversifolia Batch (15551)	2	4	5	3	9	3	3	3	3	3.89	1
L. diversifolia Ex. Mexico	2	6	9	3	7	4	4	5	4	4.89	2
L. diversifolia Ex. Veracruz	1	1	1	13	18	2	1	7	1	5.00	3
L. pallida Ex. Oaxaca	10	7	7	2	15	1	2	1	2	5.22	4
L. diversifolia Ex. Veracruz Batch (14917)	3	5	4	10	8	15	8	2	5	6.67	5

Notes: SPH = stems per hectare; PSY = psyllid damage; FBIO = foliar biomass; SBIO = stem biomass; BBIO = branch biomass; WBIO = wood biomass.

are in danger of attack by the eucalyptus longicorn beetle, *P. semipunctata* (Coleoptera: Cerambycidae), but if species such as *E. torquata* and *E. salmonophloia* are chosen, all of which are xerophytic and hence relatively resistant to the pest (see Chapter 4), only a few trees suffer light damage and pest problems are prevented. It does, however, seem to be a common practice in tropical forestry to plant the species of tree most susceptible to the local insects and conditions.

As discussed in Chapters 5 and 10, the mahogany shoot borer, *Hypsipla* spp., is one of the most serious pests in tropical forestry anywhere in the world. It attacks and damages beyond economic use many genera and species of mahogany throughout the tropical world. Its main effect is to cause young trees to fork at heights too short to produce a marketable mahogany log when finally harvested. Some trees such as the African *Khayas* seem somewhat more resistant than others, such as the Latin American *Swietenias*, which are on the whole susceptible (Perez *et al.*, 2010). Figure 6.8 shows the effect of *H. robusta* on various progenies of *Khaya ivorensis* in trials in Ghana (Ofori *et al.*, 2007), where both total tree height and height to first fork are shown to vary between progenies 2.5 years after planting. The point is that all progenies were attacked; no tree genotype is so resistant so as not to be attacked at all. The key to pest prevention success, though, is how tall can the trees be expected to grow before the main trunk forks, whether or not one of the forks quickly becomes a single dominant again and also the diameter growth achieved in a certain time. In this study, it was decided that progenies such as AM24, BB19 and BB20 were particularly suitable for use in integrated pest management (IPM) programmes. In a similar situation, this time in the Yucatan Peninsula of Mexico, Wightman *et al.* (2008) concluded that the practical application of genetic resistance in *S. macrophylla* against *H. grandella* would be to choose genotypes with good dominant heights despite pest attack incidence, to plant the trees in the best sites for mahogany and to be prepared to prune individual trees to enhance single stem dominance.

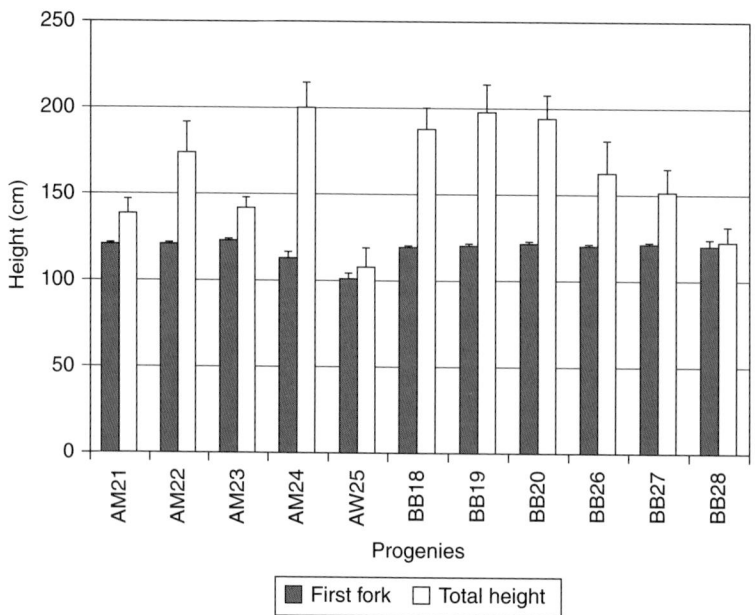

Fig. 6.8. Variations in mean total height and mean commercial height (to first fork) (± SE) among progenies of *Khaya ivorensis* in Ghana (from Ofori *et al.*, 2007, courtesy Elsevier).

Another problem linked with tree resistance is the extent to which pest incidence or damage can be expected to be reduced. In Ghana, attempts to establish plantations of *Milicia* (= *Chlorophora*) spp., valuable timber species in sub-Saharan Africa, have been hampered by the gall-forming psyllid, *Phytolyma lata* (Hemiptera: Psyllidae) (see Chapter 5). Infestations of large galls of the pest on the leaves of nursery stock and young transplants result in dieback of foliage down to woody tissue (Wagner *et al.*, 2008). In screening trials, incidence of large galls indicated that there were significant differences among individual progenies of *Milicia*, with the mean percentage of galls ranging from 15.4 to 99.6% and 39.5 to 99.9% for 1992 and 1993, respectively (Cobbinah and Wagner, 1995). However, a highly significant curvilinear relationship was shown between the incidence of dieback and relative abundance of large galls, and even the most resistant trees were likely to suffer unacceptable dieback in the nursery or just post planting. In this sort of circumstance, the best hope for tree resistance would lie in an IPM system where resistance reduces the efficiencies or densities of pests to levels where other agencies such as natural enemies might be able to exert a further influence (see Chapter 10).

A final question remains: if reliable pest resistance is obtained, how should that tree genotype be employed in tropical silviculture? Plant breeders can now supply foresters with many tree species, provenances, genotypes and so on with predictable and dependable properties. The College of Tropical Agriculture and Human Resources at the University of Hawaii, for example, can provide a *Leucaena* hybrid which has been through six cycles of selection for high forage yield and resistance to *Leucaena* psyllid, among other properties (Brewbaker, 2008), and it is obviously tempting to establish large areas of plantation with such desirable trees. However, the adaptability of insect herbivores to this novel but now ubiquitous food source must not be underestimated. A high selection pressure to adapt to overcome the resistance of the trees is presented, such that this hard-found resistance may disappear again. Probably the best practical strategy is to plant mixtures of resistant varieties if more than one is available, such that no one genotype is omnipresent. We return to the problems of planting systems and tactics in Chapter 8.

6.4.5 Resistance and biotechnology

The types of selection and breeding discussed in the previous section can take many years, even when dealing with fast-growing tropical trees. At the same time, the widespread establishment of forest plantations using progenies, provenances or even clonal material chosen for specific traits can result in narrow genetic diversity, and eventually inbreeding (Burdon and Wilcox, 2007). Biotechnology in various forms may be able to speed up breeding programmes and provide broader genetic bases (Whetten and Kellison, 2010), and though the use of genetically engineered trees in commercial forests may be some way in the future, these techniques have much more scope for gene transfers between unrelated tree species and even unrelated phyla, such as bacteria and angiosperms (Ahuja, 2009). A whole variety of forest traits can be improved using biotechnology, as shown in Table 6.5 (Sedjo, 2004), and pest and disease resistance are only two of a much wider suite of possibilities.

There are various ways of achieving such goals, such as using *in vitro* tissue cultures, somatic embryogenesis, organogenesis, micropropagated explants and so on (Merkle and Nairn, 2005), but here we will

Table 6.5. Forest traits that can be improved using biotechnology (from Sedjo, 2004).

Silviculture	Adaptability	Wood quality traits
Growth rate	Drought tolerance	Wood density
Nutrient uptake	Cold tolerance	Lignin reduction
Crown/stem	Fungal resistance	Lignin extraction
Flowering control	Insect resistance	Juvenile fibre
Herbicide		Branching

concentrate on gene transfer technologies, or to use more popular terminologies, genetic engineering or genetic modification (GM). Insect resistance, as we have discussed in Chapter 4, is partially a property of host plant chemistry, and indeed certain specific chemicals can often be identified which deter or otherwise reduce the activities of herbivores. The chemical sucrose phosphate synthase (SPS) is able to influence the levels of plant phenolics and plant nitrogen in trees such as aspen, *Populus tremula*. Higher concentrations of SPS can reduce the feeding rates of aspen defoliators like the leaf beetle *Phratora vitellinae* (Coleoptera: Chrysomelidae). Hjältén *et al.* (2007) created two genetically modified lines of aspen using a hybrid wild type as a basis; one clone expressed 1.5 times the norm and the other 4 times the norm of SPS compared with the wild type. Figure 6.9 illustrates the leaf consumption of *P. vitellinae* on the two GM lines compared with the wild type, showing that the GM line with 4 times the concentration of SPS showed significant insect antifeedant effects. SPS in itself is not a plant defensive compound, but GM poplar of this type may be useful in reducing defoliation problems, assuming that the GM crops have no other undesirable side effects.

The most common way of genetically engineering plants to resist insect attack in agriculture is the use of GM crops modified to express genes from the insecticidally active microorganism *Bacillus thuringiensis* (Bt – see also Chapter 8). A large number of arable and food crops are now available as transgenic varieties, and some countries such as Brazil and the USA plant very large areas with GM cotton and maize, for example (Speight *et al.*, 2008). Transgenic Bt trees are much less widespread, though the technology for creating them has produced some examples that work well in field trials. Take, for example, poplars, *P. nigra*, in China. Hu *et al.* (2001) carried out field trials of a transgenic variety of *P. nigra* expressing a *Cry1Ac* gene for *B.t.* var *kurstaki*, which is specific for Lepidoptera, studying the success of two species of defoliator, *Apocheima cinerarius* (Lepidoptera: Geometridae) and *Orthosia incerta* (Lepidoptera: Noctuidae) compared with that on non-transgenic trees. Not only were pest densities on transgenic trees reduced significantly but also the number of pupae, a good indication of pest densities in the next generation, was lower in transgenic plantations. Note that the number of pupae/m^2 even in the transgenic plots was an average of 20 or so, which may or may not have been sufficient to cause significant damage to trees later by newly emerging larvae. At least, though, this type of resistance system may well have a role to play as part of an IPM programme, where resistance can be combined with other forms of control such as biological or chemical (see Chapter 10).

6.5 Forest Health and the Planning Stage

Bi *et al.* (2008) present a list of recommendations to be adopted at the planning stage of a tropical forest plantation, based on experiences in south-west China, much of which we have discussed in previous sections of this chapter. Their first criterion reintroduces the concept of forest health, and the second the practice of pest and disease surveillance, i.e. the monitoring of forest health. We shall return to these topics later in the book when we discuss risk or hazard rating systems (see Chapter 9), but for now take the example of Boa (2003) in his guidebook to the state of health of trees. In particular,

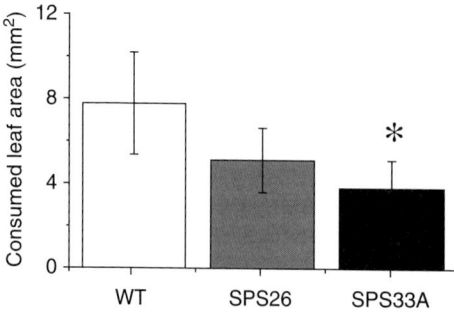

Fig. 6.9. Leaf consumption (mm^2) (±SE) by the leaf beetle *Phratora vitellinae* of two different GM lines of *Populus tremula*, compared with that on wild-type aspen. * = means sig diff @ $P < 0.05$ level (from Hjältén *et al.*, 2007, courtesy Elsevier).

Eric Boa provides a self-help document for local farmers and foresters who observe symptoms in their crops and want to get some idea at least of the cause, the severity and the future prognosis for decline, survival and possible cure. An example of best practice, Boa provides copious high-quality colour photographs under various headings to guide the user towards an identification of their tree problems. Clearly, if, with all the best planning possible, such symptoms are still detected, it may be too late to cure the problem. It is certainly important to be able to detect the differences between, say, a soil problem and a pest or pathogen; the former may be cured or prevented in the future by soil improvement or changing site, while the latter may be much more intractable. Thus, by referring to past experiences with various tree species in the locality, tree health may be maintained and pest or disease problems prevented, if at all possible. A widespread adoption of tree health diagnostic systems is urgently required.

Once some knowledge of the likely causes of tree health decline have been identified, the next questions are how to confirm the details of the problem and what to do with this information in terms of pest management, especially in a situation where access to expert advice is lacking (see also the next section). This is where expert systems may play a role. Expert systems, ES (also known as decision support systems, DSS), are a form of artificial intelligence which purports to replace the human expert with a computer (Wang *et al.*, 2009). The machine is able to provide scientific advice and technical support to any forester or pest manager who has access to the Internet (and presumably has paid the appropriate log-on fees). Such systems, once up and running, can also be used for education and training, again at a distance. They are fairly widely used for agricultural crops such as rice, and some advances in ES systems for forest pest and disease management have been produced in temperate forest situations (Kaloudis *et al.*, 2005). The CABI *Crop Protection Compendium* is probably the nearest thing to an ES for tropical forestry at the moment. Figure 6.10 shows an example of how it can be used to identify all the potential insect pests of eucalypts in Brazil, and then to single out *P. semipunctata*, for which to find details of biology, ecology, control and so on. Advance search options enable foresters and pest and disease managers to produce short (sometimes long) lists of culprits to look out for, with suggestions as to what to do if they appear. It is very important that these systems are readily available and affordable in all tropical countries that may make use of them.

6.6 Socio-economics and Forest Pest Management

As we have discussed in Chapter 1, a large part of tropical forestry takes place in the developing world. In these areas, it is possible that insect and disease management (see Chapter 10) may be of low priority, or simply unachievable for a myriad of social, educational and economic reasons. In 1991, K.S.S. Nair of the Kerala Forest Research Institute in south-west India published a crucial paper discussing the basic problems with the inception and routine use of successful insect pest management in tropical forests. His comments were based on the Indian experience, but have great relevance to most of the tropical world. In essence, Nair believes that 'to a large extent, socio-economic factors are the main constraints to successful development and adoption of IPM strategies against forest defoliators (and, undoubtedly, other types of pest too)'. He blames various factors. In India at least, all forests are government owned and forest pest control has a low priority relative to the more pressing agricultural pest problems. Nair goes on to point out that agriculture, being a private enterprise, generates more social demand than forestry. One major implication of this prioritization is the paucity of specialist forest entomologists; those that there are often have not had an opportunity for adequate training, even if basic knowledge about pest ecology and management were available. Finally, he suggests that to this latter end, scientific research in tropical forestry

(a)

Welcome to Crop Protection Compendium:

Crop Protection Compendium
The world's most comprehensive site for Crop Protection Information

Log-out

| Home | Datasheets | Abstracts Database | Library | Glossary | Search | More Resources |

Log-out
Main Menu

Advanced Datasheet Search

eucalyptus AND brazil AND insect NOT host/crop

Datasheets (22)

Sort by: relevance / date / A - Z   ◄ Previous Next ►

Showing 1 - 10 of 22

Search results for 'eucalyptus AND brazil AND insect NOT host/crop'

Phoracantha recurva
Eggs The egg is cylindrical with rounded extremities, yellow-white and 2.5 mm long. Larvae The larval body is white,

Gonipterus scutellatus
Eggs The eggs are laid in a pod or capsule containing three to 16 eggs. The pods can be found attached to both surfaces of newly

Gonipterus gibberus
Eggs The eggs are laid in greyish capsules containing one to six eggs and are attached to the leaves on both surfaces. The egg is

Phoracantha semipunctata
Eggs Elongate-oval, spindle-shaped, yellowish, 2.5-3.0 mm long, laid in groups of 30-40 in a single layer. Larva

Ctenarytaina eucalypti
Eggs These are laid singly and randomly in crevices along the leaf petioles, usually perpendicular to the plant surface. They are

Hypothenemus obscurus
Eggs Eggs are a translucent white and slightly barrel shaped. In macadamia, they are typically laid in a group of about four eggs per

Steirastoma breve
Eggs The eggs are elongate oval and slightly thickened towards one end, shining, pale yellow, translucent, and with the chorion soft and

Atta
Atta spp. have three castes of adult ants: males, females (queens) and workers. All castes pass through an immature phase, from egg to larvae and

Heliothis virescens
Eggs Eggs are subspherical with a flattened base, creamy white, developing a reddish-brown band as the embryo develops. Eggs are a uniform

Orthezia insignis
Ezzat (1956) and Green (1922) described O. insignis in detail. Body of adult female is about 1.5 mm long and 1.3 mm wide (excluding the ovisac),

◄ Previous 1 | 2 | 3 Next ►

Fig. 6.10. Extract from (a) the *Crop Protection Compendium* (b) using the eucalyptus longhorn beetle, *Phoracantha semipunctata*, as an example (CABI, 2010).

Continued

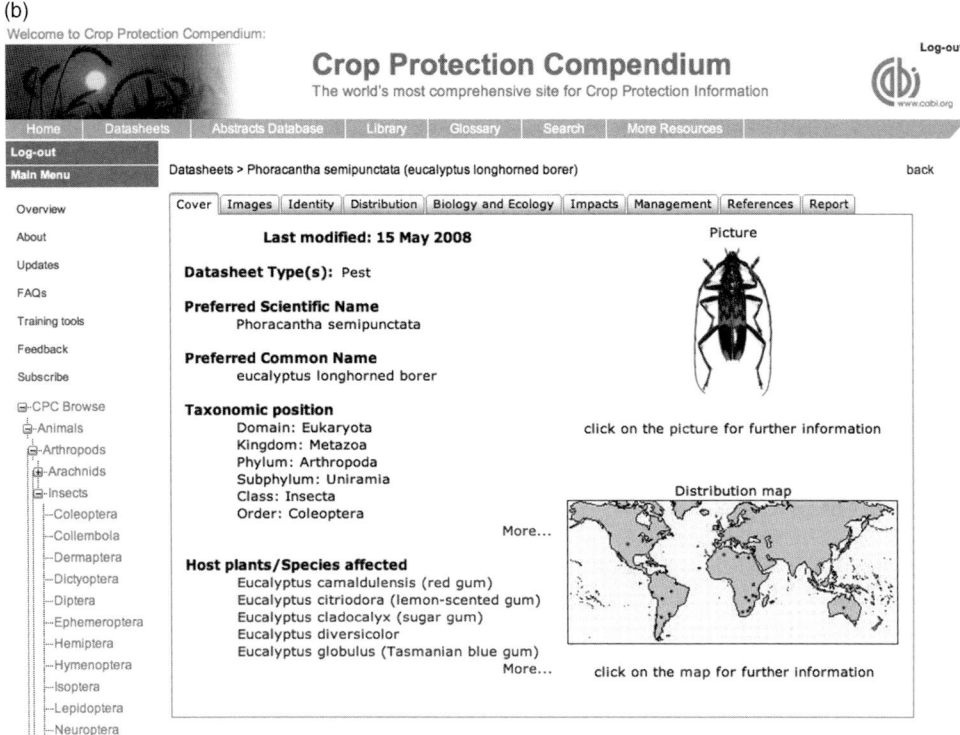

Fig. 6.10. Continued.

must be redirected into problem solving, involving multidisciplinary approaches with emphases on applied studies and the application of extension, or practical advice to workers in the field.

It must not be forgotten that local farmers and foresters in many tropical countries are in possession of a wealth of indigenous knowledge about plant health problems and their solutions (Bentley *et al.*, 2009). However, a lack of training and expertise in modern practices and technologies is still very widespread, especially when forest plantations are concerned and, undoubtedly, education is required in order to carry out best practices in pest management (Kagezi *et al.*, 2010).

Various themes repeatedly crop up, including lack of specialized training, lack of advice or extension and inadequate technology and/or funds to implement the new measures. Two examples from East Africa will illustrate these points.

What experience and expertise there is in local communities is often based on agricultural systems rather than forestry ones, and certainly in the case of agroforestry at least, the adoption of new management tactics is likely to be based on those from farming. Sileshi *et al.* (2008) interviewed farmers in Malawi, Mozambique and Zambia to discuss their perceptions of pest problems in agroforestry and to investigate their pest management tactics. Of the four major species of agroforestry tree planted in Zambia, farmers clearly felt that insect pests concerned them most. The majority of these pests were defoliating beetles, grasshoppers, moth larvae and, in particular, termites; diseases seemed to be of less concern. Despite these perceived problems, most farmers did little or nothing about the defoliators but employed mainly indigenous techniques for controlling termites, such as the use of wood ash or chopped *Euphorbia* branches in tree planting holes. Seemingly, no external pest management advice was to be had. The other example concerns a specific insect pest, the eucalyptus gall wasp, *L. invasa* (see also Chapter 5). Farmers in Uganda have been growing *Eucalyptus* woodlots for fuelwood and timber and the pest found its way to Uganda some years ago, causing substantial damage. Nyeko *et al.* (2007) questioned farmers about their problems with the gall wasp and how they handled it. In this case, some advice was available from forestry and agriculture sources in Uganda, but as Table 6.6 shows, it was variable in efficacy. Admittedly, this was a very small sample, but all manner of tactics were suggested, from the sensible but impractical in this context (plant resistant eucalypts) to the probably ineffective but dangerous (spray with chemicals). None the less, the provision of some sort of expert advice, however variable in quality, for forest pest problems is a step in the right direction and one the presence or absence of which should be taken into account at the planning stage.

Finally, in this context it must not be forgotten that in many tropical countries, communities have no access to pest management tactics, and if they did, they probably could not afford it. In dry parts of Ethiopia, for example, households have a chronic shortage of land for cultivation and tree planting is unlikely to be a high priority, despite the need for timber (Alelign *et al.*, 2011). The lack of fertilizers and pesticides exacerbates tree health and insect pest problems.

6.7 The Basis for Decision Making – Acquisition and Dissemination of Entomological Knowledge

As mentioned in the previous section, there are certain requirements for efficient management of tropical forest pests. These include: (i) surveys of major crop species for pest incidence and damage; (ii) the quantification of the impact in terms of volume losses,

Table 6.6. Advice received by farmers in Uganda about the management of *Leptocybe invasa* infestations on eucalypts (from Nyeko *et al.*, 2007).

Advice	Source	Useful, yes?	Useful, no?	NA	Total number of responses	Responses in each category (%)
Wait, we are still researching	DFD, DDA, FORRI	2	1	0	3	18.8
Spray with chemicals	DDA, FORRI	2	1	0	3	18.8
Plant resistant types of *Eucalyptus*	NFA, NFC	1	1	0	2	12.5
Cut and burn affected trees	DDA, FORRI	1	0	1	2	12.5
No chemical can control the pest	DFD	0	1	0	1	6.3
Weed properly	NFA	1	0	0	1	6.3
Plant healthy seedlings from a good source	NFA	1	0	0	1	6.3
Ensure timely planting	NFA	1	0	0	1	6.3
Apply liquid fertilizer	DFA	1	0	0	1	6.3
Beware of a disease on *Eucalyptus*	DFD	1	0	0	1	6.3
Total		11	4	1	16	100

Notes: NA = advice not yet applied; DFD = District Forest Department; DDA = District Department of Agriculture; FORRI = Forestry Resources Research Institute; NFA = National Forestry Authority; NFC = Nyabyeya Forestry College; DFA = District Farmers' Association.

Plate 1

Plate 2

Plate 3

Plate 4

Plate 5

Plate 1. Primary rainforest, Danum Valley, Sabah, Malaysia (Image courtesy of Martin Speight).
Plate 2. Primary rainforest interior, Sabah, Malaysia (Image courtesy of Martin Speight).
Plate 3. Secondary forest, Peninsular Malaysia (Image courtesy of Martin Speight).
Plate 4. *Gmelina arborea* young plantation, Kalimantan, Indonesia (Image courtesy of Martin Speight).
Plate 5. Monoculture of *Acacia mangium*, Sumatra, Indonesia (Image courtesy of Ross Wylie).

Plate 6

Plate 8

Plate 7

Plate 9

Plate 10

Plate 6. *Eucalyptus* spp. plantations in Guangxi Province, China (Image courtesy of Ross Wylie).
Plate 7. Agroforestry trial of *Leucaena leucocephela* intercropped with beans, Kenya (Image courtesy of Martin Speight).
Plate 8. *Khaya anthotheca* trees in village location, Mozambique (Image courtesy of Martin Speight).
Plate 9. Defoliation of *Casuarina cunninghamiana* by leaf beetle *Rhyparida limbatipennis* in Queensland, Australia (Image courtesy of Ross Wylie).
Plate 10. Defoliation to a stand of *Mangletia glauca* by sawflies, Vietnam (Image courtesy of Martin Speight).

Plate 11

Plate 12

Plate 11. Hurricane damage to mangrove forests, Yucatan, Mexico (Image courtesy of Martin Speight).
Plate 12. Aleppo pine, *Pinus halepensis*, under severe drought stress, killed by the bark beetle, *Pityogenes* spp., Jordan (Image courtesy of Martin Speight).

Plate 13

Plate 15

Plate 14

Plate 16

Plate 13. *Pityogenes* spp. galleries under bark of drought stressed Aleppo pine, *Pinus halepensis*, Jordan (Image courtesy of Martin Speight).
Plate 14. Larvae of the eucalyptus leaf beetle, *Paropsisterna cloelia*, Queensland, Australia (Image courtesy of CSIRO Entomology).
Plate 15. Eucalyptus snout weevil, *Gonipterus scutellatus* (Image courtesy of J. Voss).
Plate 16. Leaf-cutter ant, *Atta* spp., Costa Rica (Image courtesy of Paul Embden).

Plate 17

Plate 19

Plate 18

Plate 20

Plate 17. Defoliation by leaf-cutter ants, Costa Rica, *Atta* spp. (Image courtesy of Martin Speight).
Plate 18. Larvae of the sawfly, *Perga kirbyi*, on stem of *Eucalyptus grandis*, Queensland, Australia (Image courtesy of Ross Wylie).
Plate 19. Close-up of leaf eaten by larvae of teak defoliator, *Hyblaea puer*a, Kerala, India (Image courtesy of Martin Speight).
Plate 20. *Dendrolimus punctatus* larva on *Pinus caribaea* var *hondurensis*, Vietnam (Image courtesy of Martin Speight).

Plate 21

Plate 22

Plate 24

Plate 23

Plate 21. Larva of bagworm *Pteroma plagiophleps* feeding on *Falcataria moluccana*, Malaysia (Image courtesy of Ross Wylie).
Plate 22. Stand of *Falcataria moluccana* defoliated by bagworm *Pteroma plagiophleps* and yellow butterfly *Eurema blanda*, Sumatra, Indonesia (Image courtesy of Ross Wylie).
Plate 23. Leaf mines of *Acrocercops* spp. (Image courtesy of J. Jarman).
Plate 24. Damage caused by larvae of eucalyptus leaf tier *Strepsicrates semicanella*, Guangxi Province, China (Image courtesy of Ross Wylie).

Plate 25

Plate 26

Plate 27

Plate 28

Plate 29

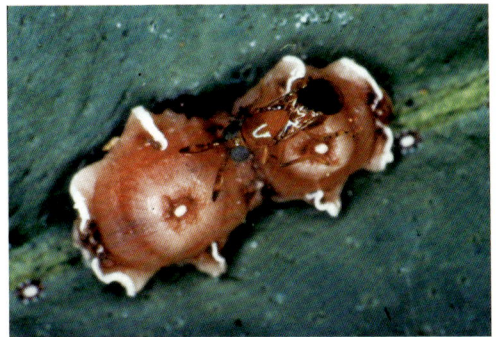

Plate 25. Pine woolly aphid, *Pineus pini*, on *Pinus radiata*, Australia (Image courtesy of J. Voss).
Plate 26. Damage to *Pinus patula* caused by pine woolly aphid, *Pineus pini*, Australia (Image courtesy of Ross Wylie).
Plate 27. Cypress aphid, *Cinara* spp., Malawi (Image courtesy of Sean Murphy).
Plate 28. *Cupressus lusitanica* damaged by cypress aphid, *Cinara* spp., Malawi (Image courtesy of Martin Speight).
Plate 29. Pink wax scale *Ceroplastes rubens*, with parasitoid, Queensland, Australia (Image courtesy of Dan Smith).

Plate 30

Plate 31

Plate 30. Bushing of *Eucalyptus robusta* caused by *Helopeltis* spp., Sumatra, Indonesia (Image courtesy of Ross Wylie).
Plate 31. Red gum lerp psyllid, *Glycaspis brimblecombei*, California, USA (Image courtesy of Jack Kelly Clark).

Plate 32

Plate 33

Plate 34

Plate 32. Damage caused by red gum lerp psyllid, *Glycaspis brimblecombei*, California, USA (photo credit Jack Kelly Clark).
Plate 33. Eggs and adults of leucaena psyllid, *Heteropsylla cubana*, Nepal (Image courtesy of Martin Speight).
Plate 34. Damage to *Eucalyptus scopairia* by the winter bronzing bug, *Thaumastocoris peregrinus*, (Image courtesy of Andrew Scales).

Plate 35

Plate 37

Plate 36

Plate 38

Plate 35. Damage to bark of *Khaya ivorensis* by larvae of *Indarbela quadrinotata*, Sabah, Malaysia (Image courtesy of Ross Wylie).
Plate 36. Adult five-spined bark beetle, *Ips grandicollis*, Queensland, Australia (Image courtesy of Luigino Doimo).
Plate 37. Galleries and larvae of *Ips grandicollis* under the bark of *Pinus elliottii*, Queensland, Australia (Image courtesy of Ross Wylie).
Plate 38. Raised welts on bark of *Eucalyptus deglupta* indicating tunnelling of the varicose borer, *Agrilus sexsignatus*, Mindanao, Philippines (Image courtesy of Ross Wylie).

Plate 39

Plate 40

Plate 41

Plate 42

Plate 39. Adult eucalyptus longicorn beetle, *Phoracantha semipunctata*, museum specimen (Image courtesy of Chris Fitzgerald).
Plate 40. Larvae and galleries of longicorn beetle, *Phoracantha semipunctata* under *Eucalyptus grandis* bark, Mozambique (Image courtesy of Martin Speight).
Plate 41. Adult *Aristobia approximator*, Vietnam (Image courtesy of Michael Cota).
Plate 42. Frass cylinders produced by the Asian ambrosia beetle, *Xylosandrus crassiusculus* (Image courtesy of Russell Mizell).

Plate 43

Plate 44

Plate 45

Plate 46

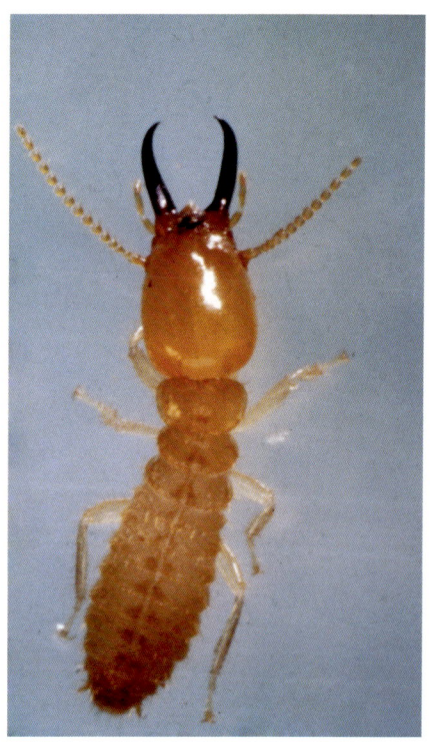

Plate 43. Gallery of Asian ambrosia beetle, *Xylosandrus crassiusculus,* with pupae (Image courtesy of Hulcr).
Plate 44. Adult female sirex woodwasp, *Sirex noctilio*, on *Pinus radiata* (Image courtesy of Michaellbbecker).
Plate 45. Larva, tunnel and frass of *Sirex noctilio*, South Africa (Image courtesy of Bernard Slippers).
Plate 46. Termite soldier, *Coptotermes acinaciformis*, Queensland, Australia (Image courtesy of B. Cowell).

Plate 47

Plate 49

Plate 48

Plate 50

Plate 47. Tunnels in stems of *Acacia mangium* produced by *Coptotermes curvignathus*, Sumatra, Indonesia (Image courtesy of Ross Wylie).
Plate 48. Patch death of *Acacia mangium* due to stem boring by *Coptotermes curvignathus*, Sumatra, Indonesia (Image courtesy of Ross Wylie).
Plate 49. Larvae of the goat moth, *Coryphodema tristi*s, in *Eucalyptus* spp., South Africa (Image courtesy of S. Lawson).
Plate 50. Damage caused by larvae of the goat moth *Coryphodema tristis* in *Eucalyptus* spp., South Africa (Image courtesy of S. Lawson).

Plate 51

Plate 52

Plate 53

Plate 54

Plate 55

Plate 51. Larva of beehole borer, *Xyleutes ceramica,* from *Gmelina arborea*, Kalimantan, Indonesia (Image courtesy of Martin Speight).
Plate 52. Pupal case and frass of beehole borer, *Xyleutes ceramica,* from *Gmelina arborea*, Kalimantan, Indonesia (Image courtesy of Martin Speight).
Plate 53. Branchlet of *Araucaria cunninghamii* infested by *Hylurdrectonus araucariae*, Papua New Guinea (Image courtesy of Barry Gray).
Plate 54. Adult pine shoot moth, *Dioryctria rubella*, Vietnam (Image courtesy of Martin Speight).
Plate 55. Mature larva of pine shoot moth, *Dioryctria rubella,* inside leader of *Pinus kesiya*, Vietnam (Image courtesy of Martin Speight).

Plate 56

Plate 57

Plate 59

Plate 58

Plate 56. Young plantation of *Pinus caribaea* var. *hondurensis* showing bushy form caused by pine shoot moth, Philippines (Image courtesy of Martin Speight).
Plate 57. Final instar larva of mahogany shoot borer, *Hypsipyla robusta,* in *Toona ciliata*, Queensland, Australia (Image courtesy of Martin Speight).
Plate 58. Webbed frass produced by larva of mahogany shoot borer, *Hypsipyla robusta*, Queensland, Australia (Image courtesy of Manon Griffiths).
Plate 59. Young *Cedrela odorata* attacked by mahogany shoot borer, *Hypsipyla grandella*, Paraguay (Image courtesy of Martin Speight).

Plate 60

Plate 61

Plate 62

Plate 63

Plate 64

Plate 60. Adult female blue gum chalcid, *Leptocybe invasa*, egg laying on *Eucalyptus* spp., Israel (Image courtesy of Zvi Mendel).
Plate 61. Exit holes in midrib gall of *Eucalyptus* spp. produced by blue gum chalcid, *Leptocybe invasa*, South Africa (Image courtesy of Jolanda Roux).
Plate 62. Leaf galls on *Eucalyptus* spp. caused by gall wasp, *Ophelimus maskelli*, Israel (Image courtesy of Zvi Mendel).
Plate 63. Adult Erythrina gall wasp, *Quadrastichus erythrinae,* Florida, USA (Image courtesy of Paul Skelley).
Plate 64. Leaf galls of *Quadrastichus erythrinae on Erythrina variegata* var. *orientalis,* Oahu, Hawaii, USA (Image courtesy of Forest and Kim Starr).

Plate 65

Plate 66

Plate 67

Plate 68

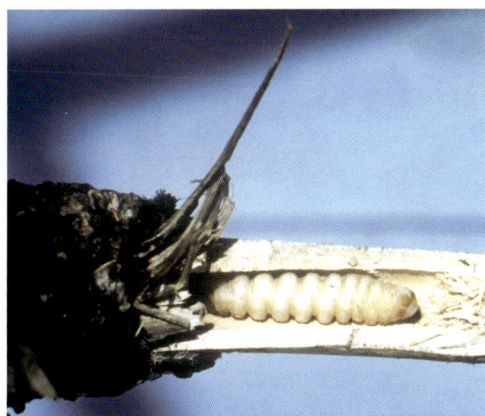

Plate 69

Plate 65. Losses in young eucalypt plantation caused by root-feeding termites, Guangxi Province, China (Image courtesy of Ross Wylie).
Plate 66. White grub, *Lepidiota stigma*, and root damage to *Eucalyptus exserta*, Guangdong Province, China (Image courtesy of Ross Wylie).
Plate 67. Gaps in young plantation of *Eucalyptus exserta* caused by root feeding of *Lepidiota stigma*, Guangdong Province, China (Image courtesy of Ross Wylie).
Plate 68. Larva of branch-pruning longicorn, *Strongylurus decoratus,* in hoop pine, Queensland, Australia (Image courtesy of G. Trinder).
Plate 69. Exit hole of longicorn, *Strongylurus decoratus*, in hoop pine, showing wood slivers plugging the pupal chamber, Queensland, Australia (Image courtesy of G. Trinder).

Plate 70

Plate 71

Plate 73

Plate 72

Plate 70. Resin canals in eucalypt stem resulting from ambrosia beetle attack, Sumatra, Indonesia (Image courtesy of Ross Wylie).
Plate 71. Sap staining associated with *Ips grandicollis* attack on fire-damaged *Pinus,* Queensland, Australia (Image courtesy of Ross Wylie).
Plate 72. Webbed frass covering tunnel of wood moth, probably *Zeuzera* sp., in young eucalypt, Sumatra, Indonesia (Image courtesy of Ross Wylie).
Plate 73. Tunnelling by the larva of a wood moth, probably *Zeuzera* sp., in this *Eucalyptus urophylla* has resulted in breakage of the leader, Sumatra, Indonesia (Image courtesy of Ross Wylie).

Plate 74

Plate 75

Plate 76

Plate 77

Plate 78

Plate 74. Tree in *Araucaria cunninghamii* seed orchard attacked by golden mealybug, *Nipaecoccus aurilanatus*, Queensland, Australia (Image courtesy of Ross Wylie).
Plate 75. Trees as part of mixed agriculture, Kandy, Sri Lanka (Image courtesy of Ross Wylie).
Plate 76. Slash and burn agriculture, Luzon, Philippines (Image courtesy of Ross Wylie).
Plate 77. Fuelwood stacked by roadside, Mozambique (Image courtesy of Martin Speight).
Plate 78. Collecting leaves and twigs for fuel in a eucalypt plantation in Guangxi Province, China (Image courtesy of Ross Wylie).

Plate 79

Plate 82

Plate 80

Plate 83

Plate 81

Plate 79. Tea crop shaded by *Grevillea robusta*, Sri Lanka (Image courtesy of Ross Wylie).
Plate 80. *Swietenia macrophylla* planted in a taungya system of village forestry, Sri Lanka (Image courtesy of Ross Wylie).
Plate 81. Roadside planting in saline soil, India (Image courtesy of Jeff Burley).
Plate 82. Sacred fig tree, Nepal (Image courtesy of Martin Speight).
Plate 83. Damaged bottle tree, Roma, Australia (Image courtesy of Ross Wylie).

Plate 74

Plate 75

Plate 76

Plate 77

Plate 78

Plate 74. Tree in *Araucaria cunninghamii* seed orchard attacked by golden mealybug, *Nipaecoccus aurilanatus*, Queensland, Australia (Image courtesy of Ross Wylie).
Plate 75. Trees as part of mixed agriculture, Kandy, Sri Lanka (Image courtesy of Ross Wylie).
Plate 76. Slash and burn agriculture, Luzon, Philippines (Image courtesy of Ross Wylie).
Plate 77. Fuelwood stacked by roadside, Mozambique (Image courtesy of Martin Speight).
Plate 78. Collecting leaves and twigs for fuel in a eucalypt plantation in Guangxi Province, China (Image courtesy of Ross Wylie).

Plate 79

Plate 82

Plate 80

Plate 83

Plate 81

Plate 79. Tea crop shaded by *Grevillea robusta*, Sri Lanka (Image courtesy of Ross Wylie).
Plate 80. *Swietenia macrophylla* planted in a taungya system of village forestry, Sri Lanka (Image courtesy of Ross Wylie).
Plate 81. Roadside planting in saline soil, India (Image courtesy of Jeff Burley).
Plate 82. Sacred fig tree, Nepal (Image courtesy of Martin Speight).
Plate 83. Damaged bottle tree, Roma, Australia (Image courtesy of Ross Wylie).

Plate 84

Plate 85

Plate 86

Plate 84. Dying *Prosopis juliflora* planted in arid site, Turkana Desert, Kenya (Image courtesy of Martin Speight).
Plate 85. Longicorn beetle larva inside branch of *Prosopis juliflora*, Turkana, Kenya (Image courtesy of Martin Speight).
Plate 86. Larvae of *Anoplophora lucipor* in *Acacia mangium* stems, Bataan, Philippines (Image courtesy of Ross Wylie).

Plate 87

Plate 88

Plate 89

Plate 87. Tunnels excavated by *Anoplophora* larvae in *Acacia mangium*, Bataan, Philippines (Image courtesy of Ross Wylie).
Plate 88. Stump of sal, *Shorea robusta*, killed by buprestid beetles, showing waterlogged mud on roots, Nepal (Image courtesy of Martin Speight).
Plate 89. Eucalypt planted on a wet site with impervious soil layer, Sumatra, Indonesia. Note drainage trenches (Image courtesy of Ross Wylie).

Plate 90

Plate 91

Plate 92

Plate 93

Plate 90. Experimental forest nursery, Mozambique (Image courtesy of Martin Speight).
Plate 91. Forest nursery showing natural forest in background, Mindanao, Philippines (Image courtesy of Ross Wylie).
Plate 92. Adult *Spodoptera litura* (museum specimen) (Image courtesy of Queensland Department of Primary Industries).
Plate 93. White grub, *Rhopaea* sp., a pest of hoop pine, *Araucaria cunninghamii,* in nurseries in Queensland, Australia (Image courtesy of Queensland Department of Primary Industries).

Plate 94

Plate 95

Plate 96

Plate 94. Variegated grasshopper, *Zonocerus variegatus*, Africa (Image courtesy of Paul Embden).
Plate 95. Root-feeding cricket, *Gryllotalpa* sp., Queensland, Australia (Image courtesy of D. Ironside).
Plate 96. Adult yellow butterfly, *Eurema hecabe* (museum specimen) (Image courtesy of Queensland Department of Primary Industries).

Plate 97

Plate 100

Plate 98

Plate 101

Plate 99

Plate 102

Plate 97. Yellow weevil, *Hypomeces squamosus,* feeding on *Falcataria* in nursery, Vietnam (Image courtesy of Martin Speight).
Plate 98. Damage to roots of hoop pine in nursery by *Aesiotes notabilis,* Queensland, Australia (Image courtesy of Queensland Department of Primary Industries).
Plate 99. Multi-stemming of *Pinus* hybrid in nursery, caused by *Nysius clevelandensis*, Queensland, Australia (Image courtesy of Chris Fitzgerald).
Plate 100. Polythene nursery pots, Zimbabwe (Image courtesy of Peter May).
Plate 101. Root curling in nursery stock of *Acacia mangium*, Sabah, Malaysia (Image courtesy of Martin Speight).
Plate 102. Deformed taproot of 8-year-old *Acacia mangium* resulting from nursery mishandling, Sabah, Malaysia (Image courtesy of Martin Speight).

Plate 103

Plate 105

Plate 104

Plate 106

Plate 103. Raised nursery beds, Sri Lanka (Image courtesy of Ross Wylie).
Plate 104. Young *Acacia mangium* showing poor pruning of lower branches, Kalimantan, Indonesia (photo credit Martin Speight).
Plate 105. Thinning of *Eucalyptus grandis* stand, Paraguay (Image courtesy of Martin Speight).
Plate 106. Dense stand of sal, *Shorea robusta,* in need of thinning, Nepal (Image courtesy of Martin Speight).

Plate 107

Plate 109

Plate 110

Plate 108

Plate 111

Plate 107. *Pinus elliottii* killed by forest fire and invaded by *Ips grandicollis*, Queensland, Australia (Image courtesy of Judy King).
Plate 108. Salvage of fire-damaged *Pinus elliottii*, Queensland, Australia (Image courtesy of Ross Wylie).
Plate 109. Watering of log piles of fire-damaged *Pinus* spp. to prevent bark beetle attack, Queensland, Australia (Image courtesy of Ross Wylie).
Plate 110. Overmature stand of *Pinus* spp. with large amounts of brash and felling debris, Malawi (Image courtesy of Martin Speight).
Plate 111. Pulp mill log pile showing rapid degradation of harvested timber, Vietnam (Image courtesy of Martin Speight).

Plate 112

Plate 113

Plate 114

Plate 115

Plate 116

Plate 112. Pupal cases of tachinid fly parasitoid inside bagworm case, Vietnam (Image courtesy of Martin Speight).
Plate 113. *Megarhyssa nortoni*, a parasitoid of the wood wasp *Sirex noctilio*, laying eggs in larva deep in the wood of *Pinus radiata,* Australia (Image courtesy of Forestry Tasmania).
Plate 114. Moth larvae killed by nucleopolyhedrovirus (NPV) (Image courtesy of Martin Speight).
Plate 115. Termite damage to the lower stem of *Eucalyptus,* South Africa (Image courtesy of Peter May).
Plate 116. Termite damage to the root of young *Acacia,* Vietnam (Image courtesy of Ross Wylie).

Plate 117

Plate 118

Plate 119

Plate 120

Plate 121

Plate 117. Mixing of nursery soil with insecticide to combat root-feeding termites, Zimbabwe (Image courtesy of Peter May).
Plate 118. Planting *Eucalyptus* seedlings with insecticide granules to kill termites, South Africa (Image courtesy of Peter May).
Plate 119. Aggregating adults of the scarab *Epholcis bilobiceps*, North Queensland, Australia (Image courtesy of Ross Wylie).
Plate 120. 'Drainpipe' pheromone trap to monitor ambrosia beetle abundance in log yard, Queensland, Australia (Image courtesy of Ross Wylie).
Plate 121. Bark beetles found in imported packaging during routine inspection at port (Image courtesy of Martin Speight).

Plate 122

Plate 125

Plate 123

Plate 126

Plate 124

Plate 127

Plate 122. *Eucalyptus* plantations, Aracruz Celulose, Brazil (Image courtesy of J. Burley).
Plate 123. Poplar plantation defoliated by larvae of *Clostera* spp. (Image courtesy of György Csoka).
Plate 124. *Eucalyptus grandis* planted on ex-soybean field sites, Paraguay (Image courtesy of Martin Speight).
Plate 125. Young *Eucalyptus grandis* showing bark chewing at base by grasshoppers, Paraguay (Image courtesy of Martin Speight).
Plate 126. Grasshopper *Baeacris* spp., Paraguay (Image courtesy of Martin Speight).
Plate 127. Base of trunk of African mahogany, *Khaya anthotheca,* that has not been attacked, Mozambique (Image courtesy of Martin Speight).

Plate 128

Plate 129

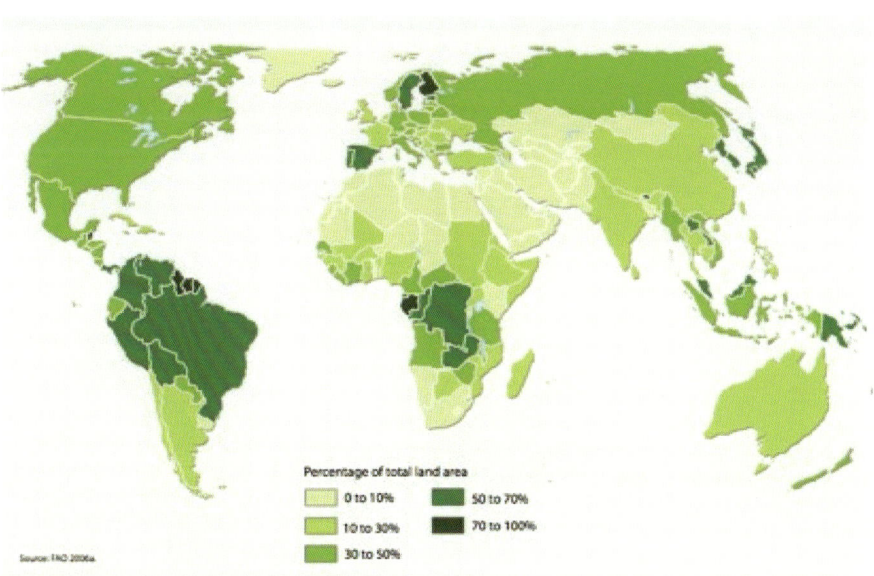

Plate 128. *Khaya anthotheca* showing multiple forking after attacks by *Hypsipyla* sp., Mozambique (Image courtesy of Martin Speight).
Plate 129. Global forest cover as a percentage of total land area (Image courtesy of UNEP, 2009).

Plate 130

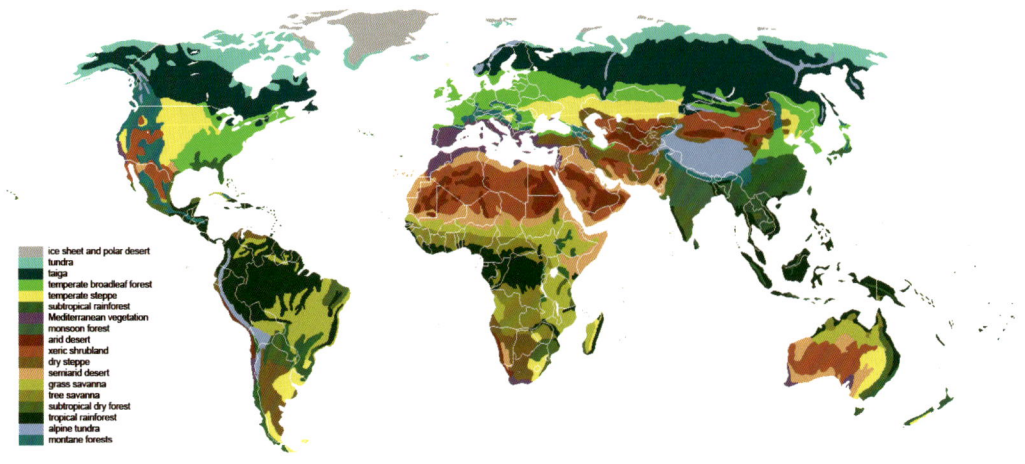

Plate 131

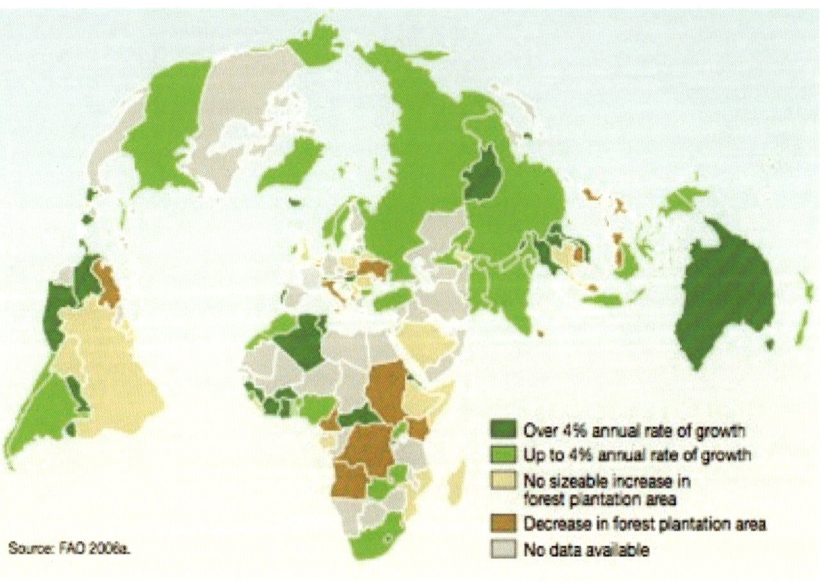

Plate 130. The major habitats (biomes) of the world, showing the distribution of tropical forests (Image courtesy of Sten Porse, licensed under the Creative Commons Attribution-Share Alike 3.0, 2011).
Plate 131. Percentage changes in the area of productive forest plantations (Image courtesy of UNEP, 2009).

deaths, degradation, etc., in terms of losses at the subsistence level, especially in agroforestry, and in money; (iii) investigations into the biology, ecology and host-tree relationships of major pest species; and (iv) the communication of these findings to foresters and entomologists in-country and also internationally. Such aims can be accomplished by: (i) suitably funded international initiatives to survey key crops for pests; (ii) funding international working groups or conferences to discuss current problems and work out collaboration for the future; (iii) disseminating findings in easily accessible publications; and (iv) providing appropriate training for local people both at home and overseas.

Many, if not all, forest insect pests are distributed far and wide and are not likely to honour national or regional borders! Hence, one country's problems are very likely to pose similar concerns in neighbouring or perhaps far-distant areas. It is simple common sense to join forces with research and development groups in any country where the same crops are grown and where insect (or fungal) pest problems are also actually or potentially present. Take, for example, the case of *Hypsipyla* spp. (Lepidoptera: Pyralidae). The larva of this moth is known variously as mahogany shoot borer, mahogany tip moth or cedar tip moth, and the genus, plus the enormous damage it exerts on *Swietenia*, *Toona*, *Cedrela* and *Khaya* spp. (see Chapters 5 and 10), is almost universal across the tropical world. Figure 6.11 illustrates its distribution. As the map shows, *Hypsipyla* species are to be found almost anywhere where mahoganies are indigenous or where attempts are made to establish plantations. Countries reporting significant damage include Australia, Columbia, Costa Rica, Cuba, Ghana, Honduras, India, Indonesia, Laos, Malaysia, Mexico, Mozambique, Pakistan, Papua New Guinea, Paraguay, Philippines, Solomon Islands, Sri Lanka, Tanzania, Thailand, Vietnam and Zimbabwe. While the taxonomy at species level is still confused, there is no doubt that if this global pest is finally to be defeated, then large-scale international collaboration is essential to: (i) combine resources and experiences and (ii) to avoid the duplication of effort. Floyd and Hauxwell (2000) provide an excellent example of how multinational and multidisciplinary collaboration may be brought to bear on shared pest problems.

The next step is putting into practice the key issues raised. This usually requires the

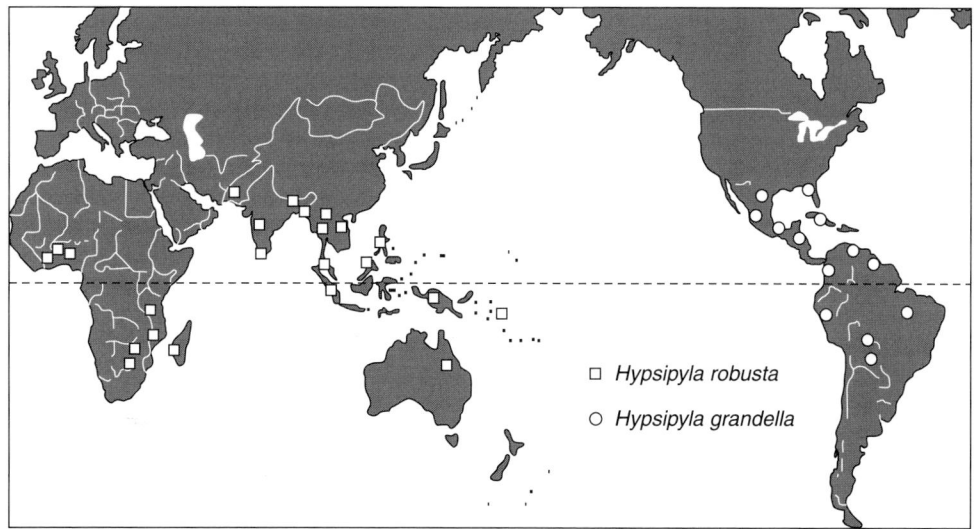

Fig. 6.11. Global distribution of reports of attacks by *Hypsipyla* spp. to mahoganies in plantations.

acquisition of considerable funding from international agencies, so that regional-wide surveys and control policies can be coordinated and carried out. In 1997, for example, a new initiative was instigated by Australian entomologists to carry out surveys of insect pests of *A. mangium* and various *Eucalyptus* species. This survey has involved forest entomologists from many South-east Asian countries, including Vietnam, Laos, Thailand, Malaysia and Indonesia. Survey tactics and protocols are standardized in each country so that the results obtained will be readily comparable and of immediate use and relevance to all concerned (Wylie *et al.*, 1998) (see Chapter 9). This example contrasts with the previous one, in that the former targeted a single pest complex, while the latter employed the host tree (crop) as the basis for collaboration.

6.8 Communication of Findings and Tactics

No survey or workshop is of any use unless the results and discussions are disseminated rapidly and efficiently to all concerned, as exemplified by the two publications mentioned in the previous section. However, reliable and extensive reference volumes are also required and it has to be said that, in the past, this facet of forest pest management has often been inadequate. Nevertheless, some prime examples of national or international publications and information retrieval systems are now available. One of the most up to date is the *Global Review of Forest Pests and Diseases* (FAO, 2009), which contains a wealth of information. Books on forest pests in various tropical countries have been appearing for years. One of the first was by Beeson, published in 1941, entitled *The Ecology and Control of Forest Insects of India and Neighbouring Countries*. This lengthy tome was of great relevance not only to the Indian situation but also, of course, to many other countries in the South Asian region. The only drawback with this type of work is its antiquity, and care must be taken by modern workers when relying on information that, at best, is over half a century old. Following Beeson's tradition, Browne published a comprehensive account of a very large number of insect and fungal pests in his *Pests and Diseases of Forest Plantation Trees: An Annotated List of the Principal Species Occurring in the British Commonwealth* in 1968. This has been the basic reference book for forest entomologists for many years, but again is desperately in need of updating. More recently, several books have been published which refer to individual countries or regions, and four examples will illustrate the scope and range of topics covered. The *Checklist of Forest Insects in Thailand* by Hutacharern and Tubtim (1995) is much as the title suggests, a long list of many forest insects found in Thailand. Reference is made to host plants and severity of damage, but there are no keys and no mention of potential management tactics. The volume is, however, of great value to entomologists in the South-east Asian region, and nothing so comprehensive has yet appeared for other parts of the tropical world.

In 1991, Wagner *et al.* published *Forest Entomology in West Tropical Africa: Forest Insects of Ghana*, with a second edition in 2008. This book provides much more detail on a smaller range of forest pests from Ghana, complete with photographs, diagrams and possible control strategies. In 1998, Elliott *et al.* published *Insect Pests of Australian Forests: Ecology and Management*. It is the first comprehensive treatment of this topic since W.W. Froggatt's books in 1923 and 1927. Importantly, in the context of this present book, it contains many references to forest insect pests in tropical and subtropical regions of Australia which have not been dealt with in Froggatt's books or elsewhere. The Tropical Agriculture Research and Training Centre (Centro Agronomico de Investigacion y Ensenanza; CATIE) in Turrialba, Costa Rica, produced the best pair of books about forest pests in Central America in 1992 (Hilje *et al.*, 1992). One volume, a handbook, provides copious information on major pests, including mites, molluscs and even vertebrates, as well as insects, nematodes and fungi. The second volume, a field guide, provides basic information about the biology of major pests, groups damage by plant parts and discusses various management techniques from silviculture to chemical control. Sadly, no more recent editions seem to be available. Then, in 2001, Speight and Wylie produced the first edition of this current book (Speight and Wylie, 2001). K.S.S. Nair published *Tropical Forest Insect Pests: Ecology, Impact and Management* in 2007, and finally, Cielsa published *Forest Entomology, A Global Perspective* in 2011. While unable to be comprehensive, or of course to deal with pests that arise after publication, all these books are to be applauded and other organizations in other countries should be encouraged to follow suit.

7

Management Systems II: Nursery Stage

7.1 Introduction

The overall objective of any nursery (Plate 90) is to produce good-quality, healthy and abundant stock at the lowest cost. It is in nurseries that forest practices most closely approach those of agriculture, and at the same time are the most different. Agriculture generally has little need for nurseries and farmers sow seeds directly into soil where the crop is to grow. Foresters, on the other hand, usually raise their seedling crop in nurseries and then plant them out elsewhere. The main reason for this is that only in nurseries can very young trees be afforded the sort of care necessary for their survival (Evans and Turnbull, 2004).

Forest nurseries occupy very small areas in comparison with the forests themselves and resemble agricultural crops in that seedling trees are planted at high density in uniform cultivated areas under intensive management (Speight and Wainhouse, 1989). This is a critical stage in production and mistakes made here may have serious consequences later in the plantation cycle. For example, if seedlings are left too long in impervious containers they become 'pot bound', coiling around inside the container. This habit continues after planting, impairing normal lateral root development and leading to instability, poor vigour and a predisposition to insect attack. With nurseries, there are both advantages and disadvantages for pest management.

7.1.1 Advantages

1. Their relatively small size makes it easier to monitor for pests and diseases than is the case in plantations. However, very few countries have specialist surveillance systems in place (see Chapter 9).
2. Even in the absence of routine surveillance, the concentration of staff at nurseries increases the likelihood that problems will be noticed earlier than they would be in most plantations and that remedial action will be taken more rapidly.
3. The combination of a high-value crop situated on a small, accessible area with an attendant labour force makes it comparatively easy to justify pesticide application.
4. Environmental side effects from any pesticide application are easier to avoid or contain over such discrete areas.

7.1.2 Disadvantages

1. Young trees in nurseries are generally the stage most nutritious and vulnerable to

insect attack. Their small size means that even numerically few insects can cause a considerable amount of damage and several plants may be destroyed by a single insect such as a cricket or caterpillar. Seedling trees have immature root systems and small energy reserves and may therefore be unable to recover from insect attack. Even if they do so, their form may be so altered as to render them useless for forest establishment (e.g. multi-leadering due to attack by sap-sucking bugs).

2. Because pest control can be justified and accomplished easily, there is a tendency towards overuse of pesticides at the slightest hint of a pest problem, when in fact the impact may be potentially minor or other control techniques may be more appropriate.

3. Nurseries in the tropics, particularly the wet tropics, are often surrounded by areas of remnant or cut-over forest which may serve as reservoirs for pest species (Plate 91) or are adjacent to agricultural or horticultural crops from which generalist feeders may migrate freely. This applies especially to temporary or 'flying' nurseries established close to the planting site.

Some of the common pest types that occur in forest nurseries, the nature of their damage and severity of impact are discussed below.

7.2 Main Pest Types in Forest Nurseries

7.2.1 Armyworms

Armyworms are the larvae of noctuid moths and are best known as pests of agricultural crops, but they also cause occasional severe damage in forest nurseries. They derive their common name from their habit of moving in enormous numbers across grass and cereal crops. Moths (Plate 92) typically lay eggs on the lower surfaces of the leaves of the host. Larvae are active at night and hide in the soil litter by day. They are defoliators and can consume completely the aerial portions of young plants. The most common genus occurring in nurseries is *Spodoptera*, which has a pan-tropical distribution. In Brazil, *S. latifascia* damages young eucalypt seedlings and is regarded as a serious pest in the region (Santos *et al.*, 1980). The species has a high fecundity, females laying up to 3000 eggs each, and the life cycle is completed in 50 days. *S. sunia* is a nursery pest of *Pinus caribaea*, *P. tropicalis* and *Casuarina* in Cuba, and losses of 40% of seedlings have been reported (Mellado, 1976; Castilla *et al.*, 2003). In Asia and Australia, *S. litura* is an important species feeding on a wide range of tree species, including teak in India, where Roychoudhury *et al.* (1995) recorded damage to about 56% of plants in one nursery, *Falcataria moluccana* in the Philippines (Braza, 1990b), *Araucaria cunninghamii* in Australia (Brown and Wylie, 1990) and *Acacia mangium* in Malaysia (Nair, 2007).

7.2.2 Cutworms

Like armyworms, cutworms are larvae of noctuid moths but are soil-inhabiting pests which typically cut off seedlings at ground level and drag them into shallow burrows in the soil to be eaten. The moths are active at night and usually lay their eggs in moist, recently cultivated soil or on the stems and leaves of plants. Young cutworm caterpillars climb plants and either skeletonize the leaves or eat small holes in them. Older caterpillars may browse on foliage but commonly cut through stems at ground level and feed on the top growth of felled plants. Caterpillars that are almost fully grown often remain underground and chew into plants at or below ground level. A single caterpillar may cut through several plant stems in one night.

Agrotis ipsilon and *A. segetum* are cosmopolitan, migrant species with a wide range of host plants. They are important pests of conifers in forest nurseries in several countries, such as Bangladesh (Baksha, 1990, 2001), India (Jha and Sen-Sarma, 2008), Zimbabwe (Mazodze, 1993) and China (Bi *et al.*, 2008). At one nursery in southern USA in 2003, cutworms destroyed over a million loblolly pine, *P. taeda*, seedlings (South and Enebak, 2006; see Table 7.2). Surveys by Ali and Chaturvedi (1996) in North Bihar, India, during 1991–1992 showed that up to 20% of seedlings of *Albizia lebbek* and *Eucalyptus tereticornis* were damaged by *A. ipsilon*, and this species also caused 10–30% mortality in eucalypt nursery stock at Qinzhou, southern China (Pang, 2003). In Central America, the most common and problematic cutworms are *A. ipsilon*, *A. malefida*, *A. repleta* and *A. subterranea* (CATIE, 1992).

7.2.3 White grubs

The larvae of some species of scarab beetles damage trees by feeding on the roots, often ring-barking and severing the stem below ground level. Usually referred to as 'white grubs' or 'curl grubs', these insects are important pests of seedlings in forest nurseries worldwide (Plate 93). They live in the soil, feeding on organic plant matter when young and later on plant roots. They migrate through the soil profile in response to temperature extremes and soil conditions, and the larval stages in some species occupy up to 2 years. In nursery beds, injury is usually first recognized when patches of previously healthy seedlings begin to exhibit drought-like symptoms, turn a faded green to brown colour and die. These seedlings are pulled out of the soil easily with a gentle tug, revealing damaged root systems.

There are numerous species of white grubs which cause problems in nurseries but some of the most frequently recorded belong to the genera *Lepidiota*, *Anomala*, *Leucopholis* and *Holotrichia*. Baksha (1990) records heavy mortality of *Hevea brasiliensis* seedlings aged 3–5 months in nurseries of rubber estates in Bangladesh due to a complex of species in the last three genera mentioned. In Sri Lanka, damage caused by the cockchafer grub of *H. serrata* in nurseries is one of the main problems in raising teak plantations, and mortalities of up to 20% have been recorded (Bandara, 1990). In India, *Schizonycha ruficollis* damaged 14–52% of teak seedlings in nurseries in Nagpur (Kulkarni *et al.*, 2007), while *H. serrata*, *H. consanguinea*, *H. rustica* and *H. mucida* have also caused significant mortalities of seedlings of teak, *A. nilotica* and *Azadirachta indica* (Ali and Chaturvedi, 1996; Kulkarni *et al.*, 2009).

7.2.4 Scarab beetles

Scarab beetle adults, as well as their larvae (white grubs), are also pests in nurseries, defoliating seedlings and sometimes girdling stems. In southern Queensland, Brown and Wylie (1990) reported that swarms of the African black beetle, *Heteronychus arator*, caused considerable losses among *P. elliottii* var. *elliottii* seedlings, damaging the aerial parts and also burrowing in the soil and ring-barking stems at or just below ground level. The outbreak is thought to have originated, in part, from fallow nursery beds sown to Gatton panic, a cultivar of *Panicum maximum*, and from grassy surrounds of the nursery. In Hong Kong, swarms of *Anomala cupripes* occasionally damage *P. massoniana* in nurseries during the summer months (Browne, 1968). Thakur and Sivaramakrishnan (1991) reported that 70–80% of the seedlings were defoliated overnight by cockchafer swarms in April–May in nurseries of *E. tereticornis*, *Dalbergia sissoo*, *Syzygium cuminii*, *Peltophorum ferrugenium* and *Santalum album* in southern India.

7.2.5 Grasshoppers and crickets

These insects, together with cutworms and white grubs, form the main pests of nursery seedlings. Grasshoppers browse the foliage of young plants and may break or sever twigs and stems. Crickets often sever stems at ground level and drag the plants into their burrows for feeding. Most of the feeding by these insects occurs at night. Grasshoppers are particularly damaging when they occur in swarms.

The variegated grasshopper, *Zonocerus variegatus* (Plate 94), which is widely distributed in tropical Africa, is primarily a pest of agricultural and agroforestry crops but also damages nursery seedlings in several countries. The insects are particularly active during the dry months and in Ghana, for example, there have been regular outbreaks every year in various localities (Wagner *et al.*, 2008). The young nymphs feed gregariously by day and roost in shrubs at night. They migrate slowly by walking rather than hopping and may devour virtually any green vegetation. In Sierra Leone, they are reported to have killed seedlings of *Terminalia ivorensis* by defoliation and browsing of the bark and young twigs (Browne, 1968). In Malawi, a related species, *Z. elegans*, is a pest of *Gmelina arborea* in nurseries, eating the growing tips and other tender tissues. Nymphs and adults of *Letana inflata* feed voraciously on the foliage of sandalwood in nurseries in India. They also lay eggs in longitudinal slits in succulent sandalwood seedling stems and the slit swells and cracks laterally, damaging the seedlings (Remadevi et al., 2005). *Valanga nigricornis* is a common pest in nurseries of *A. mangium* in Thailand and Indonesia (Nair, 2007).

As discussed in Chapter 5, *Brachytrupes* is an important genus of ground-dwelling crickets occurring in Africa and Asia, and has caused seedling mortalities in nurseries of up to 30%. Another important genus is *Gryllotalpa*, the mole crickets, which are omnivores feeding on insects and the roots of plants (Plate 95). In southern pine nurseries in the USA, adults and older nymphs eat seeds, feed on roots or cut off stems of seedlings just above the soil surface. A great deal of damage can also be caused indirectly by the tunnelling habit of mole crickets, which disturbs the soil and sometimes uproots seedlings, causing them to dry out (Bacon and South, 1989). In Venezuela, *Gryllotalpa* sp. caused losses of 20% of young *P. caribaea* seedlings in a nursery within a month of sowing (Vale et al., 1991), while in forest nurseries in Cuba *Anurogrillus* spp. are important pests of *Tabebuia angustata* (white wood), *Samanea saman* (raintree), *Cordia gerascanthus* (cordia wood) and *Cedrela odorata* (cigar box cedar) (Castilla et al., 2002).

7.2.6 Defoliating caterpillars

Species of Lepidoptera whose larvae are defoliators of nursery plants span a wide range of genera in many families, and a few examples are given here. In India, the larvae of *Pyrausta machaeralis* (Pyralidae) and *Hyblaea puera* (Hyblaeidae) are major pests of seedlings of teak. In one study by Basalingappa and Ghandi (1994), 10 out of 26 plots (each of 500 seedlings) showed 100% infestation, while the remaining 16 plots had infestations ranging from 98.6 to 99.8%. Many of the attacked plants were defoliated completely and their tops withered.

The noctuid caterpillar *Eligma narcissus* is a major pest of *Ailanthus triphysa* and *A. excelsa*, important matchwood species raised on a plantation scale in Kerala, India. Young caterpillars eat the chlorophyll tissue of leaves or the tender areas near the margin, leaving round, whitish patches or holes. Older larvae eat the marginal and other leaf tissues, often leaving only the midribs. Seedlings die after repeated defoliation (Roonwal, 1990). According to Sivaramakrishnan and Remadevi (1996), almost 100% defoliation of nursery stock of *A. triphysa* occurs annually during August–October, particularly in areas of high rainfall. In the same region, the psychid *Cryptothelia crameri* cuts down the seedlings of *Casuarina equisetifolia* and *S. album* to the level of the root collar, two to three cuttings resulting in the death of the seedlings. Yellow butterflies, *Eurema* spp. (Plate 96) produce caterpillars that are important defoliators of *F. moluccana* and *A. mangium* in nurseries in the Philippines (Braza, 1990a) and Indonesia and of *F. moluccana* in Bangladesh (Baksha, 2001). *Phalanta phalantha*, a nymphalid butterfly, is a major nursery pest of poplar, *Populus deltoides*, in India and Roychoudhury et al. (2001) report an incidence of 70–83% in nurseries at Jabalpur. The nettle caterpillar, *Darna pallivitta*, is an invasive moth in the Hawaiian Islands. It has a wide host range, having been reported to feed on over 45 plant species in 22 families. Their major economic damage comes from defoliation of ornamental nursery stock and they also pose a human health hazard due to urticating hairs that can cause painful stings and skin inflammations (Jang et al., 2009).

7.2.7 Leaf tiers or rollers

This type of damage is very common in nurseries and can sometimes be severe. Larvae of several moth groups, particularly Tortricidae and Pyralidae, web together or roll leaves as protection against predators and feed on foliage contained within the shelter.

In Central and West Africa, the pyralid moth *Lamprosema lateritalis* is widespread in the lowland rainforests of Nigeria, Ghana and the Ivory Coast. It is the only member of the genus to attack a forest tree and is the most serious pest of the valuable indigenous timber species *Pericopsis elata* (Wagner *et al.*, 2008). The first observation of the moth in Ghana was in a forest nursery in Kumasi. Eggs are laid on the upper surface of leaflets and newly emerged caterpillars combine to web two overlapping leaves together to make a nest. The caterpillars feed gregariously and skeletonize the surfaces of the leaves, which then wither. The larvae then crawl out of the damaged leaves to the nearest pair of suitable fresh leaves and repeat the process. Pupation occurs in the rolled-up, withered nests. Wagner *et al.* (1991) estimate that caterpillars emerging from one healthy, normal-sized egg batch would be capable of completely defoliating several 6- to 7-month-old *P. elata* seedlings in the nursery. Seedling loss due to repeated defoliation of *P. elata* by *L. lateritalis* was estimated at 30–40% in a period of 1 year.

Larvae of the tortricid moth *Strepsicrates* spp. (Plate 24) are serious pests of *Eucalyptus* spp. in the tropics of the eastern hemisphere (see Chapter 5). Their behaviour is somewhat similar to that of *L. lateritalis*, except that the larva is solitary and rolls a single eucalypt leaf to form a shelter in which it feeds, repeating the process when the leaf withers. Damage in one *Eucalyptus* nursery in Ghana exceeded 50% in some beds (Wagner *et al.*, 1991) and in Sabah, Chey (1996) reported infestation levels of 20%.

7.2.8 Leaf beetles, weevils and leaf-cutting ants

As with the defoliating caterpillars, there are many species of leaf beetles which chew the foliage of nursery plants, causing growth setback, deformity or mortality. Such impacts are exemplified by *Chrysomela populi*, a pest of poplar in nurseries in India and Pakistan (Khan and Ahmad, 1991; Sharma *et al.*, 2005). Eggs are laid in clusters on the lower surfaces of leaves and emerging larvae feed gregariously at first, skeletonizing the leaf and finally consuming large pieces of it. The adults similarly feed voraciously on the foliage. Attacks by *C. populi* can result in death or severe malformation of seedlings. Damage evaluation studies for another chrysomelid defoliator of *Populus* spp. in India, *Nodostoma waterhousie*, showed infestation levels ranging from less than 10% to almost 60% on various species, provenances and clones grown under nursery conditions (Singh and Singh, 1995).

Weevils such as *Myllocerus* spp. can also inflict heavy injury on nursery plants, as has been documented for these pests on *A. senegal* and *A. tortilis* in the Thar Desert of India (Vir and Parihar, 1993) and on *D. sissoo* in Bihar (Sah *et al.*, 2007). The beetles often appear suddenly in large swarms and can cause spectacular defoliation, although generally they are rated as minor pests. Another common, polyphagous weevil in Asia which defoliates dicotyledonous plants in nurseries is the gold dust yellow weevil, *Hypomeces squamosus* (Plate 97). It can also occur in large numbers and devours tender young leaves from the edge inwards but on older leaves eats only the softer tissues between the veins (Browne, 1968). The Gamari weevil *Alcidodes ludificator* is a serious pest of *G. arborea* in nurseries in northeastern India (Senthilkumar and Barthakur, 2009). In tropical and subtropical Queensland, larvae of the pine bark weevil, *Aesiotes notabilis*, cause a very different type of damage (Plate 98). They sometimes attack tubed *A. cunninghamii* seedlings in nurseries, tunnelling in the main roots, chewing away woody tissue as well as bark and finally forming cocoons at or near ground level (Brimblecombe, 1945; Elliott *et al.*, 1998).

The biology and mode of feeding of leaf-cutting ants has been discussed in Chapter 5. In addition to the damage they cause to young seedlings in the field (Marsaro *et al.*, 2004), *Atta* spp. are also serious pests in forest nurseries in South and Central America (Aguirre-Castillo, 1978). For example, studies carried out in Para State, Brazil, in 1985 to evaluate the damage caused by *Atta sexdens* to seedlings of rubber trees, *Hevea* spp., in nurseries showed that 71.3% of seedlings were attacked (Calil and Soares, 1987). In Cuba, *A. insularis* is a polyphagous pest of forest nurseries, but the vulnerable species appear to be *P. caribaea*, *P. cubensis* and *P. maestrensis* (Castilla *et al.*, 2002).

7.2.9 Psyllids

Sap-sucking psyllids can be significant pests of nursery stock, their activities resulting variously in stunting, distortion, wilting, tissue necrosis, loss of vigour and plant death.

In tropical Africa, two important genera of Psyllidae cause two very different kinds of damage. *Diclidophlebia* spp. are pests of *Triplochiton scleroxylon* in Nigeria, Ghana and the Ivory Coast. Eggs are deposited either on the principal leaf veins (*D. eastopi*) or the edges of young leaves (*D. harrisoni*) and hatch within 8 days. The nymphs live in colonies of 10–20 individuals and pass through five instar stages. Their sucking causes the periphery of the leaves to curl, roll and later turn yellow or brown. Attack can result in dieback, stunted growth, copious branching of the stems, defoliation and death. Entire plots in the nursery may be destroyed (Wagner *et al.*, 2008).

Feeding by *Phytolyma* spp., on the other hand, results in galling of plant parts. As described in Chapter 5, eggs are laid on buds, leaves and shoots of the host (*Milicia* spp.) and hatch in about 8 days. The nymphs or 'crawlers' burrow into the adjacent tissues, breaking down the epidermal cells, causing fermentation of the leaf parenchyma and the formation of a gall, which encloses the nymph. The gall eventually splits open to release the adult. These galls disrupt the plant's translocation processes, cause tip dieback and, in many cases, seedling mortality. In Ghana, 100% failures of nursery crops have been reported (Wagner *et al.*, 2008).

In India also, gall formation and stunting by psyllids has been the main problem for *Pterocarpus marsupium* and *A. odoratissima* in nurseries in Kerala (Mathew, 2005).

Spanioneura sp. caused leaf vein galls and *Arytaina* caused pouch galls on *P. marsupium* in a nursery at Peechi, resulting in leaf crinkling and severe seedling stunting. Up to 40% of seedlings were infested. At the same nursery, feeding by *Psylla oblonga* on *A. odoratissima* led to epicormic shoot formation, stunting and seedling dieback, with up to 98% of seedlings infested. Damage by these insects caused total failure of the nursery and a continuous schedule of insecticide application was found to be necessary to protect the seedlings.

7.2.10 Gall wasps

As described in Chapter 5, gall wasps most commonly cause galls on the leaf blade, midrib, petiole or stem of new shoots, resulting in deformation, stunting, premature leaf drop, dieback and sometimes death of the plant. The chalcid gall wasp *Leptocybe invasa* (Plates 60 and 61) attacks *Eucalyptus* spp. trees of all ages, from nursery stock to mature trees, but the damage is most severe on younger plants (Mendel *et al.*, 2004). In Vietnam, where the plantation industry is reliant on only a few clones of *E. urophylla* and *E. camaldulensis*, it reportedly has devastated nurseries (Thu *et al.*, 2009) and has been similarly problematic in *E. camaldulensis* nurseries in Thailand (Dell *et al.*, 2008) and Argentina (Botto *et al.*, 2010). Surveys carried out in various nurseries in Gujarat, India, revealed levels of infestation of eucalypt seedlings by *L. invasa* of between 13 and 100%, with an average of approximately 62% (Jhala *et al.*, 2010).

7.2.11 Mealybugs, scales, aphids and whiteflies

Mealybugs (Hemiptera: Pseudococcidae) are free-living insects and their bodies may be covered with fine powder, or 'meal'. The females are sap sucking but the males have no functioning mouthparts. Mealybugs can attack roots, stems, leaves and fruit of plants. In Queensland, the golden mealybug

Nipaecoccus aurilanatus is a pest of *A. cunninghamii* in nurseries. It is gregarious and can form dense colonies whose activities cause wilting and death of growing tips (Elliott et al., 1998). *Rastrococcus iceryoides* infests *A. lebbek* in India, weakening and withering the seedlings and sometimes causing death of plants (Pillai et al., 1991).

Scale insects vary greatly in appearance, although they have a similar basic life history. The young crawlers are highly mobile and seek feeding sites on the host plant. Later instar nymphs become sedentary, building their protective cover or scale and feeding more or less continually on leaf and stem tissue, using their long, very thin, flexible stylets to tap the plant sap. Their feeding can reduce tree vigour and cause mortality. In a nursery in Tamil Nadu in India, seedlings of the mangrove *Rhizophora mucronata* were infested by the scale *Aspidiotus destructor*, causing the leaves to wither and fall (Kathiresan, 1993). Five months after planting, 16% of the seedlings were dead and another 47% infested. Height and leaf numbers were reduced by more than 30%. In India, three species of coccid (*Kerria lacca, Ceroplastes actinoformis* and *Inglisia bivalvata*) have been reported as causing mortality of sandalwood plants in nurseries (Remadevi et al., 2005). Scale insects and mealybugs are serious pests on forest nursery plants such as *Khaya senegalensis, Tectona grandis, G. arborea* and *Eucalyptus* spp. in Ghana (Wagner et al., 2008).

The biology and importance to forestry of conifer aphids such as *Cinara* and *Pineus* (an adelgid) are discussed in Chapter 5. Species in both these genera are pests of nursery plants, and indeed their accidental introduction into several countries is believed to have been via infested nursery stock. Whiteflies are common nursery pests worldwide, their feeding causing leaf chlorosis, leaf shedding and reduced growth rate of the plant. The honeydew produced by nymphs leads to mould development on leaves and affects photosynthesis adversely. In India, the spiralling whitefly, *Aleurodicus dispersus*, and the babul whitefly, *Acaudaleyrodes rachipora*, are highly polyphagous, attacking important tree species both in nurseries and plantations (Sundararaj and Dubey, 2005).

7.2.12 Mirids, lygaeids and thrips

Sap-sucking mirid and lygaeid bugs 'sting' the tips of young seedlings, injecting a toxic saliva which causes tip wilt, dieback and distortion. In Queensland, the lygaeid *Nysius clevelandensis* has contributed to a serious multi-stemming problem in nursery stock of *Pinus* spp. (Plate 99) (Elliott et al., 1998). *Lygus* spp. mirids cause similar damage worldwide; for example, in one nursery in southern USA, approximately 50% of loblolly pine (*P. taeda*) seedlings were injured by *L. lineolaris* (Dixon and Fasulo, 2009). Pine seedlings severely damaged by *L. lineolaris* usually do not survive the growing season.

Thrips have rasping and sucking mouthparts and feed on soft, new growth of plants in a similar way to Hemiptera, except that penetration is relatively shallow. In *P. radiata* nurseries in Chile, *Thrips tabaci* and *Heliothrips haemorrhoidalis* caused needle crinkle and apical distortion, which resulted in production losses of 23% (Cerda, 1980). In Andhra Pradesh, India, *Frankliniella occidentalis*, commonly known as the Western flower thrips, has been associated with little leaf disease of eucalypts, which is in epidemic form in both the nursery and plantations (Kulkarni, 2010). Both adults and nymphs suck sap from the growing tips and leaves, resulting in greatly reduced leaves, pale narrow laminae, stunting, distortion, wilting of foliage and leaf drop. A mycoplasma-like organism is believed to be transmitted by the thrips, causing little leaf symptoms.

7.2.13 Shoot and stem borers

The poplar shoot borer *Eucosma glaciata* is one of the most destructive pests of young poplars in nurseries in the Western Himalayas of India (Sharma et al., 2005). Larvae of this eucosmid moth tunnel in the tender apical shoots of the plants, causing shoot death and forking. *Hypsipyla* spp. shoot borers severely damage nursery stock, as well as plantation trees of many high-value Meliaceae (*Swietenia, Cedrela, Khaya, Toona, Chukrasia*) throughout the tropics (see Chapter 5). Stem-boring scolytinae beetles have also been reported to cause considerable seedling mortality in nurseries in several countries; for example, *Hypothenemus dimorphus* on *A. auriculiformis* and *Xylosandrus crassiusculus* on *A. mangium* in Malaysia (Nair, 2007; Varma and Swaran, 2009), *Xyleborus fornicatus* on *G. arborea* in India (Mathew, 2005) and *H. birmanus* on *S. macrophylla* in Samoa and Fiji. In Sri Lanka, attack by the black twig borer *X. compactus* on seedlings of *S. macrophylla* and *K. senegalensis* has resulted in the complete failure of some nurseries (Mannakkara and Alawathugoda, 2005).

7.2.14 Termites

Termite damage is common in forest nurseries in several countries, particularly those in drier regions of Africa and India. They damage the root system below ground level by hollowing out or ring-barking the taproot, leading ultimately to the death of the seedling. In Ethiopia, *Macrotermes subhyalinus*, *Pseudacanthotermes militaris* and *Odontotermes* spp. cut the stem of nursery seedlings at ground level and *Microtermes* damages the root system (Cowie and Wood, 1989). In south India, termite damage has sometimes resulted in total loss and abandonment of nurseries (Thakur and Sen-Sarma, 2008), the main species responsible being *Microcerotermes minor*, *Odontotermes* spp. and *Microtermes obesi*. Important fodder and fuelwood tree species are attacked; Thakur (1992) records 80% mortality of *Casuarina* seedlings and 10–30% mortality of *A. nilotica*. In forest nurseries in drier areas of Bangladesh, losses due to species of *Macrotermes*, *Odontotermes* and *Microtermes* can be substantial (Baksha, 1998).

7.2.15 Pest complexes

Quite often, observed pest impacts in nurseries may be due to a combination of insect species. A good example is provided by Pillai *et al.* (1991) in their discussion of insect pest problems of *A. lebbek* in nurseries in south India. The seedlings are initially subjected to sporadic attack by *T. flavus*, which lacerate the tender leaves, causing them to wrinkle. As the seedlings put forth more leaves, the psyllids *P. hyalina* and *Acizzia indica* and the aphid *Aphis* sp. nr *craccivora* invade and commence feeding on the fresh shoots and buds, causing further curling and crinkling of leaflets. The mealybugs *R. iceryoides* and *Ferrisia virgata* then make their appearance as whitish encrustations on stems and leaves, and their feeding causes yellowing and withering of the seedlings. Copious 'honeydew' is produced by the sap-sucking insects, on which sooty mould grows, blackening the plants and reducing photosynthetic ability. This combined insect attack results in dwarfed or stunted seedlings with disfigured and discoloured foliage bunched at the top, and may lead to death of the plants. The extent of the infestation in south India nurseries ranged from 30 to 100%.

7.3 Factors Predisposing Insect Attack – Immediate and Subsequent Problems

As indicated in the introduction, nurseries are a high-value crop and a critical stage in production. Most planting programmes require trees of the right species, ready at the right time, of the right size and sturdiness and produced in sufficient numbers (Evans and Turnbull, 2004). Insect damage in the nursery can interfere with the achievement of these goals. Inappropriate nursery practices can predispose plants to insect attack, both in the seedling stage and later in the plantation cycle.

7.3.1 Root curling or binding

The use of individual, impervious containers for plants is common in the tropics, and these include black polythene bags, metal tubes, bamboo pots and tubes, veneer sleeves and tarred paper (Plate 100). Such containers generally are removed or slit at planting. Failure to do so results in root distortion, instability, stress and likely insect attack (Plate 101). The same thing can happen if seedlings are left too long in these containers in a nursery. Their roots begin to coil and this habit continues even after the tree is removed from the container and planted. Twisted and spiralled roots retard plant growth and promote strangulation and tree death (Ferreira, 1993).

Poor handling during pricking out can also result in root damage and later problems. A good example is provided by Speight (1996, 1997) in regard to patch dieback of 8-year-old *A. mangium* in Sabah, Malaysia. These trees, which should have been growing vigorously at that age, were infested heavily by the larvae of two types of bark- and wood-boring beetles, a cerambycid *Xystrocera festiva* and an unknown buprestid, which girdled the stems. These are typically secondary pests which almost invariably only attack stressed trees. Excavation of the roots of dead and infested trees showed that they had deformed and twisted root systems with no true taproot (Plate 102). Examination of tubed *A. mangium* seedlings in local nurseries showed that up to 75% had root systems which were severely curled or coiled, resulting from mishandling during pricking out by piecework labourers.

7.3.2 Poor soil mix for potting and inadequate nutrients

The best soil to use will depend on the species and what is available, but in general, a pH of 5.5–7.0 and a sandy loam texture are desirable. Evans and Turnbull (2004) provide examples of suitable soil mixtures to fill containers. Unfortunately, in many tropical countries there is poor knowledge about soils and the requirements of different plant species. Containers are sometimes filled with clayey subsoil, which bakes hard and has poor water drainage, or with very acidic or alkaline soil in which certain plant nutrients and trace elements are either leached or become insoluble (Pancel, 1993).

Production of healthy seedlings depends on an adequate supply of plant nutrients and it may be necessary to add an appropriate fertilizer. This is not always a simple matter in many tropical countries where fertilizers are imported and there may be cost and administrative difficulties. While local manures are often used in these situations, they are not always beneficial and, in one study cited by Evans (1992), growth of *P. caribaea* var. *hondurensis* was actually depressed by any mixture containing cow dung.

Exotic pines grow very poorly if not inoculated with suitable mycorrhizal fungi, since without them the roots are inefficient in nutrient uptake. Seedlings grow slowly and have chloritic, sparse foliage. Nitrogen-fixing and leguminous species may require inoculation with specific fungi such as *Rhizobium* or *Frankia*.

7.3.3 Unsuitable watering and shade regimes and poor quality water

Too much water as well as too little can stress and kill plants. In addition, the water needs to be of good quality, not alkaline or of high salinity, and free of pathogens. Shading is related to watering since lower plant and soil temperatures reduce evapotranspiration stress (Evans and Turnbull, 2004). Too much shade may cause problems for seedlings when planted out in that they are not sun-hardened and sunscald may result. Shading may affect insect behaviour as, for example, in ornamental plant nurseries in Louisiana, USA, where too much shade was found to favour feeding activity and damage by the chrysomelid beetle *Rhabdopterus picipes* (Oliver and Chapin, 1980). In some nurseries, shelter is also necessary to protect seedlings against desiccating wind, sandstorm or hail.

7.3.4 Transplant stress

Plants require conditioning before leaving the nursery in order to minimize stress at planting out. Seedlings need a balanced root:shoot ratio, need to be able to support themselves and should be hardened off. If root growth is not controlled properly, this can result in seedlings with a few, long, tenuous roots which are very unsatisfactory for planting out and have poor prospects for survival. In California, USA, attack of poplars and willows by the sesiid moth *Parantbrene robiniae* was particularly associated with trees under stress from recent transplanting (Kaya and Lindegren, 1983).

7.3.5 Damage by nursery chemicals

A wide range of chemicals is used in forest nurseries for a variety of purposes, all aimed at producing healthy, abundant stock for large-scale tree-planting programmes at the lowest unit cost. Such chemicals include fumigants for soil sterilization, herbicides, fungicides, insecticides, fertilizers, soil stabilizers and antitranspirants. Many of these chemicals can damage plants if used injudiciously, and some plants are more sensitive than others to certain chemicals. For example, in Tanzania, fungicides used to control damping-off in nurseries caused some toxicity to *P. caribaea* seedlings' pre-emergence, but *P. khasya* was less affected (Hocking and Jaffer, 1969). Combining pesticides and growth regulators may cause phytotoxic reactions in plants, such as yellowing and leaf drop (McConnell and Short, 1987). In Zimbabwe, nursery treatment of *E. camaldulensis* transplants with 10% carbosulfan to protect them against attack in the field by *M. natalensis* resulted in phytotoxicity and growth check, but field-treated plants were not similarly affected (Mazodze, 1992). Another source of accidental chemical damage to plants in the nursery is the use of the same piece of equipment for spraying insecticides as that used for spraying herbicides.

7.3.6 Silvicultural side effects

Even sound silvicultural practices can have unexpected side effects. At Toolara in subtropical Queensland, Australia, in the late 1970s, *Pinus* spp. seedlings were raised in an open-root planting system. Crop rotation was practised, and at any one time, two-thirds of the nursery was fallow and planted with a cultivar of *Panicum maximum*. The aim was to reduce build-up of *Phytophthora* root rot and aid development of good soil structure. In 1978, swarms of the scarab beetle *H. arator* attacked beds of *P. elliottii* var. *elliottii*, damaging aerial parts and ring-barking stems at or just below ground level (Brown and Wylie, 1990). The outbreak is thought to have originated in part from the fallow crop and surrounding grassed areas.

Another example of how good practice can sometimes result in a poor outcome is from North Carolina in the USA. During a hot, dry period, large numbers of the grasshoppers *Melanoplus bivittatus*, *M. femurrubrum* (which normally feed on forbs and grasses) and *Schistocerca americana* (normally feeding on weeds and broadleaf bushes and trees) fed on *P. taeda* seedlings at night in a newly planted seed orchard. Up to 52 grasshoppers per branch were recorded, and these acridids fed mainly on the branch tips of plants more than 30 cm in height. The attack was attributed to high temperatures, which encouraged the grasshoppers to feed at night on their roosting trees, and to the mowing of the grass and weeds (their normal food plants) beneath the trees (Feaver, 1985).

7.4 Management of Nursery Pests

7.4.1 Physical/mechanical control

Control of pests by physical or mechanical methods (handpicking, barriers) can be simple, effective and environmentally friendly. Handpicking is generally employed only in countries where labour is relatively cheap. In north-eastern India, control of the Gamari weevil *A. ludificator*, a pest of *G. arborea* in nurseries, is achieved by handpicking the larvae and dropping them into a tin full of water containing a little kerosene (Senthilkumar and Barthakur, 2009). A similar technique is employed for larvae of *E. narcissus*, a noctuid defoliator of *A. excelsa* in southern India. Pupal clusters of this insect are also scraped off the stems and killed by crushing (Roonwal, 1990). In Ghana, *P. elata* seedlings in the nursery are vulnerable to attack by the leaf-tying moth *L. lateritalis* (Wagner *et al.*, 2008). Because over 80% of *L. lateritalis* eggs are laid on the upper surfaces of leaves, and the eggs may be recognized easily by any unskilled labourer, it has been recommended that they be removed from seedlings during normal morning waterings. Using this mechanical method of control, mortality of nursery seedlings due to this insect can be reduced from 30–40% to 5% or less in a year.

In Bangladesh and India, nursery beds are sometimes flooded with water to force cutworms and crickets to come out of their tunnels, and the insects are then killed manually (Baksha, 1990; Sharma *et al.*, 2005). Cutworm larvae attacking *E. grandis* seedlings in a nursery in Brazil were collected manually, and this reduced the numbers of seedlings damaged by about 92% (Anjos *et al.*, 1981). Ahmad *et al.* (1996) suggest that crawling grub populations can be trapped successfully by sinking narrow-neck earthen pots half-filled with kerosenized water. In India, fresh neem twigs with leaves are used to attract adults of white grubs, which can be collected and killed in order to reduce the population (Jha and Sen-Sarma, 2008). For sap-sucking insects such as scales and mealybugs, infested branchlets can be pruned off and burned.

In several African countries, when planting-out on termite-infested sites, the plastic container bags are slit but left in place around the seedlings as a protective barrier and the seedlings are shallow-planted with part of the container above ground. However, without insecticide treatment of the potting soil this would provide virtually no barrier to termites, especially as the bags have been slit to allow root growth (Cowie *et al.*, 1989). Physical barriers sunk deep in the soil have sometimes been used around nursery beds to prevent entry of white grubs and termites (Beeson, 1941). In Nepal, simple wooden screens and mosquito-proof window mesh have been used to protect seedlings of *Pinus* spp. and *Alnus nepalensis* against a variety of pests, including grasshoppers and crickets, reducing damage in some nurseries by more than 50% (Sharpe, 1983).

Light trapping is a useful monitoring tool for forecasting pest outbreaks in nurseries, but occasionally it is also used for pest management. For example, in Papua New Guinea, light traps have been used to attract noctuid moths away from the nursery crop with the intent of reducing oviposition on the plants. Ahmad *et al.* (1996) recommend the use of light traps to capture winged termites and beetles during their emergence.

7.4.2 Biological control and biopesticides

While there are numerous examples of the occurrence in nurseries of natural enemies of pest insects, examples of the deliberate release of such agents into a nursery situation in the tropics are much fewer. This is changing in line with a general trend in pest control towards integrated pest management and away from a reliance on chemical pesticides. Predatory ladybird beetles are one of the agents most commonly employed. In Queensland, Australia, the ladybird beetle *Cryptolaemus montrouzieri*, an effective predator of a range of sap-sucking Coccoidea, has been collected in the field and released into hoop pine *A. cunninghamii* nurseries to control infestations of the golden mealybug, *N. aurilanatus*. Currently, a range of generalist natural enemies of pests of agricultural importance, including *C. montrouzieri*, is reared commercially in Queensland and can be purchased for use in forest nurseries when required. Similarly in the USA, there is a thriving industry rearing a variety of insect predators and parasitoids for the biological control of hemipteran, coleopteran and lepidopteran pests (Warner and Getz, 2008) (see also Chapter 8).

Also more common recently has been the use of entomopathogenic fungi such as *Beauveria* and *Metarhyzium*, and the bacterium *Bacillus thuringiensis* (Bt). In India, *B. brongniartii* has been used for the management of the scarab beetle *H. serrata* infesting *H. brasiliensis* seedlings in nurseries. The pathogen had been shown to spread from infected to healthy beetles during mating. Nehru *et al.* (1991) reported that the release of contaminated adults in the nursery was effective for suppressing numbers of the scarab. Various formulations of *B. thuringiensis* have been used in the control of nursery pests, for example, Baksha (2007) recommends foliar applications of Bt wettable powder to control the pyralid defoliators *Arthroschista hilaralis* and *Margaronia* sp. in nurseries in Bangladesh. Parasitic nematodes have also been used to control nursery pests such as white grubs, weevils, leaf miners and thrips, with mixed success (Georgis *et al.*, 2006). Generally, biologicals have captured only a limited share of the pesticide market, mainly because of the higher production cost compared to standard insecticides, low efficacy under unfavourable conditions, short shelf-life, short field persistence and narrow spectrum of activity (Georgis *et al.*, 2006; South and Enebak, 2006). In addition, some strains may work well in one nursery in one year but not in another year or nursery.

7.4.3 Silvicultural control

As outlined in Section 7.3, cultural practices have a major influence on pest attack in forest nurseries and can be manipulated to prevent or minimize damage. In the case of the cranberry rootworm *Rhadopterus picipes*, a chrysomelid pest of ornamental tree species in Louisiana nurseries, the timing and manipulation of non-chemical nursery practices were adjusted to minimize adult feeding injury. The plants in their containers were usually located under tree canopies to protect them during hot and cold weather. It was found that *R. picipes*, a nocturnal species which hides in litter near its hosts during the day, prefers relatively dense shade and ground cover where moist conditions prevail. Reducing the amount of shade and removing litter from the containers and nursery areas reduced feeding activity significantly (Oliver and Chapin, 1980).

The presence of weeds in and around nurseries is a commonly attributed source of pest problems. For example, Dixon and Fasulo (2009) recommend the removal of preferred host plants of the tarnished plant bug, *L. lineolaris*, from edges of loblolly pine nurseries in the southern USA and destruction of favourable overwintering sites to reduce the damage caused by this pest. In Nigeria, Akanbi (1971) reports that nectar in the flower of the weed *Tridax procumbens* is important in sustaining adults of the nymphalid butterfly *P. phalantha*, a defoliator of poplar seedlings in the nursery and field, and recommends avoiding sites where the weed is predominant or clearing these weeds. Some windbreak and ornamental tree species are alternate hosts, or harbour insects, and should not be used around nurseries.

White grubs can also be controlled by cultural methods such as the preparation of nursery beds before, and non-disturbance of the soil during, the flight period of the adult beetles, mechanical removal of grubs during bed preparation and sieving of manure before its incorporation into the soil (Sen-Sarma, 1987). Elevation of nursery stock, for example in container trays, is another way of preventing damage by soil-dwelling pests (Plate 103). Varma and Swaran (2009) reported fewer pest problems in root-trainer raised nurseries than in conventional forest nurseries in Kerala, India. The shift in potting medium from soil to inert materials like vermiculite and the placement of root-trainer blocks on stands helped to prevent damage by white grubs and termites. Wardell (1990) recommends the planting of extra seedlings to allow for termite losses, both in the nursery and after planting out, and avoiding the use of banana fibre in making pots for seedlings. He also recommends the provision of alternative sources of food (plant debris, mulch) to attract termites away from the plants, although there are conflicting views about this (Cowie *et al.*, 1989).

7.4.4 Folk medicine

For ordinary villagers or subsistence farmers, chemical pesticides are not always necessary or suitable for use. They may pose health or environmental hazards, are expensive to buy, difficult to obtain and are not always available where and when they are needed. A variety of common compounds, some with insecticidal properties and others with repellent effects, have been used to prevent or control insect attack in nurseries. For example, soapy water is often used against scale insects and mealybugs, the soap dissolving the waxy covering on these insects and exposing them to desiccation and predation. In China, green grass that has been sprinkled with sugar or vinegar is laid on the ground in nursery beds and this attracts cutworms, which are then collected and killed (Wylie and Brown, 1992). Repellent plants are said to bestow protection on nearby crops up to a distance of 1 m (Goeltenboth, 1990) and marigolds are sometimes planted in and around nurseries for this purpose. A perimeter of wood ash scattered around the base of plants or seedling beds reportedly will block walking and crawling insects, and other compounds that can be used as barriers include lime, bonemeal, powdered charcoal or any vegetable dust. Wood ash may also be mixed into nursery beds or applied as a layer below polythene planting tubes to protect seedlings from termite attack (Logan et al., 1990). Natural sprays made from garlic and pepper are powerful insect repellents. A combination of soap, water, mashed garlic bulbs and red pepper will cause caterpillars to fall to the ground, and stinging nettle tea is useful against aphids (Goeltenboth, 1990). In the traditional homestead agroforestry systems in Bangladesh, herbal pesticides such as extracts of neem leaves and cow urine are used for the control of nursery pests (Shafiqul-Islam, 2011).

7.4.5 Chemicals

As indicated in the introduction to this chapter, chemical pesticides have been widely used in nurseries because the economics are easily justifiable and the environmental side effects over such small areas can be contained. The main types of insecticides employed are stomach poisons (usually ingested as baits), contact pesticides that penetrate the body wall to cause death and fumigants, the vapours of which enter the pest's body via the spiracles. The major groups of insecticides and mode of action are listed in Table 7.1 and examples of their use in nursery pest management are discussed below, together with some approximate costs (Table 7.2).

Botanicals

These are among the earliest insecticides, their use in agriculture dating back at least two millennia in ancient China, Egypt, Greece and India (Isman, 2006). They are sometimes referred to as the 'natural insecticides' because they are derived from plants. More than 200 plant species belonging to different families and genera have been reported to contain toxic principles which are effective against insects (Meshram, 2010). Chrysanthemum flowers when dried and crushed into a dust or powder were found to have insecticidal properties and have been used for pest management for more than two centuries (Casida and Quistad, 1998). These pyrethrins are noted for their rapid knockdown action and are usually used in combination with synergists such as piperonyl butoxide. Another botanical insecticide once commonly used in nurseries is nicotine sulfate. In Queensland, Australia, a spray of nicotine sulfate with soap and water was used in the 1960s and 1970s to control golden mealybug on hoop pine seedlings (see Chapter 5). The alkaline properties of the soap were essential for activating the active ingredient in the spray (Queensland Department of Forestry, 1963). Extracts from neem seeds or leaves are often used as an alternative to more persistent and toxic chemicals (Durairaj and Muthazhagu, 2009). They have a strong antifeedant and repellent effect as well as moderate insect toxicity. Neem oil has been used in *F. moluccana* nurseries in the Philippines against the yellow butterfly, *E. blanda*, (Braza, 1992).

Table 7.1. Insecticide groups, mode of action and approximate date of introduction (after Devine and Furlong, 2007, with permission Springer-Verlag).

Primary site and mode of action	Insecticide type	Common examples	First use*
Acetylcholinesterase inhibitors Nerve action	Carbamates	Aldicarb, Bendiocarb, **Carbaryl**, Carbofuran, Carbosulfan, Methiocarb, Methomyl, **Pirimicarb**, Propoxur, Thiodicarb	1956 1969
	Organophosphates	Acephate, Chlorpyrifos, Diazinon, Dimethoate, Finitrothion, Fenthion, **Malathion**, Methamidophos, Monocrotophos, **Parathion**, Pirimiphos, Profenofos, Temephos	1950 1946
GABA-gated chloride channel antagonists Nerve action	Cyclodiene organochlorines	Aldrin, **Chlordane**, Dieldrin, Endosulfan, Endrin, gamma-HCH (Lindane), Heptachlor	1945
	Phenylpyrazoles (fiproles)	**Fipronil**	1993
Sodium channel modulators Nerve action	Organochlorine	DDT	1943
	Pyrethroids	**Allethrin**, Bifenthrin, Cylfluthrin, Cypermethrin, **Deltamethrin**, Fenvalerate, Permethrin, Resmethrin	1952 1977
	Pyrethrins	Pyrethrins **(pyrethrum)**	1850s
Nicotine acetylcholine receptor agonists/antagonists Nerve action	Neonicotinoids	Acetamiprid, **Imidacloprid**, Nitenpyram, Thiacloprid, Thiamethoxam	1991
	Nicotine	**Nicotine**	1930s
	Spinosyns	**Spinosad**	1996
Chloride channel activators Nerve action	Avermectin	**Abamectin**, Emamectin benzoate	1985
Juvenile hormone mimics and analogues Growth regulation	Juvenile hormone analogues and mimics	Hydroprene, Kinoprene, Methoprene, **Fenoxycarb**, Pyriproxyfen	1985
Selective feeding blockers Nerve action	Cryolite	**Cryolite**	1929
	Pymetrozine	**Pymetrozine**	1999
Non-specific, multi-site inhibitors	Alkyl halides	Methyl bromide and other alkyl halides	1932
	Sulfuryl fluoride	**Sulfuryl fluoride**	1959
Microbial disruptors of insect midgut membranes	*Bacillus* species	***Bacillus thuringiensis*** and subspecies (*aizawai*, *israelensis*, *kurstaki*, *sphaericus*, *tenebrionis*)	1961

Continued

Table 7.1. Continued.

Primary site and mode of action	Insecticide type	Common examples	First use*
Inhibitors of oxidative phosphorylation	Diafenthiuron	**Diafenthiuron**	1997
Energy metabolism	Chlorfenapyr	**Chlorfenapyr**	1985
Inhibitors of chitin biosynthesis	Benzolureas	Novaluron, **Diflubenzuron**, Triflumuron	1983
Growth regulation	Buprofezin	**Buprofezin**	1988
Ecdysone agonists and moulting disruptors	Diacylhydrazines	**Halofenozide**, Tubufenozide	1999
Growth regulation	Azadirachtin	**Azadirachtin**	1985
Mitochondrial complex electron transport inhibitors	Hydramethylnon	**Hydramethylnon**	1980
Energy metabolism	Rotenone	**Rotenone (Derris)**	1850s
Voltage-dependent sodium channel blockers	Indoxacarb	**Indoxacarb**	2000
Nerve action			

Note: *Dates refer to the insecticide example given in bold and are the reported dates of first registration or use.

Table 7.2. Some approximate costs of pest management in loblolly pine and slash pine nurseries (from South and Enebak, 2006, with permission Springer-Verlag).

Pest group	US$/thousand seedlings	US$/hectare	Total production cost (%)
Abiotic	0.38	500	1.0
Fungi	0.13	175	0.4
Birds and mice	0.15	200	0.4
Annual weeds	0.14	190	0.4
Insects	0.19	250	0.5
Nematodes	1.39*	3700	4.0
Total	2.38	5015	6.7

Note: *Assumes two seedling crops per fumigation with methyl bromide/chlorpicrin.

Logan et al. (1990) list several plant extracts that are reported to be toxic or repellent to termites and which are added to tree nursery irrigation systems in Malawi, India and Zimbabwe. In India, a dye extracted from the bark of *Persea macrantha* and dissolved in ethanol and teepol was found to be very effective against whiteflies in nurseries, although the individual effect of the ethanol and teepol was not assessed (Sundararaj and Dubey, 2005).

Inorganics

Inorganic or mineral insecticides have a long history of use, the Greek writer Homer recommending the use of 'pest-averting' sulfur about 3000 years ago. They still have some applications today. They generally kill slowly, a feature which makes them useful in bait formulations. At the rates they are used in commercial baits, these toxicants are not perceived by the target pests, so repellency or bait shyness generally will not occur (Bennett et al., 1997). They often have a long residual action in field situations. Common inorganics are the arsenicals and boric acid. A recommended treatment for white grubs in hoop pine nurseries in Queensland was to mix lead arsenate with sawdust, spread it over the nursery bed and dig it in just prior to sowing. Lead arsenate was also used in a bait with molasses and bran to control grasshoppers and cutworms (Queensland Department of Forestry, 1963).

Chlorinated hydrocarbons

The discovery of the insecticidal properties of DDT in 1939 revolutionized insecticide development and signalled a trend away from inorganic and botanical insecticides towards synthetic organic compounds. Until fairly recently, chlorinated hydrocarbons dominated the pesticide market, being very effective contact and stomach poisons. They affect larvae, nymphs, adults and sometimes pupae and eggs. Their persistence in the soil made them particularly useful in controlling soil-dwelling insects in the nursery such as white grubs, mole crickets and termites. They were usually applied as soil drenches or incorporated with the potting mix. Commonly used compounds included benzene hexachloride, lindane, dieldrin, aldrin, chlordane, heptachlor and endrin. It is the very persistence of these chemicals as residues in soil, animals and plant tissues, and their accumulation in food chains, that has led to their phasing-out worldwide. Methyl bromide, a fumigant also used to control soil-dwelling insects in nurseries, is now being replaced due to its ozone-depleting properties.

Organophosphates

The organophosphates were the first insecticides to replace, in some uses, the chlorinated hydrocarbons and were developed in the early 1950s. Their primary mode of action involves the inhibition of the enzyme cholinesterase in the nervous system, disrupting transmission of nerve impulses and leading to muscle spasm and death. They have contact, ingestion and fumigant action and, in contrast to the organochlorines, include several excellent systemic insecticides such as dimethoate and acephate, which have been widely used in nursery and field applications. These products are absorbed by roots and foliage and are translocated in plants to kill chewing and sucking insects. Chlorpyrifos has replaced the organochlorines as the principal soil insecticide in nurseries. Malathion is also commonly used against grasshoppers and leafhoppers and as a seed dressing to promote good germination. Most organophosphates break down quickly in the environment and are not stored for long periods in the bodies of non-target organisms, therefore posing less risk of contamination or bioaccumulation than the chlorinated hydrocarbons.

Carbamates

The first carbamates were developed in the late 1950s and through the early 1960s, and while being cholinesterase inhibitors, generally have lower mammalian toxicity than most organophosphates. Among the best known of these compounds is carbaryl, a broad-spectrum insecticide with both contact and systemic activity, which is particularly effective against beetle defoliators. Granular formulations of carbosulfan have been incorporated into potting mix in nurseries for the control of soil-dwelling pests such as subterranean termites.

Synthetic pyrethroids

Many different pyrethrin-like materials have now been produced synthetically (pyrethroids) and generally have an improved action when compared to natural pyrethrins in terms of better knockdown or longer persistence. Compounds such as cypermethrin and deltamethrin, developed in the 1970s, have been used against a wide range of pests in forest nurseries throughout the tropics, for example to control weevil defoliators in nurseries in southern China and cutworms in Brazil. Cyfluthrin and permethrin have been used to control sap-sucking insects such as aphids and adelgids, and bifenthrin is used in some forest nurseries as a soil drench or in potting mix against termites and weevils.

Novel neurotoxicants

Neuroactive compounds have been the dominant insecticides since the 1950s but resistance problems with the first generation of these insecticides prompted the search for new neurotoxicants acting at different sites to circumvent cross resistance (Casida and Quistad, 1998). Among the most effective of these new products are the contact miticide and insecticide abamectin (produced microbially) and the synthetic nicotinoids such as imidacloprid, which acts at the same site in insects as nicotine but with much higher effectiveness and safety. Spinosad, a product produced during the fermentation of the soil actinomycete, *Sacharopolyspora spinosa*, stimulates the central nervous system of the insect persistently through interaction with the nicotine acetylcholine receptors. It is particularly active against lepidopteran, dipteran and thysanopteran pests (Pineda et al., 2006). Indoxacarb, first used in 2000, interferes with sodium channels in the nerve membrane by disrupting ion transfer and the transmission of impulses between nerve cells. It is very effective against lepidopteran larvae but allows most predators and immature wasp parasites which attack these caterpillars to survive (Devine and Furlong, 2007).

Insect growth regulators

The insect growth regulators (IGRs) were at one time considered to be the third generation of insecticides (after inorganics and synthetic organics), with great potential for curtailing agricultural pesticide use (Castida and Quistad, 1998). They compete, mimic or interfere with the juvenile hormones essential for insect development (Devine and Furlong, 2007). Pyriproxyfen, which was developed in the late 1980s, is a juvenile hormone analogue that inhibits egg production and metamorphosis and primarily is active against sucking insects. Other examples are methoprene and fenoxycarb. Despite outstanding potency and apparent safety for mammals, these compounds are still limited in their application by slow action and a narrow range of sensitive stages in the life cycle.

Antitranspirants

As the name suggests, these chemicals are applied to plants for the purpose of retarding transpiration. In southern forest nurseries in the USA, they are used on pine seedbeds and appear to increase survival, especially when the plants are under stress (South and Zwolinski, 1996). Antitranspirants have a number of side effects unrelated to the reduction of transpiration. They have been used to reduce smog damage and also to lower the incidence of certain fungal diseases and insect infestations. As described in Chapter 5, gall-making *Phytolyma* spp. psyllids are serious pests of *Milicia* spp. in West Africa, particularly in nurseries (Wagner *et al.*, 2008). Systemic insecticides have been used in attempts to reduce the incidence of attack, but to no avail. However, in Nigeria, Agboola (1990) found that certain concentrations of antitranspirants prevented gall formation in seedlings. It is suggested that these chemicals, in preventing water loss from the seedlings, create an unsuitable oxygen–water balance for the development of the eggs and nymphs of *Phytolyma*. Dosage is critical, and higher concentrations of these antitranspirants kill seedlings.

7.4.6 Conclusion

Given that pesticide usage in nurseries can be justified economically and its side effects generally minimized, the environmental debate nevertheless has had considerable impact on just what chemicals are employed and in what manner. Ideally, the pesticide should act rapidly on pests yet be completely harmless to non-target organisms, should not be unduly persistent and should be cheap and readily available. As outlined here, pesticide technology is advancing rapidly and is producing safer and better-targeted insecticides. An increasing percentage of treatments are now made with the newer compounds such as abamectin, IGRs, imidacloprid and indoxacarb, which are termed 'reduced risk' insecticides (Devine and Furlong, 2007). Pesticide usage in the plantation situation is discussed in the next chapter.

8

Management Systems III: Plantation Stage

8.1 Introduction

Once the trees have been reared in the nursery, they must be planted out into the afforestation site, where some of them at least will remain until harvest, maybe decades later. During this long period, insect pest management can take several forms, which may be integrated into a package of tactics to reduce pest numbers. In this chapter, we examine the three main types of tactic employed during the plantation stage – silvicultural (or ecological) control, biological control and chemical control. A vital component of the successful management of insect pests in tropical forests concerns forest health surveillance and monitoring, as well as careful inspection and quarantine measures. For clarity, these tactics are discussed in Chapter 9, but it is important to understand that all these components need consideration and the most appropriate should be used in conjunction. Such a combination of tactics is integrated pest management (IPM), the topic of Chapter 10.

8.2 Silviculture and Pest Management

Throughout this book, we emphasize the crucial importance of prevention as the key to IPM in tropical forests (see Speight, 1997).

From the very first to the last stage of forest operations, various silvicultural or tree-husbandry techniques can be utilized to reduce or even prevent pest problems. On the contrary, the same techniques, if carried out differently, can improve conditions for insects and make outbreaks worse, or at least more likely. We shall examine the different stages in forest management and describe, where appropriate, how varying tactics can influence insect pests.

8.2.1 Planting tactics – monocultures versus polycultures

One of the most basic arguments in plantation forestry concerns the dogma which states that trees planted in monocultures (single species of trees) are more likely to suffer from pest problems than those growing in polycultures (mixed species of trees). First of all, it is important to set up some definitions. Strictly speaking, a stand of trees wherein each individual is the same species as every other one is known as monospecific; a monoculture, in fact, is a scenario where there occurs a succession of one pure stand by another of the same species (Wormald, 1992). However, if we use a much looser definition of 'monoculture', one which refers to domination of a forest by a single species of tree, then it is clear that natural monocultures are common enough phenomena, especially in temperate regions where many climax woodlands could be said to be approaching monocultures. The ecosystem, while single species dominated, still contains a wealth of plant, animal and microbe biodiversity and, indeed, a mixed age structure and genetic base of the dominant trees. Plantation monocultures, especially those which we term 'industrial', are much more extreme, where floral and faunal biodiversity is often much reduced, where the stand is even-aged (probably having been planted on the same day with the same-aged transplants), with reduced (or in the case of clonal plantings, absent) genetic diversity, and where only a few provenances are utilized.

The fundamental principle behind the potential for pest outbreaks in monocultures concerns the provision of food, breeding sites and so on, for those insect species which in the main are host-plant specialists. If, by lucky chance for the insects, a particular tree species or genotype is highly suitable and the whole forest is planted with just that tree, then resource limitation of the herbivore is minimal and epidemics can be expected. As mixtures increase and the constituent tree species become more and more dissimilar, then insects should experience increasing difficulty in locating and utilizing adjacent trees, thus reducing their performance overall in the mixed stand and helping to protect suitable trees by virtue of their being hard to find – they are, in fact, hiding among unsuitable species or genotypes.

Conclusive evidence of monocultures per se promoting pest outbreaks, and mixed cultures reducing them, has been elusive considering the universality of the dogma. Some opposing tropical examples will illustrate the problem.

In Fig. 8.1, it can be seen that there is no significant difference between the number of defoliating moth larvae, the number of dipteran galls, nor for that matter the number of fungal spots, on the leaves of the tropical tree, *Stryphnodendron microstachyum*, planted in monocultures or, alternatively, mixed in with five other species of tropical broad-leaved tree (Folgarait et al., 1995). In this instance from Costa Rica, the monoculture seemed to be no more susceptible to pests than a mixed planting.

A different example comes from the mangrove forests of Bangladesh, where the commercial but indigenous tree *Sonneratia apetala* (locally known as keora) can be attacked severely by the stem-boring moth *Zeuzera conferta* (Lepidoptera: Cossidae). Results of extensive surveys showed that an average of 32% of trees were attacked by the borer in mixed-species stands, statistically different to the 51% in monoculture plantations (Wazihullah et al., 1996).

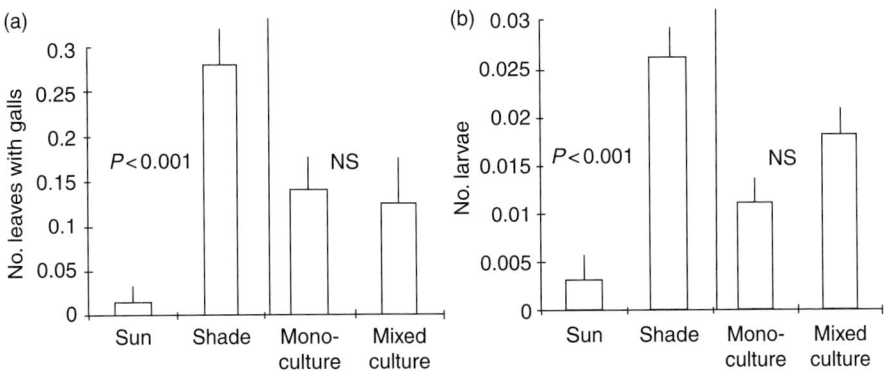

Fig. 8.1. The effect of sun or shade, monoculture or mixed culture, on two insect pests of a Costa Rican tree. (a) Cecidomyiidae (Diptera) gall former; (b) *Euclystis* spp. (Lepidoptera: Noctuidae) defoliator (from Folgarait et al., 1995).

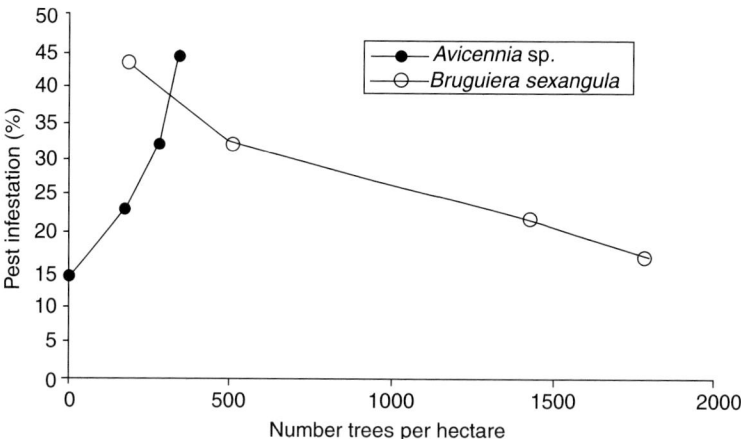

Fig. 8.2. Relationship between infestations of *Zeuzera conferta* on *Sonneratia apetala* and the occurrence of other tree species in mixed stands in Bangladesh (from Wazihullah et al., 1996).

However, the type of mixture was also found to be important. As Fig. 8.2 shows, attacks to *Sonneratia* declined as the number of *Bruguiera sexangula* (kankra) in the mixture increased, but increased with the number of *Avicennia* sp. (baen). In management terms, the best strategy would appear to be that of mixtures with kankra, assuming that this latter species remains pest free and is viable commercially. In a study in Central Panama, three native timber species, *Anacardium excelsum*, *Cedrela odorata* and *Tabebuia rosea*, were planted as either monocultures, mixed stands or mixed stands protected by insecticides and the damage inflicted by insect herbivores assessed over a 2-year period (Plath et al., 2011). Establishment of these trees in mixed stands did not have significant effects on tree survival and growth compared to pure stands. However, for one of the species, *A. excelsum*, leaf herbivory was significantly lower in unprotected mixed stands than in monocultures. In another study in Brazil, planting *Swietenia macrophylla* in a mixture with *Eucalyptus urophylla* reduced infection by the shoot borer *Hypsipyla grandela* from 71 to 25% (Neto et al., 2004), but in the Yucutan Peninsula of Mexico, *S. macrophylla* and *C. odorata* suffered similar levels of infestation by *H. grandella*, regardless of whether they were planted as monocultures or in mixtures with other species (Perez-Salicrup and Esquivel, 2008).

Jactel et al. (2005) tested the biodiversity stability theory by conducting a meta-analysis of data from a total of 54 experimental or observational studies by various authors published from 1966 to 2000 comparing pest incidence between mixed-species forest stands and pure stands. Their conclusions were:

- Tree species growing in mixed stands overall suffered less pest damage, or had lower pest populations, than pure stands. Among the 54 studies, the monoculture effect was an increase in pest damage in 39 cases and a decrease in 15.
- Three main ecological mechanisms could account for the lower damage in tree mixtures: reduced accessibility to pests of their host trees, greater impact of natural enemies and diversion from a less susceptible to a more susceptible tree species.
- The main exception to the diversity-resistance paradigm was the case of polyphagous pest insects. About half of those in the studies actually caused more damage in mixed than in pure stands. Such insects could first build up their populations on a preferred host-tree species and then spill over on to an associated host-tree species, according to the contagion process.

Nair (2007), in his own analysis of the literature and of the debate, offers some important perspectives. His first point is that we are dealing with not one but multiple hypotheses here. The overriding hypothesis is that there is a relationship between diversity and stability, such that a more diverse ecosystem is more stable. This has led to the hypothesis that natural mixed tropical forest which has a high diversity of tree species is stable and is free from pest outbreaks. The concept has been extended further to mixed forest plantations but, as Nair points out, most artificial mixtures tried in plantations consist of only two species. He regards the application of the diversity–stability principle to a simple mixed-species tree plantation as an unjustified oversimplification. In the Jactel et al. (2005) study, no distinction was drawn between naturally occurring mixed forest stands and the more simplified mixed plantations. Nair (2007) found no consistent evidence to assert that pest problems were less severe in mixed-species forest plantations than in single-species forest plantations, but did agree that in naturally occurring mixed-species stands the pest problems were less severe compared with natural single-species dominated stands (but there were exceptions). He argued that the driving force was not the stand composition but the biology of the insect species, with stand composition modifying the severity of infestation.

As seen from the above discussion, the evidence of serious losses due to pests (and diseases) resulting from forest monocultures (or, to be more precise, monospecific plantations) is inconclusive. It is possible that most observed problems may be due not to the lack of genetic diversity in the attacked stands but more to their lack of vigour (see Chapter 3), and in commercial, industrial, large-scale plantations, the benefits of monocultures may outweigh any possible disadvantages, a point conceded by Jactel et al. (2005). However, in projects such as fuelwood or agroforestry, mixed plantings may be viable.

8.2.2 Planting tactics – shading and nurse crops

There are numerous references in the literature to the benefits of shade or companion plantings in protecting desired tree crops from damage by insect pests. In West Africa, for example, *Milicia excelsa* growing under the shade of the nitrogen-fixing tree *Gliricidia sepium* had fewer galls of the psyllid *Phytolyma lata* than *M. excelsa* growing alone (Wagner et al., 2008). After 1 year, *M. excelsa* seedlings growing in 82% shade were nearly 50% taller than seedlings growing in 57% shade and 100% taller than trees growing in full sunlight. Also in Ghana, Bosu et al. (2006) found that shade from a companion planting of *Terminalia superba* appeared to reduce psyllid attack on *Milicia* species, though crop tree growth was slow. The current silvicultural recommendation for this species is that it should be planted in the partial shade of another species, in existing shade in a natural forest or alongside a fast-growing species that quickly overtops it.

In Australia's Northern Territory, Montagu and Woo (1999) studied the impact of the stem-girdling longicorn *Platyomopsis humeralis* (Coleoptera: Cerambycidae) on *Acacia auriculiformis* in clonal seed orchards at two sites. They found that attacks by this insect reduced tree height growth by 0.4–0.5 m in the 6 months following girdling, trees being most vulnerable to attack in the first year following planting. At this age, the main stem is the only stem of a size suitable for girdling, and consequently the impact is more severe than on trees with multiple branches. A control strategy was developed to counter the attack by *P. humeralis* in new plantings of *A. auriculiformis* (Montagu and Woo, 1999). The seedlings are planted into rows pretreated with herbicide. Native sorghum grasses (*Sorghum* sp.) are allowed to grow to a height of 2.5–4.0 m between the rows of acacia seedlings. By the end of the wet season, when insect numbers are at their greatest, the acacia seedlings are screened from the 'view' of the insects and consequently insect damage is reduced greatly in the first year. This effect is referred to as 'lower host-tree apparency' by Floater and Zalucki (2000). In subsequent years, when the trees have outgrown the sorghum cover crop, insect attack would occur but potentially would have less effect on tree growth than the damage in the first year. This strategy has some risks due to the higher fuel load associated with the sorghum grasses during the dry season. On experimental plots, this risk has been reduced by slashing the grasses in the early dry season. Surprisingly, 20 months after planting there appeared to be no long-term effect of insect damage on tree height or form, even when the main stem was killed by girdling (Fig. 8.3). However, the effect was obscured somewhat by the preference of the insect for attacking the larger trees at each site and the impact may very well have been greater than that shown.

Shading does not always result in a reduction in pest damage on a crop tree, as shown in Fig. 8.1 where the number of dipteran galls, defoliating moth caterpillars and fungal spots on saplings of S. microstachyum in Costa Rica was greater in the shade than in the sun (Folgarait et al., 1995). In Panama, a small stem-boring scolytine beetle, Coccotrypes rhizophorae, infests propagules and young seedlings of the mangrove Rhizophora mangle. Sousa et al. (2003) observed that C. rhizophorae attacked and killed a high proportion of the R. mangle seedlings that they had planted as part of a long-term field experiment designed to measure rates and outcomes of competition among juvenile mangroves in different forest floor environments. Within the first year of the experiment, beetles killed 72.6–89.1% of the seedlings planted in three closed-canopy understorey sites, but only 0.9–2.1% in three adjacent light gaps. These lightning-created canopy gaps afford better growth conditions than the surrounding understorey and, as importantly, provide a refuge from predation by C. rhizophorae.

Companion planting with trees such as *Azadirachta indica* (neem) and *A. excelsa* (sentang), which are sources of natural insecticides and are thought to be repellent to insect pests, has also produced mixed results. In Sabah, Malaysia, for example, mahogany (*S. macrophylla*) trees suffered serious attack by the shoot borer *H. robusta*, despite being interplanted with neem and sentang, even when the *Azadirachta* species were taller than the mahogany (Matsumoto and Kotulai, 2002).

8.2.3 Planting tactics – line or enrichment planting

Both line and enrichment planting techniques are usually applied to selectively logged secondary tropical forests, either to encourage the regeneration of indigenous tree species or, in some cases, to grow economically important species which do not do well in plantation situations. Lines or patches of natural vegetation are cleared through the secondary forest and young trees planted along them. The influence on insect pest attack in these situations is, as usual, variable.

In a moist semi-deciduous forest in Ghana, the African mahoganies *Khaya anthotheca* and *K. ivorensis* were grown under three different forest canopy shade levels: open (55% open sky), medium shade (26% open) and deep shade (11% open). *H. robusta*

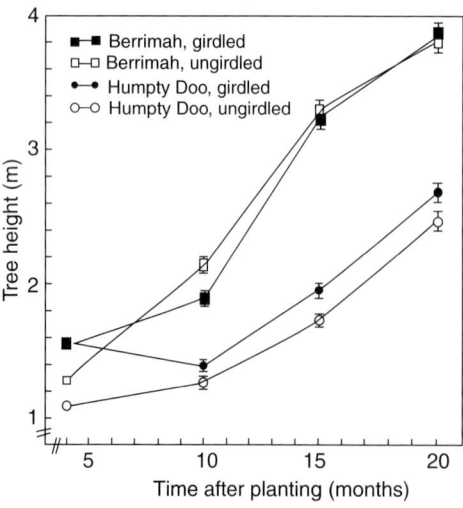

Fig. 8.3. Height of *Acacia auriculiformis* trees which either had or had not been girdled by the insect *Platyomopsis humeralis* 4 months after planting at two sites near Darwin, Australia (from Montagu and Woo, 1999).

attack on *K. anthotheca* was 85, 11 and 0% in the open, medium-shade and deep-shade treatments, respectively, and *K. ivorensis* showed similar trends. However, growth in the medium and deep shade was slow and would limit the use of this strategy for controlling *Hypsipyla* attack (Opuni-Frimpong et al., 2008a) (see also Chapter 3). In the Lae region of Papua New Guinea, Siaguru and Taurereko (1988) planted *C. odorata* in rows through logged-over rainforest, and though growth was very slow for the first year, due to heavy shade, heights of between 2 and 5 m were obtained after 15 months of establishment. No significant borer damage was recorded. It must be noted that though it is likely that *Hypsipyla* spp. are indigenous to Papua New Guinea, *C. odorata*, originating from Central America, is relatively resistant to the pest when planted in Australasia, so the results of this trial may be rather hard to explain. In the Ivory Coast (Cote d'Ivoire) of West Africa, enrichment planting of African mahoganies did not prevent serious attacks by *H. robusta* (Brunck and Mallet, 1993), and similarly in the Peruvian Amazon, planting New World mahoganies such as *Cedrela* spp. and *Swietenia* spp. in lines through native forest did nothing to reduce *H. grandella* attacks (Yamazaki et al., 1990). This is likely to be because of the high susceptibility of the trees and the extremely efficient host-finding ability of the insect (see also Chapters 5 and 10).

Some success has been achieved by planting these high-value Meliaceae in gaps in natural forests or plantations. In Sri Lanka, Mahroof et al. (2001) established field trials of *S. macrophylla* under various light conditions provided by different canopy openings of mature *S. macrophylla* and *A. auriculiformis*. The incidence of shoot borer attack 54 weeks after planting in the low shade treatment was 76% higher under mature mahogany and 31% higher under *Acacia* than under the high-shade treatment. Shadehouse experiments indicated that shading might reduce shoot borer attack by influencing both oviposition and larval development (Mahroof et al., 2002). In Brazil, Lopes et al. (2008) planted mahogany seedlings in gaps in the forest created by selective timber harvesting and recorded a low incidence of *Hypsipyla* attack and trees that were significantly taller than similar-aged natural regeneration. Nichols et al. (1998) in Ghana examined the influence of artificial gaps in tropical forest on the survival, growth and attack by the psyllid *P. lata* on *M. excelsa*. They found that insect attack occurred first and most severely in the large gaps, but spread to gaps of all sizes between the 11th and 13th months after planting. They concluded that gap sizes in the range of 10–50 m^2, where irradiances were from 30 to 60% full sunlight in forests similar to those at the study site, seemed to be most suitable for regeneration of *Milicia*.

Kirton and Cheng (2007) report ring-barking and root-debarking by termites of transplanted dipterocarp saplings (*Shorea* sp.) used in enrichment planting in logged-over lowland dipterocarp forest in Malaysia. Attack sometimes occurred on living and otherwise healthy plants, indicating that it was not necessarily secondary to other mortality factors. What was unusual in this example was that the attack occurred on dipterocarp species that were planted in their native habitat in a relatively closed-canopy forest. This throws doubt on the hypothesis that the problem arises from a lack of alternative food sources for the termites during the establishment period of young forest plantations. It also differs from the African experience, where attack has been said to cease after canopy closure and development of a leaf litter layer (Mitchell, 2002).

8.2.4 Transplanting tactics

Once the type of planting strategy has been decided on, the young trees have to be removed from the nursery and established in the plantation site. At this stage, more chemical control may be necessary to prevent serious losses from root-feeding termites (see below), but silvicultural practices may also have a role to play in reducing the risks of insect pest attack. Firstly, any stressful conditions must be avoided, not just to deter insects but also to ensure all-round vigour for successful establishment. For example, practices such as transporting nursery stock in the back of open trucks in the heat of the day for long periods of time clearly must be avoided. The age of the transplants may also be an important factor on occasion.

In Paraguay, transplants of *E. grandis* have been heavily attacked by grasshoppers (Orthoptera: Acrididae), which chew the bark just above the soil surface, causing girdling in the first 3 months after transplanting (see Chapter 10 and Plates 124–126). Even if trees are not killed outright by this damage, the first strong winds in the open planting sites cause the fragile stems to bend and break at the point of attack. Up to 70 or even 80% of young trees may be lost in this way (M. Henson and R. Davies, personal communication). It is likely that if eucalypt transplants can be kept in the nursery longer so that there is time for their lower bark to harden and toughen, then they will be less palatable to grasshoppers and able to tolerate any attacks which still occur. This tactic will, of course, require modifications of nursery plans and operations, but if successful, no further pest management should be required.

8.2.5 Pruning and brashing

Trees in a forest plantation usually have their lower branches removed manually as they grow. This is done for a variety of purposes, such as to improve stem and wood quality in some industrial crops, to provide access, to reduce fire hazard or, in agroforestry situations, to provide fodder or fuelwood (Evans and Turnbull, 2004). When low pruning involves simply breaking off small dead branches, it is called brashing. Depending on the method used, wounds may become sites of attack by a variety of insect pests that normally would not be able to invade the standing trees. Although the damage may be localized to the wound areas, care needs to be taken not to leave large wounds on the trunk or stumps of branches, both of which can cause problems (Plate 104).

In agroforestry operations in Kenya, for example, pruned trees of the multi-purpose tree *Cassia siamea* were attacked much more heavily and killed by larvae of the stem-boring moth *Xyleutes capensis* (Lepidoptera: Cossidae) than were unpruned trees (Fig. 8.4) (Mbai, 1995). While this procedure cannot be avoided, care should be taken in sites known to be infested with borers so as not to lose large numbers of trees. Other similar cases exist of pruning damage triggering borer attacks; pruning of young *Casuarina equisetifolia* on Reunion Island and of *E. deglupta* in the Philippines are reported to increase attacks by longicorn and roundhead (varicose) borers (Coleoptera: Cerambycidae and Buprestidae) (Braza, 1992; Tassin *et al.*, 1997). Studies by Kirton *et al.* (1999) in *A. mangium* plantations in Peninsular Malaysia showed that while the termite *Coptotermes curvignathus* was capable of primary attack, its entry into the wood was often facilitated by large pruning wounds, abscission scars resulting from natural pruning and damage by insect bark borers.

A good example of how silvicultural practices can be modified to counter insect attack associated with tree wounds comes from Queensland. There, attack by the pine bark weevil, *Aesiotes notabilis* (Coleoptera: Curculionidae), in plantations of hoop pine, *Araucaria cunninghamii*, and bunya pine, *A. bidwillii*, is associated mainly with injuries

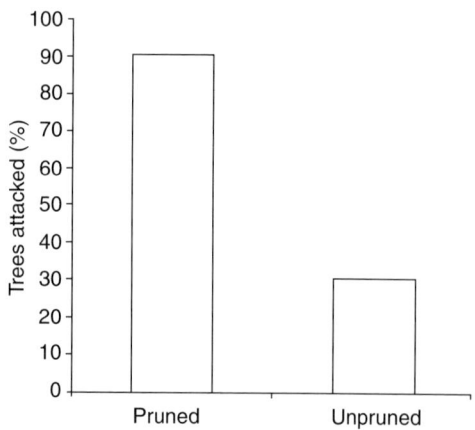

Fig. 8.4. Percentage of *Cassia siamea* trees attacked by *Xyleutes capensis* in relation to pruning activities in Kenya. Total number of trees = 400. χ^2 significantly different at $P < 0.001$ (from Mbai, 1995).

received during pruning. Damage is caused by the larvae which, after emergence from eggs laid on or near the branch stubs, tunnel in the cambial region of the bole. Larval tunnelling may girdle the stem completely or may allow the entry of secondary borers and destructive pathogens. Brimblecombe (1945) showed that adult activity and egg laying was greatest and development most rapid in wet weather during the warmer months (November–April). By carrying out pruning operations during the dry, cool months (May–August), when the insect was least active, damage by the weevil was reduced to a low level.

In Papua New Guinea, another weevil, *Vanapa oberthuri* (Coleoptera: Curculionidae), with habits similar to *A. notabilis*, is most active in stands of *A. cunninghamii* that are being thinned or pruned (Gray and Howcroft, 1970). While the number of trees killed annually by *V. oberthuri* is small compared with the total number of plantings, the extent of damage to trees at individual sites can be extremely high, with up to 51% of final crop trees killed by the beetle following pruning in one compartment (Gray and Barber, 1974). However, unlike the situation for *A. notabilis*, *V. oberthuri* is active throughout the year in the warm, wet conditions of New Guinea (Wylie, 1982b) and the only way of minimizing beetle attack is to minimize injury incurred during pruning and thinning operations.

In tropical and subtropical China, Masson pine, *Pinus massoniana*, has been widely used in reforestation programmes and is a valued source of timber, fuel and various wood products. The pine needle scale, *Hemiberlesia pitysophila* (Hemiptera: Diaspididae), a native of Taiwan and Japan, was introduced to Hong Kong accidentally on infested Christmas trees in the mid-1970s and was found on Masson pine in nearby Guangdong Province in 1982. By 1987 it had infested 315,000 ha of pines and destroyed 25% of the forest (Wilson, 1993) and, since then, the scale has killed more than half the infected trees in the province. Pruning of Masson pine's lower branches for fuel is a common practice in China. Some foresters also remove lower branches in an attempt to lessen the scale infestation, but this has the effect of concentrating the scale on a smaller crown and seems to enhance the injury. In addition, forest labourers and farmers remove the branches with knives (which leave 2–10 cm stubs on the bole) and they often injure the bark. This practice can encourage fungal infection and a recommended solution is replacement of the knife with a Meylan saw, which can remove branches as quickly as a knife and at the same time cut flush to the bole (Wilson, 1993).

Pruning is not always potentially harmful and can, on occasion, be used as part of a pest management programme. Where parts of a tree are known to be infested with insect pests, it may be possible to cut these parts away and destroy them. The chestnut gall wasp, *Dryocosmus kuriphilus* (Hymenoptera: Cynipidae), is an important pest of chestnuts in parts of China, where it causes swellings on twigs and shoots, resulting in reductions in vegetative growth and fruiting (Huang *et al.*, 1998). The pruning away of affected twigs leaving behind only healthy stems appears to be an effective way of reducing damage to individual trees, as well as to reduce pest population densities in forest stands. Similar operations have been suggested in Costa Rica for the control of the mahogany shoot borer, *Hypsipyla* spp., in *Swietenia* plantations. Stems exhibiting boring activity may be cut off and destroyed (Cornelius, 2001), allowing new or uninfested shoots to take over dominance. In some cases though, shoot borer attacks may be so severe that few if any shoots may be left after a year or two of pruning! In Australia, Collett and Neumann (2002) suggested that pruning the lower crown branches of *E. globulus* during summer could boost height growth and accelerate canopy closure, a condition which suppressed infestations by light-seeking, leaf-consuming leaf blister sawflies (*Phylacteophaga* spp.), leaf beetles (*Paropsisterna* spp. and *Paropsis* spp.) and Christmas beetles (*Anoplognathus* spp.).

The final problem is what to do with the brashed or pruned material if it is not to be utilized but instead left to lie in the plantations. This problem is discussed in Section 8.2.8.

8.2.6 Spacing and thinning

When first planted out, forest trees are arranged with certain distances between each one, taking into account the economics of stocking and the desired final number of trees in a plantation. These decisions must be coupled with the provision of a suitable habitat in which young trees can establish and flourish. As the forest grows, it is very often routine to remove a proportion of trees during thinning regimes, in order to promote the vigour of the remaining trees by reducing competition for light, water and nutrients (Plate 105). If stands are allowed to remain densely stocked (Plate 106), suppressed trees may lose vigour easily, to such a level that many secondary borer pests are able to infest them. On the other hand, trees which are spaced too widely may suffer from high levels of insolation and wind-throw, notwithstanding the point that too few mature trees on a site may not provide a viable economic return.

It has been shown in some cases that spacing may have an important influence on insect pest infestations. In Japan, the Sakhalin fir aphid, *Cinara todocola* (Hemiptera: Aphididae), is a highly destructive pest in young stands of Sakhalin fir, *Abies sachalinensis*. Figure 8.5 shows how tree spacing influences the density and mortality caused by the aphid. Trees further apart are less likely to be infested and killed by the pest, since the dispersal of wingless aphids is restricted to trees which have touching foliage (Furuta and Aloo, 1994).

Although not yet available for many tropical forest pest management programmes, associations between tree spacing and stand stocking levels have been formulated for certain serious pests in countries such as the USA. One example will illustrate how the system works. The mountain pine beetle, *Dendroctonus ponderosae* (Coleoptera: Scolytinae), attacks and kills *P. ponderosa* according to the distances between each tree in a stand and the overall density of trees, the so-called growing stock level (GSL). Figure 8.6 provides a hazard rating system developed for predicting the likelihood of beetle outbreaks. Stands with a low stock level, consisting of trees with low DBHs (diameter at breast height) and growing widely apart can be seen to be of low hazard rating, whereas those in dense stands, closely spaced and of relatively large DBH are very much at risk from beetle attack. Again, the problems of suppression and stress in the latter situations are the root cause. Similarly, in countries where the wood wasp *Sirex noctilio* has become established, overstocked stands of *Pinus* spp. are particularly susceptible

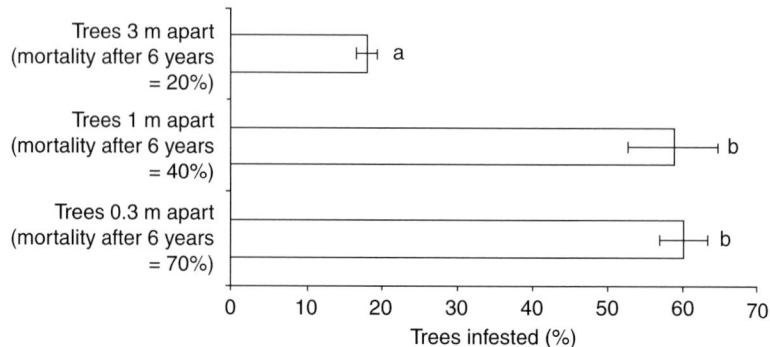

Fig. 8.5. Mean percentage of Sakhalin fir trees infested with *Cinara todocola* according to tree spacing in plantations in Japan (± SE). Means with same letter not significantly different at $P < 0.05$ level (data from Furuta and Aloo, 1994).

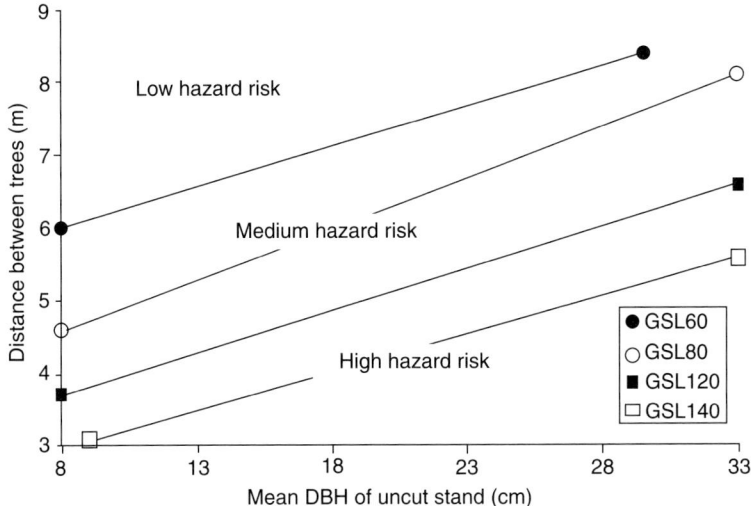

Fig. 8.6. Hazard rating system for infestations of *Dendroctonus ponderosae* in *Pinus ponderosa* according to tree spacing and growing stock level (GSL) (from Schmid et al., 1994).

to attack by this insect. Thinning is a major silvicultural tool in reducing damage caused by the pest (Gaiad et al., 2003; Dodds et al., 2007).

8.2.7 Salvage and trap logs or trees

On occasion, pest outbreaks become so severe and damaging that trees may have to be cut down and processed in various ways to protect the remaining standing trees and also to prevent timber degrade, which may render the trees unmarketable. Salvage logging is a process whereby damaged and susceptible trees are removed wholesale and treated to prevent pest attack and subsequent degrade. A good example of this comes from southern Queensland in Australia, where in 1994, around 9000 ha of *P. elliottii* plantations were damaged by forest fire during the dry season (Hood et al., 1997; Wylie et al., 1999) (Plate 107). Large numbers of fire-damaged trees were in danger of being attacked by the exotic bark beetle, *Ips grandicollis* (Coleoptera: Scolytinae), which has as a symbiont a blue stain fungus that causes discoloration and degrade to infected timber, rendering it unmarketable. Once the beetles have tunnelled under the bark of damaged trees or logs, the fungus goes to work, and salvage felling followed by storage of logs under continuous water sprinkling is required to prevent this attack (Plates 108 and 109). As Fig. 8.7 shows, the time before blue stain first appears in the timber is very short, with the first indications of staining occurring a mere 10 weeks or so after the fire. Salvage of large areas of *Pinus* had therefore to be very rapid indeed, ensuring that logs were protected from beetle invasion within a very short time (Wylie et al., 1999) (see also Chapter 5 and Fig. 5.10).

The fact that many insect pests, especially the so-called secondary pests such as bark and wood borers, preferentially infest fallen or felled trees over standing but stressed ones may be used to protect plantations at risk from this sort of attack. A number of trees in a stand may be sacrificed and used in a system known as trap logging. The concept is simple enough; a small number of trees in stands susceptible to attack by borers, during drought stress or defoliation for instance, are felled and usually piled together to produce a super-attractant stimulus for potential pests in the stand. Once the trap logs have been infested, they are debarked, treated with insecticide, or burned to destroy pests before they can emerge to

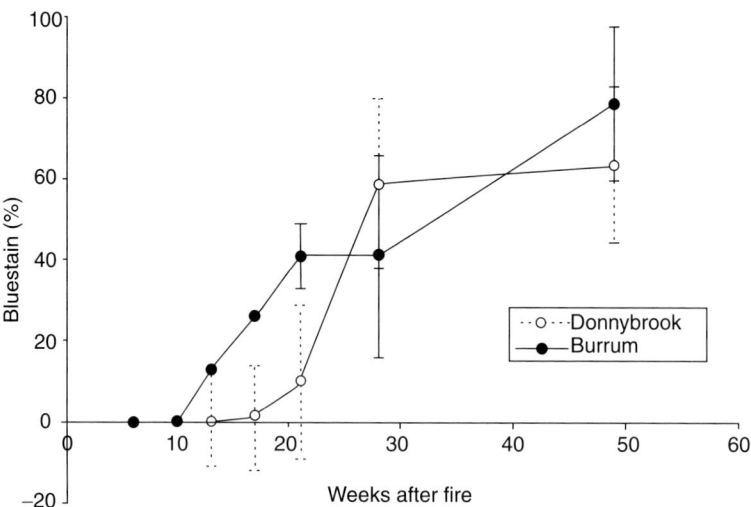

Fig. 8.7. Mean percentage bluestain in discs of *Pinus elliottii* after a forest fire in two sites in Queensland, Australia (± 95% confidence limits) (from Hood et al., 1997).

start a new generation. This system has few commercial examples in tropical forestry, but one such is the use of trap logs for the monitoring and control of the sal borer, *Hoplocerambyx spinicornis* (Coleoptera: Cerambycidae) in India (see Chapters 5 and 9). Trap trees and logs are also used in monitoring and control programmes for the wood wasp *S. noctilio* in *Pinus* spp. plantations in the southern hemisphere (Bedding and Iede, 2005). In monitoring for *Sirex*, living standing trees are stressed artificially, usually by injecting them with a herbicide, to make them attractive as an oviposition site for a female wasp and then these trees are inspected regularly for signs of attack. The primary control measure for the pest is inoculation of felled trees with the parasitic nematode *Beddingia siricidicola* which, following *Sirex* infestation of these logs, eventually will sterilize any female wasp that develops by infecting her eggs (see Chapter 5).

8.2.8 Sanitation and hygiene

The problem with pruning, brashing, thinning and harvesting is that they leave lying around in forest stands enormous amounts of potential breeding material for insect pests, unless something is done to process this debris (Plate 110). Merely stacking it in log piles on forest roads in the vicinity of plantations is asking for trouble, since pests will be attracted readily to these sites. Cursory examination of thinning and logging debris in tropical forests reveals all manner of pests, from bark beetles and ambrosia beetles (Coleoptera: Scolytinae and Platypodinae) to longicorn beetles (Coleoptera: Cerambycidae), weevils (Coleoptera: Curculionidae) and various other insect degraders of timber. The most significant problem with this sort of situation is the phenomenon of mass outbreaks, where pests which normally are unable to attack standing trees are present in such huge numbers by virtue of the provision by foresters of copious amounts of breeding material that their persistent attempts to attack vigorous, healthy trees succeed eventually by virtue of sheer weight of numbers (Speight and Wainhouse, 1989). This becomes a particularly serious problem when stands are stressed temporarily by unforeseen periods of drought or short-term defoliation events.

It is not feasible to remove all brash and thinning debris from forest plantations, but certainly efforts should be made to remove the larger material at the earliest opportunity and debark or burn it. Some examples of insect attack on trees associated with population build-up in debris are provided below.

As described in Chapter 5, the bark beetle, *I. grandicollis* (Coleoptera: Scolytinae), in Australia carries out 'feeding' and 'breeding' attacks when colonizing the bark of green to semi-green dead pine material or that of apparently healthy trees. In feeding attacks, the inner bark and outer sapwood surface are etched by large numbers of male and female adults replenishing their food reserves prior to reproduction and, as a consequence, the bark peels off. Breeding attacks are made by virgin male and female beetles and/or by fertilized females when entering the inner bark and constructing characteristic gallery systems for breeding. This insect has been mainly a secondary pest, but when large quantities of thinning slash are left on the ground, the beetle can assume a primary role. In these circumstances, feeding attacks in particular can occur on trees adjacent to the infested slash. This problem can be minimized successfully by silvicultural means such as burning or chopper rolling of fresh slash.

The bostrychid beetles *Apate monachus* and *A. tenebrans* (Coleoptera: Bostrychidae) are major forest pests in Africa, attacking a wide variety of economic species (Wagner *et al.*, 2008). In Ghana, their damage is especially evident in young plantations and taungya farms and is usually the result of maturation feeding (feeding by young adults prior to oviposition, which usually occurs on a different individual host). While their tunnelling does not kill the tree, it renders the timber valueless and may increase the tree's susceptibility to wind damage. Removal of debris from planting sites and destruction of heavily infested trees are recommended by Atuahene (1976) as control measures. *A. monachus* is also an introduced pest in Central America and the Caribbean, where it has many hosts (Rodriguez, 1981; Schabel *et al.*, 1999). In Puerto Rico, attack of mahogany, *S. macrophylla*, has been associated with a build-up of populations of this insect in slash (DeLeon, 1941). The bark borer *Scolytus major* (Coleoptera: Scolytinae) is a serious pest of *Cedrus deodara* forests in north-western India (Thapa and Singh, 1986). Localized mass attacks by this insect have occurred, killing healthy as well as unthrifty trees, and this has been linked to the rapid build-up of populations in logging debris in clear-felled areas, coupled with an insufficiency of suitable egg-laying sites. Once attack has been initiated, the insects spread out from this focal point, each generation attacking the trees nearest to those in which it was reared. As is the case for bark beetles elsewhere, drought is often a factor in these episodes. Recommended preventative measures include stand sanitation, debarking of logs immediately after felling, salvage logging of unthrifty or damaged trees and the use of trap trees.

8.2.9 Harvest

The felling of trees at the end of the forest rotation is, of course, the main aim of industrial operations. However, the same theories apply as mentioned above for commercially felled timber, i.e. they must not be allowed to accumulate insect (or pathogen) pests by allowing them to lie unprocessed in the forest or wood yard for any length of time (Plate 111). The precise period of time is hard to define. In tropical climates, timber begins to accumulate borers and pathogens at an alarming rate; a few months or even a few weeks may be all it takes for infestations to begin. Routine silvicultural techniques such as rapid removal and processing will minimize risks. Felling certainly should not take place until a guaranteed market is assured.

8.3 Biological Control

8.3.1 Introduction

Biological control has been defined as the use of natural enemies to maintain the population density of an insect pest below that at which it would exist normally in the absence of the enemies. The important concept in this definition is the term 'maintain', since the majority of biological control systems are expected to persist through time and to regulate a pest population below an economic threshold. The term 'regulation' is also crucial to the concept of biological control. As we discuss in detail in Chapter 3, regulation implies a density-dependent system, i.e. one where there is a proportional increase in mortality or other factors as the density of pests on which these factors operate increases. Only through density-dependent feedback can a biological control system be expected to persist over many generations of the pest.

The pest population is influenced both by trophic levels below it, such as food supply (quantity and quality), and by trophic levels above it, via predators, parasitoids and disease. The former is known as bottom-up regulation and the latter top-down regulation (see Chapter 4). It is crucial in any consideration of a biological control programme that the relative importance in the population dynamics of the pest of both top-down and bottom-up regulation is considered. In fact, in many cases if bottom-up regulation via resource limitation is more important than top-down, then the probability of biological control by natural enemies is low. Any forestry system which promotes food resources for insect herbivores, such as an increase in food via monocultures, stressed trees and extra breeding sites (such as log dumps, moribund trees, etc.), is likely to reduce the resource limitation which would impinge on natural pest populations and lead to outbreak situations (see Chapter 4). Under these conditions, predators, parasitoids and pathogens may be ineffective.

Natural enemies that may have potential in biological control in tropical forestry fall into three basic categories. Predators, from birds and mammals to other insects and spiders, search for their prey and eat it. Parasitoids are a special type of parasite; in fact, they appear to be a halfway house between true parasites and predators because they consume their host, in this case the pest insect, and complete the entire generation of themselves on or in one pest item. They are comprised mainly of Hymenoptera (parasitic wasps) and Diptera (parasitic flies) (Plates 112 and 113). Finally, we have the pathogens, or disease-causing organisms, which in tropical forestry at least consist mainly of viruses, protozoa, bacteria, fungi and nematodes. The likely efficiency of each of these groups has to be considered in the light of the pest's population dynamics, as mentioned above. Table 8.1. summarizes the

Table 8.1. Major types of natural enemies of insect pests which may have a role in biological control (from Speight et al., 2008).

Type of enemy	Commonly attacked pests
Predators	
Vertebrates	
Mammals (mice, voles, etc.)	Larvae and pupae of defoliators in soil
Birds	Defoliating and leaf-mining larvae
Amphibia (frogs and toads)	Larvae in soil
Fish	Aquatic insect larvae
Invertebrates	
Spiders	Small flying or crawling insects, e.g. sap feeders
Insects (beetles, bugs, lacewings, hoverflies)	Many exposed sap feeders and defoliators; soil larvae and pupae
Parasitoids	
Parasitic insects (wasps and flies)	Larvae and nymphs of many types of pests
Pathogens	
Bacteria	Most insect species attacked by one or more, often relatively species specific, pathogen groups
Fungi	
Nematodes	
Protozoa	
Viruses	

major groups of natural enemy which may be used in pest regulation. In practice, things are a little different. Globally, on all crop types, parasitoids and predators have been introduced much more frequently in classical biological control programmes compared with nematodes and pathogens (Table 8.2.) (Hajek et al., 2007). As we shall see, however, tropical and subtropical forest pest management has used just about all available enemy types at one time or another, with greater or lesser levels of success.

8.3.2 The concept of biological control

Conceptually, biological control falls into various different types. Classical biological control is where the enemy is introduced from the region where it occurs naturally into a new area where it and the pest are exotic. This is particularly the case when new crops are grown in novel countries and pests either are introduced accidentally or, alternatively, find their way through trade and lack of quarantine into the new area, where they establish quickly and become serious pests.

Augmentative biological control, on the other hand, occurs when native natural enemies are not efficient enough to maintain either native or exotic pest populations at endemic, non-economically important densities. In this case, systems such as habitat modification may be brought to bear in order to increase the density and/or the efficiency of the natural enemies occurring in the area. Inundative biological control is carried out when enemies are at low levels, or entirely absent, and very large numbers of them are bred and released into a pest-infested crop.

Many examples exist where evidence points to the regulatory ability of natural enemies. In Peninsular Malaysia, *A. mangium* is a very important short rotation forest species, which up until recently has been relatively free of serious insect pests, probably by virtue of its being exotic to that region. However, various defoliating Lepidoptera have begun to appear, such as *Spirama retorta* (Lepidoptera: Noctuidae). Defoliation rates of *Acacia* caused by the larvae of this species can reach between 20 and 30% (Sajap et al., 1997). Figure 8.8 shows the observed parasitism of *Spirama* larvae by three species of parasitoid fly (Diptera: Tachinidae). Clearly, as pest density declines, so parasitoid density increases and it is very tempting to conclude from these data that the parasitoid is responding to high densities of the moth larvae, increasing its own density at the expense of the host. In other words, this relationship may be interpreted as an example of top-down regulation. However, we really have no evidence for this conclusion; indeed, the numbers of pests might be declining for quite another reason and the parasitoids merely 'tracking' the reductions in their hosts.

We now present a series of sections which describe the successes (and failures) of biological control in tropical or

Table 8.2. Comparison of results of classical biological control programmes with different types of natural enemy (from Hajek at al., 2007).

	Parasitoids and predators	Pathogens (including nematodes)
Number of programmes	5670	131
Number (and percentage) of establishments	2008 (35.4%)	63 (48.1%)
Number of pest species	601	76
Number of enemy species	2130	45
Number of countries or islands	2130	49
Number of programmes over time	Began increasing in the 1920s, with peak numbers 1930–1939, 1950–1979	Began increasing in the 1950s, with peak numbers 1970–1989

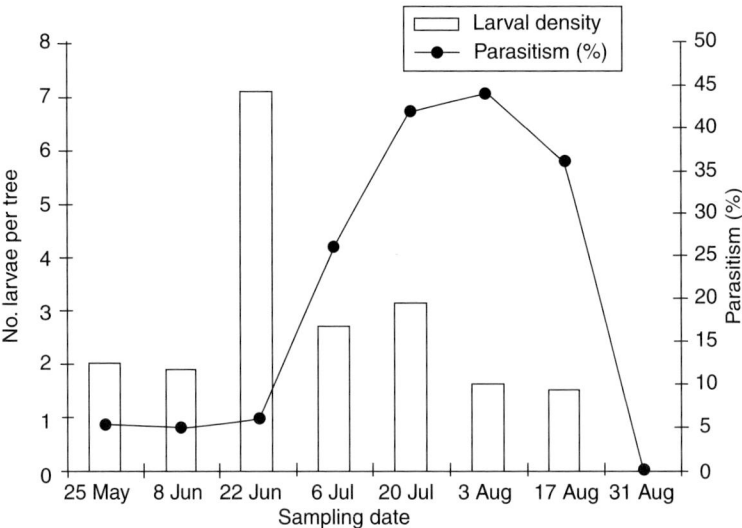

Fig. 8.8. Larval density and pooled parasitism of *Spirama retorta* on *Acacia mangium* in Malaysia (from Sajap et al., 1997, courtesy *Journal of Tropical Forest Science*).

subtropical forests, using different types of natural enemies.

8.3.3 Biological control using predators

According to van Mele (2008), predatory weaver ants in the genus *Oecophylla* were the first recorded biological control agents, dating from 304 AD in China, and they are still suggested as useful natural enemies of shoot borers and bark beetles. Their tendency to bite people as well as forest pests may be a drawback to their large-scale deployment. It has to be said that really good examples of successful biological control using predators in tropical forestry are rather difficult to find. Many attempts have been made over the years and predators have been released against all sorts of forest insect pest in numerous countries. Ladybird beetles (Coleoptera: Coccinellidae) are often reported to eat lots of sap-feeding forest pests such as psyllids and aphids (see de Oliveira et al., 2004, for example), but they do have a tendency to be cannibalistic, especially at high densities (Pervez et al., 2006). One example of classical biological control which over the years has recorded success involves the so-called ensign scale insect, *Orthezia insignis* (Hemiptera: Ortheziidae). This pest is a native of South and Central America, but has been introduced accidentally to many parts of the world, including Africa and India. It feeds on a wide variety of host plants, from ornamental shrubs and forest weeds to various indigenous tree species such as gumwood, but various economically important forest crops are also severely attacked, including *Eucalyptus* and jacaranda. The predatory ladybird, *Hyperaspis pantherina* (Coleoptera: Coccinellidae), was first collected from its native home in Mexico in 1908 and released into Hawaii, and it has been used in a variety of biological control programmes ever since. Apart from in Malawi, where some controversy arose over the outcome, the predator introductions were quite successful (Chilima and Murphy, 2000), due in part at least to the fact that, unlike many predatory insects, including ladybirds, *H. pantherina* was host specific, i.e. it would feed only on *Orthezia*. Some other attempts to control *Orthezia* species using coccinellids related to *H. pantherina* have been less successful. In Barbados, for instance, releases in 1976 and 1977 resulted in no recoveries of the introduced predators, but this might be a consequence of inadequate or non-existent post-release monitoring rather than a complete failure of the biological control programme (Booth et al., 1995). The most recent and successful example of the use of *Hyperaspis* against *Orthezia* comes from the island of St Helena (Fowler, 2004). St Helena is a small (120 km^2) tropical island isolated in the Atlantic Ocean, with much of its endemic flora disappearing, due mainly to human activities. The gumwood tree, *Commindendrum robustum*, is one of these globally endangered species, and though not a tropical forest plantation tree in the general sense of this book, it began to die from the ravages of the accidentally introduced *Orthezia* in the early 1990s. *Hyperaspis* was reared in the laboratory and first released on St Helena in 1993, producing a rapid and dramatic decline in the pest (Fig. 8.9). *Orthezia* outbreaks have ceased and the predator can no longer be detected on the island. To quote Simon Fowler, '*H. pantherina* appears to have saved the field population of a rare endemic plant from extinction'. One final difficulty with predators as biological control agents is that they usually do not discriminate between healthy prey items and those that are parasitized, so on occasion it has been found that the biological control efforts of parasitoids (see next section) are undermined by their being eaten inside their host aphids or psyllids by generalist predators co-occurring in the forest (Erbilgin et al., 2004).

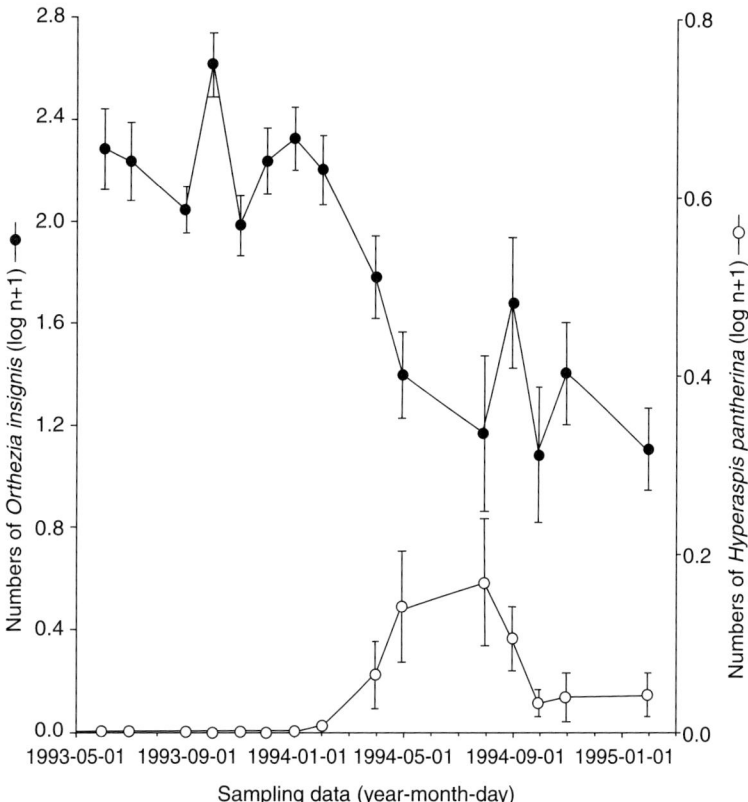

Fig. 8.9. Log-transformed mean (± SE) numbers of *Orthezia insignis* and *Hyperaspis pantherina* on shoots of initially severely and moderately infested gumwood trees on St Helena (from Fowler, 2004, courtesy Elsevier).

8.3.4 Biological control using parasitoids

Forest monocultures may not support large populations of parasitoids. For example, mixtures of *Eucalyptus* and *Acacia* in forest plantations in Australia certainly increased the abundances and richness of parasitic hymenoptera compared with monocultures (Steinbauer *et al.*, 2006), while Dall'Oglio *et al.* (2003) found the highest numbers of parasitoids in eucalypt stands in Brazil which were closest to fragments of native vegetation. Parasitoids normally are used in tropical forest pest management via inundative releases. Such is the case of the parasitic wasp, *Closterocercus chamaeleon* (Hymenoptera: Eulophidae), released against the introduced eucalyptus gall wasp, *Ophelinus maskelli* (also Hymenoptera: Eulophidae) in Israel (see Chapter 5). Protasov *et al.* (2007a) released about 12,000 adult parasitoids and found that not only were pest numbers reduced significantly within less than a year but also the enemy had spread unassisted for 120 km in the same period.

Figure 8.10 presents some of the results of a field trial on inundative biological control carried out in India against the teak defoliator moth, *Hyblaea puera* (Lepidoptera: Hyblaeidae) (Patil and Naik, 1997). *Trichogramma chilonis* (Hymenoptera: Trichogrammatidae) is a tiny parasitoid whose larvae feed and grow inside the eggs of Lepidoptera. In this field trial, varying densities of *Trichogramma* adults were released into teak plantations infested with *Hyblaea* and the resultant egg and larval populations of the pest assessed. In the figure, treatments T1 and T5 refer to the release of 100,000 and 1,000,000 parasitoids/ha, respectively, while T6 is a control where no natural enemies are released. It can be seen from the figure that both egg and larval densities were reduced in the presence of huge numbers of parasitoids, especially at peak, but the patterns in egg and larval densities mimicked each other closely as the generation of the pest progressed. In fact, the authors suggested that one of the most likely reasons for the depletion of larva shown so markedly was linked to foliage depletion – put simply, the moth larvae ate all the leaves on the trees and either died of starvation or migrated away in an attempt to find more food. Thus, in this example anyway,

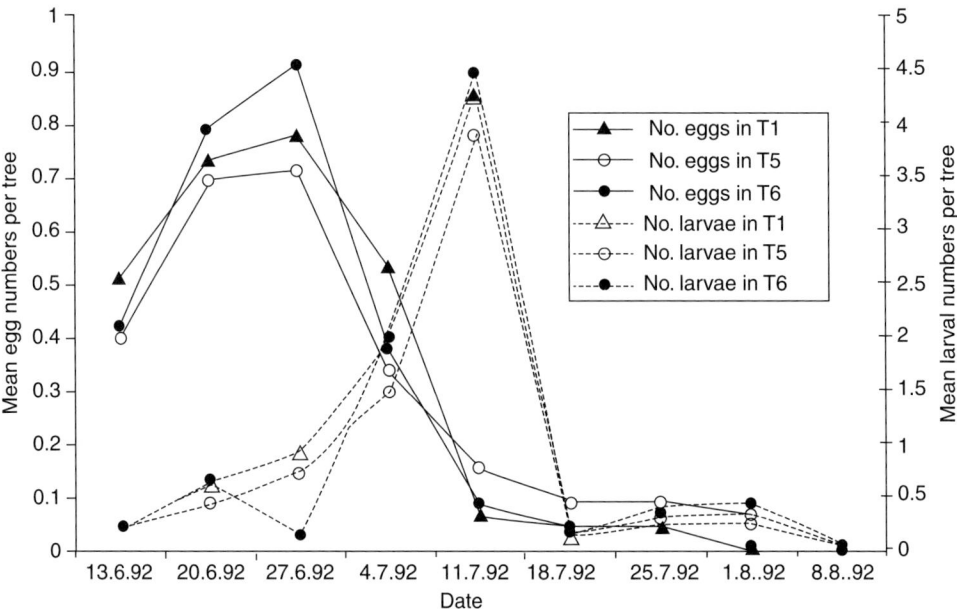

Fig. 8.10. Mean numbers of eggs and larvae of teak defoliator moth, *Hyblaea puera*, after releases of an egg parasitoid in India (from Patil and Naik, 1997, courtesy *Indian Journal of Forestry*).

even huge numbers of biological control agents did not seem to influence the population dynamics of the pest; resource limitation (bottom-up regulation) was the driving force. The amount of biological control achieved also varies enormously. Per cent mortalities can reach very high levels, but many examples exist in the literature of only moderate results. For example, Santos and de Freitas (2008) recorded between 17 and 21% parasitism of the eggs of the rubber tree lace bug, *Leptopharsa heveae* (Hemiptera: Tingidae) in Brazil, while Hanks *et al.* (2001) found that parasitism on the eucalyptus borer *Phoracantha semipunctata* reached only an average of 27% in southern California. Unless efficient IPM systems are available to deal with the remaining 75–80% survivors in these two examples, biological control cannot be claimed to be successful. Note also that any other treatments for pests and diseases, such as pesticides, may have negative effects on parasitoids (Paine *et al.*, 2011).

In 1996, Murphy estimated that the cypress aphid, then called *C. cupressi*, killed trees in southern and eastern Africa worth US$27.5 m and caused annual growth increment losses of US$9.1m (Murphy, 1996). It was introduced accidentally into parts of Africa between 1968 and 1986, and by 1990 this exotic aphid was present in Burundi, Ethiopia, Kenya, Malawi, Rwanda, South Africa, Tanzania, Uganda, Zaire, Zambia and Zimbabwe (Murphy, 1996) (see Chapter 5 and Plates 27 and 28). Since then, the aphid has appeared in other parts of the world, including South America (Sousa-Silva and Ilharco, 2001; Baldini *et al.*, 2008), where it causes serious damage to a variety of cypresses and cedars by feeding on the leaflets and injecting toxic saliva. It has become such a problem in countries such as Argentina that the characteristic red foliage symptoms it causes to trees have been termed 'cypress mortality' (El Mutjar *et al.*, 2009). The potential for classical biological control using imported natural enemies was recognized in the 1980s (Mills, 1990), and as the pest spread into new countries, this type of biological control has continued to be investigated (Montalva *et al.*, 2010).

The taxonomy of *Cinara* still requires clarification. According to Watson *et al.* (1999), aphids identified previously as *C. cupressi* appeared to belong to a species complex and, in fact, the species causing tree damage in Africa was found to be unnamed. Watson *et al.* named it *C. cupressivora*, which probably originated in a region from eastern Greece to just south of the Caspian Sea. Finally, in 2003, Remaudière and Binazzi decided that *C. cupressi* was synonymous with *C. cupressivora* (Remaudière and Binazzi, 2003). Whatever its name, the aphid is highly invasive and has a life cycle suggestive of an *r*-selected type of strategy typical of many aphids, wherein regulation by natural enemies may not be of great importance. Another factor tending towards an *r*-strategic system is the fact that, in Kenya and Malawi at least, there are no males: all reproduction is parthenogenetic and all individuals are female, thus enabling the population to be maximally reproductive. In instances such as this, pest management usually is based on host-plant resistance, essentially to slow the pest down and thus give biological control some chance to do the rest, as it were (see Chapter 6). Unfortunately, one of the most widely planted tree species in East Africa, *Cupressus lusitanica*, is one of the most susceptible to *Cinara* (Ciesla, 2003).

A regional biological control programme was established in 1991 involving the International Institute of Biological Control (IIBC), funded by the Canadian International Development Agency (CIDA) and the UK Overseas Development Administration (ODA). Classical biological control trials were based in Malawi at the Forest Research Institute of Malawi (FRIM) in Zomba (Biocontrol News and Information, 1997). In 1994, the first releases of an exotic parasitic wasp, *Pauesia juniperorum* (Hymenoptera: Aphidiidae), were carried out in Malawi (Chilima, 1995) and a widespread series of rearing programmes, releases and monitoring systems then ensued. *P. juniperorum* is a European species (Kairo and Murphy, 2005), which in Malawi has a larval period of around 8 days and a pupal period of 5–6 days (Chilima, 1996). Adult female fecundity is an average of 34 eggs during an adult lifespan of 7 days or so. Between August and

September 1994, approximately 170 mixed male and female parasitoids were released in Zomba, Malawi, in the hope that they would establish on cypress aphids. However, in 1995, recovery of *Pauesia* was very poor and it was concluded that establishment had not occurred. Because insectary facilities in Malawi were unable to rear the parasitoid locally, more shipments were made from the UK and over 1000 adult insects were released in a second wave (Chilima and Meke, 1995). Following these early aphid-season releases, 74% of all trees sampled in the November of 1995 had parasitized aphids on them, with up to nearly 50% aphid parasitism. By 1996, surveys showed that the parasitoid had dispersed up to 25 km away from the release sites and was recorded on all the ornamental cypresses and hedges inspected in and around the region (Biocontrol News and Information, 1997). Similar parasitoid releases were carried out in Mauritius in 2003 and 2004 (Alleck *et al.*, 2005), but establishment success has not been reported. Day *et al.* (2003) summarized these operations and van Driesche *et al.* (2010), looking back at these projects, reported that the declines seen in *Cinara* numbers in Kenya and Malawi were due, in part at least, to the biological control exerted by *Pauesia*. The aphid, however, continues to spread around the world, and one of the fundamental questions that any biological control programme must address is how much parasitism or predation actually is required to reduce impacts from pests to a level where the pests are no longer significant? In the case of the cypress aphid, it is felt that biological control on its own may not be efficient enough; long-term management of the cypress aphid in Africa will have to rely on a combination of silviculture and host-plant resistance, as well as biological strategies.

8.3.5 Biological control using fungi

Fungi which kill insects are known as entomopathogenic. Their big advantage over other groups of pathogens such as bacteria and viruses is their ability to infect host species via penetration of the insect cuticle (Hajek and Delalibera, 2010). To do this, they produce an invasive type of hypha that not only concentrates mechanical pressure on a localized region of the cuticle but also produces cuticle-degrading enzymes, which help to penetrate exoskeletons. Once inside the body of the host insect, fungal hyphae grow towards the body cavity, or haemocoel, where the fungus grows vegetatively, using up nutrients from the haemocoel and insect fat bodies. The infected insect thus may die of starvation, although toxins produced by the growing fungus may also debilitate and kill the host. As or just before the host dies, mycelia are produced by the fungus, which ramify through the insect's tissues and exit back through the cuticle. On the outside of the dead insect's body, spores are formed which can then be carried away by wind, water or other animals to infect a new host (Boucias and Pendland, 1998). The success of this final host-finding stage depends, of course, on the proximity of suitable insects – in pest epidemics the distance to the next host is minimized, so that infection is rapid and efficient. One fungal group, the microsporidia, has been trialled as a biological control agent of grasshoppers and locusts (Lange and Cigliano, 2010), but in tropical forestry, two fungi in particular are recognized as having great potential, white muscardine fungus, *Beauveria bassiana*, and green muscardine fungus, *Metarhizium anisopliae*. *Beauveria* has been known to be entomopathogenic for a very long time. It occurs as a ubiquitous soil pathogen and has a broad host range, which includes some fish and reptiles, and its beauty for biological control is that it can be cultured easily away from a living animal on various nutrient media. In insect hosts, it produces toxic by-products as it reproduces in the insect body and its greatest potential is in the biological control of pests which are normally hard to treat because of their way of life, such as soil insects or shoot and wood borers (Zhang *et al.*, 2011). White grubs (Coleoptera: Scarabaeidae) are notorious pests of forest nurseries (see Chapter 7) and their control is difficult since soil treatment normally is required, either by expensive and laborious sterilization or by the use of chemicals with undesirable costs and side effects. Trials in China (Table 8.3) illustrate the efficacy of *Beauveria* against these pests. It is clear that susceptibility varies with the species of pest; *Holotrichia* suffers reductions of over 90% when the soil is treated with formulations of the fungus, whereas *Blitopertha* populations, while reduced by 50% or so, are still able to cause significant damage to young trees (Li *et al.*, 1998). *Beauveria* also has potential to control defoliating insects. In China, Ding *et al.* (2004) described the movement of the pathogen through food webs in pine plantations infested with Masson's pine caterpillar, *Dendrolimus punctatus* (Lepidoptera: Lasiocampidae), while in India, Sharma and Ahmed (2004) was able to achieve up to 70% mortality of the Marwar teak defoliator, *Patialus tecomella* (Coleoptera: Curculionidae).

Metarhizium has been known to have entomopathogenic properties for over 100 years. There are many different strains of the fungus, which have been recorded from beetles (Coleoptera), Lepidoptera (moths and butterflies), Orthoptera (grasshoppers and crickets), Hemiptera (sap feeders such as aphids), Hymenoptera (bees, ants, sawflies and wasps), mosquitoes (Diptera) and termites (Isoptera) (Boucias and Pendland, 1998; Lopez and Orduz, 2003). Application systems are usually simple enough. Conidia (a type of spore) produced by the fungi can be dusted on to insects or their host plant, or sprayed in water-based formulations through standard application machinery, and many examples exist in the literature of the efficacy of both types of fungi.

One of the biggest challenges in tropical forest pest management involves termites (Isoptera) (see also Chapter 10). As described in Chapter 5, termite species can attack trees and timber all the way from the first year of life in nurseries and plantations, through the pole and mature stage to postharvest, including structural timber. Termite colonies can be vast, consisting of very many thousands of individuals, and of course, as before, they are almost all concealed. Fungal pathogens have been tested against many species of termites in many countries, and in principle, they have a set of highly desirable characteristics for the purpose of controlling pestiferous termites. For example, they are self-replicating, relatively safe to non-target organisms and small amounts of inoculum have the potential to spread throughout a termite colony, producing an epizootic (an epidemic of disease) (Jones *et al.*, 1996). In addition, the high relative humidity found within the confines of a subterranean termite colony is highly conducive to the proliferation of fungi such as *Metarhizium*, and both worker termites and alates (winged adults) can be killed with the fungus (Wright *et al.*, 2005). So, for example, Sun *et al.* (2008) worked with the wood-destroying Formosan termite, *C. formosanus* (Isoptera: Rhinotermitidae). They found that mortalities up to at least 90% could be achieved, but the results were dependent on the initial dose of fungal inoculum. In addition, the efficacy of different fungal strains varied considerably. Of course, it is one thing to kill termites when they are exposed to pathogens directly, but in practice it is vital to be able to inoculate just a small percentage of the total population and rely on them to transfer the lethal pathogen throughout the colony. Social grooming and close proximity will tend to assist this process. There are,

Table 8.3. Results of field trials in China to control white grubs in forest nurseries with the fungus *Beauveria* (from Li *et al.*, 1998).

Pest species	Level of control (%)	Infection rate (%)	Reduction in pest density (%)
Holotrichia diomphalia	67–85	56–68	69–93
Blitopertha pallidipennis	56	Not available	Not available
B. pallidipennis + *Steinernema feltiae* (nematode)	89	Not available	Not available

however, problems with this dissemination of fungi all over a termite colony. Healthy termites tend to isolate or even bury dead but infective relatives, a useful defence mechanism, but research continues into the use of fungus-infected baits for wood-destroying termites.

8.3.6 Biological control using nematodes

Infective juvenile (IJ) nematodes (the Dauer stage) in the families Steinernematidae and Heterorhabditidae are able to seek out insect pests and invade their bodies, where they grow and reproduce, releasing bacterial symbionts which kill the insects (Yu et al., 2006). In general, these nematodes have a large host range and have been used against a wide variety of insect pests in agriculture, horticulture and forestry (Georgis et al., 2006), and they have particular potential in the control of concealed pests such as wood, shoot and bark borers and soil pests, where they may be able to replace less desirable insecticides (Schulte et al., 2009). In the tropics, nematodes are used mainly to control insect pests of fruit trees (Dolinski and Lacey, 2007) and their use in biological control programmes in forestry has yet to take off, with one or two exceptions. Returning to termites, for example, Fig. 8.11 shows that various species of nematode can kill various species of pest, in laboratory trials at least (Yu et al., 2006). Notice though that some nematode species are much more effective than others, and while certain termites clearly are very susceptible, others are not. Whether this is due to differences in the relative abilities of nematodes to locate the hosts requires more detailed experimentation under proper field conditions. One of the most famous examples of the use of nematodes in forest pest management is that of the wood wasp *S. noctilio* (Hymenoptera: Siricidae) (Plates 44 and 45), whose control involves parasitoids, various silvicultural manipulations and in particular, the nematode *Beddingia.* (previously *Deladenus*) *siridicola* (Hurley et al., 2007), which renders the adult wood wasps sterile. *S. noctilio* is now abundant in many countries across the world, affecting *Pinus* spp. plantations in both temperate and subtropical regions (see Chapter 5). Collett and Elms (2009) review

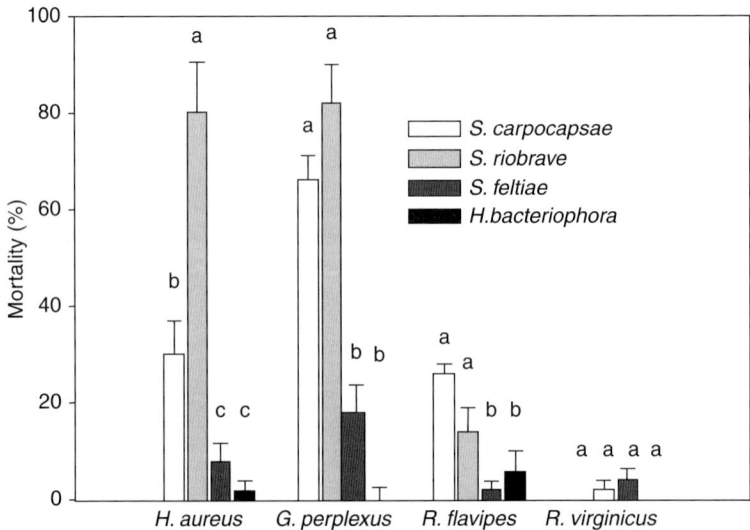

Fig. 8.11. Mortalities of the subterranean termites *Heterotermes aureus, Gnathamitermes perplexus, Reticulitermes flavipes* and *R. virginicus* after 48 h exposure to the entomopathogenic nematodes *Steinernema carpocapsae, S. riobrave, S. feltiae* and *Heterorhabditis bacteriophora* (from Yu et al., 2006, with permission Entomological Society of America).

the efficacy of biological control programmes in Australia. They report a decline in the ability of the nematode to control the pest, which seems to be related to variations in the performance of different strains of *Beddingia*.

8.3.7 Biological control using bacteria

Entomopathogenic bacteria are available for pest management in a variety of forms and activities, but the most commonly used in forest pest management is *Bacillus thuringiensis*, otherwise known as Bt. In the case of this biological control agent, we have to be fairly liberal with our definitions, since in practice, living Bt is not used to control pests, but instead a by-product of its metabolism, the so-called delta-endotoxin, is used as the active insecticidal agent. Commercial preparations of Bt do not (or at least should not) contain living bacteria at all, so this biological control agent cannot perform in a true regulatory fashion since it cannot replicate. Some would argue in fact that Bt is not a true biological control agent at all, but instead is a bioinsecticide. We will, however, include it in this section on biological control, since the delta-endotoxin is a direct product of a living organism. Bt occurs in several different basic strains or subspecies, within which there are many varieties with differing behaviours and characteristics. In tropical forestry, we are in the main interested in the subspecies *Bt kurstaki*, which predominantly kills lepidopteran larvae; in fact, the first detection of the pathogen in general was in silkworm farms at the beginning of the 20th century (Boucais and Pendland, 1998). The insecticidal endotoxin is a protein contained in crystals, which can comprise as much as 20 or 30% of the total bacterial protein in liquid broth cultures. When lepidopteran larvae ingest the toxin (and note that they must be ingested to be effective, so that any pest which feeds in a concealed way, or for that matter a sap feeder, will be unaffected), it produces a cessation of feeding and gut paralysis within a matter of hours post-ingestion. Starvation and death then ensue. Many commercial formulations of *Bt kurstaki* are now available, with trade names such as Biobit, Dipel, Delfin, Foray, Biolep, Bioasp and Lepidocide, and they are used routinely as pest control agents, especially in the developed world, in many areas of crop production including agriculture, horticulture and indeed forestry. There is little or no point at all in trying to treat bark, shoot or wood borers with Bt, since the host has to eat the toxin, but externally feeding defoliators can often be controlled successfully.

Figure 8.12 shows the results of field trials in Madhya Pradesh, India, against the teak skeletonizer moth, *Eutectona machaeralis* (Lepidoptera: Pyralidae). Both commercial formulations of *Bt kurstaki* were sprayed from the ground into the canopies of 5-year-old teak trees and the percentage mortality of larvae subsequently recorded. The figure shows that Biolep 2% was most effective, giving 77.5% mortality after 3 days, followed by Bioasp 2% and Bioasp 1%. Biolep 1% was least effective (41.7% mortality) (Meshram et al., 1997). As mentioned above, because Bt formulations are not alive as such, the entomopathogenic effects cannot be expected to persist or indeed proliferate in the environment, and hence repeated treatment is required to maintain pests at low levels, as shown in Fig. 8.13. Here, Bt was sprayed at intervals into mangrove forests on the Chinese coast against larvae of the moth *Ptyomaxia* sp. (Lepidoptera: Pyralidae), resulting in an average effectiveness of over 90% mortality (Li et al., 2007). Such treatment would, however, have to be continued for as long as the pest was abundant.

So, Bt appears to be a mainly successful alternative to heavy insecticide usage, at least for defoliators, and modern genetic engineering has produced a whole range of transgenic plants successfully that express endotoxin activity when eaten by insects, without any external treatment being required (see Chapter 6). However, serious insect pest resistance to Bt is now appearing (Gassmann et al., 2009), such that it is becoming much less effective than it used to be. So far, this problem has only really raised its ugly head in intensively treated agricultural crops, but if Bt is used indiscriminately in tropical forestry, then its future may be limited.

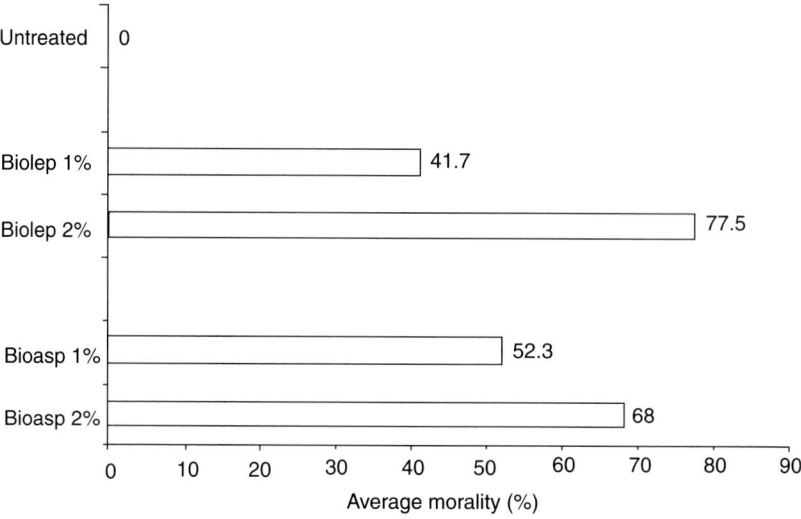

Fig. 8.12. Percentage mortality of larvae of *Eutectona machaeralis* in field trials in India with two commercial formulations of *Bacillus thuringiensis* var *kurstaki* (from Meshram et al., 1997, courtesy *Indian Forester*).

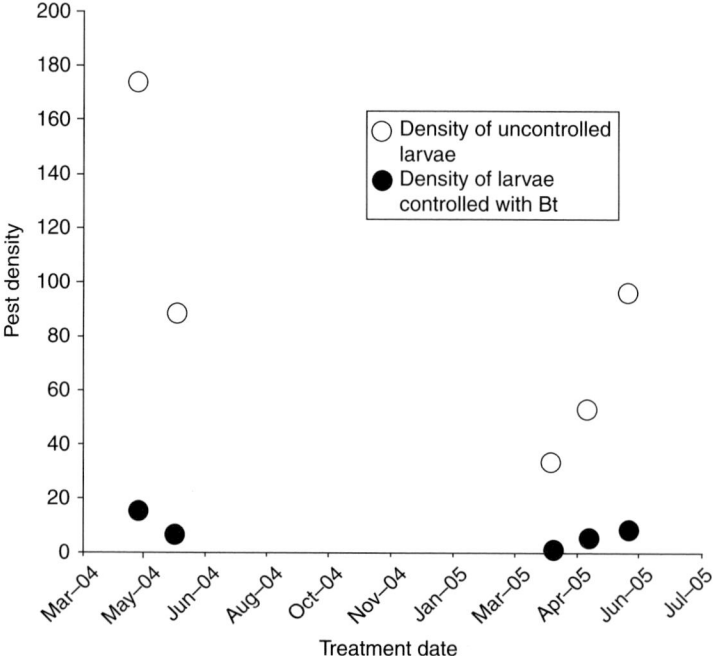

Fig. 8.13. Effect of repeated applications of *Bacillus thuringiensis* against larvae of *Ptyomaxia* in Chinese mangrove forests (data from Li et al., 2007).

8.3.8 Biological control using viruses

As with many other animals, including of course humans, insects suffer from a whole range of diseases caused by viruses and, on occasion, outbreaks of forest insects seem to decline as a result of infections by insect-specific viruses (Ilyinykh, 2011). Many viruses that kill insects are similar to those found in vertebrates, and one or two can even infect plants. The only group of viruses which are exclusive to insects are the baculoviruses, and these have great potential for the biological control of many forest insects. Baculoviruses are DNA viruses (Cory and Myers, 2003), unlike more familiar RNA viruses such as influenza and HIV, and two different basic types occur, the occluded viruses and the non-occluded viruses. Apart from some success in the biological control of palm rhinoceros beetle, non-occluded viruses have little scope, but the occluded viruses, especially the nucleopolyhedroviruses or nuclear polyhedrosis viruses (NPVs), have a potentially bright future (Plate 114). The name 'occluded' derives from the fact that the actual virus particles (virions or rods) are enclosed within a proteinaceous sheath, called a polyhedral inclusion body (PIB), which protects them from environmental conditions and enables them to persist outside the body of the host insect for long periods, especially in undisturbed habitats such as forests. Table 8.4 presents details of their biology and ecology (Speight et al., 2008). As with bacteria, viruses must be ingested (eaten) by the insect larva in order for an infection to start, but only relatively small concentrations contaminating leaf or bark material are required to start a fatal disease in the insect host. Young larvae are much more susceptible than older ones. In fact, late instars infected by an NPV might

Table 8.4. Characteristics of nucleopolyhedroviruses (NPVs) (from Speight et al., 2008).

Factor	Comments
Specificity	Completely within the phylum Arthropoda; usually within families, genera or even species; no significant effects on vertebrates detected
Occurrence	Naturally in insect populations in the field, especially in indigenous areas
Life stages infected	Larvae; eggs pupae and adults occasionally may carry inactive virus
Mode of action	Infection from contaminated substrate by ingestion only; virus attacks nuclei of gut cells in sawfly larvae and other cells (e.g. fat body) in moth larvae
Transfer between hosts	In faeces of infected but still living larvae, from disintegrating cadavers, in guts of predators (e.g. birds or beetles), may be on parasitoid ovipositors
Persistence	Environmentally stable except in UV light; persist outside the insect host on leaves or bark from generation to generation of pest, or in soil for considerable time (months or years)
Efficiency	Very high in pest epidemics via density-dependent transfer from host to host; enormous replication potential
Cost	Cheap to produce and bulk up by rearing the host species, or from field collections; complex technology not usually required beyond semi-purification
Drawbacks	Must be ingested by host insect, not plant systemic; limited or no use for borers and sap feeders
	High host specificity; rarely cross-infective from one pest species to another
	May not be available from nature
	Do not kill immediately; infected larvae may continue to damage trees for some time
	Must be applied in the same way (timing and technology) as insecticides
	Public and government distrust; tight legislation despite lack of evidence for side effects

reach the pupal stage before they die of the disease, and once they do, the pathogen ceases to function and the insect can complete its life cycle despite having picked up the virus. If death does occur in the late larval stage, host cadavers break down and release virus particles back into the environment ready to infect new hosts. Another mostly welcome trait of NPVs is their host specificity. Most of them are species specific and will not have any effect at all on even closely related insects. This means that there is no environmental impact at all from the use of such NPVs as a result of killing non-target species, and of course if they are unable to kill other insects, they certainly should be safe to use in the presence of vertebrates, including humans. Compare this with the well-known side effects of the use of synthetic chemical insecticides the world over. An added advantage is that infected insect larvae can be eaten by predators such as birds or beetles, with no effect at all on the predator. However, as the animal moves around the forest, infective NPVs reappear in the predator's faeces, producing new epicentres of disease for the target pest to encounter. The only problem with this host specificity is that most NPVs have to be discovered in the wild before they can be used – using an NPV for one forest pest against another species is usually impossible and careful searching and screening is required before a new biocontrol agent becomes available (if it ever does). Thus, though poplars (*Populus* sp.) have been grown in southern Brazil since the early 1990s, the description of an NPV from a serious poplar defoliating moth, *Condylorrhiza vestigialis* (Lepidoptera: Crambidae), was only published in 2009 (Castro *et al.*, 2009).

Figure 8.14 describes the fate of a typical NPV in the environment. Various pathways or cycles of infection exist, most of which assist in the production of a disease epizootic in the pest population (Richards *et al.*, 1998). It is clear that the potential for NPVs as biological control agents is excellent for exposed insects such as defoliators (Lepidoptera and sawflies especially) and they are being used in trials and experiments against many forest pests around the world, though with little commercial success, particularly in the tropics. In the developed world, it has to be said that there is considerable public distrust and suspicion about the use of 'germ warfare' against insect pests, since it can never be proved conclusively that pathogens such as NPVs will never cause harm to humans. However, there are situations where NPVs are being developed for tropical forest pest management and the best example, as already mentioned above, comes from Kerala in southern India against the teak defoliator moth, *H. puera* (Lepidoptera: Hyblaeidae). The impact of this pest is described in Chapter 5 and it is clear that many countries wherein teak is an important economic crop require a safe, cheap and efficient pest control system.

In Kerala, *Hyblaea* outbreaks occur almost every year, causing up to three severe defoliations per season. Relatively small trees can be sprayed from the ground (tall mature trees may be untreatable because of the prohibitive costs of spraying from fixed-wing planes or helicopters) and Nair *et al.* carried out trials with *Hp*NPV in 1993 (Nair *et al.*, 1996a). Early season (before June) applications of the virus resulted in substantial biological control of the pest, with a maximum protection of trees reaching nearly 80%. Leaf losses were reduced from nearly 50% to around 10% in March and April. Knowing the economic impact of such losses enables forest managers to assess the overall success of the control programme. The multiplication rate of *Hp*NPV is fairly typical for this type of baculovirus, and it depends on the initial dose rate, the development stage of the larval host and the time post-infection (Fig. 8.15) (Biji *et al.*, 2006). The number of PIBs per larva produced in a field mass production system averaged around 5×10^8 at death (Sudheendrakumar *et al.*, 2004). In a successful spray trial with *Hp*NPV, approximately 10^8 PIBs were used per teak tree, so that in this case, the product of one dead larva was able to provide enough virus to treat several trees.

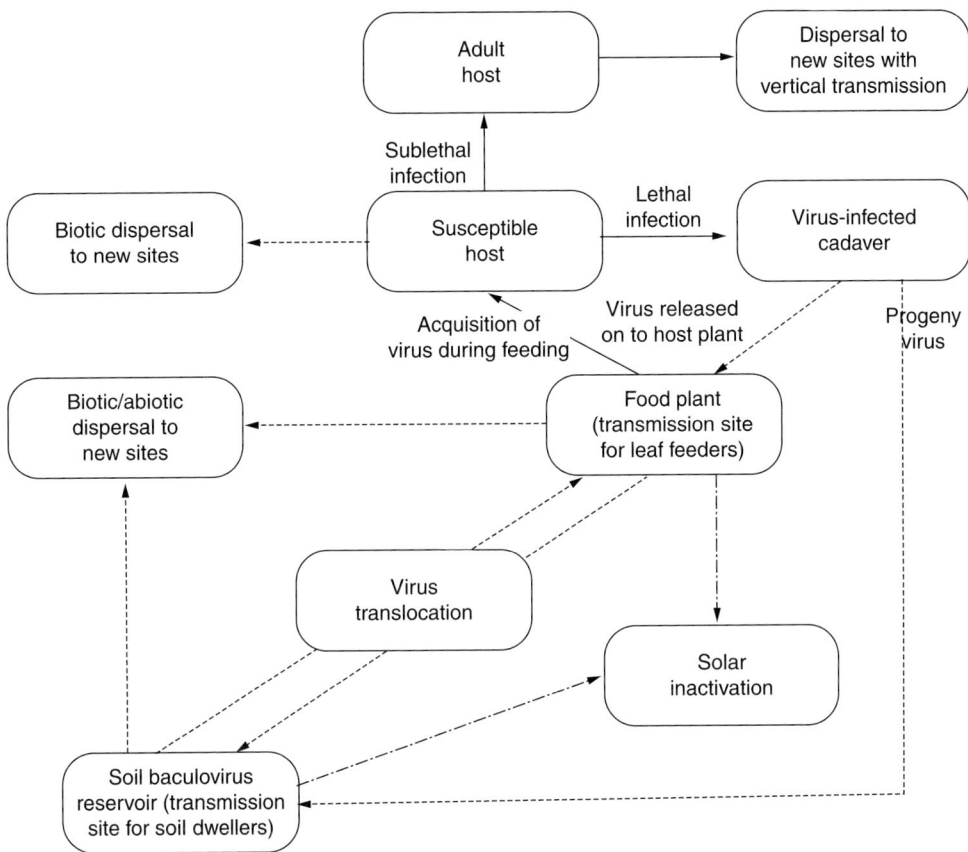

Fig. 8.14. The fate of lepidopteran nucleopolyhedroviruses in the environment. Solid arrows = NPV transmission routes; dashed arrows = NPV dispersal routes; dashed-dot arrows = NPV inactivation (from Richards et al., 1998, with permission Annual Reviews Inc).

Further work to refine and commercialize the biological control of teak defoliator using baculoviruses has continued in southern India, culminating with the production of a commercial formulation of *Hp*NPV, called 'HybCheck' (V.V. Sudheendrakumar, unpublished). The cost of this was in the region of Rs 250 (US$5)/ha in 2005. 'Control windows' have been developed which include rates and locations of larval feeding (target area), dosage–mortality relationships for all larvae (the dose), loss of virus from field factors such as ultraviolet light (attrition) and rates of coverage of foliage using selected sprayers and formulations. This sophisticated system has to be one of the best examples of biological control using pathogens in forestry.

8.3.9 The risks of biological control

In principle, biological control using the natural enemies of insects has many benefits over other control systems. It is usually self-perpetuating, environmentally benign, cheap and safe, and often requires little or no input from the grower once it is established. That it does not always work is not the fault of the scientists who do the research and development, but more due to the intractable nature of the pest–host plant–environment interactions. Foresters need to be more aware of why pest outbreaks occur in their plantations or village lots, so that basic preventative measures can be backed up by biological control tactics. So, are there any drawbacks to biological control? Some of

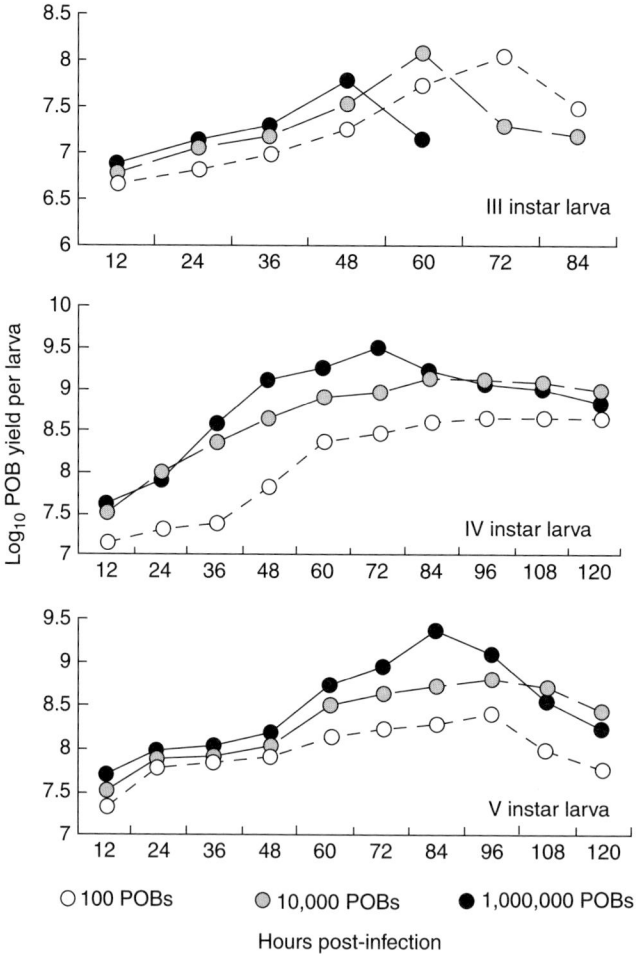

Fig. 8.15. Productivity (yield of POB/PIB) of *Hyblaea puera* NPV in laboratory cultures depending on the initial virus dose and the larval instar (from Biji *et al.*, 2006, courtesy Elsevier).

the agents we have described in the above sections are very host specific, such as most insect parasitoids and nucleopolyhedroviruses. However, some pathogens such as fungi, nematodes and bacteria are, to some extent, much more broadly infective. Some strains of *B. thuringiensis*, for example, will kill beneficial insects such as silk moths and even honeybees (Porcar *et al.*, 2008) almost as easily as their target pests, and so care must be taken before forests are treated liberally. Even with predators and parasitoids, the potential effects on non-target organisms must be considered, and the term 'biosafety' is now being used in the biological control literature (Barratt *et al.*, 2010). So far, this aspect of tropical forest pest management has been free of undesirable side effects.

8.4 Insecticides and Semiochemicals

The earliest references to the use of insecticides date back some 3000 years, although the widespread employment of chemicals to control insects is largely a development of the 19th century, starting with the successful use of Paris green against the Colorado potato beetle in the USA in 1867 (Murphy and Aucott, 1998). Insecticides generally have been used as the first line of defence in the control of insect outbreaks because: (i) they are highly effective; (ii) their effect is immediate; (iii) they can bring large insect populations under control rapidly; and (iv) they can be employed as needed. Few alternative means of control provide all these features. Serious limitations, however, are the adverse effects on human and animal health and the environment that can sometimes result from the use of these chemicals, as well as the development of insect resistance to pesticides. Worldwide, the total insecticide usage for forestry applications is a fraction of that for agricultural crops. It is generally only in nurseries, very young plantations, seed orchards and clonal hedges that such usage can be justified economically and environmental side effects contained, but there are exceptions. A classification of chemical insecticides and their use in nurseries is described in Chapter 7. Examples are provided below of their use in the planting and post-planting stages.

8.4.1 Establishment and immediate post-planting

Termites are major pests of transplanted seedlings or saplings in many parts of the tropics (Kirton and Cheng, 2007) (see also Chapter 6), but particularly in the dry regions of Africa and India (see Chapter 5). As outlined by Cowie *et al.* (1989), damage can take four forms: (i) the stem is cut near the base (e.g. by *Macrotermes* spp., *Odontotermes* spp. and *Pseudacanthotermes* spp.; Isoptera: Termitidae); (ii) attack just below the ground surface by these species extends upwards until the stem is ring-barked or downwards until the taproot tapers off and is severed; (iii) the roots are penetrated and hollowed out (e.g. by *Microtermes* spp., *Ancistrotermes* spp. and *Odontotermes* spp.; Isoptera: Termitidae); and (iv) the stem is penetrated and galleries excavated within it (Plates 115 and 116). These four modes of attack are not mutually exclusive and may grade into one another. The only effective means of preventing such attack has been by the use of persistent insecticides in the soil as a barrier around the roots and collar or by poisoning of mounds (although now, resistant tree species are being used increasingly). The guiding principle, of course, in deciding whether to apply control measures is that they are justified only when the expected loss without treatment exceeds the cost of treatment. Given the patchiness of termite attack, in most situations the cost of treating all plantings as a routine measure would be hard to justify. However, there are numerous recorded instances of particular plantation areas which are 'termite prone' and sustain repeated serious tree losses due to these pests. It is these areas that should be targeted for treatment (otherwise there would seem to be little point in replanting). One such example comes from the Guangxi Province in southern China, where young plantings of *C. citriodora*, *E. grandis* and *E. urophylla* were attacked by *O. formosanus*, *M. barneyi* and *Capritermes nitobei* (Isoptera: Termitidae), with losses of up to 73% in some areas (Wylie and Brown, 1992). Attack sometimes commenced within a few days of planting and was most severe in the first few months. Three refillings were necessary in some badly affected areas, a situation which could have been avoided by use of an insecticidal soil treatment at first refill.

Treatment of seedlings in the nursery, either by mixing insecticide with potting soil prior to filling the plastic seedling bags or by applying the liquid insecticide formulation to the seedling bags in the nursery beds is the most cost-effective and least labour-intensive method of forming a protective barrier (Nair and Varma, 1981; Cowie *et al.*, 1989) (Plate 117). When transplanting in the field, the seedlings should be planted so that the treated soil stands 2–3 cm proud of the surrounding soil surface. Incomplete removal of the plastic bag, leaving a collar of plastic around the top to keep treated soil in place, is sometimes practised. If the seedling is planted too deep, untreated soil may be washed or blown into the hole, allowing termites to breach the barrier. In India, post-planting treatment (drenching the surface soil around the plant with insecticide) confers no additional advantage, but this is essential in Africa, where the termites attack the plants at ground level, approaching the stem through the unprotected surface layer of soil (Nair, 2007).

The insecticides used most frequently for barrier treatments against termites have been the cyclodiene compounds dieldrin, aldrin, chlordane and heptachlor, because of their persistence in the soil. With the banning of these chemicals in most countries, a slow-release granular formulation of carbosulfan has been employed in Africa, Asia and South America in much the same fashion and has been found to be a very effective replacement (Canty and Harrison, 1990; Mazodze, 1992; Resende et al., 1995; Govender, 2007) (Plate 118).

Another situation where insecticides may be used at planting is to protect against attack by white grubs, which feed on the root systems of young trees, often ring-barking and severing the stems below ground. As with termites, attack is usually patchy, but it is sometimes possible to identify high-risk areas. Attack is often more severe where plantations have been established on grasslands or on land previously devoted to agriculture. In Guangdong Province in southern China, larvae of the large scarab *Lepidiota stigma* (Coleoptera: Scarabaeidae) have caused losses of up to 50% of young seedlings, necessitating refills (see Chapter 5 and Plates 66 and 67). Attack is worst in soft, sandy soils and a prophylactic treatment of slow-release carbofuran is applied at planting in these areas (Wylie and Brown, 1992). In South Africa, where there are about 106,690 ha of *A. mearnsii* plantations, white grubs are the dominant and economically most important seedling establishment pests, killing an average of 13% of wattle seedlings (Govender, 2007). Three insecticides are used against these pests; deltamethrin is applied as a drench at planting and gamma BHC or carbosulfan controlled-release granules are applied around the root plug in the planting pit. Surface applications of chlorpyrifos have been used to protect young commercial eucalypt plantations in Australia against attack by adults of the African black beetle, *Heteronychus arator*. Bulinski et al. (2006), however, have developed a cost-effective, pesticide-free approach to managing the problem, involving a nursery-applied flexible plastic mesh sleeve to cover the root mass partially. This has reduced severe damage and mortality by over 75%, with no negative effects on root formation or tree growth. Similarly, in Europe, Nordlander et al. (2009) have used a nursery-applied flexible sand coating (fine sand embedded in an acrylate dispersion) for the protection of conifer seedlings against damage by the pine weevil, *Hylobius abietis*. In field trials in commercial plantation areas, the sand coating was as effective in protecting seedlings as treatment with the insecticide imidacloprid.

A wide range of leaf-eating and sap-sucking insects can attack trees immediately post-planting. The small size of the trees means that even just a few insects can cause serious damage and the trees may not have sufficient energy reserves to recover from attack. However, their small size is an advantage when it comes to insecticidal spray application because their foliage can be reached easily from the ground by conventional spray equipment, and spraying can be targeted at individual trees or concentrated insect swarms as required. In tropical Queensland, swarms of the scarab beetle, *Epholcis bilobiceps* (Coleoptera: Scarabaeidae), attack young eucalypt plantations, completely stripping the foliage. When not actively feeding on foliage, they often congregate in large numbers on the stems of trees (Plate 119) and these concentrations can be sprayed easily with contact insecticides (Elliott et al., 1998). In *A. mearnsii* plantations in South Africa, grasshoppers (most commonly *Zonocerus elegans*) killed a maximum of about 10% of seedlings and were controlled by spraying with carbaryl or deltamethrin (Govender, 2007). Meshram and Tiwari (2003) recommend application of deltamethrin and the fungicide carbendazim for control of damage caused by the lace bug *Tingis beesoni* and the associated fungus *Hendersonula toruloidea* to young *Gmelina arborea* saplings (see Chapter 5). For insects such as leaf tiers, leaf rollers and shoot borers, whose feeding habit protects them against contact insecticides, systemic insecticides may be employed. In Costa Rica, for example, where young plantations of *Pinus* spp. have been severely attacked by the Nantucket pine tip moth, *Rhyacionia frustrana* (Lepidoptera: Tortricidae), application of dimethoate resulted in a larval mortality of 80% 8 days after treatment (Salazar, 1984). Other systemic insecticides such as phorate, monocrotophos, imidacloprid and carbofuran have also been used against various cone and seed insects in parts of Asia (Bhandari *et al.*, 2006; Nair, 2007) and the southern USA (Grosman *et al.*, 2002) (see Chapter 5).

Leaf-cutting ants (Hymenoptera: Formicidae) are one of the principal pests of plantation forestry in the New World and have been reported attacking 15 different forest tree crops, including pines, eucalypts, teak and *Gmelina* in some 16 countries (see Chapter 5 and Plates 16 and 17). It is not surprising, therefore, that a wide range of insecticides has been used against them, applied in a variety of methods as discussed by Cherrett (1986) and summarized in Table 8.5. Spraying of foraging workers with contact insecticides is only a short-term expedient and most control measures have been directed against the nest itself once this has been located. The organochlorine compounds aldrin, dieldrin, chlordane and heptachlor have often been made up as emulsions in water and poured down the nest entrances. This can be effective but is time-consuming and if the ground is dry, the liquid may not penetrate far into the nest. Gas fumigation of nests is a technique that has been employed successfully over many years using some highly toxic compounds such as carbon disulfide, hydrogen cyanide and methyl bromide. Insecticidal smoke pumped into the nest under pressure is also effective, but the machinery is expensive and the operators need special training. Because of this, small farmers tend to favour insecticidal dusts blown down nest entrances with a cheap hand pump.

All these methods require the nest to be found, are labour-intensive and entail the use of large quantities of toxicant. Toxic baits, however, do not have these shortcomings and their use began around 1957 with the manufacture of wheat flour baits containing 2% aldrin. The requirement for a bait toxicant is that it should be non-repellent to the target pest and should be sufficiently slow acting to enable it to

Table 8.5. The principal insecticides that have been used in leaf-cutting ant control, and modes of application (after Cherrett, 1986). Note that the use of the majority of these chemicals is no longer allowed.

Chemical	Method				
	Bait	Dust	Gas	Liquid	Smoke
Aldrin	X	X		X	X
BHC		X			X
Carbon disulfide			X	X	
Chlordane		X		X	X
Dieldrin	X	X		X	X
Heptachlor	X	X			
Hydramethylnon	X				
Hydrogen cyanide	X		X		
Methyl bromide			X	X	
Mirex	X				
Nonachlor	X				
Sulfur and arsenic			X		
Sulfluramid	X				

be brought back to the nest and passed around by mutual feeding before it takes full effect. Mirex, a slow-acting stomach poison developed for fire ant control, was found to be most suitable against leaf-cutting ants. Because it is persistent in the environment, toxic and possibly carcinogenic, its use has been phased out in most countries. In South America, it was incorporated in an attractant bait which was sealed inside small polythene bags (Cherrett, 1986). The ants detect the attractant through the polythene, cut the bags open and remove the bait. Sulfluramid and hydramethylnon are now used in the baits in place of mirex (Zanuncio et al., 1999; Carlos et al., 2011). Aracruz Celulose in Brazil, a leading producer of bleached sulfate eucalypt pulp for high-quality paper, has about 260,000 ha of plantations (Osland and Osland, 2007) (Plate 122), managed on a 7-year rotation, which are subject to attack by leaf-cutting ants. A three-phase control strategy is employed against these pests. Prior to planting, nests are located and baits placed around active exit holes. Immediately after planting, any nests still active are fumigated with methyl bromide. From the second month after planting until harvest, the bait bag method is used, the bags being distributed systematically through the plantations at a density of 56 bait bags/ha and also placed around individual nests (Laranjeiro, 1994).

8.4.2 Intermediate years (above reachable height to mature)

In tropical forestry, most insecticide applications are from the ground, for reasons relating to accessibility, cost and technology requirements. However, the rapid growth of many of the hardwood species used in plantations in the tropics means that the canopies of such trees are soon beyond the reach of ground-based spray equipment, and aerial spraying of pests, if economically justified and affordable, may be the best option. In India, where *Ailanthus triphysa* is planted on a large scale for the matchwood and pulp and paper industries, frequent defoliation of these plantations by caterpillars of the moth *Atteva fabriciella* (Lepidoptera: Yponomeutidae) results in retardation of growth and mortality of younger trees. Aerial spraying with malathion or endosulfan provides effective control (Jha and Sen-Sarma, 2008). Similarly, a range of insecticides has been applied aerially to control outbreaks of the teak skeletonizer *E. machaeralis* (Lepidoptera: Pyralidae) and defoliator *H. puera* (Lepidoptera: Hyblaeidae) in teak plantations in India, including endrin, malathion, fenthion, carbaryl and, more recently, *B. thuringiensis* (Jha and Sen-Sarma, 2008) (see earlier in this chapter). In a 15-year-old teak plantation in Thailand, neem extract was applied using a thermal fogger to control *H. puera*, and at higher concentrations gave 77–99% mortality of larvae in about 6 days (Eungwijarnpanya and Yinchareon, 2002). The thermal fogger and high-power sprayers had been used routinely in Thailand since 1981 to apply *B. thuringiensis* in teak seed orchards and plantations against teak defoliator.

The wattle bagworm, *Chaliopsis junodi* (Lepidoptera: Psychidae), is the most sprayed forestry pest in South Africa. Estimates of the areas sprayed annually by air are shown in Fig. 8.16 and tend to follow almost exactly the fluctuations in the bagworm population (Atkinson, 1998). Insecticides used have evolved from extremely toxic materials at high application rates (camphechlor, endrin) to the much more benign synthetic pyrethroids at extremely low rates (alphacypermethrin). Also in South Africa, plantations of *A. mearnsii* are sprayed from the air with either deltamethrin, acephate or cypermethrin to prevent outbreaks of the brown wattle mirid, *Lygidolon laevigatum* (Hemiptera: Miridae), when trapping indicates that populations could exceed the economic threshold if not treated (see earlier this chapter). As well, eucalypt plantations in that country are sprayed aerially with fenvalerate or cypermethrin to control infestations of the Eucalyptus snout weevil, *Gonipterus scutellatus* (Coleoptera: Curculionidae) (Atkinson, 1999; see Chapter 5 and Plate 15). Interestingly, in some trials with cypermethrin against this pest there was evidence that

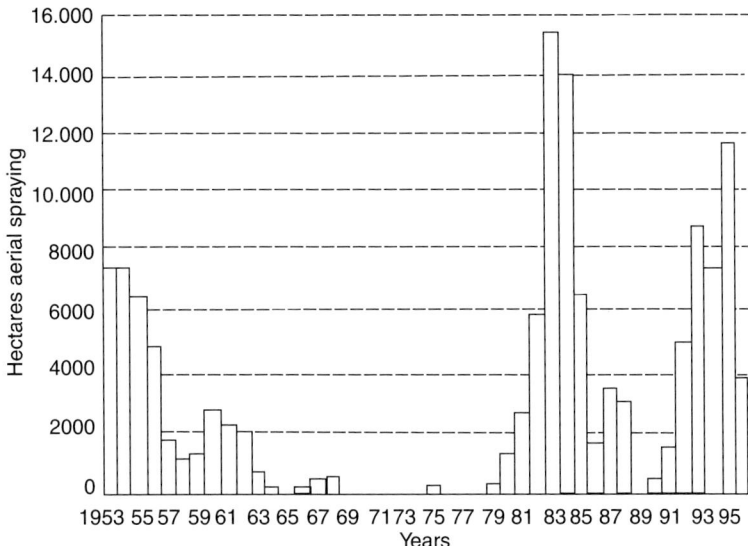

Fig. 8.16. Areas of aerial spraying annually for the wattle bagworm, *Chaliopsis junodi*, in South Africa from 1953 to 1996. These tend to follow almost exactly the fluctuations in the bagworm population (from Atkinson, 1998).

spraying might actually increase the level of egg parasitism by the mymarid wasp *Anaphes nitens* (Hymenoptera: Mymaridae) following the treatment.

Another use of insecticides in the intermediate years, and one requiring only simple methodology, is for the control of borers in stems of trees. In Mindanao, in the southern Philippines, larvae of a cossid wood moth (Lepidoptera: Cossidae), closely related to the beehole borers, and of a hepialid wood moth (possibly *Endoclita* sp.) were found infesting 3-year-old plantations of *G. arborea* being grown for use as poles and sawn timber. Surveys at three sites showed an incidence of attack ranging from 11 to 60%. Generally, there were between one and four attacks per tree (up to 14), the majority being in the lower stem. The control technique employed by local staff was to poke a cotton rag soaked with azodrin into the larval tunnel.

In Thailand, the longicorn *Aristobia horridula* (Coleoptera: Cerambycidae) is a serious pest of *Pterocarpus macrocarpus* plantations aged 4 years and older (see Chapter 5). Larval tunnelling in stems and branches causes death of young trees or makes them prone to wind-throw. Suggested control measures include painting the stems of trees with heavy oil, spraying the bark surface with insecticide during the emergence period, or injecting insecticide into holes after removing frass and wood fragments (Nair, 2007). Similarly in Bangladesh, techniques recommended for the control of the bark-eating caterpillar *Indarbela quadrinotata* (Lepidoptera: Indarbelidae) (see Chapter 5 and Plate 35), a pest of *Falcataria moluccana*, include spraying the bark with 0.1% dieldrin or malathion, injection of 5 ml of 0.1% dichlorvos into each larval hole, or plugging the larval hole with cotton soaked with kerosene. Baksha (1993) cautions that such methods, being very labour-intensive, may be impractical for large plantations but may be useful to protect high-value trees such as seed orchards and ornamental trees.

8.4.3 Mature trees

Insecticides are rarely used in mature, commercial plantations. Leaving aside any considerations of logistics, the environment or the technical difficulties of achieving adequate spray coverage through tree canopies to reach all target organisms, the main reason is that if severe insect damage were to occur, it is probably more cost-effective to salvage damaged trees, which are close to harvest anyway. Possible exceptions would be chemical usage to eradicate an exotic pest or spot treatment to contain an outbreak. Mature trees may sometimes be sprayed where the plantations have been established for reasons other than production. For example, in Taiwan in the late 1980s, outbreaks of the casuarina tussock moth, *Lymantria xylina* (Lepidoptera: Lymantriidae), occurred in plantations of *C. equisetifolia* which had been established as windbreaks in west coastal areas. Defoliation was so severe that the windbreaks lost their function. Outbreaks were controlled by helicopter spraying of a mixture of azodrin and carbaryl, sometimes in combination with *B. bassiana* (Chang, 1991).

Payne (2000) discusses the various physical, chemical and biological factors influencing aerial insecticide application to forests. A successful aerial insecticide application is one that provides the desired degree of pest control at an economic cost, with little environmental impact. Pest control products that are very specific in their activity, applied in appropriate quantities and at a time when the pest is susceptible, exposed and feeding, and with a distribution that matches that of the pest, contribute to these desired outcomes. The factors influencing the success of forestry insecticide applications differ somewhat from agricultural applications because forested treatment areas are often larger, more remote and less well defined, and the plant canopy is taller and aerodynamically rougher, making aerial applications the norm. Global positioning systems are valuable in finding and defining remote treatment areas and in guiding the pilot during the application, thereby improving efficacy and reducing off-target deposits from aerial insecticide applications to forests. The relatively tall and uneven forest canopy leads to an increased flying height to maintain pilot safety. In turn, this makes meteorological factors of somewhat greater importance in forestry applications as compared to those for agriculture (Payne, 2000).

8.4.4 Semiochemicals

As outlined by Nadel *et al.* (2011), semiochemicals are chemicals produced by one organism that affect the behaviour or physiology of another, and they are commonly used in inter- and intra-specific communication between insects and in host finding. Pheromones are used in the communication between insects of the same species, whereas allelochemicals (kairomones, allomones and synomones) act as chemical signals between different trophic levels, genera and/or species. Semiochemical research in forestry has grown exponentially over the past 40 years, most of the emphasis being on pheromones (Nadel *et al.*, 2011). They are used for the detection or monitoring of pests and are also employed in various strategies for controlling insect populations, such as mass trapping, lure and kill, lure and infect and mating disruption (El-Sayed *et al.*, 2006, 2009).

In mass trapping, suitable densities of pheromone-baited traps are deployed with the aim of pest suppression by male and/or female annihilation. Most of the work that has been carried out so far has focused on forestry pests in Europe and North America, where bark beetles have been the main targets. The largest mass-trapping programme for bark beetles was conducted in Norway and Sweden for the spruce beetle *I. typographus* during the period 1979–1982 (El-Sayed *et al.*, 2006). Approximately 600,000 pheromone-baited traps deployed in Norwegian forests caught an average of 4850 beetles/trap, or 2.9bn beetles in total. Following this, the number of attacks of living trees by the insect declined steadily, but the relative contributions of control measures and various natural factors to this decline are unknown (Eidmann, 1983). In southern California from 1989 through 1992, a 48 ha stand of the rare Torrey pine, *P. torreyana*, suffered 12% mortality due to infestation by the California fivespined ips,

I. paraconfusus (Shea and Neustein, 1995). A mass-trapping programme commenced in 1991 using aggregation pheromones in the infested parts of the stand and anti-aggregation pheromones in uninfested areas. After several seasons of trapping, mortality caused by *I. paraconfusus* was virtually eliminated. Although there were no controls, it was considered more than mere coincidence that when trapping was discontinued, mortality increased and when trapping resumed, mortality ceased. Another reported successful case of pheromone mass trapping is that of the bark beetle *I. duplicatus* in a 2000 ha 'island' of spruce in Inner Mongolia (Schlyter *et al.*, 2003). Examples from the tropics are few. In 1988, a hurricane devastated much of the island of Jamaica, damaging one-quarter of the total pine plantations, mostly *P. caribaea* (see Chapter 6). Damaged trees were colonized rapidly by bark beetles *I. calligraphus* and *I. cribicollis* (Coleoptera: Scolytinae), which could complete their life cycle within a month. Populations were monitored with multiple funnel traps baited with ipsenol and ethanol, which attracted mainly *I. cribicollis*, or a bait containing ipsdienol, ethanol and cisverbenol, which attracted mainly *I. calligraphus*. Pheromone-baited trap lines were recommended in areas where excessive quantities of blown-down timber were adjacent to stressed standing trees, the idea being to attract the beetles away from such trees and thus reduce tree mortality (Bunce and McLean, 1990). In India, Masoodi *et al.* (1990) studied the effect of disparlure, deployed in pheromone traps in poplar and willow plantations, on the suppression of mating by *L. obfuscata* (Lepidoptera: Lymantriidae). Removal of a substantial proportion of males by these adhesive-coated traps resulted in a significant suppression of mating, as shown by the egg mass count before bud burst and after leaf fall. Reduction of egg mass density was between 62 and 77%.

The lure-and-kill technique differs slightly from mass trapping in that the insect responding to the semiochemical lure is not 'entrapped' at the source of the attractant by adhesive, water or other physical device, as in mass trapping, but instead the insect is subjected to a killing agent (insecticide or insect pathogen) or sterilizing agent, which effectively eliminates it from the population after a short time (El-Sayed *et al.*, 2009).

The objective of mating disruption is to prevent the successful mating of target insects by interrupting or disrupting communication between the sexes. In this system, the responding sex is unable to find the emitting sex. Insects remain alive but disoriented during mating disruption, whereas they are removed from the population by mass trapping or lure-and-kill systems. Among the three approaches, mating disruption is the most widely used in pest management, followed by mass trapping and lure and kill (El-Sayed *et al.*, 2006). Two strategies are used for deploying synthetic pheromones over wide areas. One involves the use of widely spaced dispensers, each releasing relatively large amounts of pheromone for up to 6 months (e.g. twist-tie ropes). The other involves application of a large number of small dispensers, each releasing a small amount of pheromone over 2–4 weeks (e.g. microcapsules, beads or flakes) (Gillette *et al.*, 2006). Mating disruption using semiochemicals in the form of sex pheromones has been applied most successfully to control lepidopteran pests such as the western pine shoot borer, *Eucosma sonomana*, and the ponderosa pine tip moth, *R. zonana* (Lepidoptera: Tortricidae) in California (Gillette *et al.*, 2006). Treatments with microencapsulated pheromone disrupted orientation by both species for several weeks and resulted in a population reduction of two-thirds in each of the 2 years studied. In Virginia, forest plots treated aerially with a plastic laminated flake formulation of disparlure to disrupt gypsy moth *L. dispar* (Lepidoptera: Lymantriidae) mating were monitored for 2 years after treatment (Thorpe *et al.*, 2007). In the year of treatment, there was a greater than 98% reduction in mating success compared to controls. Mating success 1 year after treatment was reduced by 60–79%, but was not reduced significantly at 2 years after treatment. Mating disruption is a key element in the 'Slow the Spread of the Gypsy

Moth' programme in the USA. The goal of this programme is to reduce the rate of gypsy moth spread by treating and eliminating the numerous gypsy moth colonies that become established just beyond the leading edge (Sharov et al., 2002; Onufrieva et al., 2008) (Fig. 8.17).

Although semiochemical research has been undertaken for various insects in the tropics and southern hemisphere, there are few examples of this approach being used in plantation forestry (Nadel et al., 2011). The perceived high cost of discovery, development and application of semiochemicals and a lack of research capacity have been suggested as the main reasons for their minimal use in these regions. However, this situation is changing gradually, with several countries now building up their capacity in chemical ecology. The most common use of semiochemicals in the region is in the detection of forest invasive species and in monitoring the spread and population levels of various forest pests. In several Australian states, including tropical and subtropical Australia, semiochemicals are used to monitor for the presence of quarantine pests at high-risk sites around seaports and airports, as well as in forests located close to these areas (Wylie et al., 2008) (Table 8.6) (see also Chapter 9).

Australia also has an early warning system for Asian gypsy moth and other *Lymantria* species using disparlure-baited traps. In South Africa, since 2007, a national network of monitoring traps using kairomone lures, simulating stressed pine trees, has been established across the commercial forestry plantation resource (Nadel et al., 2011). Here, it is anticipated that monitoring the wood wasp *S. noctilio* ahead of the invasion front allows for early detection and swift action in the form of releasing biological control agents in newly invaded areas. A similar trapping programme is in place in Queensland for monitoring the spread of *S. noctilio*, a recent arrival in that state, employing a combination lure of α-pinene and β-pinene. Both Australia and South Africa have begun to explore the potential to use sex pheromones to control cossid wood moths that kill trees or lower timber values in commercial eucalypt plantations, such as the giant wood moth, *Endoxyla cinereus*, and the Culama wood moth, *Culama australis*, in Australia and the goat moth, *Coryphodema tristis*, in South Africa (Lawson et al., 2008; Nadel et al., 2011). Semiochemicals have been used as a deterrent to the black twig borer, *Xylosandrus compactus* (Coleoptera: Scolytinae), in an *A. koa* forest in Hawaii (Dudley et al., 2006). Zarbin et al. (2007) describe the rapid progress that has been made in research on insect pheromones in Brazil in the past decade, including identification of a pheromone of the subterranean termite *Heterotermes tenuis*, which is a serious pest in commercial plantations of *Eucalyptus* species.

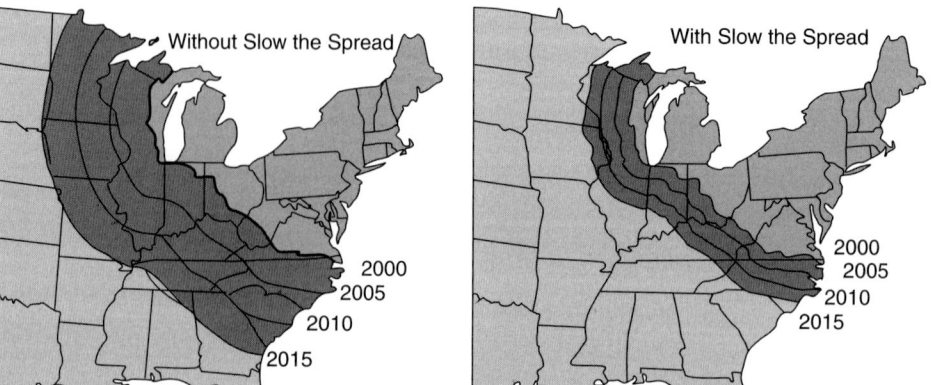

Fig. 8.17. Projected gypsy moth spread with and without the Slow the Spread Project (from Sharov et al., 2002).

Table 8.6. Forest pests and fungi target taxa and the methods used to detect them in Brisbane, Australia (from Wylie et al., 2008, with permission of *Australian Forestry*).

Target taxa	Trap	Lure

Longicorn beetles (Cerambycidae), particularly:
  Anoplophora glabripennis (Asian longhorn beetle)
  Monochamus alternatus (Japanese pine sawyer)
  Arhopalus ferus (burnt pine longicorn)
  Stromatium spp.
  Hylotrupes bajulus (European house borer)
	Panel	α-pinene + ethanol

Wood wasps (Siricidae) particularly:
  Sirex spp.
  Urocerus spp.
  Xeris spp.
	Panel	α-pinene + ethanol

Bark beetles (Scolytinae):		
Ips spp.	Panel	Ipsenol, Ips-dienol
Dendroctonus spp.	Panel	Frontalin, Exo-brevicomin
Other species, especially *Tomicus* spp., *Orthotomicus* spp. and ambrosia beetles	Panel	α-pinene + ethanol
Asian gypsy moth *Lymantria dispar* (Lymantriidae)	Delta	Disparlure
Eucalyptus rust *Puccinia psiidii*	Inspection	

Auger and powderpost beetles (Bostrichidae), especially:
  Heterobostrychus aequalis
  H. brunneus
  Lyctus africanus
  Sinoxylon spp.
	Panel	No specific lure

Some of the advantages of pheromones over insecticides are discussed by Witzgall et al. (2010). Insecticides do not achieve a long-term pest population decrease, but many studies show that continuous long-term pheromone use *does* decrease population levels of target species. This is attributable to a recovering fauna of beneficials and to an increasing efficacy of pheromones at low population densities when communication distance between sexes is increasing. Insecticide overuse also induces outbreaks of secondary pests by disturbing the natural regulation of herbivores by their antagonists. Insects with hidden, protected lifestyles, including those with underground or wood-boring larval habits, cannot be controlled easily with cover sprays of insecticides. Here, control with pheromones is advantageous, since it aims at the mobile adult life stage and functions to prevent oviposition altogether. Further examples of the uses of pheromones in insect pest monitoring are provided in Chapter 9, and Wyatt (2003) has more details of pheromone chemistry, biology and ecology.

9

Management Systems IV: Forest Health Surveillance, Invasive Species and Quarantine

9.1 Forest Health Surveillance

9.1.1 Introduction

Once a crop has been planted, whether it be agricultural, horticultural or forest tree, there is usually a need to ensure its protection, until the time of harvest and sometimes beyond, from a variety of damaging agents both natural (wildfire, hail, rainfall extremes, insects, fungi, animals) and those related to human activities (pollution, domestic stock, mechanical injury, pesticide toxicity). As described throughout this book, it is often a combination of such factors rather than a single-agent cause which underlies many crop problems. For example, in Australia, overstocking in *Pinus* plantations accompanied by drought stress can promote attack and mortality by the wood wasp *Sirex noctilio* (Collett and Elms, 2009). Evans and Turnbull (2004) outline a general approach to follow with all organic damage (fungi, insects, other animals or microorganisms); the four stages are detection, identification, analysis and action. It is in the very first of these, detection, that some significant differences between agricultural/horticultural systems and forestry systems begin to manifest themselves.

For a start, in a comparison of the minimum areas of land required in order to provide similar economic returns on investment, industrial forest plantations (IFPs), by the nature of their product, market value and long rotation times, are generally much larger than their agricultural counterparts. Crops such as wheat and sugar are broadacre but are harvested annually, while horticultural crops such as avocado and mango may take a few years to produce fruit but are of higher value per unit area. The critical mass required to make the forest industry viable in a region may also be large. For example, in southern Sumatra in Indonesia, a plantation of 170,000 ha of *Acacia mangium* has been established to supply the fibre needs for just one pulp and paper mill. Areas of natural forest that are harvested for timber are even vaster. This by itself greatly increases the difficulty of detection of health problems. Continuing this comparison of cropping systems, the greater height of forest trees within a few years after their planting and the often more rugged terrain on which they are planted make sampling difficult. Added to this is the fact that many of the forests are remote and infrequently visited, and a serious problem can pass unnoticed for a long time.

9.1.2 Purpose of routine forest health surveys

Identifying and managing threats to forests and plantations is an essential element of sound forest management (Carnegie, 2008). What constitutes a healthy forest can vary according to differing management objectives for particular forests. For example, in a forest set aside for conservation, an outbreak of a native insect pest would be regarded as a part of the normal forest dynamics and no remedial action would be taken unless such outbreak was linked in some way with human interference. In a production forest, however, any insect outbreak which threatened significant tree mortality or serious growth loss would engender prompt remedial action. In a conservation forest, stag-headed and hollow trees could be taken as a sign of a healthy ecosystem in that they provide habitat for wildlife, whereas in a forest plantation dead or dying trees are more likely to be a cause for alarm as an indicator of a potential health problem.

As defined by Carnegie (2008), forest health surveillance involves systematic surveys of forests by trained specialists, the main purposes being to: (i) detect and map outbreaks and damage by known pests or diseases; (ii) detect change in forest health over time, including the distribution and status of pests or diseases; and (iii) detect incursions of exotic pests or diseases. The principle underlying this is that early detection of a pest problem allows more scope for its management. In forests managed for conservation purposes, the emphasis is more likely to be on the detection of any unnatural biotic or abiotic factors (e.g. introduced pests, air pollution) which threaten the long-term health or vitality of the ecosystem, and on factors which are indicative of ecosystem disturbance or decline. Such surveillance, of course, would also include naturally occurring factors.

9.1.3 Requirements for effective and efficient surveillance

Forest health surveillance or detection monitoring essentially provides a 'snapshot' of the health of the forests. It includes both extensive surveys and fixed-plot monitoring, and usually entails systematic observance of a predetermined set of parameters that pertain to forest health. Data obtained build up a reference baseline and assist in the early detection of changes that call for a more detailed evaluation. In the course of such surveys, the surveillance team may be able to suggest likely causes for some of the changes observed (e.g. lightning strike as a cause of observed bark beetle attack and tree mortality).

Problems or potential problems detected during broad-scale surveillance must then be investigated by protection specialists to delineate the extent of the problem, identify the cause (if possible) and make recommendations to managers as to the appropriate course of action. Such recommendations may be to do nothing (if not a problem or not economically justified), to conduct additional targeted surveys, to monitor the occurrence closely, to take control action or to initiate detailed research on the problem. These two phases (i.e. general surveillance and follow-up investigation) encompass the four stages outlined by Evans and Turnbull (2004). Barnard et al. (1992) include an additional phase, 'intensive-site ecosystem monitoring', as part of health monitoring, the goal being to obtain a more complete understanding of the mechanisms of change in forest ecosystems. Such detailed and long-term monitoring is more commonly used in naturally occurring forests and could be regarded as being in the realm of ecosystem research. It is in a somewhat different category to the routine health surveys carried out in forest plantations.

Surveillance for forest pests is practised at varying levels of sophistication in different countries throughout the tropics. In its simplest form, it involves individual workers in the forests observing a disorder or something unusual in the course of their duties and reporting it to a superior, who takes the appropriate action. These 'eyes in the forest' are an important component of any surveillance system, but their contribution to overall probability of detection should not be overestimated by forest managers. As an example, trials in New Zealand have demonstrated that the 'efficacy' of detection by such staff is extremely low (Carter, 1989). The next level of surveillance may involve the assigning of certain staff, although non-specialist, working in a particular forest area to look for, or gather information on, disorders in that forest and to liaise with specialists. At the top level, teams of highly trained professionals are employed to conduct systematized surveillance of the forests using a range of methodologies.

For forest health surveillance to achieve its goals, it is essential to have a good backup team, otherwise monitoring is pointless. Taxonomists are required to identify correctly organisms or damage brought back by the surveillance team. This allows for an assessment of the organism's known pest status and potential to cause serious damage. Experienced staff are also required to evaluate and quantify the extent and severity of the problem, gauge impact and advise on a course of action. It is in this area of technical backup that problems are experienced in many tropical countries. For a start, in comparison with temperate regions, there are few forest entomologists and forest pathologists in the tropics, and even fewer insect or fungal taxonomists. As an example, Nair (2007) mentioned that in 2000, Indonesia, with a forested area of over 100 Mha, had only about 40 researchers in forest protection (including entomologists and pathologists). As well, the number of species in the tropics is vast and the taxonomy of many groups is poorly known. Some countries are fortunate in having large and well-curated museum collections, but many do not. Under tropical conditions, good curation is essential or collections will deteriorate rapidly, but such requirements are often financially daunting in poorer countries. Countries lacking their own taxonomic facilities have usually depended on others to service their needs, but the introduction of charges for identifications by many institutions (even if the charges are subsidized) undoubtedly has reduced the numbers of specimens submitted. All these factors combine to make identification a slow process and impair the efficient functioning of health surveillance systems, although the advent of pest image libraries and online identification keys are now improving this situation.

Other backups which are essential for a well-functioning surveillance programme are facilities and equipment for rearing or culturing collected immature organisms (since mature specimens are usually required for species identification) and systems for the efficient storage and retrieval of data. For insect rearing, sophisticated facilities such as controlled-environment rooms, while very useful, are not essential. Cages or containers for rearing generally can be made quite cheaply from local materials and kept at ambient temperatures in simple insectaries, which are shielded from direct sun and screened or otherwise protected to exclude vermin or insects that might damage specimens. An essential requirement in achieving one of the main goals of forest health monitoring, the detection of change, is an established baseline to determine if, when and where changes are occurring and to quantify those changes. This requires an efficient system of data storage and retrieval. Manual systems have now been replaced largely by computer databases, which can store and sort very large data sets rapidly and can be linked to geographic information systems (GIS) if desired. An example of a very simple output from such database systems, which is of great value to forest managers, is a map which details pest occurrence and severity (Fig. 9.1) (Wallnes, 1996). Appropriate software and hardware is generally readily available worldwide, although cost remains a limiting factor for some countries.

9.1.4 Surveillance methodologies

Methodologies used for forest health surveillance or monitoring vary according to the purpose of the surveillance and the type of forest being surveyed. In industrial forest plantations, a combination of aerial and ground surveys is commonly employed, while around ports and sawmills trapping systems may be used. Examples of these methodologies and some comments on their efficacy are provided below.

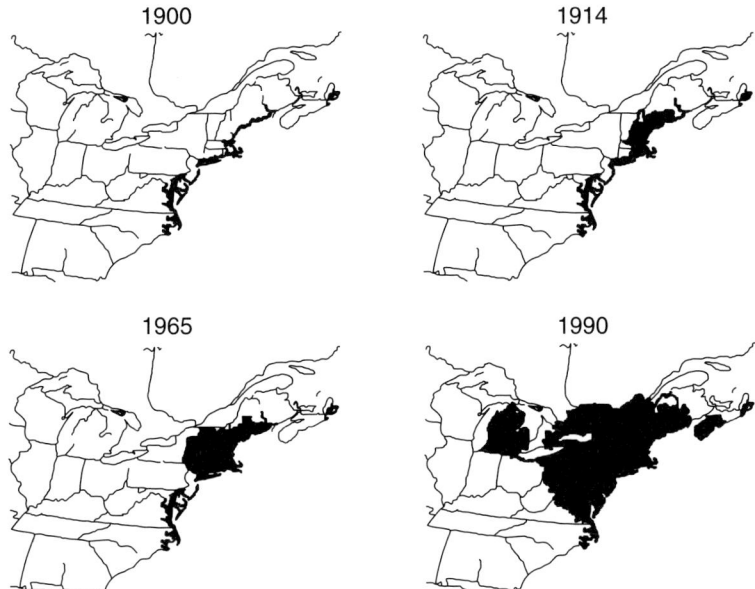

Fig. 9.1. Establishment and spread of *Lymantria dispar* in the USA (from Wallner, 1996).

Ground survey

This generally involves either drive-through or walk-through surveys, or more often a combination of both. The former are usually conducted from all-terrain (e.g. four-wheel drive) vehicles, although in northern Sumatra in Indonesia motorcycles are employed. Typically, a team of two observers will drive along roads through the plantation estate at low speed looking for symptoms of tree disorder. Routes are usually planned in advance so as to provide systematic sampling of the maximum area. Periodically, at randomly located points, observers may leave the vehicle to conduct ground inspections away from the road. When disorders are detected, the team may then sample more intensively to obtain information on the severity and extent of the problem and to collect specimens, including damage, for identification or diagnosis. Stone *et al.* (2003a,b) devised a method for assessing the effects of insect herbivory at the leaf, tree crown and stand scales for young eucalypts in the pre-canopy closure phase. Their crown damage index (CDI) is based on a visual estimate of the incidence, that is, extent of damage over the entire tree crown (as a percentage) multiplied by the average level of severity at the leaf scale (as a percentage) for the three types of leaf damage: defoliation, necrosis and discoloration. The CDI is the sum of the products of each incidence and severity and has been proposed so as to provide ground-based assessors with a generic, standardized measure of tree crown damage, which can allow comparisons to be made between plantations and districts over time.

Bulman *et al.* (1999) in New Zealand have estimated the efficiency of various pest detection survey methods. For drive-through surveys, they found that the most important factors influencing target detection were distance from the road (68% of all simulated damage was detected at road edge, 52% was detected 20 m from the road and 35% at 40 m into the stand) and driving speed – 77% of all simulated damage was detected at 15 km/h, 46% at 30 km/h and only 32% at 45 km/h. Detection was slightly better overall in the older and lower-stocked stands than the younger stands (Fig 9.2). There was also significant interaction between age/stocking and distance from the road, with better detection close to the road but worse detection further from the road in the younger, higher-stocked stands. In walk-through surveys, the most influential factor was distance; 97% of the roadside symptoms were detected, decreasing to 71% at 20 m into the stand and 47% at 40 m into the stand. Detection rates were significantly higher in the pruned stands (75%) than the unpruned stands (60%). They found that drive-through forest sampling at the slowest vehicle speed tested (15 km/h) gave detection efficiencies very similar to those obtained from walk-through sampling (Fig. 9.2). However, they caution that there is no substitute for the close-up examination of foliage and potential insect breeding sites. Insect frass on the stems or around the bases of trees, for example, is unlikely to be detected in a drive-through survey. Bulman et al. (1999) also examined the effect of using more than one observer and found that this increased the probability of detection considerably in all three types of survey tested (Fig. 9.3), as did repeat inspections by observers (Bulman, 2008).

In Australia, Wardlaw et al. (2008) measured the efficacy of aerial, roadside and ground inspection to detect nine different types of damage symptoms, ranging from very obvious (mortality and dead tops) to very cryptic (stem cankers and stem borers), each occurring at a range of incidences among five 3-year-old *Eucalyptus globulus* plantations. They found that dead tops were detected most efficiently by aerial inspection but crown symptoms produced by moderately severe insect defoliation or necrotic leaf lesions were best detected by roadside and ground inspection, both of these methods being equally efficient. Cryptic symptoms could not be detected reliably using any of the inspection platforms, even when their incidence, within small patches, was as high as 2%. They concluded that the combination of aerial and roadside inspection provided sufficient resolution to detect operationally relevant damage (i.e. damage of sufficient severity to consider remedial treatment) but was unlikely to detect damage by new incursions at a sufficiently early stage when eradication might be feasible. This finding reinforces the importance of conducting surveillance for forest invasive species at their likely points of entry, as discussed later in this chapter.

Aerial survey and remote sensing

Historically, aerial detection surveys have been of two types: visual sketch-mapping surveys and aerial photographic surveys (Pywell and Myhre, 1992). Visual sketch mapping is the technique of delineating the

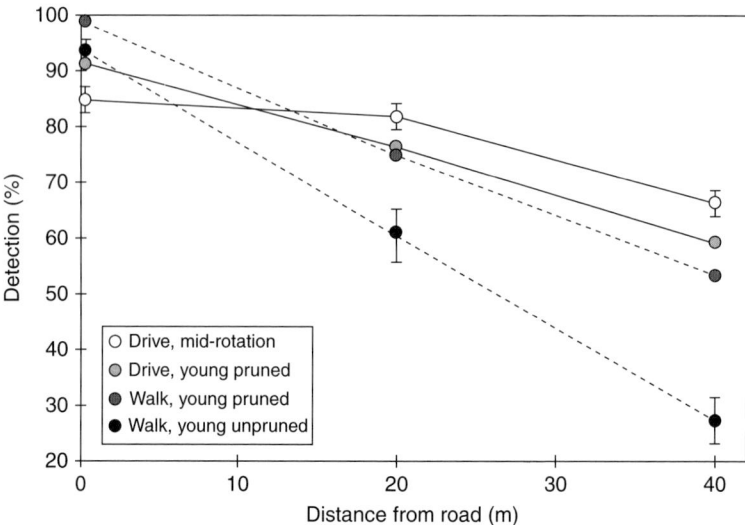

Fig. 9.2. Comparison of mean detection rates at various distances from the road for a drive-through survey at 15 km/h and a walk-through survey (from Bulman et al., 1999).

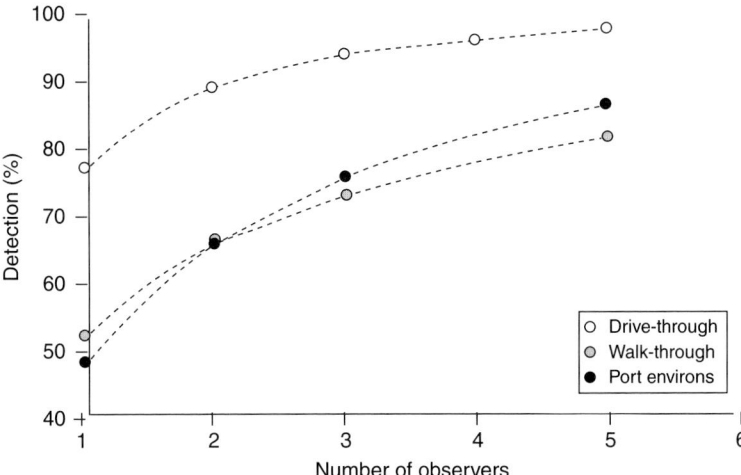

Fig. 9.3. Mean detection percentages plotted against the number of observers for drive-through, walk-through and port environs surveys (from Bulman et al., 1999).

area of pest-caused damage on to maps based on observations by forest health specialists flying in a small aircraft (Johnson and Ross, 2008). The observers look for symptoms of tree disorder, usually indicated by vegetation colour change, and map such occurrences for ground evaluation, or 'truthing'. Once the causal agent and host are identified, and if the problem is sufficiently serious and the damage characteristic, further aerial surveys may be conducted to track the severity and spread of the infestation. This technique is an efficient method of detecting and appraising recognizable pest damage over large remote forest areas, although not all countries have the necessary infrastructure. Aerial sketch mapping has been used, for example, for monitoring forest conditions in southern Brazil, primarily for assessment of damage caused by the wood wasp *S. noctilio*, monkeys, armillaria root disease and other damaging agents in pine plantations (Oliveira et al., 2006).

Aerial photography is another survey tool utilized for forest health monitoring. Colour and colour-infrared photographs have been used in the estimation of current and/or total levels of damage and mortality from pests. They provide a historical record of pest activity and have been used to monitor the rate of spread and trends of a pest over time. Most aerial photography applications in pest management can be divided into two broad classes, mapping photography and sampling photography (Pywell and Myhre, 1992). Mapping photography is a block of continuous photo coverage that can be assembled into a photo mosaic or photo map. It is usually used for mapping the total extent of a pest problem. Sampling photography is photo coverage of a small area that is representative of a large unit or type, and is used when it is not operationally feasible or cost-effective to evaluate 100% of the area of concern.

A limitation of sketch mapping is that it is highly subjective, although an assessment by Johnson and Ross (2008) shows that it is operationally acceptable for broad-scale detection and monitoring. Photography can provide more accurate information than sketch mapping, as well as a permanent record, but disadvantages are the high cost and the amount of time required in processing film and prints, interpreting photographs and transferring information from photographs to a map. The advent of digital camera technology has allowed images to be downloaded directly into the computer without the need for intermediate processing. There they can be manipulated and enhanced to aid in interpretation and can be incorporated into geographic information systems (GIS). They can be stored readily, although this requires considerable disk space. An advantage for observers using digital cameras from an aircraft or on the ground is that they can view the image immediately after it is taken and decide whether it is suitable or whether another 'photo' is required. Airborne videography has somewhat similar advantages over conventional aerial photography, although the quality of the imagery has generally not been as good.

A range of new technologies has enhanced the accuracy and efficiency of forest health surveys over the past decade (Carnegie, 2008). These include the use of GIS–GPS (global positioning system) interface tools and handheld computers to assist navigation and data collection in the field. In Queensland, Australia, laser rangefinders, linked to palmtop computers with integrated GPS, are used to enhance aerial surveys of pine plantations using fixed-wing aircraft (Ramsden et al., 2005). This system allows highly accurate spatial data to be combined with forest health descriptive data for immediate interpretation within GIS. A digital aerial sketch-mapping (DASM) system developed by the Forest Service in the USA allows users to digitize polygons directly on to a touch-screen linked to a GPS unit and computer or on to a tablet PC with an integrated GPS (Johnson and Wittwer, 2008). Improved mobile phone technology and wireless broadband has also enabled images of symptoms and damaging agents to be transmitted quickly from the field for expert opinion and action (Carnegie, 2008). In several countries, significant progress has been made in the application of digital, remotely sensed imagery to detect and classify damaged forest canopies. In Australia, for example, the airborne instruments that have received most attention have been multispectral (with few broad bandwidths) and hyperspectral (many narrow bandwidths) optical sensors (Stone and Coops, 2004; Stone and Haywood, 2006; Stone et al., 2008). Both of these instruments measure the amount of light reflected from vegetation within specific bandwidths of the electromagnetic spectrum. Any process that alters the biochemical and morphological features of leaves also influences directly the reflectance characteristics of the leaves that can be measured quantitatively (Fig. 9.4). Stone and Coops (2004) provide two examples of such use of airborne high-resolution imagery in south-eastern Australia, namely assessment of canopy decline in moist native regrowth forests associated with herbivorous insects and assessment of crown defoliation in a mature *P. radiata* plantation associated with attack by the California pine aphid, *Essigella californica*. They note that the success of the approach depends in part on a sound understanding of the progression of

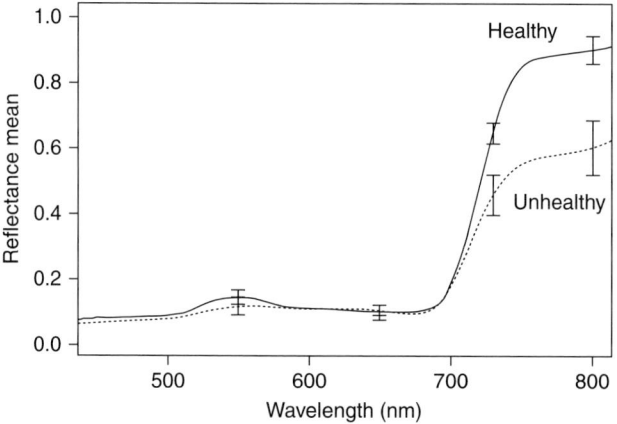

Fig. 9.4. The mean reflectance curves obtained for foliage from a *Eucalyptus paniculata* with a healthy crown compared with insect-damaged foliage from another *E. paniculata*. Vertical bars illustrate standard deviation errors of the mean of five replicates at key wavebands (from Stone et al., 2001).

symptoms at the leaf, tree crown and stand scale, especially those symptoms that influence spectral reflectance behaviour. Similarly in South Africa, a study by Ismail et al. (2007) demonstrated the potential of high-resolution digital multispectral imagery for the improved detection and monitoring of *P. patula* trees infected by *S. noctilio*.

Satellite technology has been used in the monitoring of pest outbreaks, for example in large-scale assessments of defoliation by the gypsy moth *Lymantria dispar* (Lepidoptera: Lymantriidae) in the USA (Dottavio and Williams, 1983). In the tropics, it has been used mostly for forest inventory purposes. In Malaysia, for example, it is being used to map forest types, to measure changes in forest cover due to a range of causes such as shifting cultivation, forest exploitation and urbanization, and to monitor damage caused by harvesting and extraction (Khali Aziz et al., 1992). In India, a combination of aerial photography and satellite images was used to detect a significant reduction in vegetation cover of sal (*Shorea robusta*) forests in some forest divisions (Chauhan et al., 2003). The causes were identified as deforestation, encroachment, agriculture and, in one division, severe infestation of the sal borer, *Hoplocerambyx spinicornis*. A past limitation for remote sensing, particularly in tropical regions, has been in acquiring cloud-free images, but this has been overcome by the use of microwaves, which pass straight through clouds and rain and can be used during both day and night. The spatial resolution of remotely sensed data, once too coarse for any detailed analysis, is continually improving. In 1992, the maximum resolution of individual picture units (pixels) from the best commercially available satellite data was 10 m (Khali Aziz et al., 1992); in 2011 it was 0.4 m.

With respect to aerial surveys conducted by observers in light aircraft, the probability of detection of tree disorder, which involved visible crown yellowing or browning, for various flight line spacings across a forest has been assessed by Carter (1989) and is shown in Fig. 9.5. For New Zealand, it was estimated that only 13% of potentially harmful exotic organisms would cause damage that would be visible from the air before the organism had spread so widely as to be considered ineradicable. Carter (1989) cautions, therefore, that the theoretical probability of detection of exotic pests as calculated above should be scaled down by a factor of 0.13. This again supports the need for targeted surveillance for invasives closer to their potential points of entry.

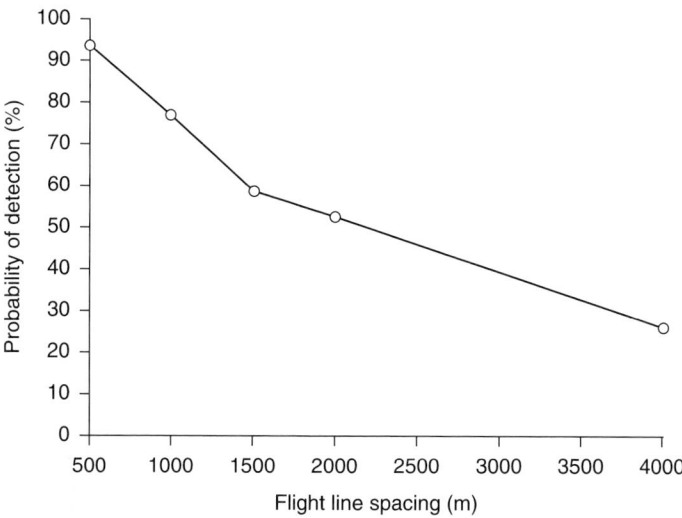

Fig. 9.5. Probability of aerial detection of visible symptoms of tree disorder at various flight line spacings (from Carter, 1989).

Trapping

Over the years, a great variety of devices or techniques have been developed for catching or trapping insects. While the intent of such devices has often been the control of insect pests, experience has shown that trapping has greatest value as a monitoring tool. Such devices may be used at ports of entry to detect the presence of imported noxious insects, to determine the spread and range of recently introduced pests in a region and to determine the seasonal appearance and abundance of insects in a locality and the need for application of control measures.

The type of trap used is governed principally by the behaviour of the insect species or group one is trying to catch. Most detection trapping is structured around the capture of flying adult insects, but some species (e.g. wood wasps) are detected through the use of trap trees or logs into which adults oviposit. The progeny which subsequently develop and/or their characteristic damage can then be identified. Traps for catching flying insects may be 'passive' (for example, suspended net traps, 'windowpane' or interception traps, water traps, sticky traps) or may lure insects from a distance (for example, light or bait traps). Some examples of detection trapping programmes that have been implemented in tropical and subtropical regions are given below and later in this chapter (see the section on hazard site surveillance).

ASIAN GYPSY MOTH – QUEENSLAND, AUSTRALIA The Asian gypsy moth, *L. dispar* (Lepidoptera: Lymantriidae), is a serious pest of forest trees and is known to feed on more than 650 plant species. It has spread from Asia to Europe, the earliest European record dating back to 1965, and in 1991 was discovered in Canada and the USA, where it was the subject of major eradication programmes. Following the finding in 1993 of viable egg masses of Asian gypsy moth on a ship that had visited Australia and New Zealand, both these countries embarked on a detection-trapping programme around designated ports, including some in tropical Queensland. Various types of pheromone trap are employed to monitor *L. dispar* in different countries. In Australia, 'Delta' traps and, more recently, 'Uni-traps' (bucket traps) baited with a sex pheromone (disparlure) are deployed around ports, usually a grid pattern of at least 40 traps at high-risk ports and a few sentinel traps in strategic areas at low-risk ports. Disparlure attracts not only *L. dispar* but also several other *Lymantria* species exotic to the region, including the Nun moth, *L. monacha*, and the Indian gypsy moth, *L. obfuscata*. Traps are inspected fortnightly during a 6-month trapping 'season' each year covering the optimum flight period of the moth and any Lepidoptera caught are sent to specialist entomologists for identification. A general contingency plan has been formulated for actions to be taken in the event of the detection of one of these exotic species.

LEPIDOPTERAN DEFOLIATORS – BRAZIL The increasing pest problems in *Eucalyptus* monocultures in Brazil led to the establishment of survey programmes in the late 1980s and early 1990s that aimed to identify and recognize the relative importance of the various eucalypt insect pests (Zanuncio et al., 2001). The main pests in these plantations were found to be defoliating insects, particularly caterpillars of moths such as *Thyrinteina arnobia* (see Chapters 5 and 10), *Stenalcidia grosica* and *Glena unipennaria* (Geometridae), *Eupseudosoma aberrans* (Arctiidae) and *Psorocampa denticulata* (Notodontidae) (Zanuncio et al., 2001, 2003, 2006; de Freitas et al., 2005). The moths of most of the lepidopteran defoliators are night active and their populations have been monitored with light traps equipped with black-light tubes and 12-V batteries placed 2 m above ground level (Pereira et al., 2001; Zanuncio et al., 2006). Trapping programmes have been conducted in several Brazilian states (Fig. 9.6) and the information obtained used to predict where and when outbreaks are likely to occur and to facilitate the timing of control measures such as the release of natural enemies.

FIVE-SPINED BARK BEETLE – QUEENSLAND, AUSTRALIA *Ips grandicollis* (Coleoptera: Scolytinae), a bark beetle pest of *Pinus* spp., was introduced accidentally into Australia from the USA in the 1940s via imported pine logs with bark on and in dunnage. For the next four decades,

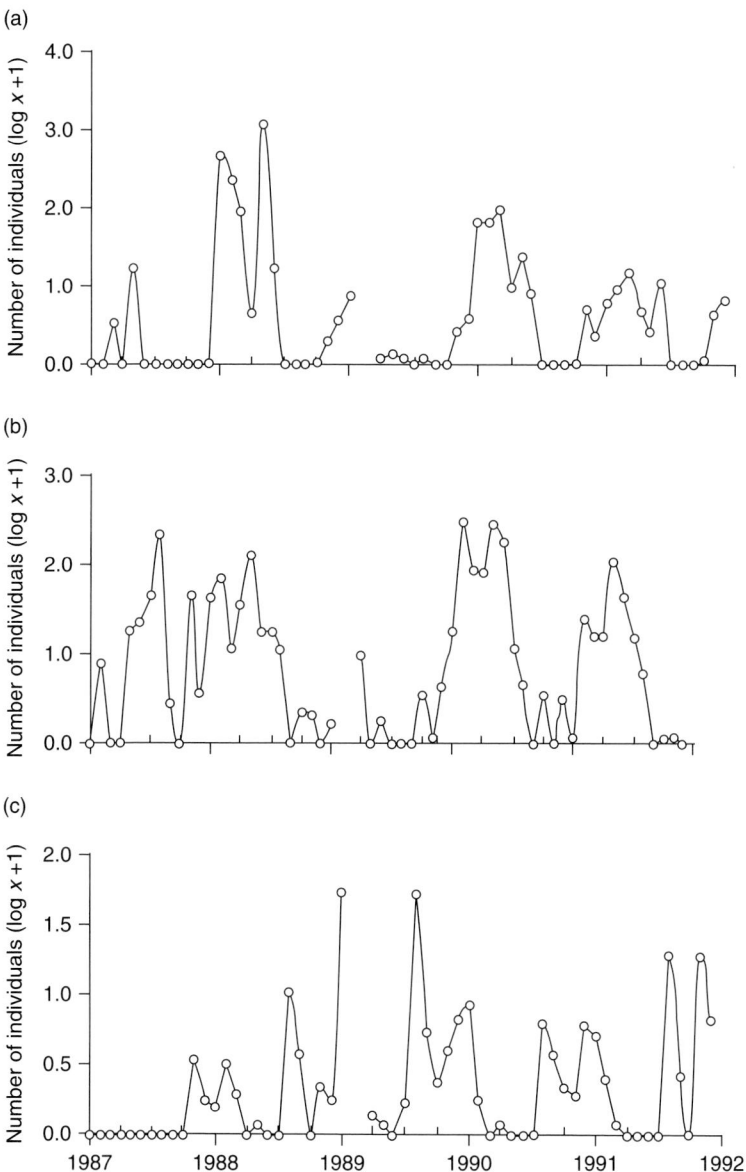

Fig. 9.6. Population fluctuation of (a) *Thyrinteina arnobia* (Geometridae) (b) *Stenalcidia* sp. (Geometridae) and (c) *Psorocampa denticulata* (Notodontidae) in the Municipality of Bom Despacho, State of Minas Gerais, Brazil, March 1987 to February 1992. Discontinued lines indicate that collections were not made on these dates (from Zanuncio et al., 2006).

its distribution was restricted to South Australia and Western Australia, but in the early 1980s it spread to the other mainland states and by 1994 had crossed the tropic of Capricorn in Queensland (Wylie et al., 1999). Its progress in Queensland has been monitored by means of 'drainpipe' traps (Plate 120) baited with a synthetic preparation combining the two bark beetle pheromones, 'ipsenol' and 'transverbenol'. Such trap systems have been shown to detect *I. grandicollis* at very low population levels. Quarantine restrictions were instituted in Queensland in 1982 to prevent the northward movement of pine logs with bark on or of pine bark which might harbour the pest. The quarantine boundaries were monitored by both ground survey of adjacent pine plantations and by pheromone trapping. The quarantine remained in place until 2009, when *I. grandicollis* was discovered in Townsville in north Queensland, believed to have 'hitch-hiked' there with a shipment of logging equipment sent from the south of the state for a post-cyclone salvage operation.

BARK BEETLES – SOUTH AFRICA   There are three exotic pine bark beetle species present in South Africa, all originating from Europe, namely *Orthotomicus erosus*, *Hylastes angustatus* and *Hylurgus ligniperda* (Coleoptera: Scolytinae). All three species feed on the inner bark and cambium of conifers, mainly *Pinus* species, and may be found simultaneously in the same pine tree (Tribe, 1991). All are vectors of both bluestain and pathogenic fungi. In their countries of origin, they are regarded as secondary pests but *O. erosus* can become primary if trees are stressed by adverse climatic conditions, and *H. angustatus* becomes a serious pest during its maturation feeding phase. Over 50% of pine seedlings in a newly planted stand may be killed by under-bark girdling of the roots and root collars by *H. angustatus*. Such a level of damage is a rare event and it is not economically viable to institute an annual spraying programme. The acceptable rate of seedling loss above which replacement becomes necessary is 15% and chemical protection through prophylactic insecticide sprays is regarded as the only effective control measure. The correct timing of such sprays is crucial. Similarly with *O. erosus*, their presence and numbers in a forest will determine the speed at which trees on a stressed site are colonized and the severity of this attack. *H. angustatus* populations are monitored by means of trap logs, while pheromone traps baited with a combination of ipsdienol, verbenone and 2-methyl-3-buten-2-ol have been used to monitor *O. erosus* populations (Tribe, 1991).

BROWN WATTLE MIRID – SOUTH AFRICA   In South Africa, wattle is an important plantation crop, with increasing demand for bark extracts for tanning, resins and adhesives. The timber is used for the production of high-quality paper, the manufacture of furniture, the production of charcoal, building poles, fencing and floors. There are approximately 112,000 ha of black wattle *A. mearnsii* in the country. The brown wattle mirid, *Lygidolon laevigatum* (Hemiptera: Miridae), causes serious damage to trees, usually those between 0.5 and 5.0 m in height (Govender and Ingham, 1998). Attacks on shoots affect tree form, resulting in witch's broom, and reduction in growth, which in turn reduces bark and timber yield. To prevent such damage, chemical control measures need to be applied early in the season before mirid populations build up to damaging levels. Plastic bottle traps coated with sticky adhesive and placed in plantations at a height of 2 m and at a density of eight traps per compartment are used to monitor mirid numbers. If eight adults are recorded on a trap in 1 week, control measures are necessary to prevent the population exceeding the economic threshold (Govender and Ingham, 1998).

SAL BORER – INDIA   The sal borer, *H. spinicornis* (Coleoptera: Cerambycidae), is a serious pest of *S. robusta* in northern India and Pakistan (Nair, 2007). It is principally a secondary pest of dying and fallen trees but during epidemics can infest and kill even healthy trees. In India, trap logs are used in the monitoring and control of the insect, taking advantage of the adult beetles' attraction to the sap (on which they feed) oozing out from the injured or wounded sal trees. The technique consists of felling a few trees and then cutting them into billets, beating up and loosening the bark. Beetles are attracted by oozing sap and take shelter underneath the loose bark, where they are collected regularly and destroyed (Roychoudhury, 1997). After every 3–4 days, the logs are cross cut again and the cut ends beaten to restore their attractiveness. A freshly cut tree remains attractive for 8–10 days. Nair (2007) rates the trap-tree operation as effective but cumbersome and research is under way to isolate the attractive components in the sal tree sap with the aim of developing synthetic lures.

9.1.5 Costs of surveillance

The costs of forest health surveillance will vary according to the purpose of the surveillance, the methodology used, the resources deployed and the technical backup required. Carnegie et al. (2008) compare the costs in the USA, Australia and New Zealand. In the USA in 2004, the cost of surveying over 180 Mha of forest using fixed-wing aircraft was US$0.025/ha. These surveys are designed mainly to detect landscape-scale outbreaks that can be mapped at a relatively small scale. In New Zealand, aerial surveys of pine plantations using fixed-wing aircraft cost NZ$0.02–0.08/ha, while in New South Wales in Australia, helicopter surveys of 90,000 ha of pine plantations cost AUS$0. 15 –0.30/ha. Compared to the USA, the New Zealand and New South Wales aerial surveys are designed to detect problems at a much higher resolution (compartment scale) using larger-scale mapping. As noted by Carnegie et al. (2008), these costs are not 'all inclusive', which can make comparisons difficult.

Speight and Wylie (2001) calculated the cost of forest health surveys of pine plantations in Queensland, Australia at AUS$1.14/ha, which included aerial and ground surveys, diagnostics and written reports. Similar costs are reported from New South Wales and New Zealand for the same services (Carnegie et al., 2008). In contrast, the cost of hardwood plantation surveys in Queensland was almost ten-fold that for softwood plantations (about AUS$10/ha), due to the small (unit size generally 20–50 ha) and disparate estates requiring more detailed ground surveys and the larger number of pests and diseases requiring processing (diagnosis) in this forest type. This cost is expected to reduce as hardwood plantation unit size and aggregation increase (Lawson et al., 2008).

Pywell and Myhre (1992) provide cost comparisons of three remote-sensing techniques discussed earlier in this chapter (Table 9.1). As can be seen, visual sketch mapping by observers in a small aircraft is the cheapest method, although videography has several advantages over this, the main ones being that it eliminates the subjective evaluation of the observer and provides a permanent record and more accurate estimates of the real extent and severity of pest damage. Both sketch mapping and videography are far cheaper than conventional aerial photography.

Carter (1989) gives a detailed breakdown of the costs of various methods of forest health surveillance in New Zealand and compares cumulative cost against probability of detection (Fig. 9.7). He concludes that maximum net benefit is achieved at survey levels which will detect 95% of all new introductions (compared to then-current levels in that country which were achieving less than 50% detection).

Table 9.1. Cost analysis of remote-sensing techniques used to define and map gypsy moth defoliation in the USA (from Pywell and Myhre, 1992).

Technique	Time (hours)	Cost of plane/ pilot per hour (US$)	Cost of mapper/ analyst per hour (US$)	Cost of film/ processing (US$)	Total cost (US$)
Aerial sketch	2.0	105	9.68	0.00	229.36
Airborne videography	1.5	250	20.90	12.95	419.30
Office sketch	2.5		10.00		25.00
					444.30
Aerial photography	2.0	250	20.90	1100.00	1641.80
Photointerpretation	40.0		10.00		400.00
					2041.80

A further example of an operational surveillance programme in the tropics and the resources required to run it is from Aracruz Celulose SA in Brazil. This company is one of the world's largest exporters of hardwood bleached pulp and has more than 260,000 ha of plantations, mostly eucalypts (Osland and Osland, 2007) (Plate 122). Its system of preventative control against target pests involves: (i) detection of primary outbreaks; (ii) outbreak evaluations; (iii) analysis of results; and (iv) definition of control strategy (Laranjeiro, 1994), and is shown schematically below (Fig. 9.8). As reported by Laranjeiro (1994), the efficacy of this system has meant that intervention was necessary on less than 0.02% of initial outbreaks detected. In the remaining cases, the pest populations returned to a balanced state due to natural control.

9.1.6 Example of survey form and surveillance guidelines

A survey form that has been used in forest health surveillance in several countries in Asia and the Pacific is shown in Fig. 9.9. The form originated from ACIAR-funded capacity-building projects on forest health in the region during the period 1997–2004. Surveys were conducted in Malaysia, Thailand, Vietnam, Indonesia, Fiji, Vanuatu, Samoa, Tonga and tropical Australia to determine the key pests occurring on tree plantations and to obtain information on their distribution and impact on plantation health and growth (Wylie et al., 1998). The surveys were conducted across a range of species, provenances, ages, geographical locations, site conditions and times of year using standardized data collection methods. The form shown was adapted from that used in Queensland in their forest health surveillance programme. The form is meant to act as a checklist for observers who are trained in the recognition of the symptoms listed. The information recorded assists in the

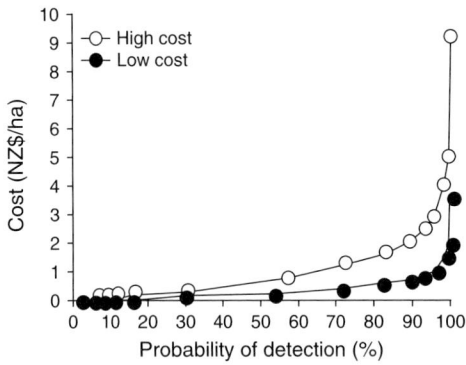

Fig. 9.7. Cumulative cost and probability of detection for high- and low-cost locations (150 and 50 km each way, to and from the forest) (from Carter, 1989).

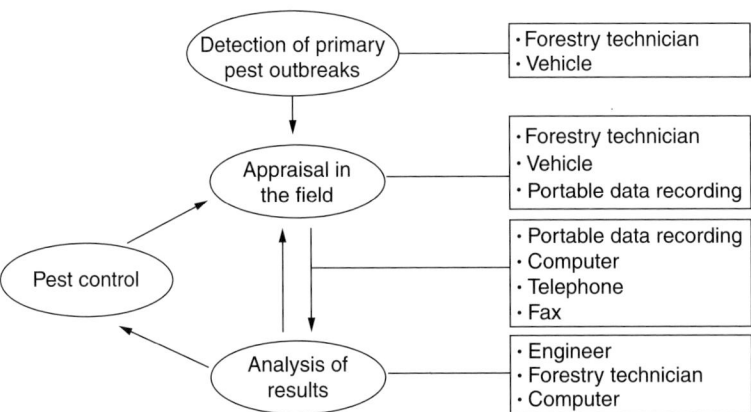

Fig. 9.8. Schematic outline of the Aracruz Celulose SA system for preventative control of insect pests of *Eucalyptus* (from Laranjeiro, 1994).

FOREST HEALTH FIELD FORM

OFFICE USE ONLY
DATE ENTERED ____/____/____
ENTERED BY _____

DATE ____/____/____   OBSERVER _____   LOCATION NO _____   OCCURRENCE NO _____

COUNTRY _____   PROVINCE/STATE _____
LOCALITY/ADDRESS _____
LOGGING AREA NAME _____   COMPARTMENT _____
COMPARTMENT AREA Ha _____   LAT/LONG _____
OWNERSHIP/ FUNDING   ☐ Private ☐ Public/Government   NAME _____

AGE & AVERAGE SIZE
Plant Date yr _____ mth _____
Height   m _____
DBH   cm _____

PRESENT STOCKING RATE
Stems/Ha _____

TREE SITUATION
☐ Commercial Plantation
☐ Native Forest
☐ Nursery
☐ Community Plots
☐ Agroforestry
☐ Amenity/Street Trees
☐ Other _____

ORIGINAL VEGETATION /SITE HISTORY
☐ Native Forest
☐ Tree Plantation
☐ Grassland
☐ No vegetation
☐ Previously agriculture
☐ Other _____

SITE & STAND TREATMENT
☐ None
☐ Regeneration
☐ Planted
☐ Herbicide
☐ Fertilised
☐ Pruned
☐ Thinned
☐ Clearfelled
☐ Stock Grazing
☐ Other _____

TREE SPECIES: _____   PROVENANCE _____   SEEDLOT _____

TREE STATUS
☐ Living
☐ Standing dead
☐ Fallen
☐ Other _____

GROWTH STAGE
☐ Seedling
☐ Sapling
☐ Pole
☐ Mature
☐ Overmature
☐ Other _____

CANOPY POSITION
☐ Dominant
☐ Codominant
☐ Suppressed
☐ Understorey

TOPOGRAPHY
☐ Gully
☐ Ridges
☐ Slope
☐ Flats
☐ Undulating

ASPECT _____

DAMAGE/SYMPTOM
☐ None
☐ Mortality
☐ Chlorosis
☐ Necrosis
☐ Skeletonising
☐ Holes
☐ Defoliation
☐ Dead Crown
☐ Wilt
☐ Dieback
☐ Witches broom
☐ Other Distortion
☐ Stunting
☐ Resinosis
☐ Oozing
☐ Swelling
☐ Cracks
☐ Ringbarking, girdle
☐ Epicormic shoots
☐ Puncture
☐ Mining
☐ Chewing
☐ Gall

☐ Tunnelling
☐ Webbing/Tying
☐ Severing
☐ Frass
☐ Other _____

DISTRIBUTION OF AFFECTED TREES
☐ Single tree
☐ Scattered
☐ Patches
☐ Widespread

AVERAGE SEVERITY / TREE
☐ Negligible
☐ Minor
☐ Moderate
☐ Severe
Severity _____ %

AREA AFFECTED _____ Ha
% OF TREES AFFECTED
_____ %
☐ Estimated
☐ Counted N=_____ out of _____

PART AFFECTED
☐ Entire Tree
☐ Stem
☐ Base of trunk
☐ Root Collar
☐ Root
☐ Sapwood
☐ Heartwood
☐ Cambium
☐ Bark
☐ New Shoot
☐ Twig
☐ Branch
☐ Leading Shoot
☐ Foliage
☐ Bud
☐ Flower
☐ Seed
☐ Other _____

☐ Upper
☐ Middle
☐ Lower
☐ Directional

POSSIBLE PREDISPOSING FACTORS
☐ Fungal
☐ Animal
☐ Mistletoe
☐ Drought
☐ Waterlogging
☐ Wind/Typhoon
☐ Hail
☐ Lightning
☐ Frost
☐ Heat/Sunshine
☐ Fire
☐ Mechanical
☐ Compaction
☐ Competition
☐ Nutrient
☐ Salt
☐ Herbicide
☐ Other _____

INSECT IDENTIFICATION
Scientific Name _____
Common Name _____
Family _____   Order _____
Identifier _____   Organisation _____

ACCESSION NO _____

Fig. 9.9. Forest health field form.

correct diagnosis of the problem and allows, for example, a cross-check on whether the observed symptoms match those which could be produced by the putative causal agent. It also provides some initial quantitative measure of the incidence and severity of the problem, which can then be followed up by more detailed surveys if the situation warrants it. Information is stored on a database and, over time, will provide a historical record of the health of particular forest areas. Such databases can be interrogated as required to produce reports useful for forest management.

Comprehensive guidelines to assist plant health scientists design surveillance programmes for detecting pests in crops, plantation forests and natural ecosystems have been funded and published by the Australian Government (McMaugh, 2005). The guidelines cover the planning of surveillance programmes for building specimen-based lists of pests, surveillance for monitoring the status of particular pests, surveillance for determining the limits of distribution of pests, surveillance for determining the presence or absence of pests in particular areas and general surveillance. Although aimed at use by developing countries in the Asia-Pacific region, these guidelines are broadly applicable. The Food and Agriculture Organization of the United Nations also has produced guidelines for surveillance as part of its International Standards for Phytosanitary Measures series of publications (FAO, 1998).

9.2 Forest Invasive Species and Quarantine

9.2.1 The need for forestry quarantine

The best way of minimizing the damage that may be caused to plantations by exotic insects is, of course, to keep the pests out in the first place. Quarantine therefore needs to be recognized by forest managers as an essential and valuable component of the overall protection effort. Exotic insects pose great risk to the stability and productivity of the forest ecosystems into which they are introduced, as a species can become established easily if it finds a suitable climate and host material in its new environment (Ciesla, 1993). In the absence of natural enemies that may regulate the insect in its native range, its numbers may increase rapidly. In addition, as host plants in the new habitat may not have been exposed previously to the introduced insect or to microorganisms of which the insect might be a vector, they may be more sensitive to injury than those in its natural range. Frequently, insects that are not considered to be major pests in their native habitats cause widespread damage when they are introduced into new areas. Good examples of this are the wood wasp *S. noctilio* in Australia, South Africa and South America, and the cypress aphid *Cinara cupressi* in Africa, both species being of only minor importance in their native Europe but devastating pests in those countries where they have become established.

In many countries, forestry quarantine is treated as just a subset of general plant quarantine, which tends to be dominated by agricultural considerations. This situation frequently leads both forestry managers and quarantine policy makers to underestimate its importance. However, forestry has many difficulties and requirements that are quite distinct from those of agriculture and which necessitate a different approach. Unlike agricultural crops, most tree crops require decades to mature and cannot be modified quickly to control or resist new pests and diseases (Wylie, 1989). Direct control measures such as spraying or dusting, which would be employed for problems in agricultural crops, are usually not logistically, environmentally or economically practicable in forests (either plantations or natural). The potential impact of some of the more serious exotic pests and diseases of forests, if introduced, could therefore be not just the loss of a few season's crops (as is often the case with agricultural pest problems) but the loss of decades of effort and investment in plantations and irreparable damage to the native flora and timber resource.

Other special difficulties for forestry relate to the early detection of exotic organisms and the sometimes limited options available for eradicative action. Included here are the vastness and isolation of much of the forest estate and the infrequency and difficulty of inspections (compared to that for intensively managed, easily accessible agricultural crops). Past experience in many countries has been that by the time an exotic pest is detected in forests it is usually well established and beyond eradication (Wylie, 1989; Wardlaw *et al.*, 2008). Some of the economic and social consequences which can result from the establishment of exotic insect species in forests have already been chronicled in Chapters 5 and 6, for example *C. cupressi* in Africa, *I. grandicollis* in Australia and *Heteropsylla cubana* in the Asia-Pacific and Africa.

9.2.2 Modes of entry of exotic insects

Common pathways of exotic pest entry into a country and examples of such introductions are given below.

Seeds

Exchange of tree seed around the world has helped in extending and diversifying forestry and in improving yields. Unfortunately, it has also helped to spread damaging tree pests from one country to another. Among the commonest insects intercepted in seed consignments are beetles (e.g. *Bruchidius* spp. in acacia seed), wasps (e.g. *Megastigmus* spp. in eucalypt and pine seed) and moths (e.g. *Tracholena* sp. in hoop pine seed). In two examples from India, Verma (1991) records the interception of the bruchid *Merobruchus columbinus* in seeds of *Samanea saman* imported from Honduras, and Verma *et al.* (1991) list *Bruchus ervi* in seeds of *A. brachystacha* imported from Australia. As discussed in Chapter 5, seed insects have importance for tree-breeding programmes and for social forestry.

Scions and nursery stock

These are very high-risk imports, as numerous examples in the literature attest. One of the most famous, or notorious, accidental introductions on scions was that of the woolly pine aphid, *Pineus boerneri*, into Zimbabwe and Kenya in 1962 on pine scions from Australia and its subsequent spread to a further six countries in Africa, mostly by the movement of infested nursery stock (Odera, 1974; Blackman *et al.*, 1995). An early example of introduction via nursery stock is the discovery of the gum tree scale, *Eriococcus coriaceus*, on plantations of *E. globulus* in the South Island of New Zealand in 1900 (Clark, 1938), which was linked to eucalypt plants imported from Australia (Lounsbury, 1917).

The Christmas tree trade has also been a vehicle for the introduction of exotic pests into several countries. For example, in Bermuda, several organisms of quarantine importance were found accompanying Pinopsida and other Christmas foliage plants imported into the country in sealed containers from various sources. Tree pests found included the spruce gall aphid, *Adelges cooleyi*, the scale *Chionaspis pinifoliae*, the balsam gall midge, *Paradiplosis tumifex*, the mite *Trisetacus quadrisetus* and the gall midge *Monarthropalus buxi* (Anon., 1997). Bonsai trees imported from Asia have been established as a proven pathway for the introduction of *Anoplophora chinensis* (Ciesla, 2004).

Hitchhiking on non-target plants

Some forestry pests have been transported accidentally from one country to another on a plant that is not a host for the species or is not a forest tree species. The eucalyptus weevil, *Gonipterus scutellatus*, was first noticed in South Africa in 1916 (Mally, 1924). South African authorities considered it extremely improbable that the insect came with eucalypt trees because the introduction of eucalypts from overseas became absolutely prohibited in 1903. The most likely pathway was thought to be as stowaways in cases of apples from Australia.

G. scutellatus is a pest of apples in Tasmania, and beetles have often been observed clinging to the stems of apples after these have been packed in cases. Large quantities of Tasmanian apples were imported into South Africa just after the Boer War (Mally, 1924). More recently, there were four interceptions of the Tasmanian eucalypt leaf beetle, *Paropsisterna bimaculata*, in the UK in 2004 in tree ferns (*Dicksonia antartica*) imported from Australia (Central Science Laboratory, 2005), while a dead specimen of the gum tree longicorn, *Phoracantha recurva*, was found in a cluster of bananas imported into Belgium in 2005, sourced from either Central or South America or Australia (Bosmans, 2006).

Cut flower trade

The transfer of pests on live plants applies as well to the cut flower trade. Leaf miners, thrips, mites and larvae of several moth species are found regularly on cut flowers, indicating the risks associated with the intercontinental flower trade (Wittenberg and Cock, 2001). The most likely pathways for dissemination of the red gum lerp psyllid, *Glycaspis brimblecombei*, and the blue gum chalcid, *Leptocybe invasa*, around the world are plants for planting or cut foliage of *Eucalyptus* from countries where these pests occur. This pathway is linked closely to air transport and one of the earliest examples was in 1928 when the airship Graf Zeppelin made its first visit to North America; seven insect species were found in bouquets decorating the cabins (Gilsen, 1948).

Logs and sawn timber

Bark beetles, longicorn beetles and wood wasps are among the most destructive pests of forest trees and can be spread all too easily through the international timber trade unless care is taken. Unprocessed logs are particularly risky imports, especially if the bark is intact, because the cambium layer is the breeding site for many species with a highly destructive potential. A good example is provided by Ciesla (1992, 1993). In April 1992, unprocessed *Pseudotsuga menziesii* and *Tsuga heterophylla* logs exported from North America into the People's Republic of China and subsequently deposited in a forested area to be used for the construction of a Buddhist temple were found to be infested by bark beetles and wood-boring beetles indigenous to western North America. Live adults of Douglas fir beetle, *Dendroctonus pseudotsugae* (Scolytinae), and live larvae of the flatheaded fir borer, *Melanophila drummondi* (Buprestidae), were recovered from the logs. *D. pseudotsugae* is the most important bark beetle pest of *P. menziesii* throughout the range of this tree in western North America, sometimes reaching epidemic levels and killing large numbers of trees. *M. drummondi* occurs throughout the same region, attacking several species of trees and capable of killing apparently healthy trees. In this particular case, the logs were deposited in a location that did not contain suitable host material for the insects and it is doubtful they would become established in the immediate area. However, in other areas of China they could well have become established.

The hazardous nature of log imports is highlighted further by figures from China, where in 27 boatloads of timber from Malaysia quarantined in 1990–1992 in Zhoushan port, Zhejiang Province, logs from 20 boatloads were found to carry termites, including nests and winged adults (Zhang and Yang, 1994). Three species found were the rubber termite, *Coptotermes curvignathus* (see Chapter 5), *C. bornensis* and *Schedorhinotermes sarawakensis*. In India, of 2000 consignments of logs and timber (mostly from the Far East) imported into Karnataka in 1989–1991, 162 were treated to control one or more of the 40 species of coleopteran pests detected (Ghodeswar *et al.*, 1992). Ambrosia beetles are generally the most common insects transported in logs and unseasoned sawn timber, and Ohno (1990a,b) records 144 species of Scolytinae and 86 species of Platypodinae in logs shipped from the island of Borneo to the Japanese port of Nagoya over the period 1982–1987.

An early example of pest spread via sawn timber is that of the gum tree longicorn,

Phoracantha semipunctata, which is believed to have been introduced into South Africa shortly before the Boer War in newly cut eucalypt railway sleepers from Australia. The insect was first observed in 1906 in a plantation near Wolseley in Cape Province, and Lounsbury (1917) noted that nearby railway sleepers of Australian origin, laid in 1898, showed tunnelling similar to that noted in the trees. *P. recurva* is thought to have entered New Zealand in a similar fashion, adults being found in Canterbury in 1873 near Australian timber imported for railway works (Miller, 1925).

Packing crates, pallets and dunnage

Timber packaging and packing materials are another common means of entry for exotic pests of forestry importance (Haack and Petrice, 2009). Such material is usually rough-sawn and often has bark remnants adhering to it, which are sufficient to harbour small numbers of adults, pupae or larvae of insects such as bark beetles (Plate 121). Haack (2006) summarized data for 8341 Coleoptera interceptions at US ports of entry from 1985 to 2000 (Table 9.2). Crating and dunnage were found to be the most common type of wood article to be infested and some of the most commonly associated products were tiles, marble and machinery. The five-spined bark beetle, *I. grandicollis*, is known to have entered Western Australia via dunnage imported from North America, and *H. ater*, *H. ligniperda* and *O. erosus* (all of European origin) have been introduced into several tropical and subtropical regions of the world via pine cargo crates (Ciesla, 1993). Larger wood-boring insects are also transported in packaging, and the wood wasps *S. juvencus* and *Urocerus* sp. are intercepted periodically in the subtropical port of Brisbane in Australia in material originating from Japan, Europe and North America (Wylie and Peters, 1987). The Asian longhorn beetle, *A. glabripennis* is believed to have entered the USA in the 1990s via wooden packaging material (Barak *et al.*, 2005).

Cargo containers

Cargo containers themselves, and not just their cargo, can serve to transport forest invasives. Gadgil *et al.* (2000) assessed contaminants on the external surfaces of 3681 shipping containers entering New Zealand. They found that 23% of containers carried quarantinable contaminants, among which were pathogenic species of *Fusarium* and a live egg mass of the gypsy moth, *L. dispar*. Most steel containers have plywood or timber floors and Stanaway *et al.* (2001) surveyed the floors of 3001 empty sea cargo containers in Brisbane, Australia, searching for live or dead insects on or in the flooring. They collected more than 7400 specimens from 1174 (39%) of the containers examined. Live insects accounted for 19% (1339) of the total specimens collected. No live infestations of timber insects were recorded, but feeding damage was detected in one floor. Insects that had the potential to infest timber were found in 104 (3.5%) of the containers inspected. Only one of these was found alive, a scolytine beetle, but 45 dead insects were found that were timber pest species exotic to Australia and of quarantine concern. They included bostrychids, curculionids, cerambycids, siricids and termites and most would have been associated with the cargo carried in the containers or with dunnage. The study shows that the wooden components of sea cargo containers are exposed constantly to timber-infesting insects from many sources. The advent of refrigerated containers for agricultural or horticultural produce may increase the risk associated with this pathway, resulting in higher survival of insects because containers are kept at constant, non-lethal temperatures throughout transport (Work *et al.*, 2005).

Air transport

Quarantine authorities have long been aware of the connection between aircraft and the dissemination of insect species from one region to another as hitchhikers in the holds or cabin areas (Dobbs and Brodel, 2004). The importance of this pathway has

Table 9.2. Summary data for 8341 Coleoptera interceptions at US ports of entry from 1985 to 2000 by insect family (from Haack, 2006, with permission of National Research Council of Canada Research Press).

Family*	No. of interceptions identified				No. identified			No. of interceptions associated with wood articles if given				Top five associated products in decreasing order
	Total	Family level only	Genus level only	Species level	Genera	Species		Crating	Dunnage	Pallets	Wood	
BOS	414	52	115	247	16	16		137	33	28	150	Tiles, woodenware, melons, machinery, marble
BUP	245	41	182	22	16	10		80	51	7	39	Tiles, marble, machinery, steel, mesquite
CER	1642	448	1048	146	79	41		390	269	57	408	Tiles, iron, ironware, machinery, marble
CUR	875	118	714	43	44	17		420	326	33	95	Tiles, steel, machinery, marble, granite
LYC	102	72	11	19	3	3		64	4	3	15	Doors, tiles, artware, bamboo, housewares
PLA	55	19	36	0	2	0		3	4	3	8	*Dracaena* plants, pineapples, bananas, woodenware, doors
SCO	5008	1547	1002	2459	40	60		2179	1841	348	601	Tiles, marble, machinery, steel, parts
Total	*8341*	*2297*	*3108*	*2936*	*200*	*147*		*3273*	*2528*	*479*	*1316*	Tiles, machinery, marble, steel, ironware

Notes: ¹Insect families: BOS = Bostrychidae; BUP = Buprestidae; CER = Cerambycidae; CUR = Curculionidae; LYC = Lyctidae; PLA = Platypodinae; SCO = Scolytinae.

increased steadily with increases both in the number of passenger and cargo flights around the world (Hulme, 2009) (Fig. 9.10) and in the speed of air travel. Studies by Dobbs and Brodel (2004) showed that almost one in four cargo aircraft arriving from Central American countries into south Florida harboured live, non-indigenous organisms of potential economic impact to US agriculture, forests and ornamentals (Figs 9.11 and 9.12). Insects intercepted on this pathway included chrysomelid leaf beetles, weevils, scarab beetles, leaf hoppers, plant bugs, termites, moths and crickets (Caton *et al.*, 2006). Extreme long-range dispersal by air travel is thought to be the main mechanism for spread of the winter bronzing bug, *Thaumastocoris peregrinus*. This hypothesis was supported by information from Brazil where, in the state of Sao Paulo, the bug was first found in *Eucalyptus* trees adjacent to two international airports in the metropolitan region of Sao Paulo city (Wilcken *et al.*, 2010). The insect may also hitchhike on the clothes of travellers or be dispersed by wind.

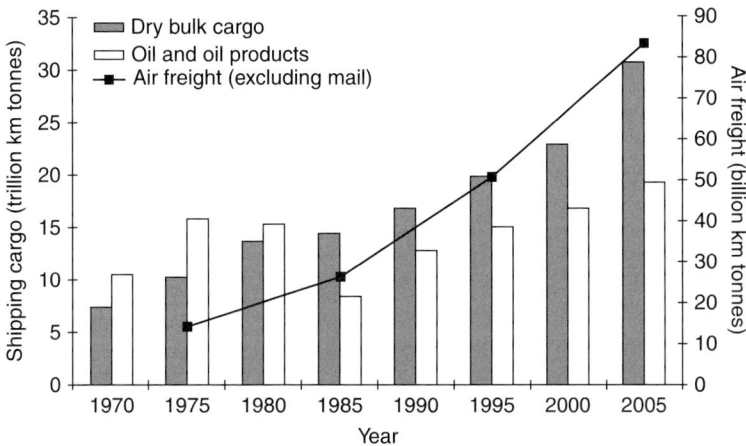

Fig. 9.10. Trends in global shipping cargo volumes and airfreight, 1970–2005. Note the three orders of magnitude difference in the scales of the left- and right-hand ordinate axes (from Hulme, 2009).

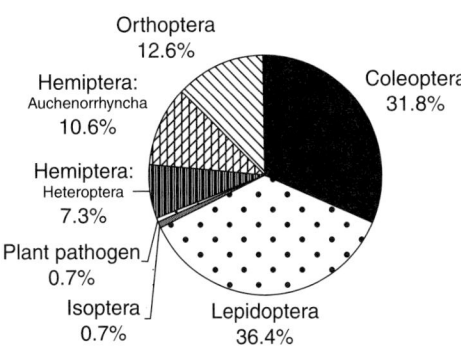

Fig. 9.11. Relative proportions of quarantine-significant taxa captured in foreign cargo aircraft arriving at Miami International Airport, 1 September 1998 to 31 August 1999 (from Dobbs and Brodel, 2004).

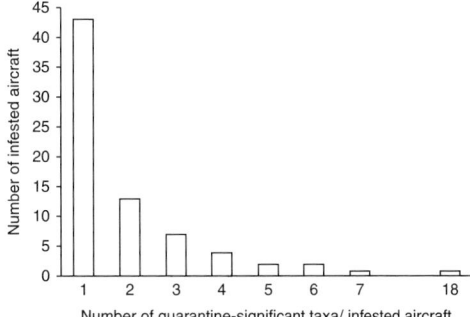

Fig. 9.12. Numbers of quarantine-significant taxa found aboard infested foreign cargo aircraft arriving at Miami International Airport, 1 September 1998 to 31 August 1999 (from Dobbs and Brodel, 2004).

Long-range dispersal by air transport may also be involved in the spread of the red gum lerp psyllid, *Glycaspis brimblecombei* (Plates 31 and 32), as evidenced in Chile where the insect was first detected in 2001 in the neighbourhood of the International Airport of Santiago on *E. camaldulensis* (Huerta *et al.*, 2010).

Wind dispersal

The dispersal of insects by aerial currents is a well-known phenomenon but it was not until the early 20th century that the first proof was obtained of long-distance transport when spruce aphids were observed in large numbers over a broad area at Spitsbergen in northern Norway. Their nearest host plants were estimated to be about 1300 km away (Elton, 1925). The first estimate of the magnitude of aerial insect population was made by Coad (1931), who found that the number of insects in a vertical column of air 2.6 km^2 and extending from 15 to 4300 m above the ground averaged 25 m throughout the year in Louisiana in the USA. In numerous studies, Hemiptera are the insect group most commonly transported long distances by wind and, among these, Aphididae is the dominant family (Holzapfel and Harrell, 1968; Hardy and Cheng, 1986). Examples of both long-distance and localized wind dispersal of forest insect pests are provided in Chapter 3.

Humans as vectors: intentional and accidental

The gypsy moth, *L. dispar*, in North America is a classic example of human-assisted introduction of an exotic pest. The European variety was brought initially into that country in 1869 for hybridization experiments with the domestic silkworm, *Bombyx mori*. A portion of the laboratory colony escaped, however, and found a suitable habitat for colonization in the surrounding oak forest, from whence they spread widely to become one of the most destructive hardwood forest insect pests in the region, defoliating an average of 1.6 Mha of forest annually in eastern USA (Ciesla, 1993). This spread was assisted by the accidental transport of life stages on recreational and commercial vehicles and on outdoor household articles (Douce *et al.*, 1994). More than a century later, North America has been involved in eradication programmes for the Asian form of the moth, which has an even more extensive host range than that of the European form. Egg masses of Asian gypsy moth were transported on ships from the Russian Far East and the newly emerged larvae blew in the wind on silk threads to shore. Since then, egg masses of the moth have been detected on machinery brought in from several countries where the insect occurs, and similar instances have been reported from other parts of the world.

The growing popularity of arthropods as pets in some places such as the USA, Europe and Japan and the increasing international trade in live insects poses quarantine risks. An example of relevance to forestry is that of the cossid moth, *Chilecomadia valdiviana*, variously known as the quince borer, carpenterworm or butterworm. It is native to Chile, where it is an emerging pest of *Eucalyptus* spp. plantations (Lanfranco and Dungey, 2001). Tens of millions of larvae of *C. valdiviana* and another species of *Chilecomadia*, *C. morrie*, are sold as fishing bait and reptile food in the USA and Europe (Thomas, 1995; Iriarte *et al.*, 1997).

In California, eucalypts are widely planted as landscape and windbreak trees, but some people in the community oppose their use, mainly because of their growth habits, facilitation of catastrophic urban wildfires and a perception that it is changing the California landscape by crowding-out or preventing the growth of native species (Paine and Miller, 2010). Between the time eucalypts were first introduced to the state in the middle of the 19th century and 1983, only two herbivorous insects, originating from Australia, had been recorded feeding on these trees. In the period between 1984 and 2008, an additional 15 species, all Australian, were recorded. The mode or route of introduction has never been established, but examination of temporal and spatial patterns suggests that the introductions are non-random processes. As outlined by Paine and Miller (2010), the hypothesis that there were intentional introductions as biocontrol agents for a perceived weed species cannot be rejected.

Tourism is also a pathway for invasive species of forestry importance. For example, in New Zealand in 2006–2007, 25% of undeclared goods seized by quarantine inspectors from arriving air passengers and crew were cut flowers or foliage (McNeill et al., 2008). Items such as tents, golf bags and baggage carried by travellers between countries have also been shown to harbour live insects or foliage (Gadgil and Flint, 1983; Liebhold et al., 2006; McCullough et al., 2006) (Fig. 9.13), as has clothing worn by travellers. Egg masses of tussock moths (Lymantriidae) are intercepted frequently on imported used vehicles at the border in New Zealand (Armstrong et al., 2003; Toy and Newfield, 2010).

9.2.3 Prevention and management options

The threat posed to plants and plant products by invasive pest species was first recognized officially in the late 1800s following a series of catastrophic pest and disease epidemics in Europe and America (Mathys and Baker, 1980). One of the earliest plant quarantine laws was passed in Germany in 1873 prohibiting the importation of plants and plant products from the USA to prevent the introduction of the Colorado potato beetle. Asia was also involved early in plant quarantine, with Indonesia legislating in 1877 to prevent the entry of coffee rust from Sri Lanka. The first international plant protection convention was established in 1881 after the introduction of grape phylloxera into Europe from America in 1865 caused major losses in the vineyards of France (Mathys and Baker, 1980). Since that time, invasive species have continued to cause significant damage, but it is not until recently that such losses have been better quantified. Pimentel et al. (2001) estimate that damages worldwide from invasive species are more than US$1.4 trillion/year, which represents nearly 5% of the world economy. For the forestry sector, in the USA alone annual losses due to invasive species total US$4.2bn (Pimentel et al., 2001). As noted by Holmes et al. (2009), biological invasions by non-native species are a by-product of economic activities and the potential costs generally are not factored into decisions about exports, imports and domestic transport of goods and people, all of which are pathways for the introduction and spread of invasive species. They suggest that the greatest economic impacts of invasive species in forests are likely to be due to the loss of non-market values (Fig. 9.14). This has prompted a range of global initiatives to address the problem. These initiatives, together with some of the key elements of pre-border, border and post-border quarantine, from a forestry perspective, are discussed below.

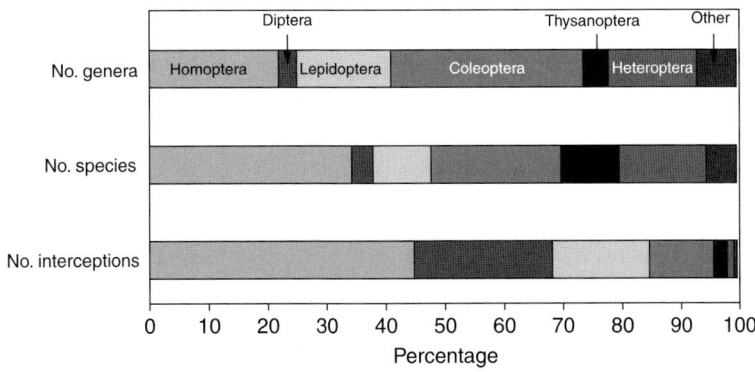

Fig. 9.13. Frequencies of different insect orders intercepted in airline baggage coming into the USA expressed as numbers of genera, numbers of species and total numbers of interceptions. The 'other' category includes Hymenoptera, Orthoptera, Isoptera and Collembola (from Liebhold et al., 2006).

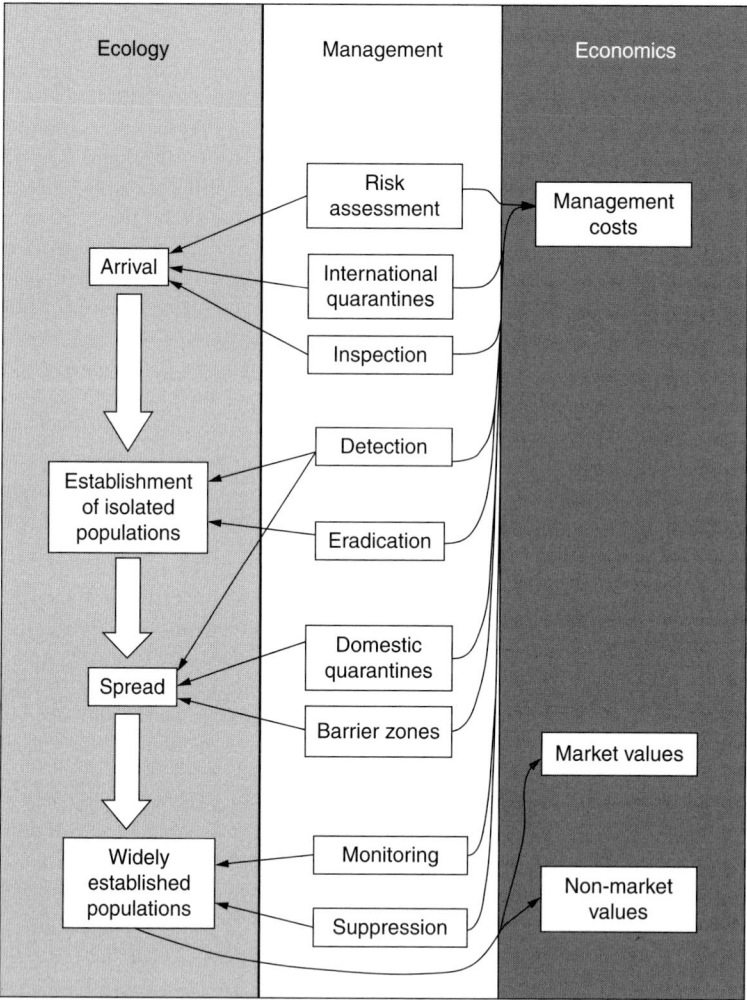

Fig. 9.14. The stages of a biological invasion are linked to management actions that can be applied at each stage; each of these management actions has economic implications (from Holmes et al., 2009).

Pest awareness

For any country endeavouring to implement forestry quarantine, knowledge of the pest species occurring elsewhere, which may pose a risk to its own forest resources should they be introduced accidentally, is important. This allows for targeting of preventative and inspection measures to exclude these pests. Such information is variously available in the literature (print or electronic), or through international and regional quarantine bodies such as the International Plant Protection Organization (IPPO) and the Pacific Plant Protection Organization (PPPO). It may also be obtainable through networks of forest protection specialists.

A recent global initiative to raise awareness of the immense costs and dangers posed by invasive species to the sustainable management of forests is the formation in 2004 of Forest Invasive Species Networks. The Asia-Pacific Forest Invasive Species Network (APFISN) is a cooperative alliance of the 33 member countries in the Asia-Pacific

Forestry Commission (APFC) – a statutory body of the Food and Agriculture Organization of the United Nations (FAO). The network focuses on inter-country cooperation that helps to detect, prevent, monitor, eradicate and/or control forest invasive species in the Asia-Pacific region. Specific objectives of the network are to: (i) raise awareness of invasive species throughout the Asia-Pacific region; (ii) define and develop organizational structures; (iii) build capacity within member countries; and (iv) develop and share databases and information. The Forest Invasive Species Network for Africa (FISNA) comprises seven African countries (Ghana, Kenya, Malawi, South Africa, United Republic of Tanzania, Uganda and Zambia). Its mandate is to coordinate the collation and dissemination of information relating to forest invasive species in sub-Saharan Africa with objectives similar to that of the APFISN.

Knowledge of indigenous fauna

Ideally, in setting up quarantine protocols to exclude exotic organisms, an ability to be able to distinguish between what is exotic and what is indigenous would seem to be of key importance. In practice, few, if any, countries would have a complete inventory of their forest insect fauna. This applies particularly in the tropics, where the number of species is large, many unidentified, and the number of entomologists few. Nevertheless, to comply with the requirements of world trade, countries need to be able to demonstrate an acceptable capacity in this regard (through their own resources or with external assistance) and an ongoing commitment to such work. This is especially the case where area freedom arrangements are sought (i.e. the ability to trade based on a certification that a particular pest does not occur in the area where the exports originated). Information about endemic and established forest organisms comes from many sources, including arthropod and pathogen collections with validated, well-curated specimens, databases, scientific literature, surveillance conducted by forest services, faunal surveys and specialist networks.

Assessing risk and selecting target pests

It should be cautioned that, when attempting to set up pest lists based on perceived risk, knowledge of the pests of much of the forest flora of the world is meagre in comparison with that for agricultural crop pests (Wylie, 1989). Coupled with this is the difficulty referred to earlier of predicting just how an organism will behave outside its native environment. Many countries have target pest lists, but again the experience has been that most of the insects that have entered and become problems are not on such lists. Despite this, knowledge of exotic pests, their behaviour and likely mode of entry is very useful in formulating generic measures to combat pest 'types' rather than individual species.

Pest risk assessment (PRA) and associated risk mitigation is the purview of biosecurity and quarantine agencies and, by its nature, focuses mainly on pre-border and border activities. It is a disciplined process that is used to predict whether or not a species is likely to become established and be invasive and to generate a relative ranking of risk (Wittenberg and Cock, 2001). Usually, scores are assigned to each of the risk components and then the scores are added or averaged to give the final result, or odds may be assigned instead of scores (Holt et al., 2006). Entire pathways may also be analysed for risk. Application of the risk assessment process, while reducing the amount of subjective judgement involved, can be labour-intensive, time-consuming and costly. For example, a risk assessment of importing unprocessed logs from Russia to the USA was estimated to cost US$500,000 (Wittenberg and Cock, 2001). However, the process also has relevance for post-border activities such as targeted surveillance and response planning. One of the issues raised by several countries attending the 2008 APFISN workshop in Vietnam on 'Risk-based targeted surveillance for forest invasive species' was the need for assistance in selecting which pests to target for post-border surveillance (FAO, 2009). Wylie (2010) proposed a nine-step guide to the selection of target pests, incorporating most elements of published international standards on pest risk assessment (EPPO, 1998; FAO, 2004), but with a slightly different emphasis and a forestry perspective. The guide and two worked examples from the Asia-Pacific region are reproduced here. Note that this is not meant as a substitute for the standard PRA, rather it is intended as a screening tool which will allow agencies involved in forest health to select a few key pests for post-border monitoring.

A nine-step guide to selection of target forestry pests for surveillance

1. WHAT DO WE WANT TO PROTECT? Most PRAs are initiated in response to requests for import of new commodities, or awareness that a new pathway has opened up or a new pest comes into prominence. Sometimes, it is because of the review or revision of a policy (FAO, 2004). Here, we commence not with the commodity but with the resource we want to protect – this may be the country's commercial forest plantations (pine or hardwood), or conservation native forests, or timber-in-service (e.g. the built environment). Ideally, surveillance would be conducted for the full range of potential pests across all resource types, but financial and other constraints may require that surveillance be prioritized.

2. WHAT EXOTIC PESTS COULD BE A THREAT TO THIS RESOURCE? Information about potential forest pest and disease threats already may reside with biosecurity or quarantine agencies in-country, and these should be consulted first. However, in several Asia-Pacific countries, the quarantine focus historically has been on agricultural pests, and forestry needs are largely unexplored. In such case, information can be sourced from international forest health or quarantine networks, the published literature, scientific meetings and the Internet.

3. DOES THE PEST HAVE THE POTENTIAL TO BE TRANSPORTED BY TRADE OR HUMAN MOVEMENT? Aspects to be considered here relate to the biology and behaviour of the organism and its potential to be transported with human assistance. In the case of an insect pest, pertinent questions are:

- What is its life cycle and behaviour?
- What are its hosts and where does it lay eggs, feed and pupate?
- Is there a resting stage?
- Is it associated with a commodity?

A longicorn beetle, for example, which is long-lived, lays its eggs in wood and whose immature stages feed and pupate inside wood, has a greater potential to be transported than a leaf beetle defoliator, which lays its eggs and spends its larval period in the soil and only feeds on tree foliage as an adult. Potential pests are not necessarily associated with a commodity. In the Russian Far East, where forests are close to the main seaports, Asian gypsy moths (*L. dispar*) are attracted to port and ship lights and lay their eggs on the superstructure of ships or on cargo. Spores of Eucalyptus rust (*Puccinia psidii*) have been found on the surfaces of cargo containers and can be transported on the clothes of travellers. The New Zealand burnt pine longicorn, *Arhopalus ferus*, is a regular hitchhiker on timber cargo being shipped to Australia (Pawson *et al.*, 2009).

4. IS THERE A PATHWAY INTO THE COUNTRY FOR THE PEST? Pathway analysis is a key component of pest risk analysis. Its purpose is to determine the likelihood of a pest arriving in a country as a result of trade in a particular commodity or human-assisted movement and the likelihood that, if it arrives, it will transfer to an appropriate local host. The four main elements of pathway analysis are listed below:

- Likelihood of trans-shipment with a commodity
  - Prevalence of pest in source area
  - Occurrence of relevant life stage associated with the commodity
  - Frequency of movement with the commodity
  - Season
  - Pest management, cultural and commercial procedures applied at the place of origin

- Likelihood of a pest surviving during transport and storage
  - Speed and conditions of transport
  - Duration of life cycle of pest in relation to transport conditions
  - Vulnerability of pest life stages to transport conditions
  - Prevalence of pest in the consignment
  - Commercial procedures

- Likelihood of a pest surviving existing pest management procedures
  - Likelihood pest will survive existing pest management procedures applied for other pests from origin to end use
  - Likelihood that pest will go undetected during inspections

- Likelihood of transfer to suitable host
  - Dispersal mechanisms including vectors
  - Number of destination points
  - Proximity of entry, transit and destination points to suitable hosts
  - Time of year of importation
  - End use of commodity
  - Risks from by-products or waste

As is evident from the list above, pathway analysis requires the gathering of a considerable amount of information from a variety of sources. Detailed knowledge of the pest in relation to the pathway is required, such as how the rate of development matches with the speed of transport (although with air transport a pest may be moved anywhere in the world within 24 h), the frequency of association with a commodity and history of past interceptions. If the pest is seasonal, then there will be periods when the risk of trans-shipment is reduced. The capability of a pest to survive adverse conditions such as low temperatures or desiccation, which it may experience on the pathway, is an important consideration, as is knowledge of how it disperses (naturally or via a vector). Information is required on the worldwide distribution of the pest and its prevalence in the source area, as well as on commercial procedures (e.g. drying, chemical application, fumigation, visual inspection) applied to the commodity at the point of origin, during transport and on arrival. This information gathering will require input from forest health specialists, biosecurity/quarantine agencies and industry in both the importing and exporting country.

5. WHAT IS THE LIKELIHOOD OF ESTABLISHMENT?

- Environmental suitability of country – climate, suitable hosts
- Geographical location of ports – proximity to suitable hosts
- Numbers and life stages of the pest present and reproductive potential

Comparison of the ecoclimatic zones of the pest's distribution with that of the country conducting the PRA is the first step. What climatic conditions, such as temperature, rainfall, relative humidity, day length, have been shown to be conducive or suppressive to the survival, development, reproduction and dispersal of the pest? It is important to keep in mind the adaptability of many invasive organisms. The wood wasp *S. noctilio*, a native of temperate Europe, is now a major pest of *Pinus* species in several subtropical regions around the world (in Australia, for example, it has taken 50 years since introduction into temperate Tasmania to spread to the subtropics of Queensland). The presence of suitable hosts in the potential recipient country is also of great importance, especially if the pest has a restricted host range. The geographical location of ports and proximity of suitable hosts will also influence the likelihood of establishment. It should be noted, however, that most major ports in a country are likely to be in close proximity to cities where, usually, a wide range of exotic tree species are used as amenity plantings, and this could facilitate establishment. Such plantings are useful in post-border surveillance. Establishment also depends greatly on the numbers and life stages of the arriving pests, whether they are likely to find a mate and their reproductive potential.

6. WHAT IS THE LIKELIHOOD OF SPREAD?

- Pest's ability for natural dispersal
- Potential for human-assisted dispersal

- Distribution and abundance of hosts
- Natural barriers

Organisms such as rust fungi, psyllids and aphids, which have short life cycles and are wind assisted, will spread rapidly on arrival and are difficult to contain. Many timber-boring insects with a long life cycle and a propensity to exploit a food resource fully before migrating are much easier to contain or eradicate. Most pests, it seems, whether short-lived or long-lived, cryptic or exposed, have potential for human-assisted dispersal. The Asian gypsy moth, *L. dispar*, is a dispersal specialist – the female moth is capable of flight distances of up to 40 km but also lays eggs on containers, cargo and motor vehicles, which assist its spread. Natural barriers such as an inhospitable region for the pest or lack of host plants may limit risk of spread.

7. WHAT ARE THE POTENTIAL CONSEQUENCES OF ESTABLISHMENT?

- Economic impact – direct damage/crop loss, costs of control or containment measures, effect on trade with other countries
- Environmental impact – damage to ecosystems or biodiversity, loss of key species, non-target effects of control measures
- Social impact – effect on workers, families, communities of loss of income/wages from affected products and imposition of area quarantine

All of the above factors will be taken into consideration in a well-constructed benefit/cost analysis. Experience with the pest in other countries may serve as a guide.

8. WHAT IS THE ABILITY TO DETECT THE PEST? As outlined by Wylie *et al.* (2008), the choice of targets for surveillance will be influenced by the ability to detect them by inspection and/or trapping. For many pests and diseases of trees, visual inspection for symptoms of disorder is the most common, and sometimes the only, practical method of detection. Tree health assessments in the vicinity of ports or high-risk sites may be undertaken periodically by trained inspectors. This method may be supplemented by the planting of 'sentinel' trees of locally important species at selected sites and then monitoring these trees at regular intervals.

For some other pest organisms affecting trees and timber, but in particular for insects, trapping is the preferred method of detection. A wide variety of devices or techniques have been developed for this purpose, the type of trap being governed mainly by the behaviour of the target insect species or group. As mentioned earlier in this chapter, most detection trapping is structured around the capture of flying adult insects using either 'passive' traps (e.g. suspended net traps, water traps, 'windowpane' traps), or luring insects from a distance (e.g. light or bait traps). For wood-boring insect species, static traps in combination with lures have been used effectively in many countries (e.g. Brockerhoff *et al.*, 2006). Lures utilizing plant-host volatiles, sex or aggregation pheromones are now commercially available for some key pest species. Spore traps have been used for the detection of some fungal pathogens.

9. WHAT IS THE CAPACITY TO ERADICATE THE PEST? The final consideration in determining which pests to target for surveillance is the ability to eradicate or contain the pest once detected. For some fungal pathogens and small insects which are dispersed rapidly by wind currents, eradication or containment is difficult, but not impossible, if detected early enough. Even if eradication or containment is considered impossible once a pest arrives, surveillance may nevertheless be conducted for such organisms to allow time to mitigate for expected impacts (for example, the deployment of resistant/tolerant tree species or clones against a pest such as the eucalypt gall insect, *Leptocybe invasa* (Nyeko *et al.*, 2009). If there is no capacity to deal with certain pests, then surveillance may be directed to other potential pests more amenable to control.

Examples of how these nine guidelines may be applied in a preliminary screening of potential target pests for the Pacific and Asia are presented in Tables 9.3 and 9.4, respectively.

Table 9.3. An example of the selection of an invasive forest pest for targeted surveillance in Fiji using the nine-step guide.

Selection criteria	Comment
1. What do we want to protect?	Fiji's mahogany plantations (*Swietenia macrophylla*), the major hardwood plantation resource totalling 50,000 ha and worth about US$200 m.
2. What exotic pests could be a threat to this resource?	The most important pests of mahogany worldwide are shoot borers *Hypsipyla* spp. which, within the pests' range, prevent the commercial growing of any of the high-value Meliaceae, including American mahoganies *Swietenia* spp., African mahoganies *Khaya* spp., cedars *Cedrela* and *Toona*, and Asian mahogany *Chukrasia*. *H. robusta* is the main pest species in Africa and the Asia-Pacific region.
3. Does the pest have the potential to be transported by trade or human movement?	The life cycle of *H. robusta* is 1–2 months, depending on climate. The moth lays its eggs on shoots, stems and leaves. Larvae burrow into new shoots and tunnel in the primary stem or branches, feeding on the pith. Pupation occurs in cocoons spun in the stem tunnels or among leaf litter and soil around the tree base. Larvae also feed on the bark, fruit and flowers of their hosts. Young plants can be infested. The pest has the potential to be transported on living host plants (family Meliaceae, mainly sub-family Swietenioidea) or via seedpods of host species.
4. Is there a pathway for the pest into the country?	The import of living plant material and seed into Fiji via international ports is regulated. Inspection of the material on arrival is likely to result in detection of the pest. Another possible pathway is via unregulated inter-island trade at a community level. The insect would be transported as eggs, larvae or pupae on living plant material or in seedpods of the host-plant species. A less likely, but possible, pathway is as a larva or pupa inside the large seed of the cannonball mangrove, *Xylocarpus granatum*, which is a host of the pest. The seeds are carried by ocean currents, can float for a month and are dispersed long distances (Smith, 1990). *H. robusta* is present in neighbouring Vanuatu, the Solomon Islands and Papua New Guinea, as well as Australia.
5. What is the likelihood of establishment?	High. Fiji has a climate suitable for the pest and host plants are distributed widely throughout its islands (Fiji has native *Xylocarpus* spp. mangroves, which are a host). The main mahogany resource is on the island of Viti Levu and there are considerable mahogany plantings within several kilometres of the principal port of Suva. The moth has high reproductive potential (a female can lay about 450 eggs over several days). The simultaneous arrival at a site of a few infested plants with late instar larvae or pupae (for example, in a consignment of nursery stock) may be sufficient to produce a mating pair of moths which could initiate establishment.
6. What is the likelihood of spread?	High. The moth is a strong flier and has impressive host-finding ability. Movement of nursery stock is also a risk.
7. What are the potential consequences of establishment?	The economic cost could be high. Forestry ranks fourth in the Fijian economy and third in foreign exchange earnings. Mahogany is a high-value timber and alternative species would not be as profitable. Line plantings as currently practised may offer some protection against the pest, but nursery stock is at risk. *Xylocarpus* mangroves are indigenous to Fiji and some ecological effects could be expected. The inability to establish new mahogany plantations would affect employment, unless an alternative species was chosen. Village plantings would also be affected, resulting in a reduction in income.
8. What is the ability to detect the pest?	A lure has been developed that will attract *H. robusta* moths and can be used in combination with static traps. The usual method of detection is by visual inspection of plants for the characteristic signs of shoot boring. Sentinel plantings can be used close to ports.
9. What is the capacity to eradicate the pest?	There is a possibility of eradication only if the pest is detected early. Once the pest is established, there are very few management options available. Enrichment or line planting in native forest such as in Fiji (and Sri Lanka) offers better prospects of a harvestable crop than open plantings.

Table 9.4. An example of the selection of an invasive forest pest for targeted surveillance in Asia using the nine-step guide.

Selection criteria	Comment
1. What do we want to protect?	*Pinus* species throughout Asia – a major plantation tree and also naturally occurring in some countries. Common species are *P. merkusii*, *P. caribaea*, *P. elliottii*, *P. massoniana*, *P. kesiya*, *P. koraiensis*, *P. roxburghii*, *P. thunbergii*, *P. yunnanensis*.
2. What exotic pests could be a threat to this resource?	There are many pest species that could be considered for surveillance by individual countries; some are exotic to the region and others (e.g. pine shoot moths and pine wilt nematode) occur in Asia but have limited distribution at present. Among pests exotic to the region, the wood wasp *S. noctilio* is a serious threat to the *Pinus* resource. It is native to Eurasia and North Africa. The female wasp lays eggs into stressed or suppressed trees, along with a phytotoxic mucus and wood decay fungus, and this combination kills the trees. It is a major pest of *Pinus*, where it has been introduced in the southern hemisphere – Australia, New Zealand, South America and South Africa – and is capable of devastating large areas of pines, causing up to 80% tree mortality.
3. Does the pest have the potential to be transported by trade or human movement?	The life cycle of the pest is approximately 1 year. Eggs are laid into the outer sapwood of standing or recently felled trees and the immature stages feed and pupate in the wood. The pest has the potential to be transported in its immature stages with *Pinus* timber, whether as a commodity or as packaging or dunnage.
4. Is there a pathway for the pest into the country?	The pest is likely to survive harvesting and milling procedures and is transported readily in pine logs and timber, as well as in solid wood packaging and dunnage. Inspection of pinewood on arrival is likely to result in detection of sirex if emergence holes and exposed tunnels of the insect are present, but otherwise would not detect immature stages in the wood. *S. noctilio* is a very common intercept in many countries, often emerging from containers in which there had been infested cargo or packaging.
5. What is the likelihood of establishment?	Analysis of the suitability of climate and hosts shows that there is a strong probability of the pest being able to establish in many parts of Asia. Container cargo and items with timber packaging are moved widely throughout most countries.
6. What is the likelihood of spread?	Adults are strong fliers capable of travelling several kilometres in search of host trees. In Australia, the annual spread of the pest is about 50 km. Human-assisted transport of infested wood is also a major factor.
7. What are the potential consequences of establishment?	Considerable timber losses can result from outbreaks of the pest. In Australia, for example, an outbreak between 1987 and 1989 killed more than 5 m *P. radiata* trees with a value of AUS$10–12 m (Haugen et al., 1990). Costs of instituting control measures can be significant and there may be severe environmental impact where native pine forests are attacked.
8. What is the ability to detect the pest?	The pest may be detected by inspection of logs, timber, dunnage or wood packaging for emergence holes or exposed tunnels. Adults may be trapped using α-pinene + β-pinene as a lure. The use of 'trap trees' is another detection method whereby selected trees are injected with an herbicide to stress the trees and make them attractive to *S. noctilio*.
9. What is the capacity to eradicate the pest?	Eradication would be attempted in the early stages of an incursion, for example, destruction of infested material and inspection/destruction of infested pine trees. Should the pest become established, biocontrol measures using the nematode *Beddingia siricidicola* and parasitoids will keep damage at a level of about 1% per annum.

The process of selecting target pests for surveillance provides a focus on pathways and helps to direct efforts towards preventing the entry of invasive species. A generic rather than species-specific approach to surveillance is preferable, but having target species is nevertheless useful. For example, in focusing on *Dendroctonus* spp. bark beetles we essentially are targeting all bark beetles, since they are likely to have a similar mode of entry.

Contingency planning

Another important aspect of preparedness is contingency planning – what to do in the event of an incursion of an exotic pest and disease. As with pest lists, preparation of very detailed plans for a specific pest may well be a fruitless exercise if, as experience shows, 'unknowns' are more likely to turn up. Generic plans for 'types' of pests are favoured. Contingency planning must be backed up by adequate resources, clear lines of responsibility and appropriate legislation to allow rapid eradicative action when required.

International treaties and standards for phytosanitary measures

The International Plant Protection Convention (IPPC) is an international treaty that was adopted in 1951, revised in 1997 and is administered within the Food and Agriculture Organization of the United Nations (Haack and Petrice, 2009). As of March 2009, there were 170 countries or multicountry contracting parties to the IPPC. The Commission on Phytosanitary Measures, the governing body of the IPPC, develops and adopts International Standards for Phytosanitary Measures through a process that includes consultation with member countries. Among recent global initiatives to contain the spread of pests of forestry importance, the International Standard for Phytosanitary Measures (ISPM) 15 for regulating wood packaging material in international trade is a major advance (FAO, 2002). It requires that all such wood packaging material be either heat treated (minimum of 56°C core temperature for 30 min) or fumigated with methyl bromide. One of the shortcomings of this standard is that the treatments are aimed at killing pest organisms that reside in the wood at the time of treatment and neither treatment has any residual effect, so reinfestation is possible, especially when bark is present (ISPM 15 does not require the elimination of bark). Haack and Petrice (2009) found that Cerambycidae and Scolytinae readily infested and developed in logs with bark after heat treatment and laid eggs in all sizes of bark patches tested down to about 25 cm^2, but did not infest control or heat-treated lumber without bark. In surveys at six US ports in 2006, 9.4% of 5945 ISPM 15-marked wood packaging material items contained bark, and 1.2% of the items with bark contained live insects of quarantine significance under the bark. While the risk of introducing quarantine pests is not entirely eliminated, there is no doubt that ISPM 15 has reduced drastically the incidence of live insects in wood packaging material.

As discussed in 9.2.2 above, many forest pests have been introduced into new locations on plants for planting. The complexity of the international horticultural plant distribution chain is shown in Fig. 9.15. There are three problems with the current phytosanitary regulatory approach: (i) it relies on inspection, but the volume of world trade has expanded far beyond inspection capacity; (ii) it focuses on addressing the risks associated with known quarantine organisms, but most pests introduced on plants for planting are previously unknown or unpredictably aggressive pests; and (iii) the IPPC prohibits requiring phytosanitary measures against unregulated pests (Britton, 2007; Campbell, 2007; IUFRO, 2007). A pathway approach to plants for planting, using IPSM 15 as an example, is in development and in April 2010 a draft ISPM was circulated for member consultation.

Barrier quarantine

The terms 'barrier quarantine' or 'border operations' are used to describe a wide range

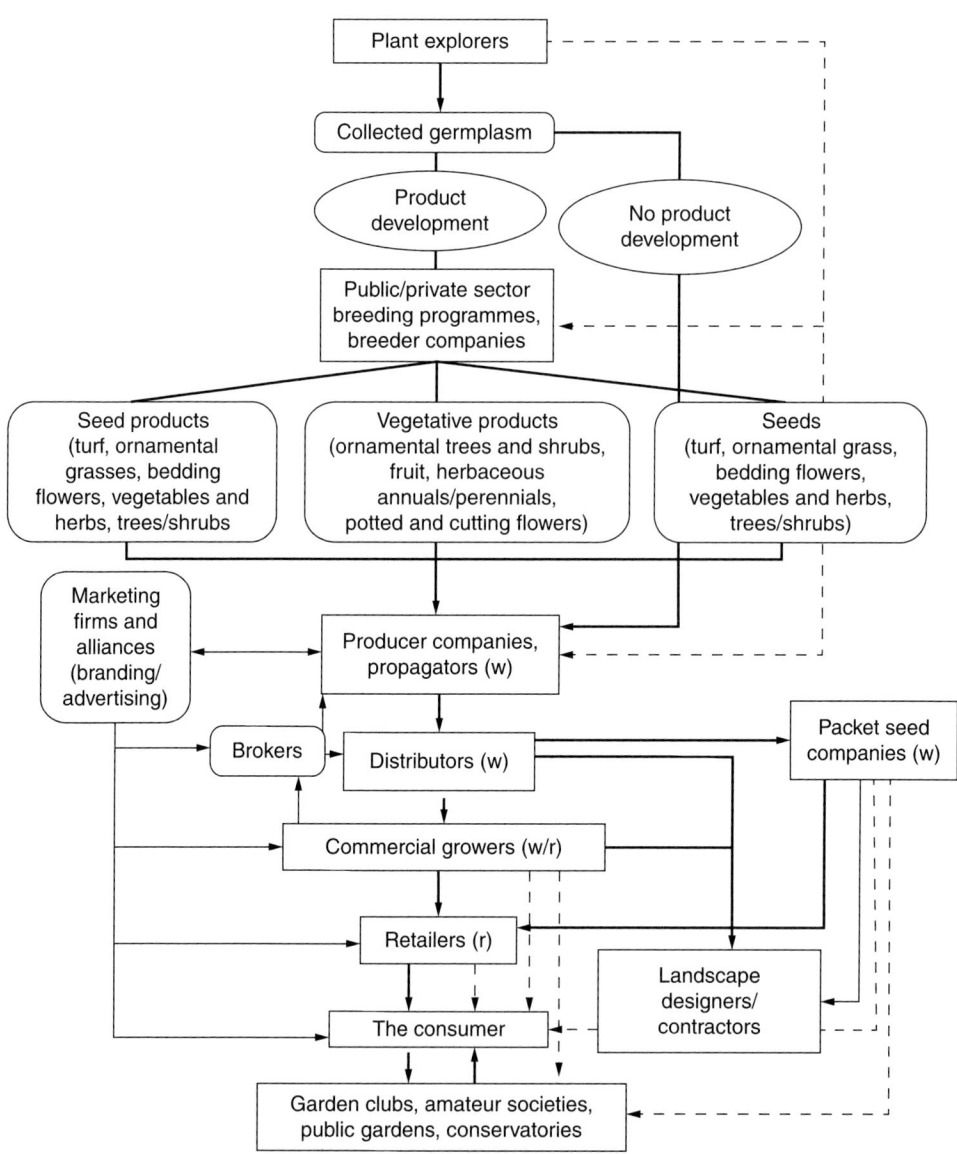

Fig. 9.15. International horticultural plant distribution chain. Key: *r* = retail; *w* = wholesale; *thick solid line* = traditional marketing conduits; *thin solid line* = information conduits; *broken line* = web-based marketing conduits (from Drew *et al.*, 2010).

of activities undertaken at airports, seaports and mail exchanges to intercept material of quarantine interest, whether introduced in accordance with quarantine procedures or as a result of illicit activity, or unintentionally. This is probably the most crucial stage of forestry quarantine, given the number of pests that can enter in lumber, wooden packaging, manufactured wooden articles and artefacts. A skilled eye is required to detect borer activity, and quarantine inspectors can benefit from specific training on forest, not just agricultural, pests. New technology is assisting in the detection of cryptic timber pests, for example, the use of odour-detection dogs and X-rays.

Interception data from national quarantine agencies are important in the development or improvement of invasion risk assessment methods, providing insight into pathways and vectors of introduction (Kenis et al., 2007). For example, McCullough et al. (2006) summarized more than 725,000 pest interceptions recorded in the Port Information Network database maintained by the US Department of Agriculture, Animal and Plant Health Inspection Service from 1984 to 2000 to examine origins, interception sites and modes of transport of non-indigenous pests. They found that about 62% of intercepted pests were associated with baggage, 30% were associated with cargo and 7% were associated with plant propagative material. Pest interceptions occurred most commonly at airports (73%), land border crossings (13%) and marine ports (9%). Insects dominated the database, comprising 73–84% of the records annually, with Homoptera, Lepidoptera and Diptera collectively accounting for over 75% of the interception records. Haack (2001), using the same database, focused on intercepted scolytine bark beetles and showed that 73% were found in solid wood packing materials, 22% in food or plants and 5% in other or unspecified materials. The products most commonly associated with scolytine-infested wood packing materials were tiles, marble, machinery, steel parts, ironware, granite, aluminium, slate and iron. Kenis et al. (2007) found that the majority of introductions of alien insects in Europe were associated with the international trade in ornamental plants.

Hazard site surveillance

As outlined by Wylie et al. (2008), hazard site surveillance is a system for post-border detection of new pest incursions targeting sites which are considered to be potentially at high risk of such introductions. The ever increasing volumes of containerized freight and competition for space at domestic ports means that many goods are being first opened at premises some distance from the port of entry, thus dispersing risk away from the main inspection point. Hazard site surveillance acts as a backstop to border control to ensure that new incursions are detected sufficiently early to allow the full range of management options, including eradication and containment, to be considered. This is particularly important for some of the more cryptic forest pests whose presence in a forest often is not discovered until populations are already high and the pest is well established (Wardlaw et al., 2008).

In setting up a hazard site surveillance programme you need to: (i) know what pests you already have; (ii) know what pests you do not want; (iii) assess the likely pathways for exotic pest entry; (iv) identify and categorize risk sites; (v) have a methodology for detection of target pests; and (vi) be able to identify what you find. All of these requirements except (iv) have already been discussed in this chapter. What constitutes a risk site for exotic forest pest incursions will vary greatly within and between different countries and cities according to such factors as trade patterns, geography, infrastructure and quarantine policy, but some generalities can be made. Self and Kay (2005), in a consultancy for the Australian Government on targeted post-border surveillance using Brisbane as an example, developed a methodology for the selection of high-hazard sites that could be used as a model. As a first step, they divided likely risk sites into four broad groupings from primary to quaternary, the primary sites being considered of highest risk. Port and international airport environs were classed as primary risk sites, and secondary risk sites included areas where containers were opened, quarantine-approved premises (QAP) and importers of raw material (timber importers). Botanic gardens, military camps and transport corridors were listed as tertiary risk sites, and forests or forest parks within city boundaries as quaternary risk sites. Primary risk sites were recommended as the first choice for trapping and inspections, and other categories could be included according to the resources available.

In determining secondary risk sites, Self and Kay (2005), with the assistance of the Australian Quarantine and Inspection Service, compiled a list of business premises involved in the importation or handling of significant volumes of high-risk items such as timber, timber packaging, wooden furniture and artefacts. Many of these sites were QAPs. Sites were assessed for risk using an arbitrary scale of 1 (low risk) to 5 (high risk) for each of five criteria, namely type of goods (likelihood of harbouring pests), volume of risk goods, cargo source (potential pest threat from country of origin and similarities of its climate and flora to that of Australia), vegetation at site (extent and type) and habitat (intensity of land development). For each site, scores for the five risk criteria were multiplied and then sites ranked according to their total score. This process aided decision making in the final selection of sites for surveillance.

Fledgling hazard site surveillance programmes have commenced in several tropical countries, e.g. Fiji, Vanuatu, Malaysia and Vietnam.

Post-entry quarantine

This is a measure to reduce further the likelihood of import of an undesirable exotic organism and is applied to high-risk products. Living plant material may be inspected on arrival and then kept in special quarantine glasshouses for several months to allow time for pest and disease symptoms to manifest themselves. Unprocessed timber may be allowed entry following barrier inspection, but there may be a requirement for subsequent inspections at the mill where the timber is being processed.

Treatment

A variety of treatments are applied worldwide for disinfestation of imported materials. Seeds are usually dusted with insecticidal powder, and living plant material is sprayed or dipped in an insecticidal solution. Fumigation is the treatment most commonly used for timber and wood products. However, care is needed in the choice of fumigant and the concentrations employed. Some fumigants are effective against the adult and larval stages of insects but do not kill the eggs; others may be effective ovicides but less successful against other stages, depending on the concentrations used. One approach to overcome this problem is to use a mixture of gas fumigants at lower dosages; for example, a low dose of methyl bromide, providing high efficacy against egg stages of forest insect pests, and sulfuryl fluoride, providing high efficacy against the other stages (Oogita *et al.*, 1998). Methyl bromide is the fumigant that has been most commonly used against timber borers and is generally very effective at the right concentrations. However, it should be noted that methyl bromide is not completely effective against borers in logs and large dimension timber when the moisture content is high. Cross (1991) has shown that it is not practical to achieve useful insecticidal doses much beyond a depth of 100 mm in 'green' (unseasoned) material using conventional tent fumigation techniques (Fig. 9.16). The use of methyl bromide is being phased out to comply with the Montreal Protocol (Barak *et al.*, 2006), although it is still the approved fumigant under ISPM 15 for wood packaging material. Other fumigants being considered under this standard are sulfuryl fluoride, phosphine and carbonyl sulfide.

Heat treatment is another measure commonly employed for disinfestation of timber. For example, coniferous sawnwood is sometimes kiln dried to eliminate possible infestations of the pinewood nematode, *Bursaphelenchus xylophilus*. This nematode is transmitted when its vector, the pine sawyer *Monochamus* spp. (Coleoptera: Cerambycidae), lays its eggs in freshly cut, felled, dying or recently dead conifers. Therefore, the nematode occasionally may be present in 'green' timber. Heat treating the timber in a kiln to a wood temperature of 56°C for 30 min is sufficient to kill all nematodes and their vector (Dwinell, 1997). This is an approved treatment for wood packaging material under ISPM 15 (FAO, 2002). A variety of other treatments are being used or tested to control wood-infesting insects, e.g. gamma radiation, X-rays, microwaves,

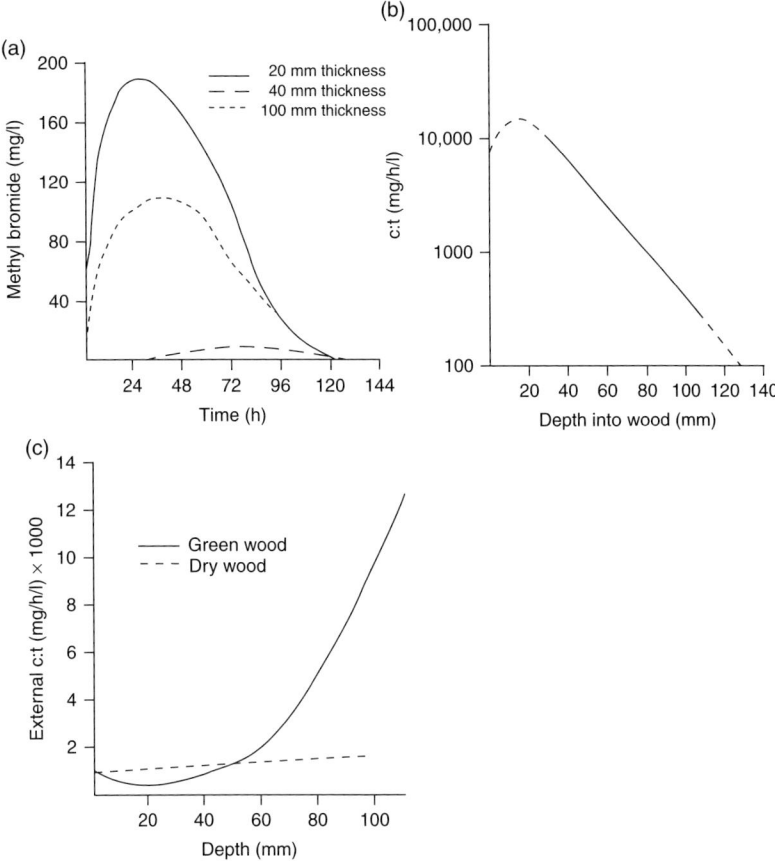

Fig. 9.16. Penetration of methyl bromide fumigant into test blocks of wood. (a) Equivalent concentration curves for the three thicknesses of test block. (b) Relationship of internal c:t (dosage = concentration × time) to external c:t held at 6224 mg/h/l. (c) c:t product to get 1000 mg/h/l at increasing depth in wood (from Cross, 1991).

infrared, electron beam treatment and chemical pressure impregnation (FAO, 2002; Fleming et al., 2003). Finally, in some cases where the usual treatments are impractical or not economically justified, infested items may be destroyed (burned).

Within-country quarantine

Once an exotic pest is detected post-barrier, quarantine restrictions are usually applied immediately within country to contain its spread while eradicative options are being assessed. Even when it becomes apparent that the pest is well established and eradication is unlikely to be successful, restrictions on the movement of high-risk material may still be imposed so as not to assist the spread of the pest to other parts of the country. In Queensland, Australia, where the exotic bark beetle *I. grandicollis* became established in the south of the state in the early 1980s, quarantine zones were declared in an effort to slow the northward spread of the pest. Movement of logs with bark on, or of bark chip, out of the zone was prohibited unless the material had been fumigated. These quarantine boundaries remained in place for 27 years until 2009, when *I. grandicollis* was discovered in Townsville in north Queensland, believed to have 'hitchhiked' there with a shipment of logging equipment sent from the south of the state for a post-cyclone salvage operation.

In the West Indies, entry of all plant products into Bermuda from the islands of Grenada and Trinidad and Tobago was prohibited in 1995 following the first observation of the mealybug, *Maconellicoccus hirsutus*, in Grenada in 1994 and subsequent widespread damage to many crops including tree crops (Jones, 1995). Another example is from North America, where the exotic bark beetle, *Tomicus piniperda*, was first detected in 1992. The insect breeds in phloem of recently cut or dying pine logs, stumps and slash. Adults feed in shoots of live pines to complete maturation. Federal and state quarantines were imposed to regulate movements of pine Christmas trees, logs and nursery stock out of infested counties (McCullough and Sadof, 1998; Haack and Poland, 2001). The final example is from Central America. The Nantucket pine tip moth, *Rhyacionia frustrana*, was first recorded in Costa Rica in 1980, where it was believed to have spread from its natural range in neighbouring Nicaragua via ornamental pines. The pest established in the north and central regions of the country and a quarantine was imposed on the movement of living pine material to the southern region with the aim of preventing the introduction of the insect to Panama and South America (Ford, 1986).

10

Integrated Pest Management (IPM)

10.1 Introduction

Continued concerns about the negative impacts of pesticides in developing countries lead to the recommendation that they should be replaced by integrated pest management (IPM) measures coupled with education and training (Atreya *et al.*, 2011). In April 2011, a literature search on ISI Web of Knowledge using the single term 'integrated pest management' yielded over 15,000 records. However, adding the word 'forest*' (the asterisk meaning a wildcard) produced only 908 references, going back to the mid-1960s. In other words, IPM would seem to be much more prevalent and commonplace in other forms of crop production, in particular agriculture and horticulture (Speight *et al.*, 2008). IPM has been defined in many ways over the years but its basic concept is simple enough. Essentially, it involves the combination of all appropriate pest management tactics into a package which reduces to tolerable levels the economic losses caused by insects. It was first conceived as a potential means to reduce the use of pesticides and to encourage the more 'ecologically friendly' use of pest natural enemies via biological control, and it was first implemented in the cotton crops of the USA, when massive doses of insecticides failed to control pests (Peshin *et al.*, 2009). As mentioned above, most success stories of IPM come, still, from agriculture. Forestry, especially in the tropics, has tended not to implement anything like the complexities found in intensive agriculture, but even then, IPM tends to be more complicated to carry out than routine spraying, for example. Its adoption, especially in situations with little expert advice and even less economic or educational infrastructures, can be problematic.

All IPM systems have several objectives in mind, i.e. more effective crop production, often at lower cost and with less environmental impact. It has been suggested that a better term for the amalgamation of all available factors to reduce or even prevent pest problems in forestry is ecological control of forest pests (ECFP) (Liang and Zhang, 2005), but in this book we choose to retain the term IPM to cover all appropriate aspects of pest management amalgamated into a functional whole. Because the definition is rather nebulous, most sets of tactics can be labelled as IPM, but in forestry at least, the most important watchword is that of prevention. Clearly, if problems never arise, then they do not need to be fixed and the preceding chapters of this book have presented detailed discussions of the multi-various ways and means that the prevention, or at least reduction, in the risks of economic damage from insect pests may be achieved. Figure 10.1 sets the IPM scene for tropical

© F.R. Wylie and M.R. Speight 2012. *Insect Pests in Tropical Forestry, 2nd Edition*
(F.R. Wylie and M.R. Speight)

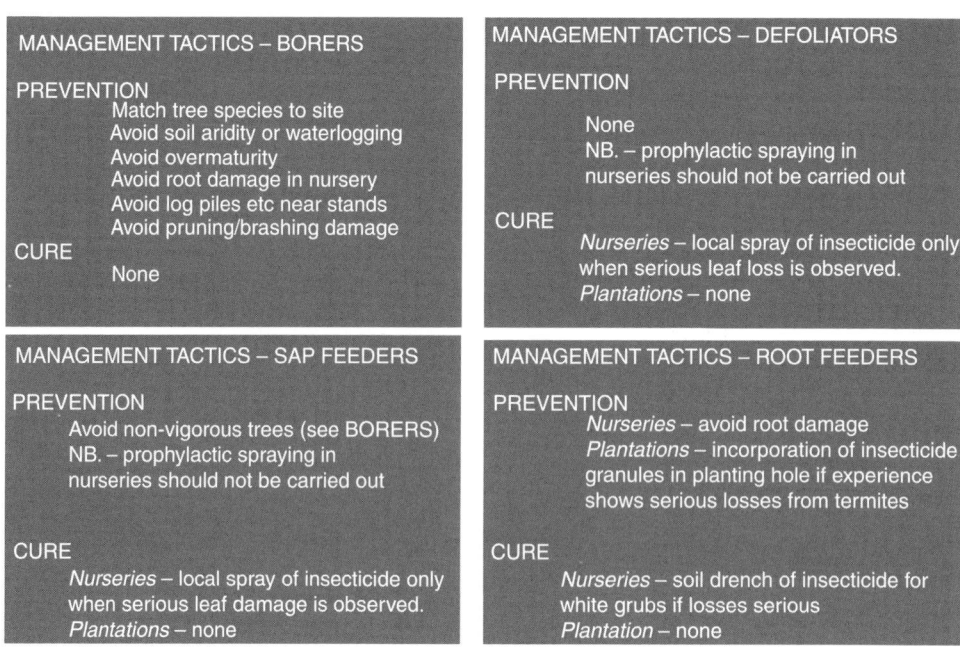

Fig. 10.1. Principal management tactics suggested for four stages in tropical forest production.

forests and suggests that, in most scenarios, prevention is the only viable method; most control tactics are inherently difficult, if not impossible, at least economically. Against this rather depressing framework, this chapter looks at the requirements of successful IPM programmes and provides examples of such systems, targeted either at specific pests in specific localities or, as a contrast, groups of tropical tree species with several, or indeed many, pest problems. It should be noted at the outset that most discussions of IPM in tropical forestry are conjectural or consist of proposals for management rather than actual, everyday strategies.

10.2 Why Do Pest Outbreaks Occur?

Figure 10.2 presents a summary of many of the basic reasons for insect pest outbreaks in tropical forestry (Speight, 1997), all of which have been covered in the preceding chapters of this book. The right-hand side of the diagram illustrates how pest problems might be lessened, while the left-hand side provides 'recipes' for disaster. We are, of course, presenting these tactics from an entomological perspective; sound and unavoidable silvicultural techniques very often have to employ risk-prone systems and then IPM will attempt to use control tactics rather than preventative ones.

10.3 Appropriate Strategies for Management

Figure 10.3 summarizes the types of management that might be employed in a general IPM 'toolbox' for tropical forestry. As can be seen in stage (a) of the diagram, IPM should contain a great deal of planning and decision making, carried out well before a tree seed is ever planted. In an ideal situation, no more pest management is required beyond careful economic and silvicultural planning, and this is why it is so important (and so often ignored) that entomology (and pathology, too) is considered fully at the very start of new forestry projects. Even so, nothing is guaranteed, so topics under stages (b) and

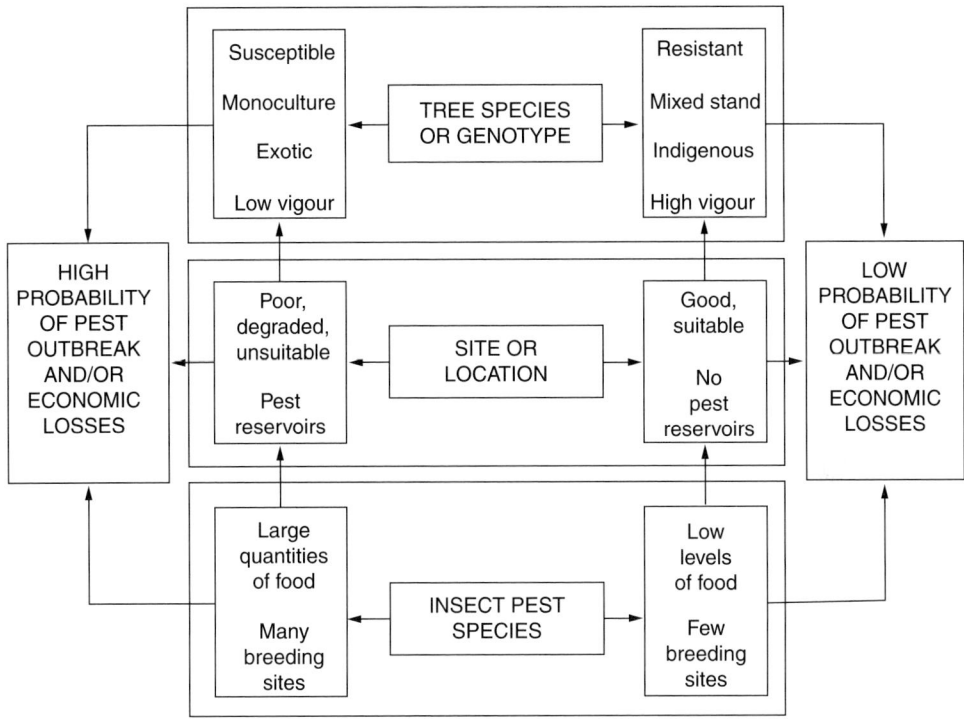

Fig. 10.2. Some major reasons for pest outbreaks in tropical forestry.

(c) must be implemented fully in an ideal IPM programme. Then, even if planning and decision making in section (a) fails to prevent pest problems completely, major losses can be predicted and duly dealt with. Stage (d), the actual hands-on manipulation of insect pest populations, comes into play once the previous stages have been implemented. Figure 10.4 presents a flow chart which suggests a structure for the implementation of the various tactics shown in Fig. 10.3. The idea is that preventative techniques are backed up by monitoring, prediction and decision making, with control tactics being rolled out only if and when required. This system may seem fairly straightforward in concept but, unfortunately, there may be drawbacks or snags at each stage. Clarke (1995) lists five separate research and management problems with the use of IPM in tropical forestry: namely (i) defining economic injury levels of pests; (ii) defining economic injury levels of pest damage; (iii) carrying out monitoring procedures; (iv) implementing management tactics; and (v) operating under tight economic constraints. Defining economic injury levels and economic damage thresholds (which may or may not equate to similar things) can be problematic, since forests are long-lived ecosystems, and a number of external factors must also be taken into account, which include company and government policy, forest economics and environmental pressure. IPM relies heavily on monitoring to identify areas where pest populations are high and economic thresholds are likely to be exceeded. In large and inaccessible areas, such monitoring may be impractical, or at least inaccurate. Finally, and perhaps most importantly, almost all tropical forestry operates within very tight economic constraints as mentioned earlier, where profit margins may well be very low, or non-existent in the case of agroforestry, for example. There simply may not be enough money to pay for monitoring staff and the various tactics required for artificial controls.

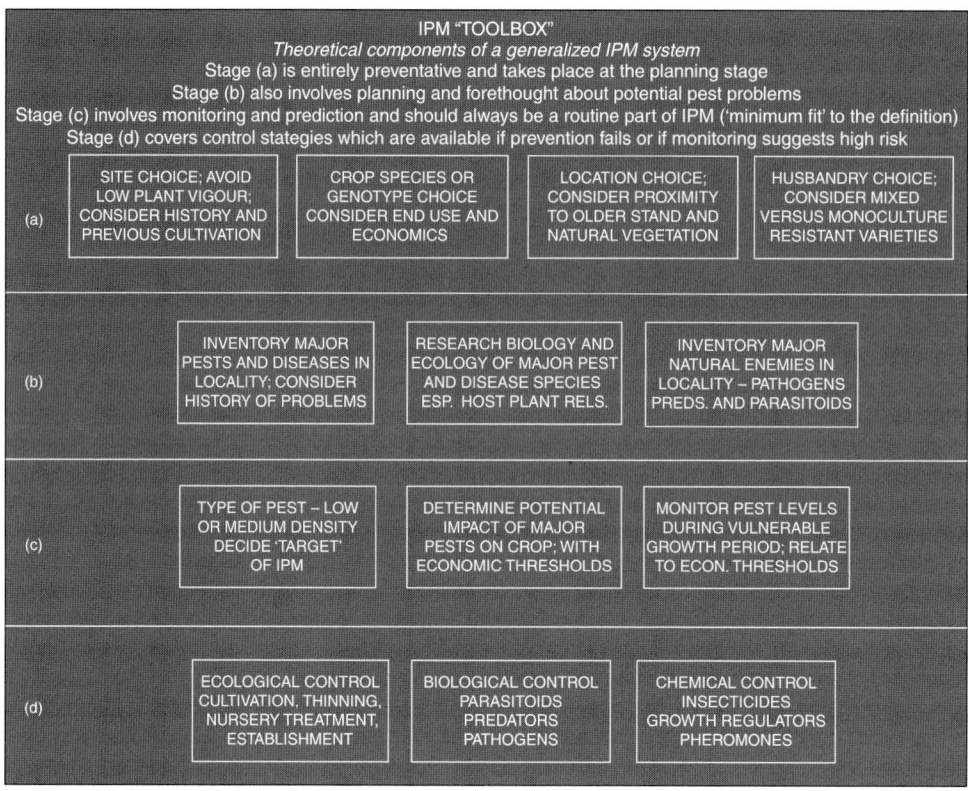

Fig. 10.3. Layers of tactics available for selection to be used in tropical forest IPM programmes – the so-called IPM Toolbox.

10.4 Advice and Extension

As is obvious from both Figs 10.2 and 10.3, IPM is complicated and requires much more expertise, planning and forethought than standard techniques of pest control, such as routine insecticide spraying, for example. To this end, advice from extension services is essential, if at all possible, where properly trained forest entomologists are on hand to provide inputs at every stage of forestry. The involvement of entomological (and pathological) extension services in appropriate strategies for all types of crop system have changed over the years. Even so, many developing tropical countries may well be unable to afford the luxury of sophisticated advice networks (see also Chapters 6 and 9), and this will be a particular problem for small-scale forest operations at a village or farm level. Nevertheless, larger-scale projects, particularly those funded by international agencies, should have no excuse not to invest in pest management expertise, preferably by well-trained local experts but, if necessary, by expatriate advisors. Even if personal visits by experts are precluded for practical or economic reasons, generally available reviews are becoming more widespread and very important. Just one example of this is provided by Mike Wingfield and Dan Robison, who published a review on the pests and diseases of *Gmelina arborea* (Wingfield and Robison, 2004). This document spans the globe wherever *Gmelina* is planted, reviewing past experiences and predicting future problems. Such articles are vital for planning new forest projects

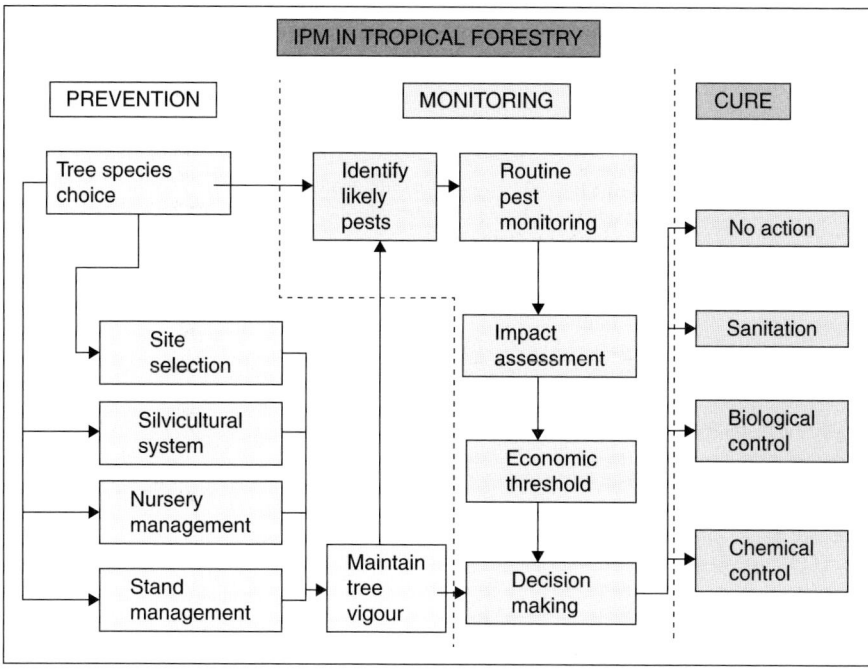

Fig. 10.4. Flow chart suggesting the order of implementation of various components of IPM in tropical forestry.

10.5 General Infrastructure

Successful and dependable IPM of tropical forest insect pests has to involve a much broader set of components as well as insect biology, ecology and impact, and Table 10.1 revisits a series of desirable external and internal contributions to the general infrastructure of IPM in the tropical forest context, first mentioned in Chapter 1. Many of these components have been alluded to in previous sections of this chapter, and the table summarizes them. Notice especially the requirement for international collaboration. As we mentioned in the latter sections of Chapter 6, insects do not acknowledge the existence of international boundaries, and experiences in one country may be of great value indeed to workers elsewhere, not just next-door neighbours but in some cases in countries widely separated geographically.

10.6 Economics and Impact Assessment

Lucas (2011), while reviewing the future of crop pest and disease management, commented that truly sustainable technologies must be 'affordable in the context of the local economy and crop value'. In other words, no amount of new and highly innovated research and development, especially at the molecular and genetic level, is of any use to tropical foresters if it is too complicated, too expensive or simply not available to them. Not only is IPM complex but also it can be expensive, and a fundamental decision must be made at the outset concerning its economic viability. Cost–benefit analyses will indicate whether the economic impact of insect pest attack

Table 10.1. Components required for efficient and successful forest pest management.

Background information	Impact data related to economic thresholds for serious pest and pathogen species, applying to local conditions
	Databases of major pest and pathogen species, their biology and ecology
	Databases of tropical tree species and their site requirements
Field assessments	Taxonomic and ecological data available for major pest and pathogen species to enable reliable identification followed by appropriate management strategies
	Ongoing research and development of economically viable, environmentally safe and technologically appropriate management tactics
	Provenance trials of important tree species in varied locations which assess the occurrence and impact of pests and pathogens
	Promotion of the use of indigenous tree species in well-matched sites and forest habitats
Infrastructure and support	Advisory and extension services available with sound knowledge of insect–tree interactions and potential control systems
	Efficient systems for reporting pest and pathogen problems to a central agency, with the capability of travel to the sites in question by trained entomologists and pathologists
	Provision of accessible plant health clinics backed up by national and/or international expertise and financial aid, where growers can obtain crop protection advice
	Incorporation of forest entomology and pathology into international aid or loan projects

in tropical forests outweighs the actual cost of management. In some cases, this may not be so, even when clear damage is occurring.

Take, for example, the situation described in Colombia by del Valle and Madrigal (1993). *Pinus patula* plantations were attacked by at least two species of defoliating moth caterpillars, *Oxydia trychiata* (Lepidoptera: Geometridae) in particular. Increment losses were assessed quantitatively over 4 years. Though some losses in diameter growth were to be expected, statistical analysis of the data showed that after 18 months post-defoliation, mean diameter growth rates were not significantly different. In fact, it was suggested that defoliated trees actually experienced enhanced growth rates due to the fertilization effects of pine needle debris and other detritus produced by insect activity in the forest canopies. Because the impact was in reality of little long-term consequence, economic analysis revealed that since the cost of IPM for defoliators in Colombia was between US$9 and US$16/ha, depending on which strategies were actually employed, then even if a reduction in pest problems of up to 90% could be achieved, IPM was not in fact an economically worthwhile operation.

It is highly desirable that such cost–benefit analyses are carried out in any afforestation project, and such an operation should be incorporated whenever possible. The crucial need for quantitative impact assessment must be addressed before any such analysis can be carried out.

10.7 Monitoring

We have discussed pest (and disease) monitoring elsewhere in this book and two examples here will suffice to highlight the requirement for such tactics to provide easily collectable information that provides a dependable prediction for pest problems in the future.

Ctenarytaina spatula (Hemiptera: Psyllidae) is a sap-feeding insect native to parts of Australia but now introduced as an exotic pest into various other tropical countries, including Brazil. This insect first appeared in Brazil in 1994, attacking drought-stressed *Eucalyptus grandis*, and it is now considered to be a potentially serious pest of many eucalypt species in plantations (de Queiroz *et al.*, 2010). A lot of work has been done on plant resistance (see Chapter 6) to the psyllid and some species of *Eucalyptus* have been found to be less susceptible than others. However, foresters and pest managers need to be able to predict the occurrence of the pest in forest stands and its likely severity. As Fig. 10.5 shows, egg counts can be used to predict the densities of nymphs, the damaging life stage of the insect, so that inspection by trained technicians can be used to form the basis of IPM tactics if required. Nurseries are one of the few forest growth stages where pest control may be really viable (see Fig. 10.1), since the trees are very small and in a very restricted area, thus allowing routine inspection and treatment if required. White grubs, the larvae of root-feeding scarabaeid beetles (see Chapter 5), are serious pests of forest nurseries in many parts of the tropics. Kulkarni *et al.* (2009) investigated the incidence of these pests in teak (*Tectona grandis*) nurseries in central India. Because of the relatively small areas of land involved, it was possible to estimate grub populations by searching through randomly selected nursery beds during and after the monsoon season (June–September) each year. Earlier work (Kulkarni *et al.*, 2007) has shown the relationship between white grub (in this case *Schizonycha ruficollis*) density and teak seedling damage (Fig. 10.6), which can be used to predict ensuing damage from early larval counts. Larval counts may also be backed up by using light traps for the night-flying adults, thus allowing nursery managers to act accordingly using chemical soil treatments to protect the highly valuable teak seedlings.

10.8 Examples of IPM in Tropical Forestry

The remaining sections of this chapter provide examples of IPM from tropical forestry. It must be borne in mind that many situations are still speculative rather than commercial and are not intended to provide practical solutions. Instead, they are presented as examples of how IPM systems may be constructed

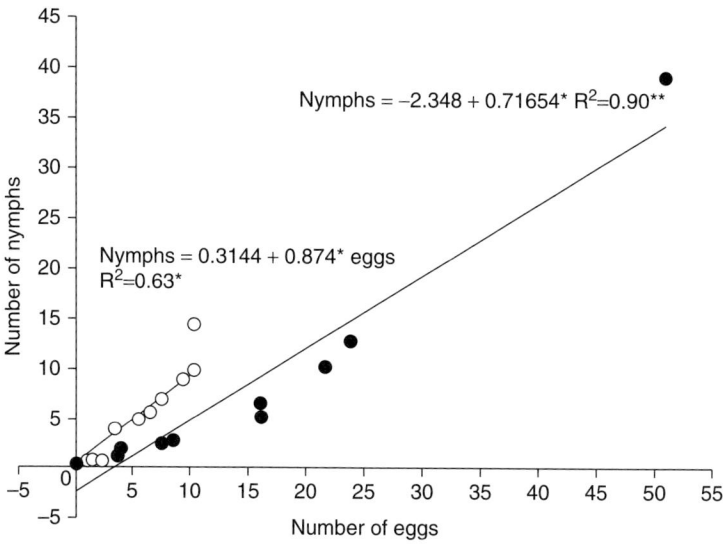

Fig. 10.5. Linear regression of the mean number of nymphs of the eucalypt-feeding psyllid *Ctenarytaina spatulata* against the mean number of eggs in winter (black circels) and spring (white circels) seasons in Brazil (from de Queiroz *et al.*, 2010, courtesy Sociedadae Brasileira de Entomologia).

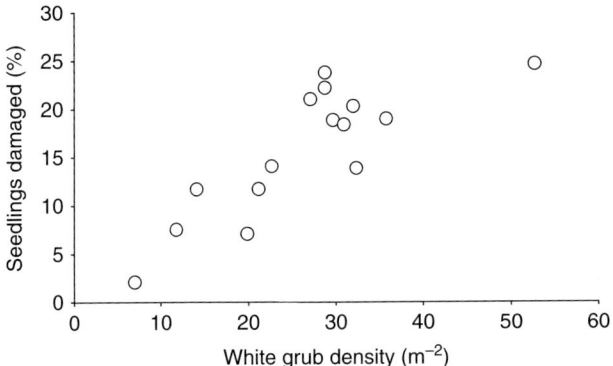

Fig. 10.6. Relationship between the density of *Schizonycha ruficollis* white grubs and damage to teak seedlings in nurseries in central India. R = 0.83, df = 13, $P < 0.01$ (from Kulkarni et al., 2007, courtesy Wiley-Blackwell).

to incorporate separate management components appropriate for the pest's biology and the crop situation. Details of all of these separate tactics have been discussed in Chapters 6–9, while many of the pests are described in detail in Chapter 5.

10.8.1 Example 1: *Dendrolimus punctatus* in pine plantations in Vietnam and China

Over the 1980s and early 1990s, hundreds of thousands of hectares of Vietnam have been planted with various pine species, including *P. massoniana*, *P. merkusii*, *P. kesiya* and *P. caribaea*. Though some early plantings have had to be replaced with eucalypts and acacias because of ruinous damage by shoot-boring Lepidoptera (Speight, 1996), large areas of pine plantations remain and continue to be established. One of their worst pest problems is defoliation by the Masson pine caterpillar, *Dendrolimus punctatus* (Lepidoptera: Lasiocampidae) (Plate 20), that can cause serious growth losses, especially to *P. massoniana* (Huang et al., 2008), which is indigenous to China. According to Billings (1991), for example, in 1987, 56,500 ha had been affected by the moth larvae, with 33,500 ha suffering from severe defoliation. Young trees may be defoliated completely, and even die as a result. Those between 7 and 15 years old are most commonly attacked, though Billings suggests that the worst affected are off-site, i.e. growing in sites to which they are not best suited, such as those located on hilltops or on shallow eroded soils, with little or no understorey vegetation. These areas may be best left to other tree species altogether, such as *Acacia mangium*, various eucalypts or the indigenous tree *Styrax*, which provides excellent timber and other raw materials. IPM of *D. punctatus* in Vietnam has in the past centred around the utilization of thousands of local people who were paid in kilogrammes of rice per day in exchange for insects cut and collected from infested pine trees. This arduous task was made all the worse by the fact that moth larvae possess severely irritating hairs that can cause skin rashes, prolonged irritation, headaches and even arthritis. In China, recourse to broad-spectrum insecticides applied through relatively old equipment was only partially successful, and new outbreaks of the pest occurred frequently. Unsurprisingly, resistance to various insecticide groups has also appeared (Wen et al., 2001).

Table 10.2 suggests a series of IPM measures which use a wide range of tactics. Monitoring with any degree of dependability is somewhat risky; pheromone trapping of adult males only works to a certain extent and though high trap captures faithfully predict larval outbreaks, some other outbreaks occur without having high male counts in pheromone

Table 10.2. Components of potential IPM programmes for the Masson pine caterpillar, *Dendrolimus punctatus*, in China and Vietnam, from the literature.

Tactic	Reference
Monitoring and prediction	
Monitoring adult densities with black light traps	Zhang *et al.* (2001)
Use sex-attractant pheromones for monitoring adult populations	Zhang *et al.* (2003)
Use airborne video to characterize defoliation intensities	Chen *et al.* (1997)
Predict damage intensity using satellite data	Zhang *et al.* (2005)
Use economic threshold of five larvae per tree	Li (2007)
Chemical control	
Aerially spray pyrethroid insecticides	Zhang *et al.* (1998)
Reduce use of synthetic insecticides	Li (2007)
Silviculture	
Enhance plant diversity in pine stands	Peng *et al.* (1996)
Plant *Pinus elliottii* instead of *P. massoniana*	Liang *et al.* (2008); He *et al.* (2007)
Mix pines with cedar or broad-leaved tree species	Wang *et al.* (2005)
Biological control – parasitoids	
Enhance populations of natural enemies by conserving understorey vegetation	CABI (2010)
Rear egg parasitoids (*Trichogramma dendrolimi*) in laboratory and release in stands	Xu *et al.* (2006)
Enhance knowledge and application of 78 parasitoid species (wasps and flies)	Xu *et al.* (2006)
Biological control – pathogens	
Apply fungus *Metarhizium anisopliae* in warm, dry conditions	Song (1997)
Apply fungus *Beauveria bassiana* in cooler, more humid conditions	Wang *et al.* (2004)
Apply bacterium *Bacillus thuringiensis* var *kurstaki* (Bt)	Chen *et al.* (1997)
Apply cytoplasmic polyhedrosis virus (CPV)	Zhao *et al.* (2004); Xiao *et al.* (2010)
Use egg parasitoids to disseminate cytoplasmic polyhedrosis virus (CPV)	Sun and Peng (2007)
Use transgenic *Pinus* species expressing Bt CRY1Ac gene	Tang and Tian (2003)
Other	
Mass trapping of adults with light traps	Zhao *et al.* (1998)
Use pupae as food for ethnic minorities	He *et al.* (1998)

traps beforehand (Zhang *et al.*, 2003). In general, outbreaks of the pest seem to be fairly erratic and hence hard to predict (Zhang *et al.*, 2008), though in some localities outbreaks appear every 3–5 years or so. Silvicultural techniques such as the use of mixed plantings seem to work as long as the goals of the forest industry can be met using less susceptible tree species. Biological control, at least on paper, seems to have most efficacy at preventing and/or controlling *D. punctatus* outbreaks. Parasitic hymenoptera and Diptera abound in the native pine forests of the region, and exotic stands can be inoculated or inundated with them. The egg parasitoid, *Trichogramma dendrolimi* (Hymenoptera: Trichogrammatidae), is particularly successful and can be reared in the laboratory with ease (Xu *et al.*, 2006). *Dendrolimus* has a well-known cytoplasmic polyhedrosis virus (CPV) associated with it, which can be bulked up in the laboratory using alternative insect host species such as the armyworm *Spodoptera exigua* (Lepidoptera: Noctuidae) (Xiao *et al.*, 2010). However, the broad host range of the virus could also be its downfall. It is reported to infect 35 species of insect in 10 families of Lepidoptera (Zhao *et al.*, 2004), including beneficial species such as silkworms, *Bombyx mori* (Lepidoptera: Saturnidae) (Hong *et al.*, 2003). For now, the most favoured pathogen is the fungus *Beauveria bassiana*, and indeed Li (2007) has called its use against

D. punctatus in China as 'one of the most successful biological control programmes in the world'. The timing and delivery of the fungus into infested stands of pine, once these have been located and identified as sufficiently serious as to warrant treatment, is still problematic however.

10.8.2 Example 2: Insect pests of eucalypt plantations in Brazil

Tropical eucalypts may yield up to 70 m^3/ha/year, with a rotation period of 7–20 years (Evans and Turnbull, 2004) and as such are planted over large areas in order to provide great quantities of timber rapidly. Brazil produced over 70 m m^3 of non-conifer timber (roundwood, pulpwood and sawlogs) in 2009 (FAO, 2009b), of which around 4 Mha consists of various species of *Eucalyptus* (Stape *et al.*, 2004). However, one of the most limiting factors is water supply (Stape *et al.*, 2010) and some species, *E. grandis* for example, is likely to be water stressed rather frequently, with consequent problems, as discussed in Chapter 3. First planted in Brazil in the early 1900s, these exotic trees have had plenty of time to 'accumulate' insect pests, both from the local indigenous fauna and also as introduced exotics. For example, Zanuncio *et al.* (2003) caught 437 species of adult Lepidoptera using light traps in plantations of *E. urophylla*, of which they rated 11 species to be primary pests and 18 species to be secondary pests. In a very similar study, this time in *E. grandis* plantations, 547 lepidopteran species were caught, with 13 and 20 primary and secondary pest species, respectively (Zanuncio *et al.*, 2006) (see also chapter 9) Flechtmann *et al.* (2001) caught 75 species of bark and ambrosia beetles (Coleoptera: Scolytinae) in *E. grandis* plantations using ethanol baited vane traps. So, there is no doubt that plantations of eucalypts in Brazil have high potential for pest problems and economic losses, which may be exacerbated as climate change effects become more intense (Ghini *et al.*, 2011).

The most important insect pests in eucalypt plantations in Brazil include *Thyrinteina arnobia* (Lepidoptera: Geometridae) (Santos *et al.*, 2000, Zanuncio *et al.*, 2003), *Stenalcidia grosica* (Lepidoptera: Geometridae) (Zanuncio *et al.*, 2006), *Euselasia eucerus* (Lepidoptera: Riodinidae) (Zanuncio *et al.*, 2009), *Gonipterus* spp. (Plate 15) (Coleoptera: Curculionidae) (Wilcken *et al.*, 2008), *Xyleborus* spp. (Coleoptera: Scolytidae) (Flechtmann *et al.*, 2001), *Syntermes* spp. and *Cornitermes* spp. (Isoptera: Termitidae) (Wilcken *et al.*, 2002), *Atta* spp. and *Acromyrmex* spp. (Plates 16 and 17) (Hymenoptera: Formicidae) (Ramos *et al.*, 2003), *Ctenarytaina spatulata* and *Glycaspis brimblecombei* (Plates 31 and 32) (Hemiptera: Psyllidae) (Santana and Burckhardt, 2007). Most of these species, such as *E. eucerus*, are native to Brazil, and some have transferred to eucalypts from their native hosts. *T. arnobia*, for example, has guava, *Psidium guajava*, as its native host but it grows significantly better on the non-coevolved eucalypts (Holtz *et al.*, 2003). A few species, such as the psyllids *C. spatulata* and *G. brimblecombei* and the weevils *Gonipterus* spp. are exotic introductions. Many of these pests can be found at the same time of the year (Fig. 10.7) (Torres *et al.*, 2006), so that overlapping pest outbreaks are to be expected, compounding the pest management problems.

As with all IPM systems, preventative measures should be used to combat insect pests in the eucalypt plantations. These involve: (i) detection of primary pest outbreaks; (ii) outbreak evaluations; (iii) analysis of results; and (iv) definitions of control strategies. Fundamental to all of this are field surveys and monitoring (Laranjeiro, 1994). Two, more detailed, examples from Brazil will illustrate these various tactics.

Field surveys are especially important for the early detection and evaluation of lepidopteran defoliators in the eucalypt stands. Once outbreaks take a hold, control becomes difficult and costly and growth losses will be unavoidable. Any effects of natural enemies are likely to be most prevalent at lower pest densities. *T. arnobia* populations need to be monitored in the Minas Gerais region of Brazil during the driest and coldest times of year, when peak outbreaks can occur (Zanuncio *et al.*, 2006), while other Lepidoptera species such as *S. grosica* should be monitored during the whole

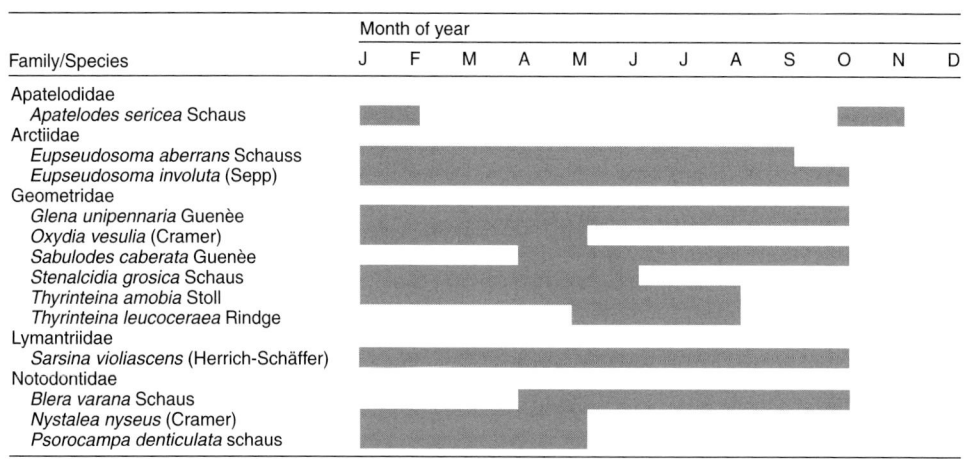

Fig. 10.7. Occurrence of 13 primary lepidopteran species defoliating *Eucalyptus* spp. in the states of Espirito Santo, Minas Gerais and São Paulo, Brazil (from Torres et al., 2006, courtesy CAB International).

growing cycle of eucalypts. *T. arnobia* had various potentially important parasitoids (Pereira et al., 2008), while *E. eucerus* had peak egg parasitism from *T. maxacalii* of nearly 50%, larval parasitism of an unidentified tachinid fly of 10% and pupal parasitism from the ichneumond, *Itoplectis* sp., of 52% (Zanuncio et al., 2009). Whether or not these biological control agents can prevent pest outbreaks most of the time will depend on other factors such as the influence of climate and host-tree factors on the lepidopteran population dynamics.

Psyllids are notorious sap-feeding pests of many tropical trees (see Chapters 5 and 6) and Brazil has seen the introduction of several species which have become serious pests in eucalypt plantations. Not only does the feeding activities of both nymphs and adults cause growth losses, loss of shoot apical dominance and dieback, sooty moulds grow on the honeydew produced by the pests, causing extra damage and yield losses (Santana and Burckhardt, 2007). Table 10.3 summarizes the various IPM tactics appropriate for psyllid control in Brazil. As mentioned above, the key to success is the reduction in susceptible hosts by choosing appropriate species and avoiding stress-inducing planting sites, by monitoring and prediction of pest occurrences and densities, by establishing self-sustaining biological control using native and introduced predators and parasitoids, and a final occasional recourse to insecticide treatment if really necessary.

10.8.3 Example 3: Insect pests of agroforestry poplars in North-west India

The tropic of Cancer is situated at 23.5°N of the equator and hence parts of north-west India are situated outside the tropics. However, forestry in this subtropical region has so many similarities and problems to that in the true tropics that we include some IPM experiences from there in this chapter. Poplars are vitally important for the rural economy as providers of shade, intercrops, multifarious timber products, fuelwood and so on, as components of agroforestry systems. Fast-growing exotic species of poplar such as *Populus deltoides* originated from the Mississippi Delta of the southern USA. Singh et al. (2004) describe the insect pest problems encountered in these agroforestry systems in Uttar Pradesh, Himachal Pradesh, Punjab and Kashmir. The most important pests of these exotic poplars in the region are larvae of the poplar leaf defoliators (Plate 123), *Clostera fulgurita* and *C. restitura* (Lepidoptera: Notodontidae) (Sangha, 2011), which can cause large-scale defoliation. Other defoliators, such as stem and shoot

Table 10.3. Various tactics employed in eucalypt plantations in Brazil to combat the psyllids *Ctenarytaina spatulata* and *Glycaspis brimblecombei* (from Santana and Burckhardt, 2007; Ferreira Filho *et al.*, 2008; Firmino-Winckler *et al.*, 2009; de Queiroz *et al.*, 2010; Silva *et al.*, 2010).

Tactic
Do not plant susceptible species such as *Eucalyptus grandis*, *E. urophylla*, *E. camaldulensis*
Avoid severe water stress of trees by site selection
Avoid continuous eucalypt plantations which facilitate the spread of the pest
Maintain natural vegetation around plantations to promote biological control
Monitor pest densities with sticky traps, manual collections or by beating foliage and branches
Monitor pest densities using egg counts related to subsequent numbers of damaging nymphs
Encourage native natural enemies (e.g. arthropod predators such as spiders and ladybirds)
Rear and introduce exotic parasitoids (e.g. *Psyllaephagus pilosus* from Australia)
Establish self-perpetuating populations of exotic parasitoids throughout eucalypt regions of Brazil
Reduce reliance on insecticides because of overlapping pest generations and side effects of toxins

borers, and root feeders, are also regular problems both in the nursery and in forest stands. Singh *et al.* (2004) describe IPM systems which include cultural methods, use of biopesticides, tolerant tree varieties, natural enemies and so on, and Table 10.4 summarizes these techniques used for different poplar pest species. Some of these tactics are more practical than others. Using resistant or tolerant trees, monitoring for pests in adjacent vegetation and alternative hosts, encouraging parasitoids and even applying defoliator-specific viruses are all sound and likely to have some efficacy. Light pruning of borer-infested trees is acceptable, though heavy pruning can seriously stress the trees and of course reduce wood biomass production significantly. Any heavy use of organophosphate and carbamate insecticides should be avoided. Clearly, to implement all these techniques will require rather sophisticated management systems, relying on the experience and expertise of local and regional growers.

10.8.4 Example 4: *Phytolyma lata* in *Milicia* plantations in West Africa

Iroko wood is an extremely valuable timber produced by the trees *Milicia excelsa* and *M. regia* in sub-Saharan Africa. The natural distributions of the two species overlap in parts of West Africa and most of the timber production in the region is from natural forests. Attempts to grow *Milicia* species in plantations are hampered severely by the so-called Iroko gallfly, a psyllid *Phytolyma lata* (Hemiptera: Psyllidae). This sap feeder attacks young trees and induces the formation of galls on buds, leaves and shoots. When the galls mature, serious dieback of terminal shoots can occur and seedlings may be killed (Cobbinah and Wagner, 1995). Most pest management systems revolve around silvicultural or genetic management of the host trees so that, for instance, the use of tree mixtures and shade planting may be useful (Bosu *et al.*, 2006). Ofori and Cobbinah (2007) set out a series of recommendations for *Phytolyma* control. Firstly, the use of insecticides is not encouraged, except in heavily attacked nurseries. For one thing, the pest is concealed inside the gall for much of its lifetime and therefore difficult to reach with chemicals and, in addition, since the pest is present all year round, routine chemical control would be likely to inundate ecosystems with undesirable toxins. Neem-based insecticides such as azodrin may, however, be acceptable. The genetic structure of *Milicia* populations and the relative resistance to the psyllid between *Milicia* species and genotypes holds promise. Cobbinah and Wagner (1995) report that some varieties or genotypes can reduce the growth and survival of the psyllid via antibiosis, others are simply tolerant of attacks, while a third group is highly susceptible. Once genetically based varietal resistance

Table 10.4. Pest management tactics for various insect pests in *Populus deltoides* agroforestry in north-western India (from Singh *et al.*, 2004).

Insect	Type of pest	Management tactics
Apriona cinerea (longhorn beetle)	Stem borer	Prune and burn attacked branches; destroy brash, storm damage and old stumps; remove alternative host plants such as *Morus* spp. and *Prunus* spp. in the locality
Eucosma glaciata (Lepidoptera)	Shoot borer	Prune and burn affected branches; foliar spray of insecticide (dimethoate) in May/June and August/September
Clostera cuprata, C. fulgurita, C. restitura, Phalantha phalantha (Lepidoptera)	Defoliators	Replace now susceptible clones of *P. deltoides* with tolerant ones; bulk up and spray *Clostera*NPV (virus); mass rear and release egg parasitoids (*Trichogramma* spp.); encourage predatory bugs (Hemiptera) in stands; spray neem oil after rains; monitor pest populations in alternate winter hosts (other poplars, *Salix* etc.) around stands
Eriosoma spp. (aphids)	Sap feeders or gallers	Encourage predatory beetles (ladybirds); remove galls manually in winter; plant resistant poplar hybrids
Holotrichia spp. (white grubs)	Root feeders	Do not weed nursery beds when adults are flying; clean cultivate and deep plough at other times; dress soil with slow release insecticide formulations
Odontotermes spp., *Coptotermes* spp. (termites)	Root and stem feeders	Apply insecticides in water or dust to soil or base of trees

has been detected, various vegetative techniques have been established to conserve and propagate resistant material for use in the field. The big risk then might involve significant reduction in *Milicia* genetic diversity. Finally, as mentioned above, there are silvicultural techniques such as mixed plantings with various other timber tree species such as *Albizia*, *Khaya*, *Terminalia* or even *Tectona* (Wagner *et al.*, 2008). Planting *Milicia* in the shade of other trees does reduce pysllid attack to some extent, but shade of course will also reduce tree growth (Bosu *et al.*, 2006). Table 10.5 presents a summary of IPM techniques compiled by Wagner *et al.* (2008).

10.8.5 Example 5: *Phoracantha* species in the Americas, Africa and Europe

The eucalyptus longicorn beetles, *Phoracantha semipunctata* and *P. recurva* (Coleoptera: Cerambycidae) (Plates 39 and 40), are now a very widespread pest of eucalypts over almost all of the tropical, subtropical and even Mediterranean regions of the world (Paine *et al.*, 2011), including Brazil, Uruguay, Paraguay, Chile, Tunisia, Morocco, Malawi, South Africa and so on. Their biology, ecology, impact and host-plant relationships have been detailed in many parts of this book, including Chapters 5 and 6. Studies of these sections would provide clear recipes for mainly preventative IPM tactics, which hinge on the use of resistant species of *Eucalyptus* on sites which are not prone to drying out. The avoidance of host-tree stress is paramount in avoiding problems.

Figure 10.8 depicts four categories of management tactic for the pests, three of which are essentially silvicultural in nature. Biological control, as discussed in Chapter 8, is the least likely to succeed, especially if the other three tactics are ignored or underplayed. Though biological control may be desirable, it seems not to be efficient enough in dealing with this low-density, mainly concealed pest. The egg parasitoid, *Avetianella longoi* (Hymenoptera: Encyrtidae), has been shown to reduce tree mortality to some extent (Wang *et al.*, 2008), while larval parasitoids such as *Syngaster lepidus* have been released in various countries. The situation has become more complicated

Table 10.5. Strategies to be considered for the IPM of *Phytolyma lata* on *Milicia* spp. in West Africa (from Wagner et al., 2008).

Strategy	*Phytolyma lata* stage affected	Type of control
Parasitism by *Trichogramma* sp.	Egg	Biological
Predation by mantid	First instar crawlers and, less so, exposed adults	Biological
Boring of all tissues followed by parasitism by hymenopteran parasitoids	Nymphs	Biological
Host selection – preliminary trials required to compare pest resistance in the two species of *Milicia*	Insect species as a whole through oviposition preference, establishment rate, growth rate, fecundity, etc.	Varietal resistance
Mixed planting	Adults	Cultural
Insecticide application (azodrin, bidrin, unden)	All stages	Chemical

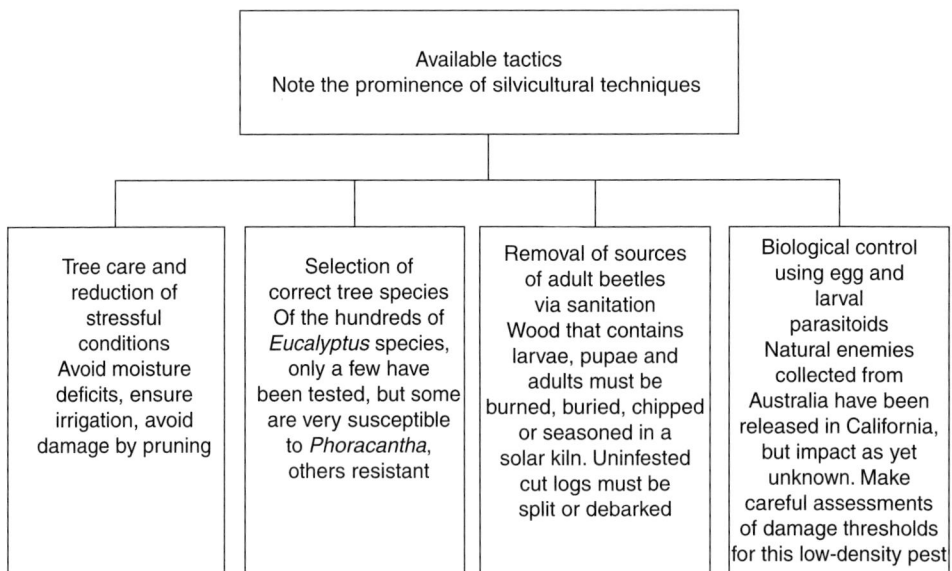

Fig. 10.8. IPM of eucalyptus longhorn beetles (from Paine et al., 1995).

with the establishment of *P. recurva*, which in places such as southern California has replaced *P. semipunctata* as the dominant pest species (Paine and Millar, 2002). Although it appears that *A. longoi* shows the same search efficiency for egg masses of both species, the rates of parasitism and rates of successful emergence are significantly lower in *P. recurva*. Thus, the parasite may be a critical driving force in the replacement of *P. semipunctata* in the environment where trees still get killed. In summary, parasitoids may both cure and cause a problem (T. Paine, 1999, personal communication).

10.8.6 Example 6: Root-feeding termites in nurseries and young forest plantations across the tropics

Termites (Isoptera: Termitidae) can do all sorts of damage to very young trees, growing trees, dead and dying trees and timber in service (i.e. wood and wood products used in construction), and we have discussed their pest status and management in various previous chapters of this book. We are mainly concerned here with the damage termites do to nursery stock and to seedlings in newly established plantations (Plates 115 and 116), and it is perhaps not surprising that these initial problems can be so severe, since forest nurseries may be thought of as agriculture crop-like in the density of plants and their susceptibility to pests and diseases, in which termites are more usually prevalent (Rouland-Lefevre, 2011). Normally, termites are thought of as being indigenous pests of exotic plant species, rather than being problematic with native tree species (Kumar and Thakur, 2011), so it is predominantly acacias and especially eucalypts which suffer most damage and death (Varma and Swaran, 2007). Just occasionally, subterranean termites such as *Odontotermes* spp. have been found to damage young indigenous tree species, so Kirton and Cheng (2007) found them attacking native dipterocarp species in the genus *Shorea* in enrichment plantings in Peninsular Malaysia. It has long been assumed that subterranean termites are most serious in plantations newly established after natural forest had been cleared, through the simple fact that their soil environment has been altered significantly by disturbance (Tsukamoto and Sabang, 2005) and they have been deprived of their normal and natural food supply, leaving them no choice but to feed on the new plantations (Sileshi *et al.*, 2009). In essence, termites in tropical forestry are usually thought of as pests on degraded sites rather than on natural ones (Kekeunou *et al.*, 2006).

So, can soil termites be managed in any real sense of IPM? In concept, there is no real difference between them and any other forest pests discussed in this book and here we mention these various potential tactics, assembled from the literature, and briefly summarize some of their drawbacks. If we take an aggregate view, the future could be said to be fairly bleak for the management of termite pests. There are too few practical, effective and ecologically and socio-economically acceptable tactics available these days, assuming we want to carry on planting exotic tree species on degraded lands.

However, Verma *et al.* (2009) have reviewed physical, chemical and biological methods as summarized in Fig. 10.9, only some of which are applicable to forest nursery and young plantation situations. The 'good old days' of long-lasting residual insecticides being used to form a chemical barrier around the roots of young trees (Wilcken *et al.*, 2002) have gone and the highly effective but otherwise undesirable organochlorine insecticides have been banned across the world (Peters and Fitzgerald, 2007), while the newer but only marginally less toxic and persistent organophosphates and carbamates have also mainly disappeared in the past few years. However, local people still view chemicals as the best and most efficient methods of termite control (Kagezi *et al.*, 2010) and undoubtedly, if we could find a commercially viable and environmentally acceptable chemical, this would still be the best way to protect our seedlings

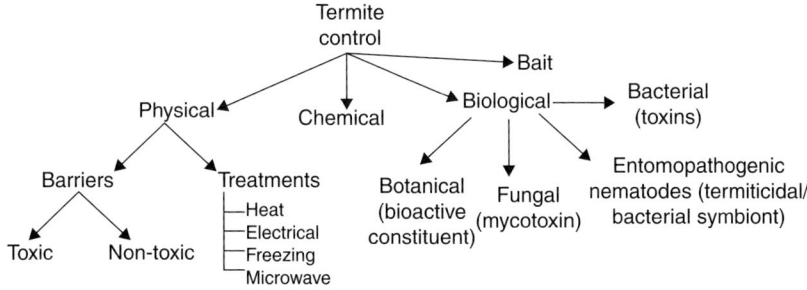

Fig. 10.9. Different types of control measures for termites (from Verma *et al.*, 2009, courtesy Elsevier).

(French and Ahmed, 2006). Plant-derived botanical insecticides may have some promise as toxins, repellants or feeding deterrants (Verma et al., 2009) and research into the performance of new products continues with compounds such as fipronil (trade name Regent or Termidor), which is a slow-acting poison used in baits (Huang et al., 2006), ivermectin, which is normally used to control internal mammalian parasites (Jiang et al., 2011), and bifenthrin, a fairly standard pyrethroid insecticide engineered to have long-lasting stability and low toxicity to non-target species by combining (immobilizing) it with silk (Guan et al., 2011). It remains to be seen whether or not chemicals such as these will ever be effective, cheap and universally available enough to be used across the world.

Most exotic tree species of any industrial use seem to be susceptible in their young stages, so there would appear to be little future in tree breeding, selection or indeed genetic modifications, which leaves biological control. Termite predators and parasitoids, though abundant in nature, are not efficient enough to prevent damaging attacks to small trees in nurseries and young plantations. Pathogens such as viruses or bacteria would be difficult to get to the main colonies of termites underground or under bark, even if strains could be found that were pathogenic to termites, which leaves fungi. Entomopathogenic fungi such as *B. bassiana* and *Metarhizium anisopliae* are well-known biological control agents of many pests, especially in humid conditions, and both are well able to kill termites (Wright et al., 2002). Because of the feeding behaviour of termites, tropholaxis (sharing of foods among social individuals) enables the distribution of fungal spores throughout a colony prior to epizootics which can kill large numbers of insects. There are many isolates and strains of pathogenic fungi which are still being tested for certain types of habitat and application, since termites as a group live in a wide variety of environments with different conditions (Wright et al., 2004). As well as being lethal pathogens, some fungi such as *M. anisopliae* can also act as deterrents, since foraging worker termites have been shown to be repelled by mulches and baits containing the fungus (Sun et al., 2008). Perhaps ironically, this behaviour may result in fewer termites coming into contact with the fungus and thus dying; the control mechanism is therefore not toxicity (pathogenicity) but repellence (Hussain et al., 2010). These variations in species, isolate, strain and environmental efficiency in termite-pathogenic fungi may make the choice of a commercial and wide-ranging effective product problematic. Until easily applicable and dependable IPM tactics for termites in tropical forestry become available, if they ever do, then foresters will have to tolerate some losses of certain tree species on certain sites and plant extra accordingly, change species, or go elsewhere.

10.8.7 Example 7: Eucalyptus grasshoppers in Paraguay

Unlike the situation in Brazil described earlier in this chapter, the neighbouring country of Paraguay has only recently begun to grow species of *Eucalyptus* on an industrial scale. Here, large areas of *E. grandis* were being established in the 1990s on old grassland or ex-agricultural sites (Plate 124), and one pest problem rapidly appeared which involved an originally unidentified species of grasshopper (Orthoptera: Acrididae) (Henson and Davies, 1999, unpublished; Speight, 1999, unpublished). In certain sites, severe losses of up to 70 or 80% were recorded in *Eucalyptus* transplants during the first 3 months after establishment. Older trees were not affected, but the younger ones had their bark chewed away close to the soil surface by grasshoppers, causing ring-barking (girdling) followed by stem breakage and tree fall (Plate 125). The grasshoppers (Plate 126) appeared to be common in the eastern region of the country only, close to the Brazilian border, and were especially prevalent in planting sites with a previous history of soybean cultivation. Even here, lines of eucalypts planted between strips of *Brachiaria* grass appeared to be much less attacked, as did trees removed from the nursery after sufficient time had been allowed for them to produce tougher, grey bark as opposed to the very soft and presumably palatable red/green bark typical of very young trees. Crucial taxonomy, both in Brazil and at the Natural History Museum in London, identified the genus as *Baeacris* (J. Marshall, personal communication), which allowed essential comparisons with the published literature on host plants, life cycles and pest status. The main member of the genus, *B. punctulatus*, is distributed fairly widely across the pampas regions of Argentina and Brazil (Cigliano *et al.*, 2000), and it is thought to be a potentially important pest of pasture and pampas grassland in South America in general (Michel and Teisaire, 1996). In Brazil at least, it is reported to be an important pest of agricultural crops, especially soybean during the summer months (November–March) (Dias-Guerra *et al.*, 2010), thus establishing the soybean connection found in the newly afforested fields in Paraguay. Four or even five generations per year may be achieved by the pest in optimal conditions, so with plenty of food and warm, wet climatic conditions there was clearly great potential for pest proliferation.

Commercially viable pest control tactics were limited. The microsporidian pathogen, *Nosema locustae*, was introduced into Argentina in the late 1970s to control grasshoppers, and field surveys in the mid-1990s found that *B. punctulatus* exhibited between 1 and 19% infection (Lange and de Wysiecki, 1996). However, biological control was not considered viable for the problem in Paraguay due to the scale and cost of the treatment likely to be required. Insecticidal treatment was also felt to be in the main ineffective, again due to scale, cost and also potential side effects, though dipping of transplants in a contact chemical might have helped. Prevention was considered the best way forward and Fig. 10.10 summarizes

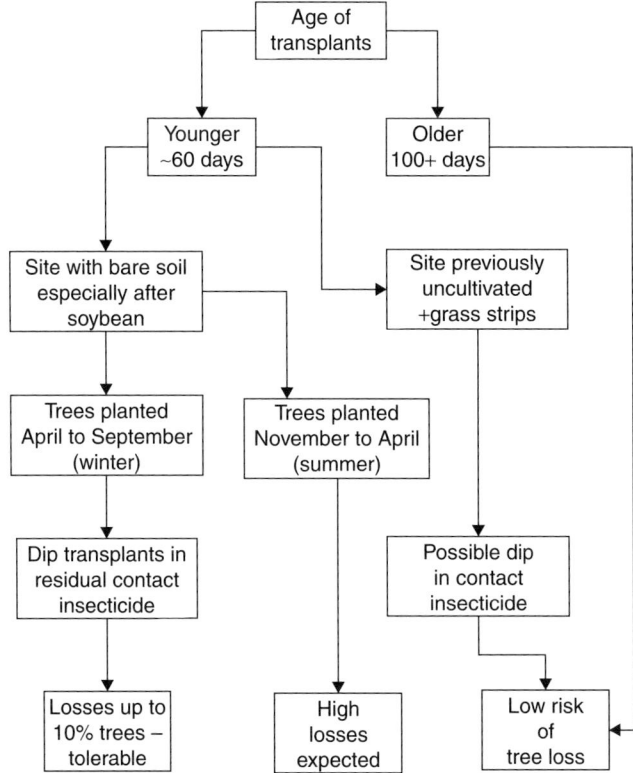

Fig. 10.10. Flow chart illustrating a speculative decision-making system for grasshopper pest management in young *Eucalyptus grandis* plantations in Paraguay (from Speight, unpublished).

some speculations about the various elements leading to grasshopper damage which have control potential. The most likely approach was thought to be the use of older, more resistant (or tolerant) transplants, established in less risk-prone sites, during times of the year when the pest was less active. The major change in tactics would involve considerable alterations to nursery practice, with consequent reductions in productivity and turnover of nursery stock.

10.8.8 Example 8: *Hypsipyla* spp. in mahogany plantations in the Old and New World

As we have discussed in various chapters in this book, the mahogany shoot borers, or tip moths, belonging to the genus *Hypsipyla* (Lepidoptera: Pyralidae) (Plates 57–59) are one of the most intractable insect pest problems in tropical forestry. Undoubtedly, if it were possible to grow *Swietenia* spp., *Cedrela* spp., *Toona* spp. or *Khaya* spp. successfully in some form of at least semi-industrial silviculture, then huge amounts of revenue could be earned by producing countries and much less pressure would be placed on natural tropical forests via the often illegal harvesting of native mahoganies. Chapters 5 and 6 provide details of the shoot borer's ecology and impact, showing that the aim of IPM for *Hypsipyla* spp. must centre around the protection of young trees from the nursery stage up until they reach around 4 or 5 years of age, when their straight main stems have reached a final marketable length. After this stage, shoot borer attack will no longer affect the shape of the trunk detrimentally, and then the main problem with maturing trees is that they are able to act as pest reservoirs for neighbouring younger and vulnerable plantations. Plates 127 and 128 show examples of African mahogany, *K. anthotheca*, illustrating the sort of tree that can be produced when not attacked by the borer and the forking shape produced after attack, respectively.

The shoot borer problem has confronted tropical foresters for as long as they have tried to grow mahoganies. Silviculture has tended in recent years to concentrate on possible methods of cultivation of mahoganies in relatively natural forests, with attempts to grow the timber in more normal (at least in terms of plantation silviculture) plantation systems (Mayhew and Newton, 1998). Whichever cultivation system is employed, attempts at *Hypsipyla* management reached their heyday in Costa Rica in the 1970s, where Grijpma and colleagues, working at the Centro Agronomico de Investigacion y Enseflanza (CATIE) in Turrialba, produced a 'shopping list' of research topics required, which included host-plant relationships and resistance, sex attractants, chemical control, insect taxonomy and biological control (Speight and Cory, 2000). A wide variety of suggestions for pest management were provided, but little actual progress was made. It is a somewhat salutary tale to realize that despite all these trials and experiments, the pest remained as serious as ever. Interest in *Hypsipyla* IPM was reignited in the 1990s, fuelled by workers in the UK and Australia, and as well as research and review publications such as those of Newton *et al.* (1993, 1998), and an international workshop was organized in Sri Lanka in 1996 (Floyd and Hauxwell, 2000). Large bodies of information are still controversial, conflicting or simply lacking. For example, we still know very little about the ecology and host-plant relationships of *Hypsipyla* species in natural forest. Until we are able to consider pest dynamics in terms of coevolved mechanisms, the underlying causes of shoot borer problems will remain a mystery. Various workers have suggested links between pest infestation, damage intensity and site conditions (see Newton *et al.*, 1998, Chapter 6). In Venezuela, Briceno-Vergara (1997) suggests that part of an IPM package for *Hypsipyla* attacking *S. macrophylla* should contain site selection of plantations with well-drained soils. This conflicts with observations, so far unpublished, for *Khaya* in Mozambique (Alves *et al.*, 1999, unpublished) and for *Swietenia* in Sabah (East Malaysia) (Speight and Chey, 1998, unpublished). In both the latter cases, trees appear straighter, taller and generally less damaged in wet valley bottoms.

Planting regimes continue to be trialled in various countries. In Martinique and French Guinea, greater success in growing mahoganies despite the ravages of *Hypsipyla* spp. are achieved using natural regeneration of *Swietenia* (Plan and Vennetier, 1998), while in Sri Lanka, extensive fieldwork is looking at shade interactions and outbreak intensity (R. Mahroof and C. Hauxwell, 1999, unpublished). In Australia, large trials, mainly of *T. ciliata*, have been established to examine genetic resistance, to revisit silvicultural techniques such as underplanting and planting and to test fertilizer–novel insecticide combinations to boost the growth of young trees and at the same time to protect them against *H. robusta* in the crucial early years. Thus far, no concrete management recommendations can be made due to the complex interactions discovered.

Additionally, the taxonomy of mahogany shoot borers is being unravelled. The idea of *H. robusta*, for example, as a single species with a distribution extending from Australia, the Pacific and Asia to Africa, has often been questioned and molecular techniques have shown that the genus *Hypsipyla* is not one but at least eleven species, four from the Americas and seven from Asia and Africa (Horak, 2001), which may explain some of the differences noticed in the behaviour and biology of the insect in different parts of its range.

Since the first edition of this book was published in 2001, work has continued around the world to investigate possible ways of either reducing the attacks of *Hypsipyla* or developing resistance or tolerance in the host trees. In 2003, Floyd *et al.* summarized the list of potential pest management methods (Table 10.6). Some of these suggestions, such as the use of pheromones, still need developing, so the major topics are: (i) genetics and host-tree susceptibility; (ii) tree manipulations (site choice, pruning, etc.); (iii) silviculture (shade, mixtures, etc.); (iv) chemical control (insecticides and repellants/deterrents); and (v) biological control (predators, parasitoids and pathogens), and we present some details of each one in turn.

Genetics and host tree susceptibility

There are many species of tree in the subfamily Swietenioidea of the family Meliaceae scattered across the tropics, most of which produce valuable hardwood timber. For example, *S. macrophylla* and *Cedrela odorata* are native to the Americas, *K. senegalensis* and *K. anthotheca* are from Africa, and *T. ciliata* is from Australasia, all of which are attacked by *Hypsipyla* virtually everywhere they are planted, except on the occasions where the insect is completely absent from an isolated locality, such as the city of Brasilia in Brazil or the island of Fiji. There has been some support for the concept that tree species growing in their native country or region are attacked more severely than exotic species (Cunningham *et al.*, 2005). However, this knowledge alone is not enough to provide a tactic which can prevent economically serious damage. As we have discussed in Chapter 6, there are compounding problems of environment versus genetics in all field trials to establish resistance or tolerance in certain tree genotypes or varieties, but much recent work has found genetic variation in attacks by shoot borer, and tree tolerance or recovery;

Table 10.6. Pest management tactics for mahogany shoot borers (from Floyd *et al.*, 2003).

Host tree resistance
Mixed species plantations and agroforestry
Biologically active compounds (kairomones and novel insecticides)
Site manipulation, particularly soil nutrients and drainage
Manipulation of the action of predators
Introduced pathogens and parasitoids
Traditional insecticides

for example, Alves (2004) in Mozambique, Ofori et al. (2007) in Ghana, Ward et al. (2008) in Mexico, Cornelius (2009) in Costa Rica, Underwood and Nikles (2009) in Australia, Soares et al. (2010) in Brazil, and so on. Unfortunately, there still seems to be no firm and dependable answer to the question of which tree species, provenance, etc. to grow where, and growers should be advised to look carefully at trials carried out in their own regions rather than to rely on research results from further afield. Also note that cloning tree genotypes that show resistance to the shoot borer may be able to produce large numbers of healthy seedlings in vitro (Pena-Ramirez et al., 2010), but how long this resistance with its very narrow genetic diversity will remain effective is uncertain.

Tree manipulations (site choice, pruning, etc.)

We discuss the crucial requirement to match tree species to planting site in Chapters 3 and 6, and the situation with the mahogany shoot borer is no exception. For example, the biggest African mahogany, K. anthotheca, specimens were found in Mozambique growing near the bottom of river valleys in forest remnants (Alves, 2004), while young K. anthotheca grow best on sandy, slightly acidic soils. Other species in other countries have similar specific requirements which should be adhered to wherever and whenever possible.

Since *Hypsipyla* larvae tunnel into and kill leading shoots, resulting in multiple leaders subsequently developing, hand pruning back to a single leader is possible. Cornelius (2001) suggests that young trees attacked by the borer can be pruned back to just below the insect's tunnel to remove all trace of the pest and then later, if required, a second pruning can produce the single leader required. When pruning is combined with genetic variations in tree form and growth, commercially acceptable trees may be produced (Cornelius and Watt, 2003), as long as there is the required manpower for this labour-intensive operation. Though not a likely strategy on a commercial scale, Perez et al. (2010) have shown that grafting relatively resistant shoot and leaf material from K. senegalensis on to relatively susceptible root stock (S. macrophylla) affects the growth and survival of pest larvae adversely.

Silviculture (shade, mixtures, etc.)

Planting mahoganies in monoculture plantations seems to be destined to failure almost everywhere it has been tried, and the use of diverse plantings using weeds, gap or line enrichment, mixed-species stands and so on seems much the better tactic. Note, however, that mixed plantings with two susceptible species such as *Swietenia* with *Cedrela* has no effect, since the insect pests seem to consider them both suitable hosts (Perez-Salicrup and Esquivel, 2008). The crucial 'trick' is to reduce the ability of the adult female shoot borer moth to find mahoganies concealed by non-suitable species, presumably a coevolutionary strategy in natural forests. Enrichment planting is a general afforestation technique wherein young trees are planted in lines or gaps in usually natural forests. The older forest provides species mixtures and shade (see below). In some ways, this system is the same as natural regeneration in forest gaps and so should be able to protect young trees from pests as they grow beyond their susceptible stage. It certainly seems to work. In Brazil, Lopes et al. (2008) succeeded in growing S. macrophylla with significantly less attacks from *Hypsipyla* and significantly taller, faster growth by planting the young trees in forest gaps created by selective logging. Once established in forest clearings, Snook and Negreros-Castillo (2004) advocate no cleaning or weeding in the plots, which also reduces borer attacks. The major drawback with these enrichment planting scenarios is, of course, their timber yield potential, which is very much lower per unit area of ground than a plantation would be if it were not for the pest.

One interesting tactic to reduce shoot borer damage with mixtures is to establish a mahogany species in a mixture with another commercial crop, as for example with C. odorata planted between rows of coffee bushes in the Yucatan Peninsular of Mexico (Navarro et al., 2004). Using provenly resistant *Cedrela* provenances within coffee plantations, it was possible to provide acceptable trees producing not only good timber but also good seed. Even though coffee bushes have a much shorter useful life than the *Cedrela* trees, the coffee should be able to provide the species mixtures needed to deter the shoot moth for the crucial early 3 or 4 years of life. Contrastingly, Plath *et al.* (2011) did not find any significant differences in height growth in young *Cedrela* planted in monoculture or mixed with two other tree species.

A final tactic that tries to combine aspects of plantation silviculture with enrichment-type planting is the use of shade. Several authors have shown that planting mahogany trees under the shade of other trees reduces shoot borer attack. Mahroof *et al.* (2002), for instance, working in Sri Lanka, found that the oviposition rate of *H. robusta* was much lower on *S. macrophylla* planted under artificial shade, as was the survival of larvae once established in shoots. Opuni-Frimpong *et al.* (2008a) carried out field trials in Ghana using different shade levels and found that, indeed, they could reduce *H. robusta* attacks on *K. anthotheca* to virtually zero under deep shade, but unfortunately tree growth was also seriously compromised. This is a clear example of the possibility of a compromise position between too much insect damage and too much tree growth suppression.

Chemical control (insecticides and repellants/deterrants)

Foresters have been trying to control mahogany shoot borers with insecticides for almost a century, with little success (Wylie, 2001). It must be remembered that the pest is archetypal low density, i.e. one attack is essentially too many, coupled with the fact that once established, the larva is concealed from external influences such as poisons (and indeed natural enemies – see next section) in its tunnel. Most contact insecticides therefore have little effect. One or two products, deltamethrin for example, have been shown to have some impact on shoot borer larvae (Mancebo *et al.*, 2002) and biopesticides such as neem extracts either kill or deter the pest (Goulet *et al.*, 2005). Other plant chemicals such as extracts of rue (*Ruta chalepensis*) have so-called phagodeterrant properties (Barboza *et al.*, 2010). It seems unlikely, however, that these products will ever be a practical solution on a large plantation. Not all chemicals are poisons, and the sex-attractant pheromones have been discussed as pest management tools in Chapter 8. *H. grandella* and *H. robusta* are listed on Pherobase (2011) as having their sex pheromones identified, but surprisingly, very little research seems to have been done on the use of these chemicals for mahogany shoot borer management since the early 1990s.

Biological control (predators, parasitoids and pathogens)

Floyd *et al.* (2003) list various species of parasitic hymenoptera that have been reared from *Hypsipyla* larvae, but none seem capable of either becoming established in mahogany plantations or of reducing the levels of pest attack to a usable level. Lim *et al.* (2008) advocate the addition of weaver ant nests to mahogany trees in Malaysia attacked by shoot moth larvae so that the predatory ants can kill and eat the pest, but again this technique is not likely to reach commercial or worldwide success. Hauxwell *et al.* (2001) have spent a lot of effort trying to identify *Hypsipyla*-specific pathogens, viruses in particular from the wild, but with little or no success. *Bacillus thuringiensis* (Bt) formulations do work to some extent (Goulet *et al.*, 2005), but only if the larvae can be exposed to the toxin, and no records are as yet available where Bt-transgenic trees have been developed and trialled against shoot borer.

So where does all this leave us? Nothing seems to be definitively successful, though the future for plantation-grown mahoganies, if there is one at all, probably lies with tree selection and breeding, silviculture and tree manipulation. Perhaps the final and conceptually best tactic would be to ignore the insect altogether and instead to allow it to do its worst, but on trees that are so tolerant that they are able to shrug off any attacks and grow a single, straight and true leader in the presence of previous borer depredations. Such super-trees have yet to be discovered or developed commercially.

Bibliography

Abbott, I. (1993) Review of the ecology and control of the introduced bark beetle *Ips grandicollis* (Eichhoff) (Coleoptera: Scolytidae) in Western Australia, 1952–1990. *CALMScience* 1, 35–46.

Abbott, I., Smith, R., Williams, M. and Voutier, R. (1991) Infestation of regenerated stands of karri (*Eucalyptus diversicolor*) by bullseye borer (*Tryphocaria acanthocera*, Cerambycidae) in Western Australia. *Australian Forestry* 54, 1–2.

Abbott, K.L. and Green, P.T. (2007) Collapse of an ant-scale mutualism in a rainforest on Christmas Island. *Oikos* 116, 1238–1246.

Abood, F., Bajwa, G.A., Ibrahim, Y.B. and Rasip, A.G.AB. (2008) A review of insect defoliators on teak with special reference to *Paliga damastesalis* Walker in Malaysia. *The Malaysian Forester* 71(1), 1–10.

Abraham, C.C., Joy, P.J. and Cheeran, A. (1978) Occurrence of *Remphan* sp. (Coleoptera: Cerambycidae: Prioninae) as a pest of live *Erythrina indica* standards in pepper plantings. *Journal of Plantation Crops* 6, 97–98.

ACIAR (2010) Establishing pest detection systems in South Pacific countries and Australia. Final Report of ACIAR Project FST/2004/053. Australian Centre for International Agricultural Research, 51 pp.

Agbidye, F.S., Ofuya, T.I. and Akindele, S.O. (2009) Marketability and nutritional qualities of some edible forest insects in Benue State, Nigeria. *Pakistan Journal of Nutrition* 8(7), 917–922.

Agboola, D.A. (1990) The effect of some antitranspirant concentrations on 2–4 month old seedlings and seedling gall formations of *Militia excelsa*. *Indian Journal of Plant Physiology* 33, 294–299.

Agounke, D., Agricola, U. and Bokonon, G.H.A. (1988) *Rastrococcus invadens* Williams (Hemiptera, Pseudococcidae), a serious exotic pest of fruit trees and other plants in West Africa. *Bulletin of Entomological Research* 78, 695–702.

Aguirre-Castillo, C. (1978) Control of the leaf-cutting ant *Atta* spp. in forest plantations. Meeting of IUFRO Working Parties S2.06.12 and S2.07.07, 'Pests and Diseases of Pines in the Tropics', 3–14 September, 'Piedras Blancas', Medillin, Columbia.

Agyeman, V.K., Ofori, D.A., Cobbinah, J.R. and Wagner, M.R. (2009) Influence of *Phytolyma lata* (Homoptera: Psyllidae) on seedling growth of *Milicia excelsa*. *Ghana Journal of Forestry* 25, 28–39.

Ahmad, M., Bhandari, R.S. and Prasad, L. (1996) Insect problems in agro-forestry tree species and their management. *Van Vigyan* 34, 84–99.

Ahmad, S., Ahmad, R. and Ghani, M.A. (1977) Notes on the incidence and biology of some parasites of the pine cone-borers in Pakistan. *Entomophaga* 22, 107–110.

Ahmed, B.M., French, J.R. and Vinden, P. (2004) Review of remedial and preventative methods to protect timber in service from attack by subterranean termites in Australia. *Sociobiology* 44, 297–312.

Ahmed, S.I. (1995) Investigations on the nuclear polyhedrosis of teak defoliator, *Hyblaea puera* (Cram) (Lep., Hyblaeidae). *Journal of Applied Entomology* 119, 351–354.

Ahmed, S.I. and Sen-Sarma, P.K. (1983) On seasonal variation in population of *Calopepla leayana* Latr. (Coleoptera: Chrysomelidae). Proceedings of the Symposium on Insect Ecology and Resource Management, pp. 83–88.

Ahmed, S.I. and Sen-Sarma P.K. (1990) Bionomics of *Calopepla leayana* Latr. Coleoptera: Chrysomelidae, a serious defoliator of *Gmelina arborea* Roxb. plantations in India. *Indian Forester* 116(1), 71–82.

Ahuja, M.R. (2009) Transgene stability and dispersal in forest trees. *Trees – Structure and Function* 23, 1125–1135.

Akanbi, M.O. (1971) The biology, ecology and control of *Phalanta phalanta* Drury (Lepidoptera: Nymphalidae), a defoliator of *Populus* spp. in Nigeria. *Bulletin of the Entomological Society of Nigeria* 3, 19–26.

Akanbi, M.O., Ashiru, M.O., Raske, A.G. and Wickman, B.E. (1991) Towards integrated pest management of forest defoliators: the Nigerian situation. Proceedings of a symposium, Towards Integrated Pest Management of Forest Defoliators, held at the 18th International Congress of Entomology, Vancouver, Canada in 1988. *Forest Ecology and Management* 39,1–4, 81–86.

Alelign, A., Yemshaw, Y., Teketay, D. and Edwards, S. (2011) Socio-economic factors affecting sustainable utilization of woody species in Zegie Peninsula, northwestern Ethiopia. *Tropical Ecology* 52, 13–24.

Alfaro, R.I., Humble, L.M., Gonzalez, P., Villaverde, R. and Allegro, G. (2007) The threat of the ambrosia beetle *Megaplatypus mutatus* (Chapuis) (=*Platypus mutatus* Chapuis) to world poplar resources. *Forestry* 80(4), 471–479.

Ali, M.S. and Chaturvedi, O.P. (1996) Major insect pests of forest trees in North Bihar. In: Nair, K.S.S., Sharma, J.K. and Varma, R.V. (eds) *Impact of Diseases and Insect Pests in Tropical Forests*. Proceedings of the IUFRO Symposium, 23–26 November 1993, Peechi, India, pp. 464–467.

Allan, R., Morris, A. and Carlson, C. (2000) Survival and early growth effects of some re-establishment practices with *Pinus patula*. *Southern African Forestry Journal* 187, 29–36.

Alleck, M., Seewooruthun, S.I. and Ramlugun, D. (2005) Cypress aphid status in Mauritius and trial releases of *Pauesia juniperorum* (Hymenoptera: Braconidae), a promising biocontrol agent. MAS 2005, Food and Agricultural Research Council, Réduit, Mauritius, pp. 316–321.

Allen, W.A. and Rajotte, E.G. (1990) The changing role of extension entomology in the IPM era. *Annual Review of Entomology* 35, 379–397.

Almeida, W.R., Wirth, R. and Leal, I.R. (2008) Edge-mediated reduction of phorid parasitism on leaf-cutting ants in a Brazilian Atlantic forest. *Entomologia Experimentalis et Applicata* 129, 251–257.

Alves, M.T.A.F. (2004) The potential for the domestication of *Khaya anthotheca* in Central Mozambique. Unpublished DPhil thesis, University of Oxford.

Ananthakrishnan, T.N., Gopichandran, R. and Gurusubramanian, G. (1992) Influence of chemical profiles of host plants on the infestation diversity of *Retithrips syriacus*. *Journal of Biosciences Bangalore* 17, 483–489.

Anderson, C. and Lee, S.Y. (1995) Defoliation of the mangrove *Avicennia marina* in Hong Kong – cause and consequences. *Biotropica* 27(2), 218–226.

Anderson, J.B. and Brower, L.P. (1996) Freeze-protection of overwintering monarch butterflies in Mexico: critical role of the forest as a blanket and an umbrella. *Ecological Entomology* 21, 107–116.

Andrew, R.L., Wallis, I.R., Harwood, C.E. and Foley, W.J. (2010) Genetic and environmental contributions to variation and population divergence in a broad-spectrum foliar defence of *Eucalyptus tricarpa*. *Annals of Botany* 105, 707–717.

Angulo, A.O. and Olivares, T.S. (1991) *Chilecomadia valdiviana* (Philippi) (Lepidoptera: Cossidae) associated with *Ulmus glabra* Hudson variety (Laud.) Rehder ("*Ulmus pendula*") in the VIII region (Conception, Chile). *BOSQUE* 12(1), 67–68.

Anjos, N. dos, Ludwig, A., Santos, G.P. and Moreira, J.F. (1981) Efficiency of three chemical products and one mechanical measure for the control of cutworms attacking seedlings of *Eucalyptus grandis* Hill ex Maiden. *Revista Arvore* 5, 218–223.

Anjos, N. dos, Santos, G.P. and Zanuncio, J.C. (1987) The eucalyptus defoliator *Thyrinteina arnobia* (Stoll 1782) (Lepidoptera: Geometridae). Boletim Technico, Empresa de Pesquisa Agropecuariade Minas Gerais, No. 25.

Annamalai, R., Surendran, C., Thirumurthi, S. and Gunasekharan, V. (1996) Insect damage to neem in Tamil Nadu, India. In: Nair, K.S.S., Sharma, J.K. and Varma, R.V. (eds) *Impact of Diseases and Insect Pests in Tropical Forests*. Proceedings of the IUFRO Symposium, 23–26 November 1993, Peechi, India, pp. 349–355.

Anon. (1997) Unwelcome visitors to the island at Christmas. Monthly Bulletin of the Department of Agriculture, Fisheries and Parks, Bermuda 68, 8 pp.

Armstrong, A.J., Van, H.H.J., Scott, D.F. and Milton, S.J. (1996) Are pine plantations 'inhospitable seas' around remnant native habitat within south-western cape forestry areas? *South African Forestry Journal*, 1–10.

Armstrong, K.F., McHugh, P., Chinn, W., Frampton, E.R. and Walsh, P.J. (2003) Tussock moth species arriving on imported used vehicles determined by DNA analysis. *New Zealand Plant Protection* 56, 16–20.

Atkinson, P.R. (1998) Aerial spraying trials with alternative insecticides and application methods for the wattle bagworm, *Chaliopsis junodi* (Heylaerts) (Lepidoptera: Psychidae). ICFR Bulletin Series No. 06/98.

Atkinson, P.R. (1999) Eucalyptus snout beetle, *Gonipterus scutellatus* Gyll., and its control in South Africa through biological, cultural and chemical means. ICFR Bulletin Series No. 01/99.

Atkinson, P.R. and Laborde, R.M. de (1993) Estimates of the extent of, and the potential losses from, infestations of common forestry pests in South Africa. ICFR Bulletin Series No. 19/93.

Atkinson, T.H., Foltz, J.L., Wilkinson, R.C. and Mizell, R.F. (2005) Asian ambrosia beetle, granulate ambrosia beetle (no official common name), *Xylosandrus crassiusculus* (Motschulsky) (Insecta: Curculionidae: Scolytinae). University of Florida IFAS Extension EENY131.

Atreya, K., Sitaula, B.K., Johnsen, F.H. and Bajracharya, R.M. (2011) Continuing issues in the limitations of pesticide use in developing countries. *Journal of Agricultural and Environmental Ethics* 24, 49–62.

Atuahene, S.K.N. (1972) The major entomological problems facing Ghana's reforestation program. Proceedings of the Seventh World Forestry Congress, 14–18 October, 1972, Buenos Aires, Argentina, pp. 1587–1590.

Atuahene, S.K.N. (1976) Incidence of *Apate* spp. (Coleoptera: Bostrychidae) on young forest plantation species in Ghana. *Ghana Forestry Journal* 2, 29–35.

Austin, A.D., Quicke, D.L.J. and Marsh, P.M. (1994) The hymenopterous parasitoids of eucalypt longicorn beetles, *Phoracantha* spp. (Coleoptera: Cerambycidae) in Australia. *Bulletin of Entomological Research* 84, 145–174.

Austin, G.T., Haddad, N.M., Mendez, C., Sisk, T.D., Murphy, D.D., Launer, A.E. and Ehrlich, P.R. (1996) Annotated checklist of the butterflies of the Tikal National Park Area of Guatemala. *Tropical Lepidoptera* 7, 21–37.

Austin, M.T., Sorensson, C.T., Brewbaker, J.L., Sun, W. and Shelton, H.M. (1995) Forage dry matter yields and psyllid resistance of thirty-one leucaena selections in Hawaii. *Agroforestry Systems* 31, 211–222.

Averill, R.D., Wilson, L.F. and Fowler, R.F. (1982) Impact of the redheaded sawfly (Hymenoptera: Diprionidae) on young red pine plantations. *Great Lakes Entomologist* 15, 65–91.

Axmacher, J.C., Brehm, G., Hemp, A., Tuente, H., Lyaruu, H.V.M., Mueller-Hohenstein, K. and Fiedler, K. (2009) Determinants of diversity in afrotropical herbivorous insects (Lepidoptera: Geometridae): plant diversity, vegetation structure or abiotic factors? *Journal of Biogeography* 36, 337–349.

Azevedo-Ramos, C., de Carvalho, O. Jr, and do Amaral, B.D. (2006) Short-term effects of reduced-impact logging on eastern Amazon fauna. *Forest Ecology and Management* 232, 26–35.

Bacon, C.G. and South, D.B. (1989) Chemicals for control of common insect and mite pests in southern pine nurseries. *Southern Journal of Applied Forestry* 13, 112–116.

Baksha, M.W. (1990) Some major forest insect pests of Bangladesh and their control. Bulletin No. 1, Forest Entomology Series, Forest Research Institute, Chittagong, Bangladesh, 19 pp.

Baksha, M.W. (1991) Incidence of bark-eating caterpillar on moluccana koroi (*Paraserianthes falcataria*). *Bangladesh Journal of Forest Science* 20, 44–48.

Baksha, M.W. (1993) Control of bark-eating caterpillar of moluccana koroi (*Paraserianthes falcataria*). *Bangladesh Journal of Forest Science* 22, 1–5.

Baksha, M.W. (1997) Biology, ecology and control of gamar defoliator, *Calopepla leayana* Latr. (Chrysomelidae: Coleoptera) in Bangladesh. *Bangladesh Journal of Forest Science* 26(2), 31–36.

Baksha, M.W. (1998) Attack of termites in forest nurseries and plantations and their management. *Bangladesh Journal of Forest Science* 27(1), 49–54.

Baksha, M.W. (2001) Important Pests of Forest Nurseries of Bangladesh and their Management. Bulletin 7, Forest Entomology Series, Bangladesh Forest Research Institute, Chittagong, Bangladesh, 17 pp.

Baksha, M.W. (2007) Major forest nursery pests and diseases in Bangladesh. Sixth meeting of the IUFRO Working Party 7.03.04 (Diseases and insects in forest nurseries). *Communicationes Instituti Forestalis Bohemicae* 23, 24–28.

Baksha, M.W. and Crawley, M.J. (1998) Population dynamics of teak defoliator, *Hyblaea puera* Cram. (Lep., Hyblaeidae) in teak plantations of Bangladesh. *Journal of Applied Entomology-Zeitschrift fur Angewandte Entomologie* 122, 79–83.

Balachander, M., Remadevi, O.K., Sasidharan, T.O. and Bai, N.S. (2009) Infectivity of *Metarhizium anisopliae* (Deuteromycotina: Hyphomycetes) isolates to the arboreal termite *Odontotermes* sp. (Isoptera: Termitidae). *International Journal of Tropical Insect Science* 29, 202–207.

Baldini, A., Oltremari, J. and Holmgren, A. (2008) Effect of *Cinara cupressi* on *Austrocedrus chilensis* following chemical control. *Ciencia e Investigacion Agraria* 35, 341–350.

Bandara, G.D. (1990) Chemical control of cockchafer grub (*Holotrichia serrata*) in teak nurseries. *Sri Lanka Forester* 19, 47–50.

Barak, A.V., Wang, Y., Xu, L., Rong, Z., Hang, X. and Zhan, G. (2005) Methyl bromide as a quarantine treatment for *Anoplophora glabripennis* (Coleoptera: Cerambycidae) in regulated wood packing material. *Journal of Economic Entomology* 98(6), 1911–1916.

Barak, A.V., Wang, X., Yuan, P., Jin, X., Liu, Y., Lou, S. and Hamilton, B. (2006) Container fumigation as a quarantine treatment for *Anoplophora glabripennis* (Coleoptera: Cerambycidae) in regulated wood packing material. *Journal of Economic Entomology* 99(3), 664–670.

Barboza, J., Hilje, L., Duron, J., Cartin, V. and Calvo, M. (2010) Phagodeterrence by a crude extract of common rue (*Ruta chalepensis*, Rutaceae) and its partitions on *Hypsipyla grandella* (Lepidoptera: Pyralidae) larvae. *Revista de Biologia Tropical* 58, 1–14.

Barnard, J.E., Radloff, D.L., Loomis, R.C. and Space, J.C. (1992) Forest health monitoring: taking the pulse of America's forests. In: Wood, G. and Turner, B. (eds) *Integrating Forest Information Over Space and Time*. IUFRO Conference, 13–17 January 1992, Canberra, Australia, pp. 343–348.

Barratt, B.I.P., Howarth, F.G., Withers, T.M., Kean, J.M. and Ridley, G.S. (2010) Progress in risk assessment for classical biological control. *Biological Control* 52, 245–254.

Barrion, A.T., Aguda, R.M., Litsinger, J.A., Barrion, R., Withington, D. and Brewbaker, J.L. (1987) The natural enemies and chemical control of the leucaena psyllid, *Heteropsylla cubana* Crawford (Hemiptera: Psyllidae), in the Philippines. Proceedings of a workshop on biological and genetic control strategies for the leucaena psyllid, November 3–7. *Leucaena Research Reports* 7, 45–49.

Barton, K.E. and Koricheva, J. (2010) The ontogeny of plant defense and herbivory: characterizing general patterns using meta-analysis. *American Naturalist* 175, 481–493.

Basalingappa, S. and Ghandi, M.R. (1994) Infestation of the seedlings of *Tectona grandis* by the lepidopteran larvae of *Hapalia machaeralis* (Pyralidae) and *Hyblaea puera* (Hyblaeidae). *Journal of Ecobiology* 6, 67–68.

Basset, Y. (2001) Invertebrates in the canopy of tropical rain forests – how much do we really know? *Plant Ecology* 153, 87–107.

Basset, Y., Hammond, P.M., Barrios, H., Holloway, J.D. and Miller, S.E. (2003) Vertical stratification of arthropod assemblages. In: Basset, Y., Novotny, V., Miller, S.E. and Kitching R.L. (eds) *Arthropods of Tropical Forests*. Cambridge University Press, pp. 17–27.

Bassett, M.A., Baumgartner, J.B., Hallett, M.L., Hassan, Y. and Symonds, M.R.E. (2011) Effects of humidity on the response of the bark beetle *Ips grandicollis* (Eichhoff) (Coleoptera: Curculionidae: Scolytinae) to synthetic aggregation pheromone. *Australian Journal of Entomology* 50, 48–51.

Bassus, W. (1974) On the biology and ecology of *Dendrolimus punctatus* Walk. (Lepidoptera: Lasiocampidae), the most important pest of pines in the Democratic Republic of Vietnam. *Archiv fur Phytopathologie und Pflanzenschutz* 10, 61–69.

Batista-Pereira, L.G., Marques, E.N., Da Silva, M.J., Groke-Junior, P.H. and Pereira-Neto, S.D. (1995) Mortality rate of *Thyrinteina arnobia* (Stoll, 1782) (Lepidoptera: Geometridae) by parasitoids and entomopathogens. *Revista Arvore* 19, 396–404.

Baylis, W.B.H. and Barnes, R.D. (1989) International provenance trials of *Pinus caribaea* var. *bahamensis*. *Forest Genetic Resources Information*, pp. 24–25.

Bayliss, J., Makunga, S., Hecht, J., Nangoma, D. and Bruessow, C. (2007) Saving the island in the sky: the plight of the Mount Mulanje cedar *Widdringtonia whytei* in Malawi. *Oryx* 41(1), 64–69.

Becerra, J. and Ezcurra, E. (1986) Glandular hairs in the *Arbutus xalapensis* complex in relation to herbivory. *American Journal of Botany* 73, 1427–1430.

Bechtold, W.A., Hoffard, W.H. and Anderson, R.L. (1992) Summary report: forest health monitoring in the South, 1991. General Technical Report SE-81, Southeastern Forest Experiment Station, US Department of Agriculture Forest Service, Asheville, North Carolina, 40 pp.

Beck, J. and Kitching, I.J. (2009) Drivers of moth species richness on tropical altitudinal gradients: a cross-regional comparison. *Global Ecology and Biogeography* 18, 361–371.

Beck, J., Brehm, G. and Fiedler, K. (2011) Links between environment, abundance and diversity of Andean moths. *Biotropica* 43(2), 208–217.

Becker, V.O. (1973) Some Microlepidoptera associated with *Pinus* in Central America (Pyralidae and Tortricidae). *Turrialba* 23, 104–106.

Beckerman, A.P., Benton, T.G., Ranta, E., Kaitala, V. and Lundberg, P. (2002) Population dynamic consequences of delayed life-history effects. *Trends in Ecology and Evolution* 17, 263–269.

Bedding, R.A. and Iede, E.T. (2005) Application of *Beddingia siricidicola* for *Sirex* woodwasp control. In: Grewal, P.S., Ehlers, R.U. and Shapiro-Ilan, D.I. (eds) *Nematodes as Biocontrol Agents*. CAB International, Wallingford, UK, pp. 385–400.

Beeson, C.F.C. (1941) *The Ecology and Control of the Forest Insects of India and Neighbouring Countries.* Vasant Press, Dehra Dun, India.

Beeson, C.F.C. and Bhatia, B.M. (1939) On the biology of the Cerambycidae (Coleoptera). *Indian Forest Records (Entomology)* 5(1), 1–234.

Belghazi, B., Ezzahiri, M., Ouassou, A., Achbah, M., el Yousfi, S.M. and Beighazi, T. (2008) Behaviour of some eucalyptus clones in the Maamora forest (Morocco). *Cahiers Agricultures* 17, 11–22.

Bell, H.L. (1979) The effects on rain-forest birds of plantings of teak, *Tectona grandis*, in Papua New Guinea. *Australian Wildlife Research* 6, 305–318.

Belmont, R.A. and Habeck, D.H. (1983) Parasitoids of *Dioryctria* spp. (Pyralidae: Lepidoptera) coneworms in slash pine seed production areas in north Florida. *The Florida Entomologist* 66, 399–407.

Belsky, J. and Siebert, S. (1995) Managing rattan harvesting for local livelihoods and forest conservation in Kerinci-Seblat National Park, Sumatra. *Selbyana* 16(2), 212–222.

Benedick, S., Hill, J.K., Mustaffa, N., Chey, V.K., Maryati, M., Searle, J.B., Schilthuizen, M. and Hamer, K.C. (2006) Impacts of rain forest fragmentation on butterflies in northern Borneo: species richness, turnover and the value of small fragments. *Journal of Applied Ecology* 43, 967–977.

Bennett, G.W., Owens, J.M. and Corrigan, R.M. (1997) *Truman's Scientific Guide to Pest Control Operations*, 5th edn. Advanstar Communications, Cleveland, Ohio.

Bentley, J.W., Boa, E.R., Kelly, P., Harun-Ar-Rashid, M., Rahman, A.K.M., Kabeere, F. and Herbas, J. (2009) Ethnopathology: local knowledge of plant health problems in Bangladesh, Uganda and Bolivia. *Plant Pathology* 58, 773–781.

Bernays, E.A. (1991) Evolution of insect morphology in relation to plants. *Philosophical Transactions of the Royal Society London B, Biological Sciences* 333, 257–264.

Bernays, E.A. and Chapman, R.F. (1994) *Host-plant Selection by Phytophagous Insects.* Contemporary Topics in Entomology, Vol. 2. Chapman and Hall, London.

Berryman, A.A. (1986) *Forest Insects: Principles and Practice of Population Management.* Plenum Press, New York.

Berryman, A.A. (1987) The theory and classification of outbreaks. In: Barbosa, P. and Schultz, J.C. (eds) *Insect Outbreaks.* Academic Press, San Diego, California, pp. 3–29.

Berti Filho, E. (1974) *Biologia de Thyrinteina arnobia (Stol, 1782) (Lepidoptera: Geometridae) e observacoes sobre a ocorrencia de inimigos naturais.* ESALQ/USP, Piracicaba, Brazil.

Bevan, D. and Jones, T. (1971) Report on the beetle *Hylastes angustatus* in pine (*Pinus* spp.) plantations of Swaziland and adjacent areas of the Republic of South Africa. Forestry Commission UK, Research and Development Paper No. 84.

Bhandari, R.S., Pande, S., Joshi, M.C., Zaidi, S.M.H., Rawat, J.M.S. and Kumar, V. (2006) Control of cone worm, *Dioryctria abietella* Denis and Schiffermueller (Lepidoptera: Pyralidae) in seed production areas of spruce, *Picea smithiana* Boiss by systemic insecticides. *Indian Forester* 132(8), 1041–1046.

Bhowmick, A.K. and Vaishampayan, S.M. (1986) Observations on the activity of teak defoliator *Hyblaea puera* Cramer on teak (*Tectona grandis*) influenced by the movement of monsoon. *Journal of Tropical Forestry* 2, 27–35.

Bi, H., Simpson, J., Eldridge, R., Sullivan, S., Li, R-W., Xiao, Y-G., *et al.* (2008) Survey of damaging pests and preliminary assessment of forest health risks to the long term success of *Pinus radiata* introduction in Sichuan, southwest China. *Journal of Forestry Research (Harbin)* 19, 85–100.

Bigger, M. (1980) *Hyblaea puera* on teak. Forest Pests of the Solomon Islands, No. 5. Forestry Division, Ministry of Natural Resources, Honiara.

Bigger, M. (1985) The effect of attack by *Amblypelta cocophaga* China (Hemiptera: Coreidae) on growth of *Eucalyptus deglupta* in the Solomon Islands (Western Pacific Ocean). *Bulletin of Entomological Research* 75, 595–608.

Bigger, M. (1988) The insect pests of forest plantation trees in the Solomon Islands. Solomon Islands' Forest Record, No. 4. ODNRI, Chatham, UK.

Biji, C.P., Sudheendrakumar, V. and Sajeev, T.V. (2006) Influence of virus inoculation method and host larval age on productivity of the nucleopolyhedrovirus of the teak defoliator, *Hyblaea puera* (Cramer). *Journal of Virological Methods* 133, 100–104.

Billings, R.F. (1991) The pine caterpillar *Dendrolimus punctatus* in Vietnam – recommendations for integrated pest-management. *Forest Ecology and Management* 39, 97–106.

Billings, R.F. and Richter, A.M. (1990) Life history parameters of bark beetles (Coleoptera: Scolytidae) attacking West Indian pines in the Dominican Republic. *Florida Entomologist* 73, 591–603.

Billings, R.F., Raske, A.G. and Wickman, B.E. (1991) The pine caterpillar *Dendrolimus punctatus* in Vietnam; recommendations for integrated pest management. Proceedings of a symposium, Towards Integrated

Pest Management of Forest Defoliators, held at the 18th International Congress of Entomology, Vancouver, Canada in 1988. *Forest Ecology and Management* 39, 97–106.

Biocontrol News and Information (1997) *Cypress Aphid Parasitoid Established in Malawi*. Vol. 18(2). CAB International, Wallingford, UK.

Biocontrol News and Information (1999) *Brighter Future for Conifer Forests: Pauesia in East Africa*. Vol. 20(2). CAB International, Wallingford, UK.

Bisiaux, F., Peltier, R. and Mulielie, J.-C. (2009) Industrial plantations and agroforestry for the benefit of populations on the Bateke and Mampu plateaux in the Democratic Republic of the Congo. *Bois et Forets Des Tropiques*: 21–32.

Blackman, R.L. and Eastop, V.F. (1994) *Aphids on the World's Trees*. CAB International, Wallingford, UK.

Blackman, R.L., Watson, G.W. and Ready, P.D. (1995) The identity of the African pine woolly aphid: a multidisciplinary approach. *EPPO Bulletin* 25(1–2), 337–341.

Boa, E.R. (1995) *A Guide to the Identification of Diseases and Pests of Neem (Azadirachta indica)*. RAP Publication 1995/41. FAO, Bangkok.

Boa, E. (2003) *An Illustrated Guide to the State of Health of Trees: Recognition and Interpretation of Symptoms and Damage*. FAO, Rome, pp. 49.

Boa, E.R., Grodzki, W., Knizek and Forster, B. (1998) Disease surveys of agroforestry trees in tropical countries. Methodology of forest insect and disease survey in Central Europe. Proceedings, First Workshop of the IUFRO WP 7.03.10, Ustron-Jaszowiec, Poland, 21–24 April, pp. 9–15.

Booth, R.G., Cross, A.E., Fowler, S.V. and Shaw, R.H. (1995) The biology and taxonomy of *Hyperaspis pantherina* (Coleoptera: Coccinellidae) and the classical biological control of its prey, *Orthezia insignis* (Homoptera: Ortheziidae). *Bulletin of Entomological Research* 85, 307–314.

Booth, T.H. (1996) The development of climatic mapping programs and climatic mapping in Australia. In: Booth, T.H. (ed.) *Matching Trees and Sites*. Proceedings of an International Workshop held in Bangkok, Thailand, 27–30 March 1995. ACIAR Proceedings No. 63, pp. 38–42.

Boreham, G.R. (2006) A survey of cossid moth attack in *Eucalyptus nitens* on the Mpumalanga Highveld of South Africa. *South African Forestry Journal* 206, 23–26.

Bosmans, B. (2006) *Phoracantha recurva* (Coleoptera: Cerambycidae) found in a cluster of bananas. *Phegea* 34, 105–106.

Bossart, J.L. and Opuni-Frimpong, E. (2009) Distance from edge determines fruit-feeding butterfly community diversity in afrotropical forest fragments. *Environmental Entomology* 38, 43–52.

Bossel, H. and Krieger, H. (1994) Simulation of multi-species tropical forest dynamics using a vertically and horizontally structured model. *Forest Ecology and Management* 69, 123–144.

Bosu, P.P., Cobbinah, J.R., Nichols, J.D., Nkrumah, E.E. and Wagner, M.R. (2006) Survival and growth of mixed plantations of *Milicia excelsa* and *Terminalia superba* 9 years after planting in Ghana. *Forest Ecology and Management* 233, 352–357.

Botto, E.N., Aquino, D., Lolacono, M. and Pathauer, P. (2010) Presence of *Leptocybe invasa* Fisher & La Salle, the 'eucalyptus gall wasp' in Argentina. Institute of Microbiology and Agricultural Zoology IPM Newsletter No. 16.

Boucias, D.G. and Pendland, J.C. (1991) Attachment of mycopathogens to cuticle: the initial event of mycoses in arthropod hosts. In: Cole, G.T. and Hoch, H.C. (eds) *The Fungal Spore and Disease Initiation in Plants and Animals*. Xxv+555p. Plenum Press: New York, London, pp. 101–128.

Boucias, D.G. and Pendland, J.C. (1998) *Principles of Insect Pathology*. Kluwer Academic Publishers, Dordrecht.

Braby, M.F. (1995) Reproductive seasonality in tropical satyrine butterflies: strategies for the dry season. *Ecological Entomology* 20, 5–17.

Bradley, A. and Millington, A. (2008) Agricultural land-use trajectories in a cocaine source region: Chapare, Bolivia. Land-Change Science in the Tropics: Changing Agricultural Landscapes, pp. 231–250.

Brady, N.C. (1996) Alternatives to slash-and-burn: a global imperative. Special issue: alternatives to slash and burn agriculture. *Agriculture, Ecosystems and Environment* 58(1), 3–11.

Braker, E. and Chazdon, R.L. (1993) Ecological, behavioural and nutritional factors influencing use of palms as host plants by a neotropical forest grasshopper. *Journal of Tropical Ecology* 9, 183–197.

Bray, R.A. and Woodroffe, T.D. (1988) Resistance of some *Leucaena* species to the leucaena psyllid. *Tropical Grasslands* 22, 11–16.

Bray, R.A. and Woodroffe, T.D. (1991) Effect of the leucaena psyllid on yield of *Leucaena leucocephala* cv Cunningham in south-east Queensland. *Tropical Grasslands* 25, 356–357.

Braza, R.D. (1987) Growth and survival of varicose borer (*Agrilus sexsignatus*) larvae in Philippines provenance of bagras (*Eucalyptus deglupta*). *Malaysian Forester* 50, 235–238.

Braza, R.D. (1988a) Biology of the varicose borer, *Agrilus sexsignatus* (Fisher), on bagras, *Eucalyptus deglupta* Blume. *Philippine Entomologist* 7, 351–358.
Braza, R.D. (1988b) Extent of infestation by shoot-pruner beetles, *Callimetopus* sp. in a *Paraserianthes falcataria* plantation. *Nitrogen Fixing Tree Research Reports* 6, 61–62.
Braza, R.D. (1989) Parasitoids of immature stages of *Agrilus sexsignatus* (Fisher) (Coleoptera: Buprestidae) attacking *Eucalyptus deglupta* Blume in Suriago del Sur. *Philippine Entomologist* 7, 479–483.
Braza, R.D. (1990a) Important pests and their control in PICOP forest nurseries. *PCARRD Monitor* 18, 88–110.
Braza, R.D. (1990b) New records of major insect pests attacking *Paraserianthes falcataria* in the Philippines. *Nitrogen Fixing Tree Research Reports* 8, 147–148.
Braza, R.D. (1991) Important pests of bagras trees and their control. *PCARRD Monitor* 19, 8–9.
Braza, R.D. (1992) Neem oil for the control of forest pests. *PCARRD Monitor* 20, 6–7.
Braza, R.D. (1993) Bee-hole borer is a destructive pest of mangium. *PCARRD Monitor* 21, 6–7.
Brehm, G. (2007) Contrasting patterns of vertical stratification in two moth families in a Costa Rican lowland rain forest. *Basic and Applied Ecology* 8(1), 44–54.
Brenes-Arguedas, T., Coley, P.D. and Kursar, T.A. (2009) Pests vs. drought as determinants of plant distribution along a tropical rainfall gradient. *Ecology* 90, 1751–1761.
Brennan, E.B., Gill, R.J., Hrusa, G.F. and Weinbaum, S.A. (1999) First record of *Glycaspis brimblecombei* (Moore) (Homoptera: Psyllidae) in North America: initial observations and predator associations of a potentially serious new pest of *Eucalyptus* in California. *Pan-Pacific Entomology* 75, 55–57.
Brewbaker, J.L. (2008) Registration of KX2-Hawaii, Interspecific-Hybrid *Leucaena. Journal of Plant Registrations* 2, 190–193.
Briceno-Vergara, A.J. (1997) Preliminary approach towards an integrated management program for *Hypsipyla grandella* (Zeller) in western Venezuela. *Revista Forestal Venezolana* 31, 41–51.
Brienco, A.J. (1987) *Bolax palliatus*, the leaf skeletonizer of guamo. *Revista Forestal Venezolana* 31, 71–74.
Brimblecombe, A.R. (1945) The biology, economic importance and control of the pine bark weevil *Aesiotes notabilis* Pasc. *Queensland Journal of Agricultural Science* 2, 1–88.
Brimblecombe, A.R. and Heather, N.W. (1965) Occurrence of the kauri coccid, *Conifericoccus agathidis* Brimblecombe (Homoptera: Monophlebiidae) in Queensland. *Journal of the Entomological Society of Queensland* 4, 83–85.
Britton, K.O. (2007) What you can do to help improve regulation of the plants for planting pathway. Proceedings of the Sudden Oak Death Third Science Symposium. General Technical Report PSW-GTR-214, pp. 141–145.
Brockerhoff, E.G., Jones, D.C., Kimberley, M.O., Suckling, D.M. and Donaldson, T. (2006) Nationwide survey for invasive wood-boring and bark beetles (Coleoptera) using traps baited with pheromones and kairomones. *Forest Ecology and Management* 228, 234–240.
Brockerhoff, E.G., Jactel, H., Parrotta, J.A., Quine, C.P. and Sayer, J. (2008) Plantation forests and biodiversity: oxymoron or opportunity? *Biodiversity and Conservation* 17, 925–951.
Brown, B.N. and Wylie, F.R. (1990) Diseases and pests of Australian forest nurseries: past and present. Paper presented at the first meeting of IUFRO Working Party S2.07-09 (Diseases and Insects in Forest Nurseries). Victoria, British Columbia, Canada, 22–30 August 1990.
Brown, C. (2001) *Future Production from Forest Plantations*. Forest Plantations Thematic Papers, Working Paper FP/13, FAO, Rome.
Brown, K.S., Yokum, B.P., Riegel, C. and Carroll, M.K. (2007) New parish records of *Coptotermes formosanus* (Isoptera: Rhinotermitidae) in Louisiana. *Florida Entomologist* 909(3), 570–572.
Brown, S. and Lugo, A.E. (1990) Tropical secondary forests. *Journal of Tropical Ecology* 6, 1–32.
Brown, V.K. and Lawton, J.H. (1991) Herbivory and the evolution of leaf size and shape. *Philosophical Transactions of the Royal Society of London B, Biological Sciences* 333, 265–272.
Browne, D.J., Peck, S.B. and Ivie, M.A. (1993) The longhorn beetles (Coleoptera Cerambycidae) of the Bahama Islands with an analysis of species–area relationships, distribution patterns, origin of the fauna and an annotated species list. *Tropical Zoology* 6, 27–53.
Browne, F.G. (1968) *Pests and Diseases of Forest Plantation Trees: An Annotated List of the Principal Species Occurring in the British Commonwealth*. Clarendon Press, Oxford, UK.
Brunck, F. and Mallet, B. (1993) Problems relating to pests attacking mahogany. *Bois et Forets des Tropiques* 9–28.
Brush, S. and Stabinsky, D. (eds) (1996) *Indigenous People and Intellectual Property Rights*. Island Press, Washington, DC, pp. 167–185.
Bryan, H. (1989) Control of the tarnished plant bug at the Carters nursery. *Tree Planters' Notes* 40, 30–33.

Bulinski, J., Matthiessen, J.N. and Alexander, R. (2006) Development of a cost-effective, pesticide-free approach to managing African black beetle (*Heteronychus arator*) in Australian eucalyptus plantations. *Crop Protection* 25, 1161–1166.

Bulman, L.S. (2008) Pest detection surveys on high-risk sites in New Zealand. *Australian Forestry* 71(3), 242–244.

Bulman, L.S., Kimberley, M.O. and Gadgil, P.D. (1999) Estimation of the efficiency of pest detection surveys. *New Zealand Journal of Forestry Science* 29, 102–115.

Bunce, H.W.F. and McLean, J.A. (1990) Hurricane Gilbert's impact on the natural forests and *Pinus caribaea* plantations of Jamaica. *Commonwealth Forestry Review* 69, 147–155.

Burckhardt, D., Lozada, P.W. and Diaz, W.B. (2008) First record of the red gum lerp psyllid *Glycaspis brimblecombei* (Hemiptera: Psylloidea) from Peru. *Bulletin de la Societe Entomologique Suisse* 81, 83–85.

Burdon, R.D. and Wilcox, P.L. (2007) Population management: potential impacts of advances in genomics. *New Forests* 34, 187–206.

Burkey, T.V. (1993) Edge effects in seed and egg predation at two neotropical rainforest sites. *Biological Conservation* 66, 139–143.

Burnham, R.J. (1994) Patterns in tropical leaf litter and implications for angiosperm paleobotany. *Review of Palaeobotany and Palynology* 81, 99–113.

Burrows, D.W., Balciunas, J.K. and Edwards, E.D. (1996) Herbivorous insects associated with the paperbark *Melaleuca quinquenervia* and its allies: V. Pyralidae and other Lepidoptera. *Australian Entomologist* 23, 7–16.

Buschke, F.T., Watson, M. and Seama, M.T. (2011) The partitioning of macroinvertebrate diversity across multiple spatial scales in the upper Modder River System, South Africa. *African Journal of Ecology* 49, 81–90.

Button, G. (2007) *Thaumastocoris peregrinus*. Forestry Facts. Available at http://www.nctforest.com

Bybee, L.F., Millar, J.G., Paine, T.D., Campbell, K. and Hanlon, C.C. (2004a) Effects of temperature on fecundity and longevity of *Phoracantha recurva* and *P. semipunctata* (Coleoptera: Cerambycidae). *Environmental Entomology* 33(2), 138–146.

Bybee, L.F., Millar, J.G., Paine, T.D., Campbell, K. and Hanlon, C.C. (2004b) Seasonal development of *Phoracantha recurva* and *P. semipunctata* (Coleoptera: Cerambycidae) in Southern California. *Environmental Entomology* 33(5), 1232–1241.

CABI (1999) *Forestry Compendium*. CAB International, Wallingford, UK.

CABI (2005) *Forestry Compendium* (CD version). CAB International, Wallingford, UK.

CABI (2010) *Forestry Compendium*. CAB International, Wallingford, UK.

CABI (2010) *Crop Protection Compendium*. CAB International, Wallingford (www.cabicompendium.org/cpc).

CABI/EPPO (2005) *Nipaecoccus viridis* (Newstead). Distribution Maps of Plant Pests No. 446. CAB International, Wallingford, UK.

CABI/EPPO (2007a) *Leptocybe invasa*. Distribution Maps of Plant Pests No. 698. CAB International, Wallingford, UK.

CABI/EPPO (2007b) *Phoracantha semipunctata*. Distribution Maps of Plant Pests No. 272. CAB International, Wallingford, UK.

CABI/EPPO (2010) *Gonipterus scutellatus* Gyllenhal (Coleoptera: Curculionidae). Distribution Maps of Plant Pests No. 344, 2nd revision. CAB International, Wallingford, UK.

Cagnolo, L., Valladares, G., Salvo, A., Cabido, M. and Zak, M. (2009) Habitat fragmentation and species loss across three interacting trophic levels: effects of life-history and food-web traits. *Conservation Biology* 23, 1167–1175.

Caldeira, M.C., Fernandez, V., Tome, J. and Preira, J.S. (2002) Positive effect of drought on longicorn borer survival and growth on eucalyptus trunks. *Annals of Forest Science* 59, 99–106.

Calil, A.C.P. and Soares, W.O. (1987) Damage caused by leaf-cutter ants (*Atta sexdens*) in rubber (*Hevea* spp.) nurseries. *Boletim da Faculdade de Ciencias Agrarias do Para, Brasil* No. 16, 23–30.

Camargo, J.L.C. and Kapos, V. (1995) Complex edge effects on soil moisture and microclimate in central Amazonian forest. *Journal of Tropical Ecology* 11, 205–221.

Campbell, F.T. (2007) Curbing introductions of forest insects and diseases on nursery stock. USDA Interagency Research Forum – GTR-NRS-P-28, pp. 9–11.

Candy, S.G., Elliott, H.J., Bashford, R. and Greener, A. (1992) Modelling the impact of defoliation by the leaf beetle, *Chrysophtharta bimaculata* (Coleoptera: Chrysomelidae), in height growth of *Eucalyptus regnans*. *Forest Ecology and Management* 54, 69–87.

Cannicci, S., Burrows, D., Fratini, S., Smith, T.J., Offenberg, J. and Dahdouh-Guebas, F. (2008) Faunal impact on vegetation structure and ecosystem function in mangrove forests: a review. *Aquatic Botany* 89, 186–200.

Canty, C. and Harrison, P.G. (1990) Transfer of controlled release insecticide technology for protection of forest seedlings from termite attack. Proceedings of the Australian Institute of Agricultural Science National Conference, 23–26 May, 1989. AIAS Occasional Publication No. 54, pp. 183–189.

Carlos, A.A., Rodrigues, A., Forti, L.C., Passador, M.M. and Sierra, J.F. (2011) Filamentous fungi found in *Atta sexdens rubropilosa* colonies after treatment with different toxic bait formulations. *Journal of Applied Entomology* 135, 326–331.

Carne, P.B. (1969) On the population dynamics of the eucalypt-defoliating sawfly *Perga affinis affinis* Kirby (Hymenoptera). *Australian Journal of Zoology* 17, 113–141.

Carne, P.B. and Taylor, K.L. (1978) Insect pests. In: Hillis, W.E. and Brown, A.G. (eds) *Eucalypts for Wood Production*. CSIRO, Canberra, pp. 155–168.

Carnegie, A.J. (2008) A decade of forest health surveillance in Australia: an overview. *Australian Forestry* 71(3), 161–163.

Carnegie, A.J., Cant, R.G., Eldridge, R.H. (2008) Forest Health Surveillance in New South Wales, Australia. *Australian Forestry* 71(3), 164–176.

Carnegie, A.J., Stone, C., Lawson, S. and Matsuki, M. (2005) Can we grow certified eucalypt plantations in subtropical Australia? An insect management perspective. *New Zealand Journal of Forestry Science* 35(2/3), 223–245.

Carnegie, A.J., Matsuki, M., Haugen, D.A., Hurley, B.P., Ahumada, R., Klasmer, P., Sun, J.H. and Iede, E.T. (2006) Predicting the potential distribution of *Sirex noctilio* (Hymenoptera: Siricidae), a significant exotic pest of *Pinus* plantations. *Annals of Forest Science* 63, 119–128.

Carpintero, D.L. and Dellape, P.M. (2006) A new species of *Thaumastocoris* Kirkaldy from Argentina (Heteroptera: Thaumastocoridae). *Zootaxa* 1228, 61–68.

Carroll, S.P. and Boyd, C. (1992) Host race radiation in the soapberry bug: natural history with the history. *Evolution* 46, 1052–1069.

Carson, M.J. and Burdon, R.D. (1991) Relative advantages of clonal forestry and vegetative multiplication. In: Miller, J.T. (comp./ed.) *Proceedings of the New Zealand Forest Research Institute/New Zealand Forest Products Forests Limited Clonal Forestry Workshop*, 1–2 May 1989, Rotorua, New Zealand. Bulletin No. 160, N.Z. Forest Research Institute, pp. 41–43.

Carter, P.C.S. (1989) Risk assessment and pest detection surveys for exotic pests and diseases which threaten commercial forestry in New Zealand. *New Zealand Journal of Forestry Science* 19, 353–374.

Casida, J.E. and Quistad, G.B. (1998) Golden age of insecticide research: past, present, or future? *Annual Review of Entomology* 43, 1–16.

Castilla, R.A.L., Casanova, A.D., Rivero, C.G., Escoto, H.C. and Issasi, N.T. (2002) Forest nursery pest management in Cuba. National Proceedings: Forest and Conservation Nursery Associations 1999, 2000 and 2001. USDA Forest Service Proceedings RMRS-P-24, September 2002.

Castilla, R.A.L., Rivero, C.G., Casanova, A.D., Escoto, H.C., Vera, A.F., Garcia, A., Varela Lao, Y., Berrios, M. Triguero, I. and Vila Maria, I. (2003) Actualization of catalogue of insects and microorganisms harmful to forest trees in Cuba. *Fitosanidad* 7(2), 3–9.

Castro, M.E.B., Ribeiro, Z.M.A., Santos, A.C.B., Souza, M.L., Machado, E.B., Sousa, N.J. and Moscardi, F. 2009 Identification of a new nucleopolyhedrovirus from naturally-infected *Condylorrhiza vestigialis* (Guenee) (Lepidoptera: Crambidae) larvae on poplar plantations in South Brazil. *Journal of Invertebrate Pathology* 102, 149–154.

CATIE (1992) *Forest Pests in Central America: Field Guide*. Tropical Agriculture Research and Training Centre (CATIE), Turrialba, Costa Rica.

Caton, B.P., Dobbs, T.T. and Brodel, C.F. (2006) Arrivals of hitchhiking insect pests on international cargo aircraft at Miami International airport. *Biological Invasions* 8, 765–785.

Cavalcanti, C. (1996) Brazil's new forests bring profit and pain. *People and the Planet* 5, 14–16.

Central Science Laboratory (2005) *Pest Risk Analysis for Chrysophtharta bimaculata (Oliver)*. Central Science Laboratory, York, UK.

Cerda, L.A. (1980) Thysanoptera in *Pinus radiata* in Chile. *Turrialba* 30, 113–114.

Chaloner, W.G., Scott, A.C. and Stephenson, J. (1991) Fossil evidence for plant–arthropod interactions in the Paleozoic and Mesozoic. *Philosophical Transactions of the Royal Society of London B, Biological Sciences* 333, 177–186.

Champagne, D.E., Koul, O., Isman, M.B., Scudder, G.G.E. and Towers, G.H.N. (1992) Biological activity of limonoids from the Rutales. *Phytochemistry* 31, 377–394.

Chandrasekhar, N., Sajeev, T.V., Sudheendrakumar, V.V. and Banerjee, M. (2005) Population dynamics of the teak defoliator (*Hyblaea puera* Cramer) in Nilambur teak plantations using randomly amplified gene encoding primers (RAGEP). *BMC Ecology* 5, 1.

Chang, Y.C. (1991) Integrated pest management of several forest defoliators in Taiwan. *Forest Ecology and Management* 39, 65–72.

Chapman, J.D. (1994) Notes on Mulanje cedar – Malawi's national tree. *Commonwealth Forestry Review* 73, 235–242.

Chauhan, P.S., Porwal, M.C., Sharma, L. and Negi, J.D.S. (2003) Change detection in sal forest in Dehradun Forest Division using remote sensing and geographical information system. *Journal of the Indian Society of Remote Sensing* 31(3), 211–218.

Chen, I.C., Shiu, H.J., Benedick, S., Holloway, J.D., Chey, V.K., Barlow, H.S., Hill, J.K. and Thomas, C.D. (2009) Elevation increases in moth assemblages over 42 years on a tropical mountain. *Proceedings of the National Academy of Sciences of the United States of America* 106, 1479–1483.

Chen, S., Chen, E., Suo, Q., Duan, Z., Hu, C., Chen, S.W., et al. (1997) *Dendrolimus* virus resources and their application in Yunnan. *Chinese Journal of Biological Control* 13, 122–124.

Chen, Z.Q. and Wu, S.X. (1984) Preliminary observations on *Hyblaea puera* Cramer. *Insect Knowledge* 21, 161–163.

Cherrett, J.M. (1986) The biology, pest status and control of leaf-cutting ants. *Agricultural Zoology Reviews* 1, 1–37.

Cherrett, J.M. and Peregrine, D.J. (1976) A review of the status of leaf-cutting ants and their control. Association of Applied Biologists: Proceedings of the Association of Applied Biologists. Tropical Pests. *Annals of Applied Biology* 84, 124–128.

Chey, V.K. (1994) Comparison of biodiversity between plantation and natural forests in Sabah using moths as indicators. Unpublished DPhil thesis, University of Oxford.

Chey, V.K. (1996) Forest pest insects in Sabah. Sabah Forest Record No. 15, Sabah Forest Department, Sandakan, Borneo.

Chey, V.K., Holloway, J.D. and Speight, M.R. (1997) Diversity of moths in forest plantations and natural forests in Sabah. *Bulletin of Entomological Research* 87, 371–385.

Chey, V.K., Holloway, J.D., Hambler, C. and Speight, M.R. (1998) Canopy knockdown of arthropods in exotic plantations and natural forest in Sabah, north-east Borneo, using insecticidal mist-blowing. *Bulletin of Entomological Research* 88, 15–24.

Chilima, C.Z. (1991) Termite control in young *Eucalyptus* plantations in Malawi using controlled-release insecticides. *Commonwealth Forestry Review* 70, 237–247.

Chilima, C.Z. (1995) Cypress aphid control: first African release of *Pauesia juniperorum*. FRIM Newsletter Forestry Research Institute of Malawi, 2.

Chilima, C.Z. (1996) Establishment of the cypress aphid biological control agent, *Pauesia juniperorum*, in Malawi. Forest Research Institute of Malawi Research Leaflet No. 10, Pest Series 2.

Chilima, C.Z. and Leather, S.R. (2001) Within-tree and seasonal distribution of the pine woolly aphid *Pineus boerneri* on *Pinus kesiya* trees. *Agricultural and Forest Entomology* 3, 139–145.

Chilima, C.Z. and Meke, G.S. (1995) Biological control of conifer aphids in Malawi: an update on developments since 1991. In: Allard, G.B. and Kairo, M.T.K. (eds) *Biological Control of Conifer Aphids in Africa. Country Updates and Progress Reports, 1991–1993*. CAB International Institute of Biological Control (IIBC), Kenya, pp. 37–46.

Chilima, C.Z. and Murphy, S.T. (2000) The introduction of *Pauesia juniperorum* (Stary) (Hymenoptera: Braconidae) into Malawi for the biological control of *Cinara* sp. (Homoptera: Aphididae). In: Proceedings of the XXI IUFRO World Congress, Kuala Lumpur, Malaysia.

Choo, T.G.T. (1982) Preliminary finding on the biology of *Autoserica rufocuprea* (Blanchard) sensu Brenske (Coleoptera: Melolonthinae) in Singapore. *Gardens Bulletin* 35, 51–63.

Chown, S.L., Scholtz, C.H., Klok, C.J., Joubert, F.J. and Coles, K.S. (1995) Ecophysiology, range contraction and survival of a geographically restricted African dung beetle (Coleoptera: Scarabaeidae). *Functional Ecology* 9, 30–39.

Chung, A.Y.C., Eggleton, P., Speight, M.R., Hammond, P.M. and Chey, V.K. (2000) The diversity of beetle assemblages in different habitat types in Sabah, Malaysia. *Bulletin of Entomological Research* 90, 475–496.

Chung, F., Chung, F.J., Soejarto, D.D., Rivier, L., Gyllenhaal, C. and Famsworth, N.R. (1996) Interests and policies of the state of Sarawak, Malaysia regarding intellectual property rights for plant derived drugs. *Intellectual Property Rights, Naturally Derived Bioactive Compounds and Resource Conservation* 51, 201–204.

Ciesla, W.M. (1991a) Cypress aphid: a new threat to Africa's forests. *Unasylva* 42, 51–55.

Ciesla, W.M. (1991b) Cypress aphid, *Cinara cupressi*, a new pest of conifers in eastern and southern Africa. *FAO Plant Protection Bulletin* 39, 82–93.

Ciesla, W.M. (1992) Introduction of bark beetles and wood borers into China in coniferous logs from North America. *FAO Plant Protection Bulletin* 40, 154–158.

Ciesla, W.M. (1993) Recent introductions of forest insects and their effects: a global view. *FAO Plant Protection Bulletin* 41, 3–13.

Ciesla, W.M. (2003) European woodwasp: a potential threat to North America's conifer forests. *Journal of Forestry* 101(2), 18–23.

Ciesla, W.M. (2004) Wood and wood products as pathways for introduction of exotic bark beetles and woodborers. *Crop Protection Compendium* (www.cabi.org, accessed 1/6/12).

Ciesla, W.M. (2011) Forest Entomology: A Global Perspective. Wiley-Blackwell.

Cigliano, M.M., de Wysiecki, M.L. and Lange, C.E. (2000) Grasshopper (Orthoptera: Acrididae) species diversity in the pampas, Argentina. *Diversity and Distributions* 6(2), 81–91.

Clark, A.F. (1938) A survey of the insect pests of eucalypts in New Zealand. *New Zealand Journal of Science and Technology* 19(12), 750–761.

Clarke, A.R. (1995) Integrated pest management in forestry: some difficulties in pursuing the holy-grail. *Australian Forestry* 58, 147–150.

Cleary, D.F.R., Genner, M.J., Koh, L.P., Boyle, T.J.B., Setyawati, T., de Jong, R. and Menkem, S.B.J. (2009) Butterfly species and traits associated with selectively logged forest in Borneo. *Basic and Applied Ecology* 10, 237–245.

Clissold, F.J., Sanson, G.D., Read, J. and Simpson, S.J. (2009) Gross vs. net income: how plant toughness affects performance of an insect herbivore. *Ecology* 90, 3393–3405.

Coad, B.R. (1931) Insects captured by airplane are found at surprising heights. *Yearbook of the United States Department of Agriculture*. USDA, pp. 320–323.

Cobbinah, J.R. (1986) Factors affecting the distribution and abundance of *Phytolyma lata* (Homoptera: Psyllidae). *Insect Science and its Application* 7, 111–115.

Cobbinah, J.R. and Wagner, M.R. (1995) Phenotypic variation in *Milicia excelsa* to attack by *Phytolyma lata* (Psyllidae). *Forest Ecology and Management* 75, 147–153.

Coley, P.D. (1998) Possible effects of climate change on plant/herbivore interactions in moist tropical forests. *Climatic Change* 39, 455–472.

Coley, P.D., Bateman, M.L. and Kursar, T.A. (2006) The effects of plant quality on caterpillar growth and defense against natural enemies. *Oikos* 115, 219–228.

Collett, N.G. and Elms, S. (2009) The control of sirex wood wasp using biocontrol agents in Victoria, Australia. *Agricultural and Forest Entomology* 11, 283–294.

Collett, N.G. and Neumann, F.G. (2002) Effects of simulated chronic defoliation in summer on growth and survival of blue gum (*Eucalyptus globulus* Labill.) within young plantations in northern Victoria. *Australian Forestry* 65(2), 99–106.

Colwell, R.K., Brehm, G., Cardelus, C.L., Gilman, A.C. and Longino, J.T. (2008) Global warming, elevational range shifts, and lowland biotic attrition in the wet tropics. *Science* 322, 258–261.

Connahs, H., Aiello, A., van Bael, S. and Rodriguez-Castaneda, G. (2011) Caterpillar abundance and parasitism in a seasonally dry versus wet tropical forest of Panama. *Journal of Tropical Ecology* 27, 51–58.

Connor, E.F. and Beck, M.W. (1993) Density-related mortality in *Cameraria hamadryadella* (Lepidoptera: Gracillariidae) at epidemic and endemic densities. *Oikos* 66, 515–525.

Connor, E.F. and Taverner, M.P. (1997) The evolution and adaptive significance of the leaf-mining habit. *Oikos* 79, 6–25.

Cornelius, J.P. (2001) The effectiveness of pruning in mitigating *Hypsipyla grandella* attack on young mahogany (*Swietenia macrophylla* King) trees. *Forest Ecology and Management* 148, 287–289.

Cornelius, J.P. (2009) The utility of the predictive decapitation test as a tool for early genetic selection for *Hypsipyla* tolerance in big-leaf mahogany (*Swietenia macrophylla* King). *Forest Ecology and Management* 257, 1815–1821.

Cornelius, J.P. and Watt, A.D. (2003) Genetic variation in a *Hypsipyla*-attacked clonal trial of *Cedrela odorata* under two pruning regimes. *Forest Ecology and Management* 183, 341–349.

Cornelius, M.L. and Osbrink, W.L.A. (2010) Effect of soil type and moisture availability on the foraging behavior of the formosan subterranean termite (Isoptera: Rhinotermitidae). *Journal of Economic Entomology* 103, 799–807.

Cory, J.S. and Myers, J.H. (2003) The ecology and evolution of insect baculoviruses. *Annual Review of Ecology Evolution and Systematics* 34, 239–272.

Costa, V.A., Berti Filho, E., Wilcken, C.F., Stape, J.L., La Salle, J. and Teixeira, L.D. (2008) *Eucalyptus* gall wasp, *Leptocybe invasa* Fisher & La Salle (Hymenoptera: Eulophidae) in Brazil: new forest pest reaches the New World. *Revista de Agricultura* 83(2), 136–139.

Cowie, R.H. and Wood, T.G. (1989) Damage to crops, forestry and rangeland by fungus-growing termites (Termitidae: Macrotermitinae) in Ethiopia. *Sociobiology* 15, 139–153.

Cowie, R.H., Logan, J.W.M. and Wood, T.G. (1989) Termite (Isoptera) damage and control in tropical forestry with special reference to Africa and Indo-Malaysia: a review. *Bulletin of Entomological Research* 79, 173–184.

Cowles, R.S. and Downer, J.A. (1995) Eucalyptus snout beetle detected in California. *California Agriculture* 49, 38–40.

Craig, T.P., Wagner, M.R., McCullough, D.G., Frantz, D.P., Raske, A.G. and Wickman, B.E. (1991) Effects of experimentally altered plant moisture stress on the performance of *Neodiprion* sawflies. Proceedings of a symposium, Towards Integrated Pest Management of Forest Defoliators, held at the 18th International Congress of Entomology, Vancouver, Canada in 1988. *Forest Ecology and Management* 39, 247–261.

Cross, D.J. (1991) Penetration of methyl bromide into *Pinus radiata* wood and its significance for export quarantine. *New Zealand Journal of Forestry Science* 21, 235–245.

Culliney, T.W., Beardsley, J.W. and Drea, J.J. (1988) Population regulation of the Eurasian pine adelgid (Homoptera: Adelgidae) in Hawaii. *Journal of Economic Entomology* 81, 142–147.

Cunningham, S.A. and Floyd, R.B. (2006) *Toona ciliata* that suffer frequent height-reducing herbivore damage by a shoot-boring moth (*Hypsipyla robusta*) are taller. *Forest Ecology and Management* 225, 400–403.

Cunningham, S.A., Floyd, R.B., Griffiths, M.W. and Wylie, F.R. (2005) Patterns of host use by the shoot-borer *Hypsipyla robusta* (Pyralidae: Lepidoptera) comparing five Meliaceae tree species in Asia and Australia. *Forest Ecology and Management* 205, 351–357.

Currie, G.A. (1937) Galls on Eucalyptus trees: a new type of association between flies and nematodes. *Proceedings of the Linnaean Society of New South Wales* 62, 147–174.

D'Amico, V. and Elkinton, J.S. (1995) Rainfall effects on transmission of gypsy moth (Lepidoptera: Lymantriidae) nuclear polyhedrosis virus. *Environmental Entomology* 24, 1144–1149.

da Costa, F.V., Neves, F.S., Silva, J.O. and Fagundes, M. (2011) Relationship between plant development, tannin concentration and insects associated with *Copaifera langsdorffii* (Fabaceae). *Arthropod–Plant Interactions* 5, 9–18.

da Silva, P.G., Hagen, K.S. and Gutierrez, A.P. (1992) Functional response of *Curinus coeruleus* (Col.: Coccinellidae) to *Heteropsylla cubana* (Horn.: Psyllidae) on artificial and natural substrates. *Entomophaga* 37, 555–564.

Daane, K.M., Sime, K.R., Dahlsten, D.L., Andrews, J.W. Jr and Zuparko, R.L. (2005) The biology of *Psyllaephagus bliteus* Riek (Hymenoptera: Encyrtidae), a parasitoid of the red gum lerp psyllid (Hemiptera: Psylloidea). *Biological Control* 32, 228–235.

Daily, G.C. and Ehrlich, P.R. (1996) Nocturnality and species survival. *Proceedings of the National Academy of Sciences of the United States of America* 93, 11709–11712.

Dall'Oglio, O.T., Zanuncio, J.C., Azevedo de Freitas, F. and Pinto, R. (2003) Hymenopteran parasitoids collected in an *Eucalyptus grandis* plantation and in a native vegetation area in Ipaba, State of Minas Gerais, Brazil. *Ciencia Florestal* 13, 123–129.

Das, G.M. and Sengupta, N. (1960) On the biology of *Rhesala moestalis* Walker (Lep.: Noctuidae), a serious pest of nursery and young shade trees in tea of north-east India. *Indian Agriculturist* 4, 95–103.

Das, S.C. (1984) Resurgence of tea mosquito bug, *Helopeltis theivora* Waterh., a serious pest of tea. *Two and a Bud* 31, 36–39.

Das, S.C., Mukherjee, S., Gope, B. and Satyanarayana, G. (1982) Termites in Cachar – an ecological study. *Two and a Bud* 29, 3–4.

Davidson, J. (1996) Developing Plantgro plant files for forest trees. In: Booth, T.H. (ed.) *Matching Trees and Sites*. Proceedings of an International Workshop, Bangkok, Thailand, 27–30 March 1995. ACIAR Proceedings No. 63, pp. 93–96.

Davies, K.A. and Giblin-Davis, R.M. (2004) The biology and associations of *Fergusobia* (Nematoda) from the *Melaleuca leucadendra*-complex in eastern Australia. *Invertebrate Systematics* 18, 291–319.

Davis, A.L.V. and Dewhurst, C.F. (1993) Climatic and biogeographical associations of Kenyan and northern Tanzanian dung beetles (Coleoptera: Scarabaeidae). *African Journal of Ecology* 31, 290–305.

Day, R.K., Kairo, M.T.K., Abraham, Y.J., Kfir, R., Murphy, S.T., Mutitu, K.E., et al. (2003) Biological control of homopteran pests of conifers in Africa. In: Neuenschwander, P., Borgemeister, C. and Langewald, J. (eds) *Biological Control of IPM Systems in Africa*. CAB International, Wallingford, UK.

de Bach, P. (ed.) (1964) *Biological Control of Insect Pests and Weeds*. Chapman and Hall, London,

de Freitas, F.A., Zanuncio, T.V., Zanuncio, J.C., Conceicao, P.M., Fialho, M.C.Q. and Bernardino, A.S. (2005) Effect of plant age, temperature and rainfall on Lepidoptera insect pests collected with light traps in a *Eucalyptus grandis* plantation in Brazil. *Annals of Forest Science* 62, 85–90.

de Little, D.W., Elliott, H.J., Madden, J.L. and Bashford, R. (1990) Stage-specific mortality in two field populations of immature *Chrysophtharta bimaculata* (Oliver) (Coleoptera: Chrysomelidae). *Journal of the Australian Entomological Society* 29, 51–55.

de Oliveira, M.L. (1999) Seasonality and daily activity of Euglossinae bees (Hymenoptera, Apidae) in terra firme forest in Central Amazonia. *Revista Brasileira de Zoologia* March 16, 83–90.

de Oliveira, N.C., Wilcken, C.F. and de Matos, C.A.O. (2004) Biological cycle and predation of three coccinellid species (Coleoptera, Coccinellidae) on giant conifer aphid *Cinara atlantica* (Wilson) (Hemiptera, Aphididae). *Revista Brasileira de Entomologia* 48, 529–533.

de Queiroz, D.L., Zanol, K.M.R., Oliveira, E.B., dos Anjos, N. and Majer, J. (2010) Feeding and oviposition preferences of *Ctenarytaina spatulata* Taylor (Hemiptera, Psyllidae) for *Eucalyptus* spp. and other Myrtaceae in Brazil. *Revista Brasileira de Entomologia* 54, 149–153.

de Souza, R.M., dos Anjos, N. and Mourao, S.A. (2009) *Apate terebrans* (Pallas) (Coleoptera: Bostrychidae) attacking Neem trees in Brazil. *Neotropical Entomology* 38, 437–439.

del Valle, J.I. and Madrigal, A. (1993) Economic impact of damage by defoliators in plantations of *Pinus patula*. *Turrialba* 43, 119–126.

DeLeon, D. (1941) Some observations on forest entomology in Puerto Rico. *Caribbean Forester* 2, 160–163.

Dell, B., Hardy, G. and Burgess, T. (2008) Health and nutrition of plantation eucalypts in Asia. *South Forest* 70, 131–138.

Dennill, G.B. (1985) The effect of the gall wasp *Trichilogaster acaciaelongifoliae* (Hymenoptera: Pteromalidae) on reproductive potential and vegetative growth of the weed *Acacia longifolia*. *Agriculture, Ecosystems and Environment* 14, 53–61.

Devine, G.J. and Furlong, M.J. (2007) Insecticide use: contexts and ecological consequences. *Agriculture and Human Values* 24, 281–306.

Dhakal, L.P., Jha, P.K. and Kjaer, E.D. (2005) Mortality in *Dalbergia sissoo* (Roxb.) following heavy infection by *Aristobia horridula* (Hope) beetles: will genetic variation in susceptibility play a role in combating declining health? *Forest Ecology and Management* 218, 270–276.

Di-Iorio, O.R. (2004) Exotic species of Cerambycidae (Coleoptera) introduced in Argentina. Part 1. The Genus *Phoracantha* Newman, 1840. *Agrociencia* 38, 503–515.

Diabangouya, M. and Gillon, Y. (2001) Adaptation of *Helopeltis schoutendeni* Reuter (Heteroptera: Miridae) to planted eucalyptus in Congo: damage and control. *Bois et Forets des Tropiques* 267(1), 6–20.

Dias-Guerra, W., de Oliveira, P.C. and Barrientos-Lozano, L. (2010) Life history and population dynamics of *Baeacris punctulatus* (Thunberg 1824) (Orhtoptera: Acrididae) in the State of Mato Grosso, Brazil. *Journal of Orthoptera Research* 19(2), 333–340.

Dibog, L., Eggleton, P., Norgrove, L., Bignell, D.E. and Hauser, S. (1999) Impacts of canopy cover on soil termite assemblages in an agrisilvicultural system in southern Cameroon. *Bulletin of Entomological Research* April 89, 125–132.

Didham, R.K. and Springate, N.D. (2003) Determinants of temporal variation in community structure. In: Basset, Y., Novotny, V., Miller, S.E. and Kitching, R. (eds) *Arthropods of Tropical Forests*. Cambridge University Press, Cambridge, pp. 28–39.

Didham, R.K., Ghazoul, J., Stork, N.E. and Davis, A.J. (1996) Insects in fragmented forests: a functional approach. *Trends in Ecology and Evolution* 11, 255–260.

Didham, R.K., Hammond, P.M., Lawton, J.H., Eggleton, P. and Stork, N.E. (1998) Beetle species responses to tropical forest fragmentation. *Ecological Monographs* 68, 295–323.

Diekmann, M., Sutherland, J.R., Nowell, D.C., Morales, F.J. and Gillard, G. (eds) (2002) FAO/ IPGRI Technical Guidelines for the Safe Movement of Germplasm No. 21. *Pinus* spp. Food and Agriculture Organisation of the United Nations, Rome/International Plant Genetic Resources Institute, Rome.

Ding, D., Li, Z., Fan, M. and Wang B. (2004) Host transfer of *Beauveria bassiana* populations in pine stand ecosystems and impact of its genetic diversity on sustainable control. *Yingyong Shengtai Xuebao* 15(12), 2315–2320.

Diodato, L. and Venturini, M. (2007) Presence of red gum lerp psyllid (*Glycaspis brimblecombei*, Hemiptera, Psyllidae), a pest of *Eucalyptus* species in Santiago del Estero, Argentina. *Quebracho* 14, 84–89.

Dix, M.E. (1996) Pest management in agroforestry systems: worldwide challenges in the 21st century. *Journal of Forestry* 94, 8–12.

Dixon, W.N. and Fasulo, T.R. (2009) Tarnished plant bug, *Lygus lineolaris* (Palisot de Beauvois) (Insecta: Hemiptera: Miridae). Institute of Food and Agricultural Sciences Extension, University of Florida. EENY-245, 5 pp.

Dobbs, T.T. and Brodel, C.F. (2004) Cargo aircraft as a pathway for the entry of nonindigenous pests into South Florida. *Florida Entomologist* 87(1), 65–78.

Doccola, J.J., Smith, S.L., Strom, B.L., Medeiros, A.C. and von Allmen, E. (2009) Systemically applied insecticides for treatment of Erythrina Gall Wasp, *Quadrastichus erythrinae* Kim (Hymenoptera: Eulophidae). *Arboriculture and Urban Forestry* 35(4), 173–181.

Dodds, K.I., Cooke, R.R. and Gilmore, D.W. (2007) Silvicultural options to reduce pine susceptibility to attack by a newly detected invasive species, *Sirex noctilio*. *Northern Journal of Applied Forestry* 24(3), 165–167.

Doganlar, M. (2005) Occurrence of *Leptocybe invasa* Fisher & La Salle, 2004 (Hym.: Chalcidoidea: Eulophidae), on *Eucalyptus camaldulensis* in Turkey, with a description of the male sex. *Zoology in the Middle East* 35, 112–114.

Doganlar, M. and Mendel, Z. (2007) First record of the Eucalyptus Gall Wasp *Ophelimus maskelli* and its parasitoid, *Closterocerus chamaeleon*, in Turkey. *Phytoparasitica* 35(4), 333–335.

Dolinski, C. and Lacey, L.A. (2007) Microbial control of arthropod pests of tropical tree fruits. *Neotropical Entomology* 36, 161–179.

Don, A., Schumacher, J. and Freibauer, A. (2011) Impact of tropical land use change on soil organic carbon stocks – a meta-analysis. *Global Change Biology* 17(4), 1658–1670.

Donald, D.G.M. (1989) Resistance of seedling progeny of *Pinus pinaster* and *P. radiata* to the pine woolly aphid *Pineus pini* L. *South African Forestry Journal* 150, 20–23.

Dottavio, C.L. and Williams, D.L. (1983) Satellite technology: an improved means for monitoring forest insect defoliation. *Journal of Forestry* 81, 30–34.

Douce, G.K., Hamilton, R.D. and Clement, G.L. (1994) 1992 Gypsy moth programs in the southeast. *Journal of Entomological Science* 29, 381–397.

Drake, D.W. (1974) Fungal and insect attack of seeds in unopened *Eucalyptus* capsules. *Search* 5, 444.

Drake, V.A. (1994) The influence of weather and climate on agriculturally important insects: an Australian perspective. *Australian Journal of Agricultural Research* 45, 487–509.

Drake, V.A. and Farrow, R.A. (1989) The 'aerial plankton' and atmospheric convergence. *Trends in Ecology and Evolution* 4, 381–385.

Drew, J., Andersen, N. and Andow, D. (2010) Conundrums of a complex vector for invasive species control: a detailed examination of the horticultural industry. *Biological Invasions* 12, 2837–2851.

Duarte, A., Menendez, J.M., Fernandez, A. and Martinez, J. (1992) Use of a biopreparation of *Metarhizium anisopliae* strain Nina Bonita in plantations of *Pinus caribaea* for control of *Rhyacionia frustrana*. *Revista Baracoa* 22, 17–23.

Dudley, N., Stein, J.D., Jones, T. and Gillette, N. (2006) Semiochemicals provide a deterrent to the black twig borer *Xylosandrus compactus* (Coleoptera: Curculionidae: Scolytinae). Proceedings 17th US Department of Agriculture Interagency Research Forum on Gypsy Moth and Other Invasive Species. USDA Forest Service General Technical Report NRS-P-10, p. 34.

Duke, N.C. (2002) Sustained high levels of foliar herbivory of the mangrove *Rhizophora stylosa* by a moth larva *Doratifera stenosa* (Limacodidae) in north-eastern Australia. *Wetlands Ecology and Management* 10, 403–419.

Dunbar, C.S. and Wagner, M.R. (1992) Bionomics of *Neodiprion gillettei* (Hymenoptera: Diprionidae) on *Pinus ponderosa*. *Annals of the Entomological Society of America* 85, 286–292.

Durairaj, S. and Muthazhagu, S. (2009) Biological effects of neem. In: Singh, K.K. (ed.) *Neem: A Treatise*. I.K International Pvt. Ltd., pp. 286–298.

Duryea, M.L., Kampf, E., Littell, R.C. and Rodriguez-Pedraza, C.D. (2007) Hurricanes and the urban forest: II. Effects on tropical and subtropical tree species. *Arboriculture and Urban Forestry* 33, 98–112.

Dvorak, W.S. (2004) World view of *Gmelina arborea*: opportunities and challenges. *New Forests* 28, 111–126.

Dwinell, L.D. (1997) The pinewood nematode: regulation and mitigation. *Annual Review of Phytopathology* 35, 153–166.

Edward, E., Chamshama, S.A.O. and Mugasha, A.G. (2006) Growth performance of lesser-known *Leucaena* species/provenances at Gairo inland plateau, Morogoro, Tanzania. *Southern African Forestry Journal*, 53–62.

Eggleton, P. and Bignell, D.E. (1995) Monitoring the response of tropical insects to change in the environment: the trouble with termites. In: Harrington, R. and Stork, N.E. (eds) *Insects in a Changing Environment*. Academic Press, London, pp. 473–497.

Eggleton, P., Bignell, D.E., Sands, W.A., Waite, B., Wood, T.G. and Lawton, J.H. (1995) The species richness of termites (Isoptera) under differing levels of forest disturbance in the Mbalmayo Forest Reserve, southern Cameroon. *Journal of Tropical Ecology* 11, 85–98.

Eichhorn, M.P., Fagan, K.C., Compton, S.G., Dent, D.H. and Hartley, S.E. (2007) Explaining leaf herbivory rates on tree seedlings in a Malaysian rain forest. *Biotropica* 39, 416–421.

Eidmann, H.H. (1983) Management of the spruce bark beetle *Ips typographus* in Scandanavia using pheromones. In: *10th International Congress of Plant Protection*, Vol. 3. Proceedings of a conference, Brighton, UK, 20–25 November 1983, pp. 1042–1050.

El Mujtar, V., Covelli, J., Delfino, M.A. and Grau, O. (2009) Molecular identification of *Cinara cupressi* and *Cinara tujafilina* (Hemiptera: Aphidae). *Environmental Entomology* 38(2), 505–512.

El Yousfi, M. (1989) The principles of control of *Phoracantha semipunctata* Fabr. *Boletin de Sanidad Vegetal, Plagas* 15, 129–137.

El Yousfi, M. (1992) Possibilities of control of *Phoracantha semipunctata* (F.) by selection of *Eucalyptus* species. *Boletin de Sanidad Vegetal, Plagas* 18, 735–743.

El-Sayed, A.M., Suckling, D.M., Wearing, C.H. and Byers, J.A. (2006) Potential of mass trapping for long-term pest management and eradication of invasive species. *Journal of Economic Entomology* 99(5), 1550–1564.

El-Sayed, A.M., Suckling, D.M., Byers, J.A., Jang, E.B. and Wearing, C.H. (2009) Potential of 'lure and kill' in long-term pest management and eradication of invasive species. *Journal of Economic Entomology* 102(3), 815–835.

Elliott, H.J., Bashford, R. and Greener, A. (1993) Effects of defoliation by the leaf beetle, *Chrysophtharta bimaculata*, on growth of *Eucalyptus regnans* plantation in Tasmania. *Australian Forestry* 56, 22–26.

Elliott, H.J., Ohmart, C.P. and Wylie, F.R. (1998) *Insect Pests of Australian Forests: Ecology and Management*. Inkata Press, Melbourne, Australia, 214 pp.

Elton, C.S. (1925) The dispersal of insects to Spitsbergen. *Transactions of the Royal Entomological Society of London* 73(1–2), 289–299.

EPPO (1998) Guidelines on pest risk analysis: check-list of information required for pest risk analysis (PRA). European and Mediterranean Plant Protection Organisation Standards PM 5/1 (1).

EPPO (2010) *Xylosandrus crassiusculus* (Coleoptera: Scolytidae) (http://www.eppo.org/QUARANTINE/Alert_List/insects/xylosandrus_crassiusculus.htm).

Epstein, P.R. (1995) Emerging diseases and ecosystem instability: new threats to public health. *American Journal of Public Health* 85, 168–172.

Erasmus, M.J. and Chown, S.L. (1994) Host location and aggregation behaviour in *Hylastes angustatus* (Herbst) (Coleoptera: Scolytidae). African *Entomology* 2, 7–11.

Erbilgin, N., Dahlsten, D.L. and Chen, P.Y. (2004) Intraguild interactions between generalist predators and an introduced parasitoid of *Glycaspis brimblecombei* (Homoptera: Psylloidea). *Biological Control* 31, 329–337.

Ernst, W.H.O., Tolsma, D.J. and Decelle, J.E. (1989) Predation of seeds of *Acacia tortilis* by insects. *Oikos* 54, 294–300.

Ernst, W.H.O., Decelle, J.E., Tolsma, D.J. and Verweij, R.A. (1990) Lifecycle of the bruchid beetle *Bruchidius uberatus* and its predation of *Acacia nilotica* seeds in a tree savanna in Botswana. *Entomologia Experimentalis et Applicata* 57, 177–190.

Erskine, P.D., Lamb, D. and Bristow, M. (2006) Tree species diversity and ecosystem function: can tropical multi-species plantations generate greater productivity? *Forest Ecology and Management* 233, 205–210.

Erthal, M. and Tonhasca, A. (2000) Biology and oviposition behaviour of the phorid *Apocephalus attophilus* and the response of its host, the leaf-cutting ant *Atta laevigata*. *Entomologia Experimentalis et Applicata* 95, 71–75.

Erwin, T.L. (1982) Tropical forests: their richness in Coleoptera and other arthropod species. *Coleopterists Bulletin* 30, 14–19.

Espinoza, J.A (2004) Within-tree density gradients in *Gmelina arborea* in Venezuela. *New Forests* 28, 309–317.

Eungwijarnpanya, S. and Yincharoen, S. (2002) Control of teak defoliator, *Hyblaea puera* Cramer (Lepidoptera: Hyblaeidae), by thermal fogger application of neem extract. In: Hutacharern, C., Napompeth, B., Allard, G. and Wylie, F.R. (eds) *Pest Management in Tropical Forest Plantations*. FAO Forestry Research Support Programme for Asia and the Pacific, Bangkok, pp. 123–125.

Evans, J. (1992) *Plantation Forestry in the Tropics: Tree Planting for Industrial, Social, Environmental, and Agroforestry Purposes*. Clarendon Press, Oxford, UK.

Evans, J. and Turnbull, J.W. (2004) *Plantation Forestry in the Tropics: The Role, Silviculture, and Use of Planted Forests for Industrial, Social, Environmental, and Agroforestry Purposes*. Oxford University Press, pp. 467.

Ezcurra, E., Gomez, J.C. and Becerra, J. (1987) Diverging patterns of host use by phytophagous insects in relation to leaf pubescence in *Arbutus xalapensis* (Ericaceae). *Oecologia* 72, 479–480.

FABI (2007) Blue gum chalcid. Pest Alert. Forestry and Agriculture Biotechnology Institute, Tree Protection Cooperative Programme (http://www.fabinet.up.ac.za/tpcp/Leptocybe_alert).

Faizal, M.H., Prathapan, K.D., Anith, K.N., Mary, C.A., Lekha, M. and Rini, C.R. (2006) *Erythrina* gall wasp *Quadrastichus erythrinae*, yet another invasive pest new to India. *Current Science* 90(8), 1061–1062.

FAO (1997) *State of the World's Forests*. Food and Agriculture Organization of the United Nations, Rome, 202 pp.

FAO (1998) International standards for phytosanitary measures: guidelines for surveillance. Publication No. 6. Food and Agriculture Organisation of the United Nations, Rome.

FAO (2002) International standards for phytosanitary measures: guidelines for regulating wood packaging material in international trade. Publication No. 15. Food and Agriculture Organisation of the United Nations, Rome.

FAO (2004) International standards for phytosanitary measures: pest risk analysis for quarantine pests, including analysis of environmental risks and living modified organisms. Publication No. 11. Food and Agriculture Organisation of the United Nations, Rome.

FAO (2007) *Heteropsylla cubana* Crawford, 1914. Forest Pest Species Profile. FAO, Rome. (www.fao.org/forestry/13564-1-0.pdf)

FAO (2009a) Asia-Pacific Forestry Week: Forestry in a changing world. A summary of events of the first Asia-Pacific Forestry Week, Hanoi, Vietnam, 21–26 April 2008. RAP Publication 2009/04, 116 pp.

FAO (2009b) Global review of forest pests and diseases. FAO Forestry Paper 156, pp. 222.

FAO (2010) Global forest resources assessment. Forestry Paper 163. Food and Agriculture Organization of the United Nations, Rome.

Farley, K.A. (2007) Grasslands to tree plantations: forest transition in the Andes of Ecuador. *Annals of the Association of American Geographers*, 97(4): 755–771.

Fasoranti, J.O. and Ajiboye, D.O. (1993) Some edible insects of Kwara State, Nigeria. *American Entomologist* 39, 113–116.

Faveri, S.B., Vasconcelos, H.L. and Dirzo, R. (2008) Effects of Amazonian forest fragmentation on the interaction between plants, insect herbivores, and their natural enemies. *Journal of Tropical Ecology* 24, 57–64.

Feaver, M.N. (1985) Grasshoppers (Orthoptera: Arrcrididae) damage to pine seedlings at night in a seed orchard. *Florida Entomologist* 68, 694–696.

Febles, G., Ruiz, T., Torres, V., Achan, G., Noda, A. and Alonso, J. (2008) Preliminary studies on the resistance of tropical tree species to the pest *Atta insularis* (Guer) in Cuba. *Cuban Journal of Agricultural Science* 42, 405–409.

Feller, I.C. (1995) Effects of nutrient enrichment on growth and herbivory of dwarf red mangrove (*Rhizophora mangle*). *Ecological Monographs* 65, 477–505.

Fernandes, M.E.B., Nascimento, A.A.M. and Carvalho, M.L. (2009) Effects of herbivory by *Hyblaea puera* (Hyblaeidae: Lepidoptera) on litter production in the mangrove on the coast of Brazilian Amazonia. *Journal of Tropical Ecology* 25, 337–339.

Ferreira Filho, P.J., Wilcken, C.F., de Oliveira, N.C., Ferreira do Amaral Dal Pogetto, M.H. and Vianna Lima, A.C. (2008) Population dynamics of red gum lerp psyllid, *Glycaspis brimblecombei* (Moore, 1964) (Hemiptera: Psyllidae) and its parasitoid, *Psyllaephagus bliteus* (Hymenoptera: Encyrtidae), in *Eucalyptus camaldulensis* plantation. *Ciencia Rural* 38, 2109–2114.

Ferreira, M.C. (1993) Forest nursery cultural practices: impact on pests affecting seedlings. In: Perrin, R. and Sutherland, J.R. (eds) *Diseases and Insects in Forest Nurseries*. Proceedings of IUFRO Conference, Dijon, France, 3–10 October 1993. INRA, Paris, pp. 163–170.

Field, S.P., Momuat, E.O., Ayre, S.R. and Bamualim, A. (1992) Developing land rehabilitation technologies compatible with farmer needs and resources. *Journal of the Asian Farming Systems Association* 1, 385–394.

Filho, O.P., Dorval, A., Schimdt, V., Berti Filho, E. and Moura, R.G. (2010) Biology of *Thyrinteina arnobia* on leaves of *Eucalyptus camaldulensis*, *Corymbia citridora*, *Psidium guajava* and *Vernonia condensata*. *Revista de Agricultura (Piracicaba)* 85(1), 1–14.

Filip, V. and Dirzo, R. (1985) Life table of the bagworm *Malacosoma incurvum* var. *aztecum* Neumogen (Lepidoptera: Lasiocampidae) in Xochimilco, D.F. Mexico. *Folia Entomologica Mexicana* 31–45.

Firmino-Winckler, D.C., Wilcken, C.F., de Oliveira, N.C. and de Matos, C.A.O. (2009) Red gum lerp psyllid *Glycaspis brimblecombei* Moore (Hemiptera, Psyllidae) biology in *Eucalyptus* spp. *Revista Brasileira de Entomologia* 53(1), 144–146.

Flechtmann, C.A.H., Ottati, A.L.T. and Berisford, C.W. (2001) Ambrosia and bark beetles (Scolytidae: Coleoptera) in pine and eucalypt stands in southern Brazil. *Forest Ecology and Management* 142, 183–191.

Fleming, M.R., Hoover, K., Janowiak, J.J., Fang, Y., Wang, X., Liu, W.M., Wang, Y.J., Hang, X.X., Agrawal, D., Mastro, V.C., Lance, D.R., Shield, J.E. and Roy, R. (2003) Microwave irradiation of wood packing material to destroy the Asian longhorned beetle. *Forest Products Journal* 53(1), 46–52.

Fletcher, L.E. (2007) Vibrational signals in a gregarious sawfly larva (*Perga affinis*): group coordination or competitive signalling? *Behavioural Ecology and Sociobiology* 61, 1809–1821.

Floater, G.J. and Zalucki, M.P. (2000) Habitat structure and egg distribution in the processionary caterpillar *Ochrogaster lunifer*: lessons for conservation and pest management. *Journal of Applied Ecology* 37, 87–99.

Floyd, R. and Hauxwell, C. (2000) Proceedings of an International Workshop on *Hypsipyla* Shoot Borers of the Meliaceae, Kandy, Sri Lanka, 1996, ACIAR.

Floyd, R.B., Hauxwell, C., Griffiths, M., Horak, M., Sands, D.P.A., Speight, M.R., Watt, A.D. and Wylie, F.R. (2003) Taxonomy, ecology, and control of *Hypsipyla* shoot borers of Meliaceae. *Big-leaf Mahogany: Genetics, Ecology and Management* 159, 381–394.

Foggo, A. and Speight, M.R. (1993) Root damage and water stress: treatments affecting the exploitation of the buds of common ash *Fraxinus excelsior* L., by larvae of the ash bud moth *Prays fraxinella* Bjerk. (Lep., Yponomeutidae). *Oecologia Heidelberg* 96, 134–138.

Foggo, A., Speight, M.R. and Gregoire, J.C. (1994) Root disturbance of common ash, *Fraxinus excelsior* (Oleaceae), leads to reduced foliar toughness and increased feeding by a folivorous weevil, *Stereonychus fraxini* (Coleoptera, Curculionidae). *Ecological Entomology* 19, 344–348.

Foggo, A., Ozanne, C.M.P., Hambler, C. and Speight, M.R. (2000) Edge effects in forest canopies: status and threats. In: Linsenmair, K.E., Davis A.J. and Speight, M.R. (eds) *Tropical Forest Canopies. Plant Ecology* 153(1–2), 347–359.

Foggo, A., Ozanne, C.M.P., Speight, M.R. and Hambler, C. (2001) Edge effects and tropical forest canopy invertebrates. *Plant Ecology* 153, 347–359.

Folgarait, P.J., Marquis, R.J., Ingvarsson, P., Braker, H.E. and Arguedas, M. (1995) Patterns of attack by insect herbivores and a fungus on saplings in a tropical tree plantation. *Environmental Entomology* 24, 1487–1494.

Follett, P.A. and Roderick, G.K. (1996) Genetic estimates of dispersal ability in the leucaena psyllid predator *Curinus coeruleus* (Coleoptera: Coccinellidae): implications for biological control. *Bulletin of Entomological Research* 86, 355–361.

Ford, L.B. (1978) The status of pine insects in Costa Rica. Meeting of IUFRO Working Parties S2.06.12 and S2.07.07, 'Pests and Diseases of Pines in the Tropics', 'Piedras Blancas', Medillin, Columbia, 3–14 September 1978.

Ford, L.B. (1986) The Nantucket pine tip moth. *Turrialba* 36, 245–248.

Fowler, H.G. and Robinson, S.W. (1979) Field identification and relative pest status of Paraguayan leaf-cutting ants. *Turrialba* 29, 11–16.

Fowler, S.V. (2004) Biological control of an exotic scale, *Orthezia insignis* Browne (Homoptera: Ortheziidae), saves the endemic gumwood tree, *Commidendrum robustum* (Roxb.) DC. (Asteraceae) on the island of St. Helena. *Biological Control* 29, 367–374.

Fox, L.R. and Morrow, P.A. (1983) Estimates of damage by herbivorous insects on eucalyptus trees. *Australian Journal of Ecology* 8, 139–147.

Fox, L.R. and Morrow, P.A. (1992) Eucalypt responses to fertilization and reduced herbivory. *Oecologia* 89, 214–222.

Franceschi, V.R., Krokene, P., Christiansen, E. and Krekling, T. (2005) Anatomical and chemical defences of conifer bark against bark beetles and other pests. *New Phytologist* 167(2), 353–357.

Fredericksen, T.S., Hedden, R.L. and Williams, S.A. (1995) Susceptibility of loblolly pine to bark beetle attack following simulated wind stress. *Forest Ecology and Management* 76, 95–107.

Freer-Smith, P. and Carnus, J.M. (2008) The sustainable management and protection of forests: analysis of the current position globally. *Ambio* 37, 254–262.

French, J.R.J. and Ahmed, B. (2006) Future termite control requires partnerships between industry, government and people (Isoptera). *Sociobiology* 48, 599–620.

Froggatt, W.W. (1923) *Forest Insects of Australia*. Government Printer, Sydney, Australia.

Froggatt, W.W. (1927) *Forest Insect and Timber Borers*. Government Printer, Sydney, Australia.

Fryer, J.H. (1996) Climatic mapping for eucalypts in Central America. In: Booth, T.H. (ed.) *Matching Trees and Sites*. Proceedings of an International Workshop, Bangkok, Thailand, 27–30 March 1995. ACIAR Proceedings No. 63, pp. 50–55.

Fujisaka, S., Dumanski, J., Pushparajah, E., Latham, M. and Myers, R. (1991) Thirteen reasons why farmers do not adopt innovations intended to improve the sustainability of upland agriculture. *Evaluation for Sustainable Land Management in the Developing World* 2, 509–522.

Furuta, K. and Aloo, I.K. (1994) Between-tree distance and spread of the Sakhalin fir aphid (*Cinara todocola* Inouye) (Horn., Aphididae) within a plantation. *Journal of Applied Entomology* 117, 64–71.

Futuyama, D.J. and Mitter, C. (1996) Insect–plant interactions: the evolution of component communities. *Philosophical Transactions of the Royal Society of London B Biological Sciences* 351, 1361–1366.

Futuyama, D.J. and Mitter, C. (1997) Insect–plant interactions: the evolution of component communities. *Plant Life Histories: Ecology, Phylogeny and Evolution*, pp. 253–264.

Gadgil, P.D. and Flint, T.N. (1983) Assessment of the risk of introduction of exotic forest insects and diseases with imported tents. *New Zealand Journal of Forestry* 28(1), 58–67.

Gadgil, P.D., Bulman, L.S., Crabtree, R., Watson, R.N., O'Neil, J.C. and Glassey, K.L. (2000) Significance to New Zealand Forestry of contaminants on the external surfaces of shipping containers. *New Zealand Journal of Forestry Science* 30(3), 341–358.

Gaiad, D.C.M., Figueiredo, F.A., Oliveira, E.B. and Penteado, S.R.C. (2003) Evolution of *Sirex noctilio* infestation in relation to diameter distribution of *Pinus taeda*. *Floresta* 33(1), 21–29.

Gajaseni, J. (1992) Socioeconomic aspects of Taungya. In: Jordan, C.F., Gajaseni, J. and Watanabe, H. (eds) *Taungya: Forest Plantations with Agriculture in Southeast Asia*. CAB International, Wallingford, UK.

Gara, R.I., Sarango, A. and Cannon, P.G. (1990) Defoliation of an Ecuadorian mangrove forest by the bagworm, *Oiketicus kirbyi* Guilding (Lepidoptera: Psychidae). *Journal of Tropical Forest Science* 3, 181–186.

Garg, V.K. and Tomar, D.S. (2008) Preferential feeding of bark eating caterpillar, *Indarbela quadrinotata* (Walker) on horticultural and forest plant species. *Indian Forester* 134(4), 579–580.

Garraway, E. (1986) The biology of *Ips calligraphus* and *Ips grandicollis* (Coleoptera: Scolytidae) in Jamaica. *Canadian Entomologist* 118, 113–121.

Gassmann, A.J., Fabrick, J.A., Sisterson, M.S., Hannon, E.R., Stock, S.P., Carriere, Y. and Tabashnik, B.E. (2009) Effects of pink bollworm resistance to *Bacillus thuringiensis* on phenoloxidase activity and susceptibility to entomopathogenic nematodes. *Journal of Economic Entomology* 102, 1224–1232.

Ge, Q.J., Guan, L.Q., Li, Z.Z., Xue, X.Q. and Zhou, X.G. (1988) A study on the control threshold of *Dendrolimus punctatus* Walker. *Journal of Nanjing Forestry University* 94–100.

Gebeyehu, S., Hurley, B.P. and Wingfield, M.J. (2005) A new lepidopteran insect pest discovered on commercially grown *Eucalyptus nitens* in South Africa. *South African Journal of Science* 101, 26–28.

Geiger, C.A. and Gutierrez, A.P. (2000) Ecology of *Heteropsylla cubana* (Homoptera: Psyllidae): psyllid damage, tree phenology, thermal relations and parasitism in the field. *Population Ecology* 29(1), 76–86.

Geiger, C.A., Napompeth, B. and Van Den Beldt, R. (1995) An update on the status of the leucaena psyllid in South-east Asia. In: Shelton, H.M., Piggin, C.M. and Brewbaker, J.L. (eds) *Leucaena – Opportunities and Limitations*. ACIAR Proceedings No. 57, pp. 125–128.

Georgis, R., Koppenhofer, A.M., Lacey, L.A., Belair, G., Duncan, L.W., Grewal, P.S., Samish, M., Tan, L., Torr, P. and vanTol, R.W. (2006) Successes and failures in the use of parasitic nematodes for pest control. *Biological Control* 38, 103–123.

Ghani, M.A. and Cheema, M.A. (1973) Biology, ecology and behaviour of principal natural enemies of major insect pests of forest trees in Pakistan. Miscellaneous Publication No. 4, Commonwealth Institute of Biological Control, Commonwealth Agricultural Bureau, UK, 100 pp.

Ghini, R., Bettiol, W. and Hamada, E. (2011) Diseases in tropical and plantation crops as affected by climate changes: current knowledge and perspectives. *Plant Pathology* 60, 122–132.

Ghodeswar, K.S., Meenakshikutty, C., Thippanna, G.M., Varkey, K.J., Sivagnanam, V., Arunachalam, A.S., et al. (1992) The interceptions of insect pests and fungi on imported wooden logs through the ports of New Mangalore and Kawar (Karnataka) from 1989 to 1991. *Plant Protection Bulletin Faridabad* 44, 1–5.

Ghosh, S.N. (1993) Effect of Eucalyptus (*Eucalyptus tereticornis*) plants as intercrop in the cashew plantation – a case study in West Bengal. *Cashew* 7, 17–19.

Gillette, N.E., Stein, J.D., Owen, D.R., Webster, J.N. and Mori, S.R. (2006) Pheromone-based disruption of *Eucosma sonomana* and *Rhyacionia zozana* (Lepidoptera: Tortricidae) using aerially applied microencapsulated pheromone.

Canadian Journal of Forest Research 36(2), 361–368.

Gilsen, T. (1948) Aerial plankton and its conditions of life. *Biological Reviews* 23(2), 109–126.

Girardi, G.S., Gimenez, R.A. and Braga, M.R. (2006) Occurrence of *Platypus mutatus* Chapuis (Coleoptera: Platypodidae) in a brazilwood experimental plantation in southeastern Brazil. *Neotropical Entomology* 35(6), 864–867.

Glenday, J. (2008) Carbon storage and carbon emission offset potential in an African riverine forest, the lower Tana river forests, Kenya. *Journal of East African Natural History* 97, 207–223.

Glover, N. and Heuveldop, J. (1985) Multipurpose tree trials in Acosta-Puriscal, Costa Rica. *Nitrogen Fixing Tree Research Reports* 3, 4–6.

Goeltenboth, F. (1990) *Subsistence Agriculture Improvement: Manual for the Humid Tropics*. Wau Ecology Institute Handbook No. 10.

Goodyer, G.J. (1985) Cutworm Caterpillars. Agfact AE40. New South Wales Department of Agriculture.

Goolsby, J.A., Makinson, J. and Purcell, M. (2000) Seasonal phenology of the gall-making fly *Fergusonina* sp. (Diptera: Fergusoninidae) and its implications for biological control of *Melaleuca quinquenervia*. *Australian Journal of Entomology* 39, 336–343.

Goolsby, J.A., Burwell, C.J., Makinson, J. and Driver, F. (2001) Investigation of the biology of Hymenoptera associated with *Fergusonina* sp. (Diptera: Fergusoninidae), a gall fly of *Melaleuca quinquenervia*, integrating molecular techniques. *Journal of Hymenoptera Research* 10(2), 163–180.

Gorlushkina, V.P. (1987) Resistance factors of *Pinus sylvestris* L. plantations to pests of buds and shoots. *Lesovedenie* 65–71.

Gotoh, T. (1994) Insect borers of some valuable timber species in Thailand. *Tropical Forestry* 30, 30–37.

Gotoh, T., Kotulai, J.R. and Matsumoto, K. (2003) Stem borers of teak and yemane in Sabah, Malaysia, with analysis of attacks by the teak beehole borer (*Xyleutes ceramica* Wlk.). *Jarq-Japan Agricultural Research Quarterly* 37, 253–261.

Gotoh, T., Eungwijarnpanya, S., Yincharoen, S., Choldumrongkul, S., Nakamuta, K., Pholwicha, P., *et al.* (2007) Emergence, oviposition and larval behaviors in the teak beehole borer (*Xyleutes ceramica* Wlk.) in northern Thailand (Lepidoptera: Cossidae). *Jarq-Japan Agricultural Research Quarterly* 41, 307–314.

Gottlieb, O.R. and Borin, M.R.D.M.B. (1995) The diversity of plants: Where is it? Why is it there? What will it become? *Anais da Academia Brasileira de Ciencias* 66, 55–82.

Goulet, E., Rueda, A. and Shelton, A. (2005) Management of the mahogany shoot borer, *Hypsipyla grandella* (Zeller) (Lepidoptera: Pyralidae), through weed management and insecticidal sprays in 1- and 2-year-old *Swietenia humilis* Zucc. plantations. *Crop Protection* 24, 821–828.

Govender, P. (2007) Status of seedling establishment pests of *Acacia mearnsii* De Wild. (Mimosaceae) in South Africa. *South African Journal of Science* 103, 141–148.

Govender, P. and Ingham, D. (1998) Practical recommendations for monitoring the brown wattle mirid, *Lygidolon laevigatum* Reuter (Hemiptera: Miridae). ICFR Bulletin Series No. 09/98.

Govender, P. and Wingfield, M.J. (2005) Overview on the entomological research in the Mediterranean forest ecosystems of South Africa. In: Lieutier, F. and Ghaioule, D. (eds) *Entomological Research in Mediterranean Forest Ecosystems*. Institut National De La Recherche Agronomique, Paris, 275 pp.

Gowda, A.N.S. and Narayana, R. (1998) Spike disease of sandal (*Santalum album* L.): a patho-physiological study. In: Radomiljac, H.S., Ananthapadmanabho, H.S., Welbourn, R.M. and Satyanarayana, R. (eds) *Sandal and its Products*. Proceedings of an International Seminar, 18–19 December 1997. ACIAR Proceedings No. 84, Bangalore, India, pp. 175–180.

Gray, B. (1974) Sex ratio, relative size, and order of attack of adults of *Hylurdrectonus araucariae* (Coleoptera: Scolytidae). *Annals of the Entomological Society of America* 67, 144–145.

Gray, B. (1975) Distribution of *Hylurdrectonus araucariae* Schedl (Coleoptera: Scolytidae) and progress of outbreak in major hoop pine plantations in Papua New Guinea. *Pacific Insects* 16, 383–394.

Gray, B. (1976) Infestation, susceptibility and damage to *Araucaria* plantations in Papua New Guinea by *Hylurdrectonus araucariae* Schedl (Coleoptera: Scolytidae). *Bulletin of Entomological Research* 66, 695–711.

Gray, B. and Barber, I.A. (1974) Studies on *Vanapa oberthuri* Pouillaude (Coleoptera: Curculionidae), a pest of hoop pine plantations in Papua New Guinea. *Zeitschrift für Angewandte Entomologie* 76, 394–405.

Gray, B. and Buchter, J. (1969) Termite eradication in *Araucaria* plantations in New Guinea. *Commonwealth Forestry Review* 48, 201–207.

Gray, B. and Howcroft, N.H.S. (1970) Notes on the incidence, attack, associated insects and control of *Vanapa oberthuri* Pouillaude (Coleoptera: Curculionidae), a pest of hoop pine, *Araucaria cunninghamii* D. on plantations in New Guinea. *Zeitschrift für Angewandte Entomologie* 66, 248–256.

Gray, B. and Lamb, K.P. (1975) Biology of *Hylurdrectonus araucariae* Schedl (Coleoptera: Scolytidae), a pest of hoop pine plantations in Papua New Guinea. *Bulletin of Entomological Research* 65, 21–32.

Greenslade, P. (2008) Climate variability, biological control and an insect pest outbreak on Australia's Coral Sea islets: lessons for invertebrate conservation. *Journal of Insect Conservation* 12, 333–342.

Griffiths, M. (1997) The biology and host relations of the red cedar tip moth, *Hypsipyla robusta* Moore (Lepidoptera: Pyralidae) in Australia. PhD thesis, Department of Entomology, University of Queensland, St Lucia.

Griffiths, M.W. (2001) The biology and ecology of *Hypsipyla* shoot borers. In: Floyd, RB. and Hauxwell, C. (eds) *Hypsipyla shoot borers in Meliaceaae*. Proceedings of an International Conference held at Kandy, Sri Lanka, 20–23 August 1996. ACIAR Proceedings No. 97, pp. 74–80.

Grijpma, P. (1973) Proceedings of the First Symposium on Integrated Control of *Hypsipyla*, Turrialba, Costa Rica, 5–12 March 1973.

Grimbacher, P.S. and Stork, N.E. (2009) How do beetle assemblages respond to cyclonic disturbance of a fragmented tropical rainforest landscape? *Oecologia* 161, 591–599.

Grissell, E.E. (2006) A new species of *Megastigmus* Dalman (Hymenoptera: Torymidae), galling seed capsules of *Eucalyptus camaldulensis* Dehnhardt (Myrtaceae) in South Africa and Australia. *African Entomology* 14(1), 87–94.

Grosman, A.H., van Breemen, M., Holtz, A., Pallini, A., Rugama, A.M., Pengel, H., Venzon, M., Zanuncio, J.C., Sabell's, M.W. and Tanssen, A. (2005) Searching behaviour of an omnivorous predator for novel and native host plants of its herbivores: a study on arthropod colonization of eucalyptus in Brazil. *Entomologia Experimentalis et Applicata* 116, 135–142.

Grosman, D.M., Upton, W.W., McCook, F.A. and Billings, R.F. (2002) Systemic insecticide injections for control of cone and seed insects in loblolly pine seed orchards – 2 year results. *Southern Journal of Applied Forestry* 26(3), 146–152.

Guan, Y.-Q., Chen, J.M., Li, Z.B., Feng, Q.L. and Liu, J.-M. (2011) Immobilisation of bifenthrin for termite control. *Pest Management Science* 67, 244–251.

Guedes, R.N.C., Zanuncio, T.V., Zanuncio, J.C. and Medeiros, A.G.B. (2000) Species richness and fluctuation of defoliator Lepidoptera populations in Brazilian plantations of *Eucalyptus grandis* as affected by plant age and weather factors. *Forest Ecology and Management* 137, 179–184.

Guillade, A.C. and Folgarait, P.J. (2011) Life-history traits and parasitism rates of four phorid species (Diptera: Phoridae), parasitoids of *Atta vollenweideri* (Hymenoptera: Formicidae) in Argentina. *Journal of Economic Entomology* 104(1), 32–40.

Gul, H. and Chaudhry, M.I. (1992) Some observations on natural enemies of poplar borers in Pakistan. *Pakistan Journal of Forestry* 42, 214–222.

Haack, R.A. (2001) Intercepted Scolytidae (Coleoptera) at U.S. ports of entry: 1985–2000. *Integrated Pest Management Reviews* 6, 253–282.

Haack, R.A. (2006) Exotic bark- and wood-boring Coleoptera in the United States: recent establishments and interceptions. *Canadian Journal of Forest Research* 36, 269–288.

Haack, R.A. and Paiz-Schwartz, G. (1997) Bark beetle (Coleoptera: Scolytidae) outbreak in pine forests of the Sierra de las Minas biosphere reserve, Guatemala. *Entomological News* 108, 67–76.

Haack, R.A. and Petrice, T.R. (2009) Bark- and wood-borer colonization of logs and lumber after heat treatment to ISPM specifications: the role of residual bark. *Journal of Economic Entomology* 102(3), 1075–1084.

Haack, R.A. and Poland, T.M. (2001) Evolving management strategies for a recently discovered exotic forest pest: the pine shoot beetle, *Tomicus piniperda* (Coleoptera). *Biological Invasions* 3, 307–322.

Haack, R.A., Billings, R.F. and Richter, A.M. (1990) Life history parameters of bark beetles (Coleoptera: Scoltytidae) attacking West Indian pines in the Dominican Republic. *Florida Entomologist* 73, 591–603.

Hackett, C. (1988) *Matching Plants and Land: Development of a Broadscale System from a Crop Project for Papua New Guinea*. Natural Resources Series No. 11. CSIRO Division of Water and Land Resources, Melbourne, Australia, 82 pp.

Haddad, N.M. and Baum, K.A. (1999) An experimental test of corridor effects on butterfly densities. *Ecological Applications* May 9, 623–633.

Hadlington, P. and Hoschke, F. (1959) Observations on the ecology of the phasmatid *Ctenomorphodes tessulatus* (Gray). *Proceedings of the Linnaean Society of New South Wales* 84, 146–159.

Haines, M.W., Nikles, D.G. and Spidy, T. (1988) A multiple-source approach to selection in Honduras Caribbean pine seed orchard establishment for the Northern Territory of Australia. *Commonwealth Forestry Review* 67, 35–40.

Haines, W.P., Heddle, M.L., Welton, P. and Rubinoff, D. (2009) A recent outbreak of the Hawaiian Koa Moth, *Scotorythra paludicola* (Lepidoptera: Geometridae), and a review of outbreaks between 1892 and 2003. *Pacific Science* 63, 349–369.

Hajek, A.E. and Delalibera, I. (2010) Fungal pathogens as classical biological control agents against arthropods. *Biocontrol* 55, 147–158.

Hajek, A.E., McManus, M.L. and Delalibera, I. (2007) A review of introductions of pathogens and nematodes for classical biological control of insects and mites. *Biological Control* 41, 1–13.

Halbert, S.E., Gill, R. and Nisson, J.N. (2003) *Eucalyptus* psyllid, *Blastopsylla occidentalis* Taylor and red gum lerp psyllid, *Glycaspis brimblecombei* Moore (Insecta: Hemiptera: Psyllidae). University of Florida, IFAS Extension EENY-306, 4 pp.

Hambler, C. (2004) *Conservation*. Cambridge University Press, Cambridge, UK.

Hambler, C. and Speight, M.R. (1995) Biodiversity conservation in Britain: science replacing tradition. *British Wildlife* 6, 137–148.

Hambler, C. and Speight, M.R. (2004) Extinction rates and butterflies. *Science* 305, 1563–1563.

Hanks, L.M. and Denno, R.F. (1993) Natural enemies and plant water relations influence the distribution of an armored scale insect. *Ecology Washington DC* 74, 1081–1091.

Hanks, L.M., Paine, T.D. and Millar, J.G. (1991) Mechanisms of resistance in *Eucalyptus* against larvae of the eucalyptus longhorned borer (Coleoptera: Cerambycidae). *Environmental Entomology* 20, 1583–1588.

Hanks, L.M., Millar, J.G. and Paine, T.D. (1995a) Biological constraints on host-range expansion by the wood-boring beetle *Phoracantha semipunctata* (Coleoptera: Cerambycidae). *Annals of the Entomological Society of America* 88, 183–188.

Hanks, L.M., Paine, T.D., Millar, J.G. and Horn, J.L. (1995b) Variation among Eucalyptus species in resistance to eucalyptus longhorned borer in Southern California. *Entomologia Experimentalis et Applicata* 74, 185–194.

Hanks, L.M., Paine, T.D. and Millar, J.G. (1996) Tiny wasp helps protect eucalypts from eucalyptus longhorned borer. *California Agriculture* 50, 14–16.

Hanks, L.M., Millar, J.G., Paine, T.D., Wang, Q. and Paine, E.O. (2001) Patterns of host utilization by two parasitoids (Hymenoptera: Braconidae) of the eucalyptus longhorned borer (Coleoptera: Cerambycidae). *Biological Control* 21, 152–159.

Hanks, L.M., Paine, T.D. and Millar, J.G. (2005) Influence of the larval environment on performance and adult body size of the wood-boring beetle *Phoracantha semipunctata*. *Entomologia Experimentalis et Applicata* 114, 25–34.

Hanula, J.L., Meeker, J.R., Miller, D.R. and Barnard, E.L. (2002) Association of wildfire with tree health and numbers of pine bark beetles, reproduction weevils and their associates in Florida. *Forest Ecology and Management* 170, 233–247.

Hardi, T.W.H., Husaeni, E.A., Darwiati, W. and Nurtjahjawilasa (1996) A study of the morphology and morphometrics of adults of *Xystrocera festiva* Pascoe. *Buletin Penelitian Hutan* 604, 39–48.

Hardy, A.C. and Cheng, L. (1986) Studies in the distribution of insects by wind currents. III. Insect drift over the sea. *Ecological Entomology* 11, 283–290.

Haribal, M. and Renwick, J.A. (1996) Oviposition stimulants for the monarch butterfly: flavonol glycosides from *Asclepias curassavica*. *Phytochemistry Oxford* 41, 139–144.

Harsh, N.S.K. and Joshi, K.C. (1993) Loss assessment of *Albizia lebbek* seeds due to insect and fungus damage. *Indian Forester* 119, 932–939.

Harsh, N.S.K., Jamaluddin and Tiwari, C.K. (1992) Top dying and mortality in provenance trial plantations of *Gmelina arborea*. *Journal of Tropical Forestry* 8, 55–61.

Hashim, N.R. and Hughes, F.M.R. (2010) The responses of secondary forest tree seedlings to soil enrichment in Peninsular Malaysia: an experimental approach. *Tropical Ecology* 51, 173–182.

Haugen, D.A., Bedding, R.A., Underdown, M.G. and Neumann, F.G. (1990) National strategy for control of *Sirex noctilio* in Australia. *Australian Forest Grower* 13(2), 8 pp. (special liftout).

Hauxwell, C., Vargas, C. and Opuni-Frimpong, E. (2001) Entomopathogens for control of *Hypsipyla* spp. In: Floyd, R.B. and Hauxwell, C. (eds) *Hypsipyla Shoot Borers in Meliaceae*. Proceedings of an International Workshop, Kandy, Sri Lanka 20–23 August 1996. ACIAR Proceedings No. 97, pp. 131–139.

Hawes, J., Motta, C.D., Overal, W.L., Barlow, J., Gardner, T.A. and Peres, C.A. (2009) Diversity and composition of Amazonian moths in primary, secondary and plantation forests. *Journal of Tropical Ecology* 25, 281–300.

Hawkeswood, T.J. (1992) Review of the biology, host plants and immature stages of the Australian Cerambycidae (Coleoptera): Part 1. Parandrinae and Prioninae. *Giornale Italiano di Entomologia* 6, 207–224.

He, J., Zhang, R., Tong, Q., Lu, N., Niu, J., He, J.Z., et al. (1998) Investigation on using pine caterpillars (*Dendrolimus* spp.) as food in Yunnan minor nationality areas. *Forest Research* 11, 396–401.

He, Z., Coa, H.-Z., Zeng, J.-P., Liang, Y.-Y., Han, R.-D. and Ge, F. (2007) Feeding preference of *Dendrolimus punctatus* Walker (Lepidoptera: Lasiocampidae) on pines *Pinus massoniana* and *P. elliottii*. *Acta Entomologica Sinica* 50(2), 125–135.

Heather, N.W. (1965) Occurrence of Cleptidae parasites in eggs of *Ctenomorphodes tessulatus* (Gray) (Phasmida: Phasmidae) in Queensland. *Journal of the Entomological Society of Queensland* 4, 86–87.

Heather, N.W. and Schaumberg, J.B. (1966) Plantation problems of kauri pine in south east Queensland. *Australian Forestry* 30, 12–19.

Helium, K. and Sullivan, F. (1990) Symbiosis between insects and the seeds of *Sesbania grandiflora* Desv. near Muak lek, Thailand. *Embryon* 3, 37–39.

Hellin, J.J. and Gomez, R.R. (1991) The nursery performance of thirty-nine *Leucaena* seedlots in Honduras. *Leucaena Research Reports* 12, 23–25.

Hengxiao, G., McMillin, J.D., Wagner, M.R., Zhou, J., Zhou, Z. and Xu, X. (1999) Altitudinal variation in foliar chemistry and anatomy of yunnan pine, *Pinus yunnanensis*, and pine sawfly (Hym., Diprionidae) performance. *Journal of Applied Entomology-Zeitschrift fur Angewandte Entomologie* 123, 465–471.

Hernandez, J.V. and Jaffe, K. (1995) Economic damage caused by populations of the ant *Atta laevigata* (F. Smith) in plantations of *Pinus caribaea* Mor. and elements for pest management. *Anais da Sociedade Entomologica do Brasil* 24, 287–298.

Hernandez-Stefanoni, J.L., Dupuy, J.M., Tun-Dzul, F. and May-Pat, F. (2011) Influence of landscape structure and stand age on species density and biomass of a tropical dry forest across spatial scales. *Landscape Ecology* 26, 355–370.

Hesami, S.H., Ale Mansour, H. and Seyed Ebrahimi, S. (2005) Report of *Leptocybe invasa* (Hym.: Eulophidae), gall wasp of *Eucalyptus camaldulensis* with notes on biology in Shiraz vicinity. *Journal of Entomological Society of Iran* 24(2), 99–107.

Heu, R.A., Tsuda, D.M., Nagamine, W.T. and Suh, T.H. (2005) Erythrina Gall Wasp, *Quadrastichus erythrinae* Kim (Hymenoptera: Eulophidae). State of Hawaii Department of Agriculture, New Pest Advisory No. 05–03.

Heu, R.A., Tsuda, D.M., Nagamine, W.T., Yalemar, J.A. and Suh, T.H. (2006) Erythrina Gall Wasp, *Quadrastichus erythrinae* Kim (Hymenoptera: Eulophidae). State of Hawaii Department of Agriculture, New Pest Advisory No. 05-03.

Hilje, Q.L., Araya, F.C., Scorza, R.F. and Viquez, C.M. (1992) Forest pests in Central America: handbook. Technical Series: Centro Agtonomico Tropical de Investigacion y Ensenanza.

Hill, C.J. (1996) Habitat specificity and food preferences of an assemblage of tropical Australian dung beetles. *Journal of Tropical Ecology* 12, 449–460.

Hill, J.K., Hamer, K.C., Lace, L.A. and Banham, W.M.T. (1995) Effects of selective logging on tropical forest butterflies on Buru, Indonesia. *Journal of Applied Ecology* 32, 754–760.

Hill, J.K., Griffiths, H.M. and Thomas, C.D. (2011) Climate change and evolutionary adaptations at species' range margins. *Annual Review of Entomology* 56, 143–159.

Hjalten, J., Lindau, A., Wennstrom, A., Blomberg, P., Witzell, J., Hurry, V., et al. (2007) Unintentional changes of defence traits in GM trees can influence plant–herbivore interactions. *Basic and Applied Ecology* 8, 434–443.

Hocking, D. and Jaffer, A.A. (1969) Damping-off in pine nurseries: fungicidal control by seed pelleting. *Commonwealth Forestry Review* 48, 355–363.

Hoddle, M.S. (2010) Red Gum Lerp Psyllid, *Glycaspis brimblecombei* (Hemiptera: Psyllidae). Centre for Invasive Species Research (http://cisr.ucr.edu/red_gum_lerp_psyllid.html).

Hodkinson, I.D. (2009) Life cycle variation and adaptation in jumping plant lice (Insecta: Hemiptera: Psylloidea): a global synthesis. *Journal of Natural History* 43, 65–179.

Hodkinson, I.D. and Casson, D.S. (1987) A survey of food-plant utilization by Hemiptera (Insecta) in the understorey of primary lowland rain forest in Sulawesi, Indonesia. *Journal of Tropical Ecology* 3, 75–85.

Hoebeke, E.R., Haugen, D.A. and Haack, R.A. (2005) *Sirex noctilio*: discovery of a Palearctic siricid woodwasp in New York. *Newsletter of the Michigan Entomological Society* 50, 24–25.

Hoffman, J.D., Aguilar-Amuchastegui, N. and Tyre, A.J. (2010) Use of simulated data from a process-based habitat model to evaluate methods for predicting species occurrence. *Ecography* 33, 656–666.

Hollis, D. (1973) African gall bugs of the genus *Phytolyma* (Hemiptera: Psylloidea). *Bulletin of Entomological Research* 63, 143–154.

Holmes, T.P., Aukema, J.E., Von Holle, B., Liebhold, A. and Sills, E. (2009) Economic impacts of invasive species in forests. *The Year in Ecology and Conservation Biology* 1162, 18–38.

Holt, J., Black, R. and Abdallah, R. (2006) A rigorous yet simple quantitative risk assessment method for quarantine pests and non-native organisms. *Annals of Applied Biology* 149, 167–173.

Holtz, A.M., de Oliveira, H.G., Pallini, A., Venzon, M., Zanuncio, J.C., Oliveira, C.L., Marinho, J.S., and Rasado, M. (2003) Performance of *Thyrinteina arnobia* Stoll (Lepidoptera: Geometridae) on eucalyptus and on guava: Isn't the native host a good host? *Neotropical Entomology* 32, 427–431.

Holzapfel, E.P. and Harrell, J.C. (1968) Transoceanic dispersal studies of insects. *Pacific Insects* 10(1), 115–153.

Hong, J.J., Duan, J.L., Zhao, S.L. and Peng, H.Y. (2003) The sensitivity of *Bombyx mori* larvae to *Dendrolimus punctatus* cytoplasmic polyhedrosis virus. *Acta Entomologica Sinica* 46, 409–416.

Hong, Y., Booth, T.H. and Zuo, H. (1996) GREEN – a climatic mapping program for China and its use in forestry. In: Booth, T.H. (ed.) *Matching Trees and Sites*. Proceedings of an International Workshop, Bangkok, Thailand, 27–30 March 1995. ACIAR Proceedings No. 63, pp. 24–29.

Hood, I., Ramsden, M., Del Dot, T. and Self, M. (1997) *Rigidoporus lineatus* (Pers.) Ryvarden in fire salvaged logs stored under water sprinklers in south east Queensland. *Material und Organismen Berlin* 31, 123–143.

Horak, M. (2001) Current status of the taxonomy of *Hypsipyla* Ragonot (Pyralidae: Phycitinae). In: Floyd, R.B. and Hauxwell, C. (eds) *Hypsipyla Shoot Borers in Meliaceae*. Proceedings of an International Workshop held at Kandy, Sri Lanka, 20–23 August 1996. ACIAR Proceedings No. 97, pp. 69–73.

Horn, S. and Horn, G.N. (2006) New host record for the Asian ambrosia beetle, *Xylosandrus crassiusculus* (Motschulsky) (Coleoptera: Curculionidae). *Journal of Entomological Science* 41(1), 90–91.

House, S., Nester, M., Taylor, D., King, J. and Hinchley, D. (1998) Selecting trees for the rehabilitation of saline sites in south-east Queensland. Technical Paper No. 52, Queensland Forestry Research Institute.

Howard, F.W. and Meerow, A.W. (1993) Effect of mahogany shoot borer on growth of West Indies mahoganies in Florida. *Journal of Tropical Forest Science* 6, 201–203.

Howard, J.J. (1990) Infidelity of leafcutting ants to host plants: resource heterogeneity or defense induction? *Oecologia* 82, 394–401.

Howlader, M.A. (1992) Host range, suitability of host plants as food, and seasonal abundance of the bagworm moth, *Pteroma plagiophleps* Hamps. (Lepidoptera: Psychidae) in Bangladesh. *Bangladesh Journal of Zoology* 20, 177–183.

Hsiao, K.J. (1981) The use of biological agents for the control of the pine defoliator, *Dendrolimus punctatus* (Lepidoptera: Lasiocampidae), in China. *Protection Ecology* 2, 297–303.

Hu, J.J., Tian, Y.C., Han, Y.F., Li, L. and Zhang, B.E. (2001) Field evaluation of insect-resistant transgenic *Populus nigra* trees. *Euphytica* 121, 123–127.

Huang, H.J., Chen, B.X., Sun, Y.R., Liu, J.M. and Ouyang, L.Z. (1998) The occurrence regularity of *Dryocosmus ruriphilus* Yasumatsu and its control. *China Fruits* 2, 33, 36.

Huang, J.H. and Chen, L.Z. (1994) Analysis of species diversity of the forest vegetation in Dongling Mountain, Beijing. *Acta Botanica Sinica* 36, 178–186.

Huang, J.S. (1995) Study on the occurrence and integrated management of *Euzophera batangensis*. *Scientia Silvae Sinicae* 31, 421–427.

Huang, L.L., Wang, G.H., He, Z. and Ge, F. (2008) Effect of pine foliage damage on the incidence of larval diapause in the pine caterpillar *Dendrolimus punctatus* (Lepidoptera: Lasiocampidae). *Insect Science* 15, 441–445.

Huang, P.Y., Fang, Y.W., Huang, J., Lin, S.M. and Wang, H.Y. (2005) A new alien invasive species, *Quadrastichus erythrinae*, in mainland China. *Chinese Bulletin of Entomology* 42(6), 731–733.

Huang, Q.Y., Lei, C.L. and Xue, D. (2006) Field evaluation of a fipronil bait against subterranean termite *Odontotermes formosanus* (Isoptera: Termitidae). *Journal of Economic Entomology* 99, 455–461.

Huber, J.T., Mendel, Z., Protasov, A. and La Salle, J. (2006) Two new Australian species of *Stethynium* (Hymenoptera: Mymaridae), larval parasitoids of *Ophelimus maskelli* (Ashmead) (Hymenoptera: Eulophidae) on *Eucalyptus*. *Journal of Natural History* 40(32–34), 1909–1921.

Huerta, A., Faundez, M. and Araya, J.E. (2010) Susceptibility of *Eucalyptus* spp. to an induced infestation of red gum lerp psyllid *Glycaspis brimblecombei* Moore (Hemiptera: Psyllidae) in Santiago, Chile. *Ciencia e Investigacion Agraria* 37(2), 27–33.

Hulcr, J., Beaver, R.A., Puranasakul, W., Dole, S.A. and Sonthichai, S. (2008) A comparison of bark and ambrosia beetle communities in two forest types in Northern Thailand (Coleoptera: Curculionidae: Scolytinae and Platypodinae). *Environmental Entomology* 37(6), 1461–1470.

Hulme, P.E. (2009) Trade, transport and trouble: managing invasive species pathways in an era of globalization. *Journal of Applied Ecology* 46, 10–18.

Hunter, M.D. (2001) Multiple approaches to estimating the relative importance of top-down and bottom-up forces on insect populations: Experiments, life tables, and time-series analysis. *Basic and Applied Ecology* 2, 295–309.

Hurley, B.P., Slippers, B. and Wingfield, M.J. (2007) A comparison of control results for the alien invasive woodwasp, *Sirex noctilio*, in the southern hemisphere. *Agricultural and Forest Entomology* 9, 159–171.

Hurley, B.P., Slippers, B., Croft, P.K., Hatting, H.J., van der Linde, M., Morris, A.R., Dyer, C. and Wingfield, M.J. (2008) Factors influencing parasitism of *Sirex noctilio* (Hymenoptera: Siricidae) by the nematode *Deladenus siricidicola* (Nematoda: Neotylenchidae) in summer rainfall areas of South Africa. *Biological Control* 45, 450–459.

Hussain, A., Tian, M.Y., He, Y.R., Bland, J.M. and Gu, W.X. (2010) Behavioral and electrophysiological responses of *Coptotermes formosanus* Shiraki towards entomopathogenic fungal volatiles. *Biological Control* 55, 166–173.

Huston, M.A. (1994) *Biological Diversity: The Coexistence of Species on Changing Landscapes*. Cambridge University Press, Cambridge, UK, 681 pp.

Hutacharern, C. (1978) The situation of insects and diseases of pines in Thailand. Meeting of IUFRO Working Parties S2.06.12 and S2.07.07, 'Pests and Diseases of Pines in the Tropics', 'Piedras Blancas', Medillin, Columbia, 3–14 September, 1978.

Hutacharern, C. (1990) Forest insect pests in Thailand. Proceedings of the IUFRO Workshop on Pests and Diseases of Forest Plantations, June, 1988, Bangkok, pp. 75–80.

Hutacharern, C. (1995) Stem borer: a threat to *Pterocarpus macrocarpus* Kurz plantations. *Malaysian Forester* 58, 93–94.

Hutacharern, C. and Panya, S.E. (1996) Biology and control of *Aristobia horridula* (Hope) (Coleoptera: Cerambycidae), a pest of *Pterocarpus macrocarpus*. Proceedings of IUFRO Symposium on 'Impact of Diseases and Insect Pests in Tropical Forests', Peechi, India, 23–26 November 1993, pp. 392–397.

Hutacharern, C. and Sabhasri, B. (1985) Insect pests of *Eucalyptus camaldulensis* Dehn. in the community woodlot in Thailand. *Zeitschrift für Angewandte Entomologie* 99, 170–174.

Hutacharern, C. and Tubtim, N. (1995) *Checklist of Forest Insects in Thailand*. Office of Environmental Policy and Planning, Bangkok.

Iede, E.T. and Machado, D.C. (1989) Insect pests of erva-mate (*Ilex paraguariensis*) and their control. *Boletim de Pesquisa Florestal* 18–19, 51–60.

Ignatov, I.I., Janovec, J.P., Centeno, P., Tobler, M.W., Grados, J., Lamas, G. and Kitching, I.J. (2011) Patterns of richness, composition, and distribution of sphingid moths along an elevational gradient in the Andes-Amazon region of Southeastern Peru. *Annals of the Entomological Society of America* 104, 68–76.

Ilyinykh, A. (2011) Analysis of the causes of declines in Western Siberian outbreaks of the nun moth *Lymantria monacha*. *Biocontrol* 56, 123–131.

Imo, M. (2009) Interactions amongst trees and crops in taungya systems of western Kenya. *Agroforestry Systems* 76, 265–273.

Impson, F.A.C., Kleinjan, C.A., Hoffmann, J.H. and Post, J.A. (2008) *Dasineura rubiformis* (Diptera: Cecidomyiidae), a new biological control agent for *Acacia mearnsii* in South Africa. *South Africa Journal of Science* 104(7–8), 247–249.

Ingwell, L.L. and Preisser, E.L. (2011) Using citizen science programs to identify host resistance in pest-invaded forests. *Conservation Biology* 25, 182–188.

Intachat, J. (1995) Assessment of moth diversity in natural and managed forests in Peninsular Malaysia. Unpublished DPhil thesis, University of Oxford.

Intachat, J. and Kirton, L.G. (1997) Observations on insects associated with *Acacia mangium* in Peninsular Malaysia. *Journal of Tropical Forest Science* 9, 561–564.

Intari, S.E. (1978) Some important pests of forest trees in Indonesia. Proceedings of the Eighth World Forestry Congress, Jakarta, 16–28 October, 1978. No. FID-0-11.

Intari, S.E. (1990) Effects of *Neotermes tectonae* Damm attack on the quality and quantity of teak timber in the Kebonharjo Forest Division, Central Java. *Bulletin Penelitian Hutan* 530, 25–35.

Intari, S.E. and Natawiria, D. (1973) White grubs in forest tree nurseries and young plantations. Laporan, Lembaga Penelitian Hutan No. 167.

Intari, S.E. and Ruswandy, H. (1986) Preferences of *Dioryctria* sp. to three species of pines. *Buletin Penelitian Hutan* 479, 44–48.

Ipute, S. (1996) The status of *Canarium* as forest timber trees in Solomon Islands. In: Stevens, M.L., Bourke, R.M. and Evans, B.R. (eds) *South Pacific Indigenous Nuts*. Proceedings of a Workshop, 31 October – 4 November 1994, Le Lagon Resort, Port Vila, Vanuatu. ACIAR Proceedings No. 69, 176 pp.

Irianto, R.S.B., Matsumoto, K. and Mulyadi, K. (1997) The yellow butterfly species of the genus *Eurema* Hubner causing severe defoliation in the forestry plantations of *Albizzia*, *Paraserianthes falcataria* (L.) Nielsen, in the western part of Indonesia. *JIRCAS Journal* No. 4, 41–49.

Iriarte, J.A., Feinsinger, P. and Jaksic, F.M. (1997) Trends in wildlife use and trade in Chile. *Biological Conservation* 81, 9–20.

Ishino, M.N., de Sibio, P.R. and Rossi, M.N. (2011) Leaf trait variation on *Erythroxylum tortuosum* (Erythroxylaceae) and its relationship with oviposition preference and stress by a host-specific leaf miner. *Austral Ecology* 36, 203–211.

Ismail, R., Mutanga, O. and Bob, U. (2007) Forest health and vitality: the detection and monitoring of *Pinus patula* trees infected by *Sirex noctilio* using digital multispectral imagery. *Southern Hemisphere Forestry Journal* 69(1), 39–47.

Isman, M.B. (2006) Botanical insecticides, deterrents, and repellents in modern agriculture and an increasingly regulated world. *Annual Review of Entomology* 51, 45–66.

Itioka, T. and Inoue, T. (1996) The consequences of ant-attendance to the biological control of the red wax scale insect *Ceroplastes rubens* by *Anicetus beneficus*. *Journal of Applied Ecology* 33, 609–618.

Itioka, T. and Yamauti, M. (2004) Severe drought, leafing phenology, leaf damage and lepidopteran abundance in the canopy of a Bornean aseasonal tropical rain forest. *Journal of Tropical Ecology* 20, 479–482.

ITTO (2009) *Annual Review and Assessment of the World Timber Situation*. International Tropical Timber Organization.

IUCN (2010) Ecology of *Quadrastichus erythrinae*. International Union for Conservation of Nature, Invasive Species Specialist Group, Global Invasive Species Database (http://www.issg.org/database/species/ecology.asp?si=965&fr=1&sts=&%20ang=EN&).

IUFRO (2007) Recommendation of a Pathway Approach for Regulation of Plants for Planting. A Concept Paper from the IUFRO Unit on Alien Invasive Species and International Trade (http://www.forestry-quarantine.org/Documents/IUFRO-ConceptPaper-%20Plants-Planting.pdf).

Ivory, M.H. (1977) Preliminary investigations of the pests of exotic forest trees in Zambia. *Commonwealth Forestry Review* 56, 47–56.

Ivory, M.H. and Speight, M.R. (1993) Pests and diseases. In: Pancel, L. (ed.) *Tropical Forestry Handbook*, Vol. II. Springer Verlag, Berlin.

Jacob, J.P. and Balu, A. (2007) Biochemical and cuticular variation in teak clone leaves and resistance to teak defoliator. *Indian Forester* 133, 197–205.

Jactel, H., Brockerhoff, E. and Duelli, P. (2005) A test of the biodiversity-stability theory: meta-analysis of tree species diversity effects on insect pest infestations, and re-examination of responsible factors. In: Scherer-Lorenzen, M., Korner, Ch. and Schulze, E.-D. (eds) *Forest Diversity and Function: Temperate and Boreal Systems*. Ecological Studies 176, 235–262.

Jaffe, K. (1986) Control of *Atta* spp. in pine tree plantations in the Venezuelan Llanos. In: Vander Meer, R.K. and Lofgreen, C.S. (eds) *Fire Ants and Leaf-cutting Ants: Biology and Management*. Westview Press, Boulder, Colorado, pp. 409–416.

Jaffe, K. and Vilela, E. (1989) On nest densities of the leaf-cutting ant *Atta cephalotes* in tropical primary forest. *Biotropica* 21, 234–236.

Jain, N.C. (1996) Growing pest problems in *Prosopis cineraria* in arid zones of India. In: Nair, K.S.S., Sharma, J.K. and Varma, R.V. (eds) *Impact of Diseases and Insect Pests in Tropical Forests*. Proceedings of the IUFRO Symposium, 23–26 November 1993, Peechi, India, pp. 507–510.

Jalali, S.K. and Singh, S.P. (1992) Biology and feeding potential of *Curinus coeruleus* (Mulsant) and *Chrysoperla carnea* (Stephens) on subabul psyllid, *Heteropsylla cubana* Crawford. *Journal of Insect Science* 5, 89–90.

Jalali, S.K. and Singh, S.P. (1993) Life-table studies on *Curinus coeruleus* Mulsant, an exotic predator of *Heteropsylla cubana* Crawford. *Journal of Insect Science* 6, 281–282.

Jamal, A. (1994) Major insect pests of gum arabic trees *Acacia senegal* Willd. and *Acacia seyal* L. in Western Sudan. *Journal of Applied Entomology* 117, 10–20.

Jang, E.B., Siderhurst, M.S., Conant, P. and Siderhurst, L.A. (2009) Phenology and population radiation of the nettle caterpillar, *Darna pallivitta* (Moore) (Lepidoptera: Limacodidae) in Hawai'i. *Chemoecology* 19, 7–12.

Javaregowda and Prabhu, S.T. (2010) Susceptibility of *Eucalyptus* species and clones to gall wasp, *Leptocybe invasa* Fisher & La Salle (Eulophidae: Hymenoptera) in Karnataka. *Karnataka Journal of Agricultural Science* 23(1), 220–221.

Jeeva, V., Saravanan, S., Devaraj, P. and Lakshmidevi, R. (1998) Malady and remedy of sandal cultivation in farmlands and private lands – an overview. In: Radomiljac, H.S., Ananthapadmanabho, H.S., Welbourn, R.M. and Satyanarayana, R. (eds) *Sandal and its Products*. Proceedings of an International Seminar, 18–19 December 1997, Bangladore, India. ACIAR Proceedings No. 84, pp.16–18.

Jha, L.K. and Sen-Sarma, P.K. (eds) (2008) *A Manual of Forest Extension Education*. APH Publishing: New Delhi, 400 pp.

Jhala, R.C., Pater, Z.P. and Shah, A.H. (1988) Pests of milk tree (*Manilkara hexandra*), a rootstock for sapodilla (*Manilkara achras*). *Indian Journal of Agricultural Sciences* 58, 730–731.

Jhala, R.C., Chauhan, N.R., Patel, M.G. and Bharpoda, T.M. (2009) Infestation of invasive gall inducer, *Leptocybe invasa* Fisher & La Salle (Hymenoptera: Eulophidae) in nurseries of *Eucalyptus* in Middle Gujarat, India. *Insect Environment* 14(4), 191–192.

Jhala, R.C., Patel, M.G. and Vaghela, N.M. (2010) Effectiveness of insecticides against blue gum chalcid, *Leptocybe invasa* Fisher & La Salle (Hymenoptera: Eulophidae), infesting eucalyptus seedlings in middle Gujarat, India. *Karnataka Journal of Agricultural Science* 23(1), 84–86.

Jiang, Z., Wu, W., Wei, J., Chen, L. and Mo, J. (2011) Evaluation of Ivermectin for the control of *Odontotermes formosanus* and *Macrotermes barneyi* (Isoptera: Termitidae) in a seedling nursery. *Sociobiology* 57, 355–360.

Johnson, E.W. and Ross, J. (2008) Quantifying error in aerial survey data. *Australian Forestry* 71(3), 216–222.

Johnson, E.W. and Wittwer, D. (2008) Aerial detection surveys in the United States. *Australian Forestry* 71(3), 212–215.

Jones, D.L. and Elliot, W.R. (1986) *Pests, Diseases and Ailments of Australian Plants*. Macarthur Press, Paramatta, Australia.

Jones, J.E. (1995) Quarantine alert! *Monthly Bulletin of the Department of Agriculture, Fisheries and Parks, Bermuda* 66(11), 106.

Jones, O.T. (1994) The current use of pheromones and other semiochemicals in the integrated management of insect pests. *Pesticide Outlook August* 1994, 26–31.

Jones, W.E., Grace, J.K. and Tamashiro, M. (1996) Virulence of seven isolates of *Beauveria bassiana* and *Metarhizium anisopliae* to *Coptotermes formosanus* (Isoptera: Rhinotermitidae). *Environmental Entomology* 25, 481–487.

Jordan, G.J., Potts, B.M. and Clark, A.R. (2002) Susceptibility of *Eucalyptus globulus* ssp. *globulus* to sawfly (*Perga affinis* ssp. *insularis*) attack and its potential impact on plantation productivity. *Forest Ecology and Management* 160, 189–199.

Jorge Pena-Ramirez, Y., Juarez-Gomez, J., Gomez-Lopez, L., Jeronimo-Perez, J.L., Garcia-Shesena, I., Gonzalez-Rodriguez, J.A., *et al.* (2010) Multiple adventitious shoot formation in Spanish Red Cedar (*Cedrela odorata* L.) cultured in vitro using juvenile and mature tissues: an improved micropropagation protocol for a highly valuable tropical tree species. *In Vitro Cellular and Developmental Biology-Plant* 46, 149–160.

Joseph, K.J. and Venkitesan, T.S. (1995) Predation of the subabul psyllid *Heteropsylla cubana* by the dragonfly *Pantala flavescens* in nature. *Entomon* 20, 273–275.

Joshi, K.C., Roychoudhury, N., Kulkarni, N. and Sambath, S, (2006) Sal heartwood borer in Madhya Pradesh. *Indian Forester* 132, 799–808.

Jost, L. (2007) Partitioning diversity into independent alpha and beta components. *Ecology* 88, 2427–2439.

Junker, R.R., Itioka, T., Bragg, P.E. and Bluethgen, N. (2008) Feeding preferences of phasmids (Insecta: Phasmida) in a Bornean dipterocarp forest. *Raffles Bulletin of Zoology* 56, 445–452.

Junqueira, L.K., Diehl, E. and Berti Filho, E. (2008) Termites in eucalyptus forest plantations and forest remnants: an ecological approach. *Bioikos, Campinas* 22(1), 3–14.

Kagezi, G.H., Kaib, M., Nyeko, P. and Brandl, R. (2010) Pest status and control options for termites (Isoptera) in the Luhya Community of Western Kenya. *Sociobiology* 55, 815–829.

Kairo, M.T.K. and Murphy, S.T. (1999) Temperature and plant nutrient effects on the development, survival and reproduction of *Cinara* sp. nov., an invasive pest of cypress trees in Africa. *Entomologia Experimentalis et Applicata* 92, 147–156.

Kairo, M.T.K. and Murphy, S.T. (2005) Comparative studies on populations of *Pauesia juniperorum* (Hymenoptera: Braconidae), a biological control agent for *Cinara cupressivora* (Hemiptera: Aphididae). *Bulletin of Entomological Research* 95, 597–603.

Kalia, S. and Lal, R.R. (1999) Insect pests of *Dalbergia sissoo* Roxb. at and around Jabalpur. *Advances in Forestry Research in India* 20, 190–202.

Kalia, S. and Lal, R.R. (2000) Efficacy of three varietal toxins of *Bacillus thuringiensis* tested against some important forest insect pests of multipurpose forest tree species. *Indian Forester* 126(1), 62–66.

Kaloudis, S., Anastopoulos, D., Yialouris, C.P., Lorentzos, N.A. and Sideridis, A.B. (2005) Insect identification expert system for forest protection. *Expert Systems with Applications* 28, 445–452.

Kalshoven, L.G.E. (1953) Important outbreaks of insect pests in the forests of Indonesia. *Transactions of the 9th International Congress of Entomology* 2, 229–234.

Kamata, N. and Igarashi, M. (1995) Relationship between temperature, number of instars, larval growth, body size, and adult fecundity of *Quadricalcarifera punctatella* (Lepidoptera: Notodontidae): cost–benefit relationship. *Environmental Entomology* 24, 648–656.

Kamnerdratana, P.Y. (1987) The economically significant insect pests of trees and timber in Thailand. In: de Guzman, E.D. and Nuhamara, S.T. (eds) *Forest Pests and Diseases in Southeast Asia*. Biotrop Special Publication No. 26, Bogor, Indonesia.

Kaplan, I., Dively, G.P. and Denno, R.F. (2009) The costs of anti-herbivore defense traits in agricultural crop plants: a case study involving leafhoppers and trichomes. *Ecological Applications* 19, 864–872.

Kasenene, J.M. and Roininen, H. (1999) Seasonality of insect herbivory on the leaves of *Neoboutonia macrocalyx* in the Kibale National Park, Uganda. *African Journal of Ecology* 37, 61–68.

Kasno and Husaeni, E.A. (2002) An integrated control of sengon stem borer in Java. In: Hutacharern, C., Napompeth, B., Allard, G. and Wylie, F.R. (eds) *Pest Management in Tropical Forest Plantations*. Forestry Research Support Program for Asia and the Pacific, FAO, Bangkok, pp. 127–131.

Kathiresan, K. (1993) Dangerous pest on nursery seedlings of *Rhizophora*. *Indian Forester* 119, 1026.

Kawabe, Y. and Ito, K. (2003) Disease and insect pest damage in afforested areas on acid sulphate soil in the Mekong Delta, Vietnam. *Tropical Forestry* 57, 25–33.

Kaya, H.K. and Lindegren, J.E. (1983) Parasitic nematode controls western poplar clearwing moth. *California Agriculture* 37(3–4), 31–32.

Kekeunou, S., Messi, J., Weise, S. and Tindo, M. (2006) Insect pests' incidence and variations due to forest landscape degradation in the humid forest zone of Southern Cameroon: farmers' perception and need for adopting an integrated pest management strategy. *African Journal of Biotechnology* 5, 555–562.

Kenis, M., Rabitsch, W., Auger-Rozenberg, M.-A. and Roques, A. (2007) How can alien species inventories and interception data help us prevent insect invasions? *Bulletin of Entomological Research* 97, 489–502.

Kent, D. (1995) Eucalypt sawflies (*Pergadorsalis* and *Pergagrapta* spp.). State Forests of New South Wales, Forest Protection Research Division Series No. E5.

Kessler, M., Abrahamczyk, S., Bos, M., Buchori, D., Putra, D.D., Gradstein, S.R., Hohn, P., Kluge, J., Orend, F., Pitopang, B., Saleh, S., Schulze, C.H., Spom, S.G., Steffan-Dewenter, I., Tijtroedirdjo, S.S. and Tschmtke, T. (2011) Cost-effectiveness of plant and animal biodiversity indicators in tropical forest and agroforest habitats. *Journal of Applied Ecology* 48, 330–339.

Kettler, J.S. (1995) A new insect pest of *Erythrina poeppigiana* in Costa Rica. *Nitrogen Fixing Tree Research Reports* 13, 51–53.

Khaemba, B.M. and Wanjala, F.M.E. (1993) Some aspects of the biology of *Cinara cupressi* Buckton (Homoptera: Lachnidae) when bred under laboratory condition. *Insect Science and its Application* 14, 693–695.

Khali Aziz, H., Azman, H., Ismael, H. and Chong, P.F. (1992) Better and efficient forest management through remote sensing. *The Malaysian Forester* 55, 158–166.

Khan, H.R., Kumar, S. and Prasad, L. (1988) Studies on seasonal activity of some agro-forestry insect pests by light trap. *Indian Forester* 114, 215–229.

Khan, M.A. and Ahmad, D. (1991) Comparative toxicity of some chlorinated hydrocarbon insecticides to poplar leaf beetle, *Chrysomela populi* L. (Coleoptera: Chrysomelidae). *Indian Journal of Forestry* 14, 42–45.

Khan, T.N. (1993) Biology and ecology of *Plocaederus obesus* Gahan (Coleoptera: Cerambycidae): a comparative study. *Proceedings of the Zoological Society Calcutta* 46, 39–49.

Khanmirzaei, A., Kowsar, S.A. and Sameni, A.M. (2011) Changes of selected soil properties in a floodwater-irrigated *Eucalyptus* plantation in the Gareh Bygone Plain, Iran. *Arid Land Research and Management* 25, 38–54.

Khuhro, R.D., Nizamani, S.M., Jiskani, M.M., Abbasi, Q.D. and Solangi, G.S. (2005) New host records of Asian ambrosia beetle (AAB) in Sindh Pakistan. *Pakistan Journal of Agriculture, Agricultural Engineering and Veterinary Sciences* 21(1), 43.

Khumbongmayum, A.D., Khan, M.L. and Tripathi, R.S. (2006) Biodiversity conservation in sacred groves of Manipur, northeast India: population structure and regeneration status of woody species. *Biodiversity and Conservation* 15, 2439–2456.

Kidd, N.A.C. (1994) Resource deprival as an anti-herbivore strategy in plants, with particular reference to aphids. Proceedings of the 4th International Symposium on Aphids held in the Czech Republic, August 30 – September 3 1993. *European Journal of Entomology* 91, 53–56.

Kim, I.K., Delvare, G. and La Salle, J. (2004) A new species of *Quadrastichus* (Hymenoptera: Eulophidae): a gall-inducing pest on *Erythrina* (Fabaceae). *Journal of Hymenoptera Research* 13(2), 243–249.

Kim, I.-K., Mendel, Z., Protasov, A., Blumberg, D. and La Salle, J. (2008) Taxonomy, biology, and efficacy of two Australian parasitoids of the eucalyptus gall wasp, *Leptocybe invasa* Fisher & La Salle (Hymenoptera: Eulophidae: Tetrastichinae). *Zootaxa* 1910, 1–20.

King, S.R., Carlson, T.J. and Moran, K. (1996) Biological diversity, indigenous knowledge, drug discovery, and intellectual property rights. In: Brush, S.B. and Stabinsk, D. (eds) Valuing Local Knowledge, Island Press: Washington DC, pp 167–185.

Kirsten, F. and Tribe, G. (1995) The biological control of *Phoracantha semipunctata* and *Phoracantha recurva* (Coleoptera: Cerambycidae) in South Africa. In: IUFRO XX World Congress, 6–12 August 1995, Tampere, Finland.

Kirton, L.G. and Cheng, S. (2007) Ring-barking and root debarking of dipterocarp saplings by termites in an enrichment planting site in Malaysia. *Journal of Tropical Forest Science* 19, 67–72.

Kirton, L.G., Brown, V.K. and Azmi, M. (1999) The pest status of the termite *Coptotermes curvignathus* in *Acacia mangium* plantations: incidence, mode of attack and inherent predisposing factors. *Journal of Tropical Forestry Science* 11(4), 822–831.

Klein, B.C. (1989) Effects of forest fragmentation on dung and carrion beetle communities in Central Amazonia (Brazil). *Ecology* 70, 1715–1725.

Kleinschmit, J., Khurana, D.K., Gerhold, H.D. and Libby, W.J. (1993) Past, present and anticipated applications of clonal forestry. In: Ahuja, M.R. and Libby, W.J. (eds) *Clonal Forestry II: Conservation and Application*. Springer-Verlag, pp. 9–41.

Kliejunas, J.T., Tkacz, B., Burdsall, H.H. Jr, DeNitto, G.A., Eglitis, A., Haugen, D.A., et al. (2001) Pest risk assessment of the importation into the United States of unprocessed *Eucalyptus* logs and chips from South America. General Technical Report FPL-GTR-124. USDA Forest Service, Forest Products Laboratory, Madison, Wisconsin, 134 pp.

Koh, L.P. and Sodhi, N.S. (2010) Conserving Southeast Asia's imperiled biodiversity: scientific, management, and policy challenges. *Biodiversity and Conservation* 19, 913–917.

Koh, L.P., Miettinen, J., Liew, S.C. and Ghazoul, J. (2011) Remotely sensed evidence of tropical peatland conversion to oil palm. *Proceedings of the National Academy of Sciences of the United States of America* 108, 5127–5132.

Koricheva, J., Larsson, S. and Haukioja, E. (1998) Insect performance on experimentally stressed woody plants: a meta-analysis. *Annual Review of Entomology* 43, 195–216.

Krell, F.T. (1996) Historical biogeography of *Temnorhynchus* species (Insecta: Coleoptera: Scarabaeidae: Dynastinae). *Zoologischer Anzeiger* 234, 209–226.

Krishna, K. and Weesner, F.M. (1970) *Biology of Termites*, Vol. II. Academic Press, New York.

Krokene, P. (1994) How bark beetles are able to kill trees. *Fauna (Oslo)* 47, 242–253.

Kruess, A. and Tscharntke, T. (1994) Habitat fragmentation, species loss, and biological control. *Science Washington DC* 264, 1581–1584.

Kulkarni, H.D. (2010) Little leaf disease of eucalyptus by thrips, *Frankliniella occidentalis* (Pergande). *Karnataka Journal of Agriculture Sciences* 23(1), 203–206.

Kulkarni, N., Chandra, K., Wagh, P.N., Joshi, K.C. and Singh, R.B. (2007) Incidence and management of white grub, *Schizonycha ruficollis* on seedlings of teak (*Tectona grandis* Linn.f.). *Insect Science* 14, 411–418.

Kulkarni, N., Paunikar, S., Joshi, K.C. and Rogers, J. (2009) White grubs, *Holotrichia rustica* and *Holotrichia mucida* (Coleoptera: Scarabaeidae) as pests of teak (*Tectona grandis* L.f.) seedlings. *Insect Science* 16, 519–525.

Kumar, A., Matharoo, A.K., Singh, S. and Singh, A.N. (2006) Planting stock improvement in *Gmelina arborea*. *Indian Forester* 132, 691–699.

Kumar, N.S., Sarma, S. and Barthakur, N.D. (2010) Effect of abiotic factors on the population of *Calopepla leayana* Latr. (Coleoptera: Chrysomelidae) on *Gmelina arborea* (Roxb.) (Verbenaceae). *Indian Forester* 136(4), 487–496.

Kumar, S. and Thakur, R.K. (2011) Termite problems in forest nurseries and plantation and their management. *Journal of Experimental Zoology India* 14, 329–332.

Kursar, T.A. and Coley, P.D. (1991) Nitrogen content and expansion rate of young leaves of rain forest species: implications for herbivory. *Biotropica* 23, 141–150.

Kursar, T.A. and Coley, P.D. (2003) Convergence in defense syndromes of young leaves in tropical rainforests. *Biochemical Systematics and Ecology* 31, 929–949.

Kuusipalo, J., Adjers, G., Jafarsidik, Y., Otsamo, A., Tuomela, K. and Vuokko, R. (1995) Restoration of natural vegetation in degraded *Imperata cylindrica* grassland: understorey development in forest plantations. *Journal of Vegetation Science* 6, 205–210.

La Salle, J., Arakelian, G., Garrison, R.W. and Gates, M.W. (2009a) A new species of invasive gall wasp (Hymenoptera: Eulophidae: Tetrastichinae) on blue gum (*Eucalyptus globulus*) in California. *Zootaxa* 2121, 35–43.

La Salle, J., Ramadan, M. and Kumashiro, B.R. (2009b) A new parasitoid of the Erythrina Gall Wasp, *Quadrastichus erythrinae* Kim (Hymenoptera: Eulophidae). *Zootaxa* 2083, 19–26.

Labandeira, C.C. and Phillips, T.L. (1996) Insect fluid-feeding on Upper Pennsylvanian tree ferns (Palaeodictyoptera, Marattiales) and the early history of the piercing-and-sucking functional feeding group. *Annals of the Entomological Society of America* 89, 157–183.

Labandeira, C.C., Dilcher, D.L., Davis, D.R. and Wagner, D.L. (1994) Ninety-seven million years of angiosperm-insect association: paleobiological insights into the meaning of coevolution. *Proceedings of the National Academy of Sciences of the United States of America* 91, 12278–12282.

Lack, D. (1976) *Island Birds*. Blackwell Scientific, Oxford.

Lale, N.E.S. (1998) Neem in the conventional Lake Chad Basin area and the threat of Oriental yellow scale insect (*Aonidiella onentalis* Newstead) (Homoptera: Diaspididae). *Journal of Arid Environments* 40, 191–197.

Lamb, D. (2011) *Different Types of Reforestation. Regreening the Bare Hills: Tropical Forest Restoration in the Asia-Pacific Region*. Springer, Dordrecht, The Netherlands, pp. 135–155.

Lambin, E.F. and Meyfroidt, P. (2011) Global land use change, economic globalization, and the looming land scarcity. *Proceedings of the National Academy of Sciences of the United States of America* 108, 3465–3472.

Lamien, N., Tigabu, M., Dabire, R., Guinko, S. and Oden, P.C. (2008) Insect (*Salebria* sp.) infestation and impact on *Vitellaria paradoxa* CF Gaertn. fruit production in agroforestry parklands. *Agroforestry Systems* 72, 15–22.

Lamprey, H.F., Halevy, G. and Makacha, S. (1974) Interactions between *Acacia*, bruchid seed beetles and large herbivores. *East African Wildlife Journal* 12, 81–85.

Landis, D.A. and Werling, B.P. (2010) Arthropods and biofuel production systems in North America. *Insect Science* 17, 220–236.

Landsberg, J. and Wylie, F.R. (1983) Water stress, leaf nutrients and defoliation: a model of dieback in rural eucalypts. *Australian Journal of Ecology* 8, 96–100.

Lanfranco, D. and Dungey, H.S. (2001) Insect damage in *Eucalyptus*: a review of plantations in Chile. *Austral Ecology* 26, 477–481.

Lange, C.E. and Cigliano, M.M. (2010) Prevalence and infection intensity of the biocontrol agent *Paranosema locustae* (Microsporidia) in field-collected, newly-associated hosts (Orthoptera: Acrididae: Melanoplinae). *Biocontrol Science and Technology* 20, 19–24.

Lange, C.E. and de Wysiecki, M.L. (1996) The fate of *Nosema locustae* (Microsporidia: Nosematidae) in Argentine grasshoppers (Orthoptera: Acrididae). *Biological Control* 7, 24–29.

Lanier, G.N. (1987) The validity of *Ips cribicollis* (Eich.) (Coleoptera: Scolytidae) as distinct from *I. grandicollis* (Eich.) and the occurrence of both species in Central America. *Canadian Entomologist* 119, 179–187.

Lapis, E.B. (1985a) Geographical and altitudinal distribution of the six-spined engraver beetle (*Ips calligraphus*) (Germar), (Coleoptera: Scolytidae) in the Philippines. *Sylvatrop* 10, 211–217.

Lapis, E.B. (1985b) Evaluation of insecticides for the control of the pine shoot moth *Dioryctria rubella* Hamps. in the Philippines. *Sylvatrop* 10, 77–85.

Lapis, E.B. and Borden, J.H. (1993) Olfactory discrimination by *Heteropsylla cubana* (Homoptera: Psyllidae) between susceptible and resistant species of Leucaena (Leguminosae). *Journal of Chemical Ecology* 19, 83–90.

Laranjeiro, A.L. (1994) Integrated pest management at Aracruz Celulose. *Forest Ecology and Management* 65, 45–52.

Larsson, S. (1989) Stressful times for the plant stress–insect performance hypothesis. *Oikos* 56, 277–283.

Larsson, S. and Bjorkman, C. (1993) Performance of chewing and phloem-feeding insects on stressed trees. *Scandinavian Journal of Forest Research* 8, 550–559.

Laurance, W.F. (1991) Edge effects in tropical forest fragments: application of a model for the design of nature reserves. *Biological Conservation* 57, 205–220.

Laurance, W.F. (2007) Have we overstated the tropical biodiversity crisis? *Trends in Ecology and Evolution* 22, 65–70.

Lavallee, R., Albert, P.J. and Mauffette, Y. (1994) Influence of white pine watering regimes on feeding preferences of spring and fall adults of the white pine weevil *Pissodes strobi* (Peck). *Journal of Chemical Ecology* 20, 831–847.

Lawson, S.A. (2007) Eucalypts as source and sink for invasive pests and diseases. IUFRO Unit 7.03.12 Monograph. In: Evans, H. and Osazako, T. (eds) *Alien Invasive Species and International Trade*. Forest Research Institute, Warsaw, pp. 133–138.

Lawson, S.A., Moore, C.J., Slippers, B., Hurley, B.P. and Wingfield, M.J. (2008) Isolation of sex pheromones of cossid wood moth (Lepidoptera: Cossidae) pests of Eucalypt plantations in Australia and South Africa. International Union of Forest Research Organisations Working Group 7.3.00. Recent Advances in Forest Entomology, 1–6 July 2008, Pretoria, South Africa.

Lawton, J.H., Bignell, D.E., Bolton, B., Bloemers, G.F., Eggleton, P., Hammond, P.M., Hodda, M., Holt, R.D., Larssen, T.B., Mawdseleg, N.A., Stosk, N.E., Stivastara, D.S. and Watt, A.D. (1998) Biodiversity inventories, indicator taxa and effects of habitat modification in tropical forest. *Nature London* 391, 72–76.

Leather, S.R. (1993) Influence of site factor modification on the population development of the pine beauty moth (*Panolis flammea*) in a Scottish lodgepole pine (*Pinus contorta*) plantation. *Forest Ecology and Management* 59, 207–223.

Lee, Y.K., Lee, D.K., Woo, S.Y., Park, P.S., Jang, Y.H. and Abraham, A.R.G. (2006) Effect of *Acacia* plantations on net photosynthesis, tree species composition, soil enzyme activities, and microclimate on Mt. Makiling. *Photosynthetica* 44, 299–308.

Lemos, R.N.S., Crocomo, W.B., Forti, L.C. and Wilcken, C.F. (1999) Feeding selectivity and influence of leaf age of *Eucalyptus* spp. for *Thyrinteina arnobia* (Lepidoptera: Geometridae). *Pesquisa Agropecuaria Brasileira* 34(1), 7–10.

Li, G., Zan, Q., Zhao, S., Xiao, Y., Wang, Y., Xu, H., et al. (2007) Biological characters of *Ptyomaxia* sp. and the toxicity and effectiveness of *Bacillus thuringiensis* against its larvae. *Chinese Journal of Applied and Environmental Biology* 13, 50–54.

Li, H.M., Xiao, H., Peng, H., Han, H.X. and Xue, D.Y. (2006) Potential global range expansion of a new invasive species, the Erythrina Gall Wasp, *Quadrastichus erythrinae* Kim (Insects: Hymenoptera: Eulophidae). *The Raffles Bulletin of Zoology* 54(2), 229–234.

Li, L.Z., Zhou, X.S., Cui, Y.S., Yan, J., Yang, H.P., Song, Y.F., et al. (1998) Study of using entomopathogenic fungi *Beauveria brongniartii* to control white grubs in forest nursery. *Journal of Northeast Forestry University* 26, 33–36.

Li, Z. (2007) Fifty years of application of entomogenous fungi in China. *Journal of Anhui Agricultural University* 34(2), 203–207.

Li, Z., Alves, S.B., Roberts, D.W., Fan, M., Delalibera, I., Jr Tang, J., et al. (2010) Biological control of insects in Brazil and China: history, current programs and reasons for their successes using entomopathogenic fungi. *Biocontrol Science and Technology* 20, 117–136.

Liang, C.F. (1991) Experiments on the effects of dimilin III for controlling pine caterpillar and its influence on parasitic natural enemies. *Natural Enemies of Insects* 13, 171–174.

Liang, C.J., Zhao, L., Huang, J.Y., Meng, M.Q., Yang, X.H. and Li, J.J. (1999) Study on aerial equipment for spraying pesticide and techniques of monitoring. II. Controlling *Dendrolimus punctatus* by aerial equipment for spraying pesticide. *Forest Research* 12, 74–78.

Liang, J. and Zhang, X.-Y. (2005) Ecological control of forest pest: a new strategy for forest pest control. *Journal of Forestry Research (Harbin)* 16, 339–342.

Liang, Y.-Y., Li, H.-G., He, Z. and Ge, F. (2008) Effect of alternative hosts on the growth and development of masson pine caterpillar, *Dendrolimus punctatus*. *Chinese Bulletin of Entomology* 45(5), 739–743.

Lieberman, D. and Lieberman, M. (1984) The causes and consequences of synchronous flushing in a dry tropical forest. *Biotropica* 16, 193–201.

Liebhold, A., Kamata, N. and Jacob, T. (1996) Cyclicity and synchrony of historical outbreaks of the beech caterpillar, *Quadricalcarifera punctatella* (Motschulsky) in Japan. *Researches on Population Ecology Kyoto* 38, 8794.

Liebhold, A.M., Work, T.T., McCullough, D.G. and Cavey, J.F. (2006) Airline baggage as a pathway for alien insect species invading the United States. *American Entomologist* 52, 48–54.

Lim, G.T., Kirton, L.G., Salom, S.M., Kok, L.T., Fell, R.D. and Pfeiffer, D.G. (2008) Mahogany shoot borer control in Malaysia and prospects for biocontrol using weaver ants. *Journal of Tropical Forest Science* 20(3), 147–155.

Linzmeier, A.M. and Ribeiro-Costa, C.S. (2008) Seasonality arid temporal structuration of Alticini community (Coleoptera, Chrysomelidae, Galerucinae) in the Araucaria Forest of Parana, Brazil. *Revista Brasileira De Entomologia* 52, 289–295.

Liu, G.X. and Liang, Z.C. (1993) Root rot of slash pine and its relation to defoliation caused by *Dendrolimus punctatus*. *Journal of South China Agricultural University* 14, 109–113.

Liu, H.Q., Zhang, C.H., Cai, J., Shi, L., Li, L. and Chen, Y.P. (1998) The study of tropical climate (Jinghong, Yunnan) influence on white wax scale. *Forest Research* 11, 508–512.

Liu, W.C., Bonsall, M.B. and Godfray, H.C.J. (2007) The form of host density-dependence and the likelihood of host-pathogen cycles in forest-insect systems. *Theoretical Population Biology* 72, 86–95.

Liu, X. (1997) Occurrence and control of termites in ancient and valuable trees in Guangzhou. *Natural Enemies of Insects* 19(4), 169.

Loch, A.D. and Floyd, R.B. (2001) Insect pests of Tasmanian blue gum, *Eucalyptus globulus globulus*, in south-western Australia: history, current perspectives and future prospects. *Austral Ecology* 26(5), 458–466.

Logan, J.W.M., Cowie, R.H. and Wood, T.G. (1990) Termite (Isoptera) control in agriculture and forestry by non-chemical methods: a review. *Bulletin of Entomological Research* 80, 309–330.

Loganathan, J. and David, P.M.M. (1999) Rainfall: a major factor leading to outbreak of teak (*Tectona grandis*) defoliator, *Hyblaea puera* Cramer (Lepidoptera: Hyblaeidae) in commercial teak plantation. *Indian Forester* 125, 316–320.

Longo, S., Palmeri, V. and Sommariva, D. (1993) Activity of *Avetianella longoi*, egg-parasitoid of *Phoracantha semipunctata* in Southern Italy. *Redia* 76, 223–239.

Lopes, J.C.A., Jennings, S.B. and Matni, N.M. (2008) Planting mahogany in canopy gaps created by commercial harvesting. *Forest Ecology and Management* 255, 300–307.

Lopez, E. and Orduz, S. (2003) *Metarhizium anisopliae* and *Trichoderma viride* for control of nests of the fungus-growing ant, *Atta cephalotes*. *Biological Control* 27, 194–200.

Lounsbury, C.P. (1917) The *Phoracantha* beetle: a borer pest of *Eucalyptus* trees. Local Series, Division of Entomology, Department of Agriculture, Pretoria, South Africa, 24, 1–10.

Louzada, J., Gardner, T., Peres, C. and Barlow, J. (2010) A multi-taxa assessment of nestedness patterns across a multiple-use Amazonian forest landscape. *Biological Conservation* 143, 1102–1109.

Loyttyniemi, K. (1991) Biology and control of *Phoracantha* beetles in exotic eucalypt plantations with special reference to Zambia. In: Schonau, A.P.G. (ed.) *IUFRO Symposium on Intensive Forestry: The Role of Eucalypts*, P2.02-01 Productivity of Eucalypts, Durban, South Africa, 2–6 September, 1991, Proceedings Vol. 2, pp. 768–775.

Lu, H.L., Zhu, L.L., Jiang, X.G., Gui, G.R. and Zhu, D.J. (1997) An investigation into the occurrence and control of *Dendrolimus punctatus* in the Dongjin forests, Jiangsu. *Journal of Jiangsu Forestry Science and Technology* 24, 32–33.

Lucas, J.A. (2011) Advances in plant disease and pest management. *Journal of Agricultural Science* 149.

Lugo, A.E. (1995) Management of tropical biodiversity. *Ecological Applications* 5, 956–961.

Lugo, A.E., Farnworth, E.G., Pool, D., Jerez, P. and Kaufman, G. (1973) The impact of the leaf cutter ant *Atta colombica* on the energy flow of a tropical wet forest. *Ecology* 54, 1292–1301.

Luna, R.K., Kaushal, P. and Banyal, R. (2006) Study on mortality of Kikar (*Acacia nilotica*) in Punjab. *Indian Forester* 132, 281–296.

Ma, W.Y., Peng, J.W., Wang, X.L. and Zuo, Y.X. (1989) Biology and population fluctuations of *Casinaria nigripes* Gravenhorst. *Acta Ecologica Sinica* 9, 27–34.

Mabberley, D.J. (1990) *The Plant Book*. Cambridge University Press, Cambridge, UK.

Mabberly, D.J. (2008) *Mabberly's Plant-Book*, 3rd edition. Cambridge University Press, Cambridge, UK, 1040 pp.

MacArthur, R.H. and Wilson, E.O. (1967) *The Theory of Island Biogeography*. Princeton University Press, Cambridge, Massachusetts.

McCann, K.S. (2000) The diversity–stability debate. *Nature* 405, 228–233.

McClure, M.S. (1990) Cohabitation and host species effects on the population growth of *Matsucoccus resinosae* (Homoptera: Margarodidae) and *Pineus boerneri* (Homoptera: Adelgidae) on red pine. *Environmental Entomology* 19, 672–676.

McConnell, D.B. and Short, D.E. (1987) Combining pesticides and growth regulators may cause phytotoxic reactions in plants. *American Nurseryman* 166, 170–178.

McCullough, D.G. and Sadof, C.S. (1998) Evaluation of an integrated management and compliance program for *Tomicus piniperda* (Coleoptera: Scolytidae) in pine Christmas tree fields. *Journal of Economic Entomology* 91, 785–795.

McCullough, D.G., Work, T.T., Cavey, J.F., Liebhold, A.M. and Marshall, D. (2006) Interceptions of nonindigenous plant pests at US ports of entry and border crossings over a 17-year period. *Biological Invasions* 8, 611–630.

McHowa, J.W. and Ngugi, D.N. (1994) Pest complex in agroforestry systems: the Malawi experience. *Forest Ecology and Management* 64, 277–284.

McMillin, J.D. and Wagner, M.R. (1995) Effects of water stress on biomass partitioning of ponderosa pine seedlings during primary root growth and shoot growth periods. *Forest Science* 41, 594–610.

McNeill, M.R., Payne, T.A. and Bewsell, D.T. (2008) Tourists as vectors of potential invasive alien species and a strategy to reduce risk. Re-creating tourism. New Zealand Tourism and Hospitality Research Conference. Hanmer Springs, 3–5 December 2008, 12 pp.

Madoffe, S.S. and Austara, O. (1990) Impact of pine woolly aphid, *Pineus pini* (Macquart) (Horn., Adelgidae), on growth of *Pinus patula* seedlings in Tanzania. *Journal of Applied Entomology* 110, 421–424.

Madoffe, S.S. and Austara, O. (1993) Abundance of the pine woolly aphid, *Pineus pini*, in *Pinus patula* stands growing on different sites in the Sao Hill district, Tanzania. *Commonwealth Forestry Review* 72, 118–121.

Maeto, K., Noerdjito, W.A., Belokobylskij, S.A. and Fukuyama, K. (2009) Recovery of species diversity and composition of braconid parasitic wasps after reforestation of degraded grasslands in lowland East Kalimantan. *Journal of Insect Conservation* 13, 245–257.

Mahroof, R.M., Hauxwell, C. and Edirisinghe, J.P. (2001) Variation in pattern of attack by *Hypsipyla robusta* (Moore) and recovery of seedlings following simulated *Hypsipyla* damage on *Swietenia macrophylla* (King) grown under different light conditions. *Ceylon Journal of Science, Biological Sciences* 26, 35–49.

Mahroof, R.M., Hauxwell, C., Edirisinghe, J.P., Watt, A.D. and Newton, A.C. (2002) Effects of artificial shade on attack by the mahogany shoot borer, *Hypsipyla robusta* (Moore). *Agricultural and Forest Entomology* 4, 283–292.

Mailu, A.M., Khamala, C.P.M. and Rose, D.J.W. (1978) Evaluation of pine woolly aphid damage to *Pinus patula* and its effect on yield in Kenya. *East African Agricultural and Forestry Journal* 43, 259–265.

Mailu, A.M., Khamala, C.P.M. and Rose, D.J.W. (1980) Population dynamics of pine woolly aphid, *Pineus pini* (Gmelin) (Hemiptera: Adelgidae), in Kenya. *Bulletin of Entomological Research* 70, 483–490.

Malcolm, J.R. (1994) Edge effects of central Amazonian forest fragments. *Ecology Tempe* 75, 2438–2445.

Maleque, M.A., Maeto, K. and Ishii, H.T. (2009) Arthropods as bioindicators of sustainable forest management, with a focus on plantation forests. *Applied Entomology and Zoology* 44, 1–11.

Mally, C.W. (1924) The *Eucalyptus* snout-beetle (*Gonipterus scutellatus*, Gyll.). *Journal of the Department of Agriculture, Union of South Africa* 9, 415–442.

Mancebo, F., Hilje, L., Mora, G.A. and Salazar, R. (2002) Biological activity of two neem (*Azadirachta indica* A. Juss., Meliaceae) products on *Hypsipyla grandella* (Lepidoptera: Pyralidae) larvae. *Crop Protection* 21, 107–112.

Mangoendihardjo, S., Mahrub, E. and Warrow, J. (1990) Endemic natural enemies of the leucaena psyllid in Indonesia. In: Napompeth, B. and MacDicken, K.G. (eds) *Leucaena Psyllid: Problems and Management*. Proceedings of an International Workshop, Bogor, Indonesia, 16–21 January 1989. Winrock International.

Manian, S. and Udaiyan, K. (1992) Altitudinal distribution of the Lantana lace bug *Teleonemia scrupulosa* Stal. in the Anaimalai hills (western Ghats), India. *Tropical Pest Management* 38, 93–95.

Mannakkara, A. and Alawathugoda, R.M.D. (2005) Black twig borer *Xylosandrus compactus* (Eichhoff) (Coleoptera: Scolytidae) damage for forest tree species. *Sri Lanka Forester* 28, 65–75.

Mariau, D. (1999) (ed.) *Integrated Pest Management of Tropical Perennial Crops*. Science Publishers, Enfield, New Hampshire.

Marsaro, A.L., Souza, R.C., Della Lucia, T.M.C., Fernandes, J.B., Silva, M.F.G.F. and Vieira, P.C. (2004) Behavioural changes in workers of the leaf-cutting ant *Atta sexdens rubropilosa* induced by chemical components of *Eucalyptus maculata* leaves. *Journal of Chemical Ecology* 30(9), 1771–1780.

Martinez, A.J., Lopez-Portillo, J., Eben, A. and Golubov, J. (2009) Cerambycid girdling and water stress modify mesquite architecture and reproduction. *Population Ecology* 51, 533–541.

Martinez, E.J.M. (1982) *Phoracantha semipunctata* [on eucalypts] in southwestern Spain. Summary of the control programme using trap trees. *Boletin de la Estacion Central de Ecologia* 11, 57–69.

Martinez, G. and Bianchi, M. (2010) First record in Uruguay of the bronze bug, *Thaumastocoris peregrinus* Carpintero and Dellape, (Heteroptera: Thaumastocoridae). *Agrociencia* 14(1), 15–18.

Martinez-Ramos, M., Anten, N.P.R. and Alcalde-Sanchez, M. (2009) Defoliation and ENSO effects on vital rates of an understorey tropical forest palm. *Journal of Ecology* 97(5), 1050–1061.

Marturano, R. and Vergara, A.J.B. (1997) *Thyrinteina arnobia* Stoll, (Lepidoptera: Geometridae) defoliating *Eucalyptus* spp. in eastern Venezuela. 1. Life cycle. *Revista Forestal Venezolana* 31(41–2), 119–127.

Masoodi, M.A., Trali, A.R. and Bhat, A.M. (1990) Supression of *Lymantria obfuscata* Walker by sex pheromone trapping of males. *Indian Journal of Entomology* 52, 414–417.

Massad, T.J., Chambers, J.Q., Rolim, S.G., Jesus, R.M. and Dyer, L.A. (2011) Restoration of pasture to forest in Brazil's Mata Atlantica: the roles of herbivory, seedling defences and plot design in reforestation. *Restoration Ecology* 19, 257–267.

Mathew, G. (1986) Insects associated with forest plantations of *Gmelina arborea* Roxb. in Kerala, India. *Indian Journal of Forestry* 9, 308–311.

Mathew, G. (1987) Cossid pests of plantation crops in India and the prospects of their management. *Journal of Coffee Research* 17, 137–140.

Mathew, G. (1989) Natural enemies of the bagworm, *Pteroma plagiophleps* Hampson (Lepidoptera: Psychidae) in Kerala (India). *Entomon* 14, 335–338.

Mathew, G. (1993) Nursery pests of indigenous tree species and their management in Kerala, India. In: Perrin, R. and Sutherland, J.R. (eds) *Diseases and Insects in Forest Nurseries*. Proceedings of IUFRO Conference in Dijon, France, 3–10 October, 1993. INRA, Paris, pp.145–151.

Mathew, G. (2005) Nursery pest problems on some native tree species in Kerala province, India. *Working Papers of the Finnish Forest Research Institute* 11, 31–36.

Mathews J.H. (2006) Assemblages of plants and arthropods associated with *Acacia senegal* inside and outside plantations. Unpublished DPhil thesis, University of Oxford, UK.

Mathur, R.N. (1979) Biology of *Tingis (Caenotingis) beesoni* Drake (Heteroptera: Tingitidae). *Indian Forest Bulletin* No. 276.

Mathys, G. and Baker, E.A. (1980) An appraisal of the effectiveness of plant quarantines. *Annual Review of Phytopathology* 18, 85–101.

Matsumoto, K. and Irianto, R.S.B. (1998) Adult biology of the albizia borer, *Xystrocera festiva* Thomson (Coleoptera: Cerambycidae), based on laboratory breeding, with particular reference to its oviposition schedule. *Journal of Tropical Forest Science* 10, 367–378.

Matsumoto, K. and Kotulai, J.R. (2002) Field tests on the effectiveness of *Azadirachta* companion planting as a shoot borer repellent to protect mahogany. *JIRCAS Journal* 10, 1–8.

Matsumoto, K., Irianto, R.S.B. and Kitajima, H. (2000) Biology of the Japanese green-lined albizzia longicorn, *Xystrocera globosa* (Coleoptera: Cerambycidae). *Entomological Science* 3(1), 33–42.

Mattson, W.J. (1980) Herbivory in relation to plant nitrogen content. *Annual Review of Ecology and Systematics* 11, 119–161.

Mayhew, J.E. and Newton, A.C. (1998) *The Silviculture of Mahogany*. Institute of Ecology and Resource Management, University of Edinburgh, UK.

Mazodze, R. (1992) Field application of carbosulfan for the protection of eucalypt transplants from termite attack. *South African Forestry Journal* 162, 21–25.

Mazodze, R. (1993) An overview of forest nursery diseases and insects in Zimbabwe. In: Perrin, R. and Sutherland, J.R. (eds) *Diseases and Insects in Forest Nurseries*. Proceedings of IUFRO Conference in Dijon, France, 3–10 October, 1993. INRA, Paris.

Mbai, M.L.O. (1995) Investigation of arthropods associated with agroforestry in Machakos, Kenya. Unpublished DPhil thesis, University of Oxford, UK.

McMaugh, T. (2005) Guidelines for plant pests in Asia and the Pacific. *Australian Centre for International Agricultural Research Monograph 119*, 192 pp.

Medina, G.S., Covas, F.G., Abreu, E., Ingles, R. and Gaud, S.M. (1987) The insects of Nispero (*Manilkara zapota* (L.) P. van Rogen) in Puerto Rico. *Journal of Agriculture of the University of Puerto Rico* 71, 129–132.

Meeker, J.R. (2008) Southern pine coneworm, *Dioryctria amatella* (Hulst) (Insecta: Lepidoptera: Pyralidae). University of Florida Institute of Food and Agricultural Sciences (IFAS) Extension. EENY325, 6 pp.

Mehlig, U. and Menezes, M.P.M. (2005) Mass defoliation of the mangrove tree *Avicennia germinans* by the moth *Hyblaea puera* (Lepidoptera: Hyblaeidae) in equatorial Brazil. *Ecotropica* 11, 87–88.

Meisel, J.E. (2006) Thermal ecology of the neotropical army ant *Eciton burchellii*. *Ecological Applications* 16, 913–922.

Mellado, B. (1976) *Spodoptera sunia* (Lepidoptera: Noctuidae: Amphipyrinae). Determination of the life cycle in the laboratory. *Baracoa* 6(3–4), 13–18.

Mendel, Z., Protasov, A., Fisher, N. and La Salle, J. (2004) Taxonomy and biology of *Leptocybe invasa* gen. & sp. n. (Hymenoptera: Eulophidae), an invasive gall inducer on *Eucalyptus*. *Australian Journal of Entomology* 43, 51–63.

Mendelsohn, R. and Balick, M.J. (1995) The value of undiscovered pharmaceuticals in tropical forests. *Economic Botany* 49, 223–228.

Menken, S.B.J. (1996) Pattern and process in the evolution of insect-plant associations: *Yponomeuta* as an example. *Entomologia Experimentalis et Applicata* 80, 297–305.

Mercer, C.W.L. (1990) Prospects for integrated pest management in forestry in Papua New Guinea. *Brighton Crop Protection Conference, Pests and Diseases*, Vol. 1, pp. 385–390.

Merkle, S.A. and Nairn, C.J. (2005) Hardwood tree biotechnology. *In Vitro Cellular and Developmental Biology-Plant* 41, 602–619.

Merrifield, L.E. and Howcroft, N.H.S. (1975) *Ceroplastes rubens* Maskell damage of *Pinus caribaea* Morelet with notes on the scale's preference of certain clones of host material (Hemiptera: Coccidae). *Silvae Genetica* 24, 110–113.

Meshram, P.B. (2010) Role of some biopesticides in management of some forest insect pests. *Journal of Biopesticides* 3, 250–252.

Meshram, P.B. and Tiwari, C.K. (2003) Lace bug *Tingis beesoni* (Drake) causing top dying of *Gmelina arborea* (Linn.) Hi Tech plantations and evaluation of certain pesticides against the bug. *Journal of Applied Zoological Researches* 14(2), 200–203.

Meshram, P.B., Husen, N. and Joshi, K.C. (1993) A new report of ambrosia beetle, *Xylosandrus compactus* Eichhoff (Coleoptera: Scolytidae) as a pest of African mahogany, *Khaya* sp. *Indian Forester* 120, 58–61.

Meshram, P.B., Joshi, K.C. and Sarkar, A.K. (1994) Relative resistance of certain clones of *Tectona grandis* to teak leaf skeletonizer, *Eutectona machaeralis* Walk. (Lepidoptera: Pyralidae). *Indian Forester* 120, 58–61.

Meshram, P.B., Bisaria, A.K., Shamila, K. and Kalia, S. (1997) Efficacy of Bioasp and Biolep – a microbial insecticide against teak skeletonizer, *Eutectona machaeralis* Walk. *Indian Forester* 123, 1202–1204.

Meshram, P.B., Pande, P.K. and Banerjee, S.K. (2001) Impact of pest problems in *Gmelina arborea* Linn. plantations in Western Maharashtra. *Indian Forester* 127(12), 1377–1386.

Messing, R.H., Noser, S. and Hunkeler, J. (2009) Using host plant relationships to help determine origins of the invasive Erythrina gall wasp, *Quadrastichus erythrinae* Kim (Hymenoptera: Eulophidae). *Biological Invasions* 11, 2233–2241.

Michel, A. and Teisaire, E.S. (1996) Chronology of the normal embryonic development of *Baeacris punctulatus* (Thunberg) (Orthoptera: Acrididae). *Revista Chilena de Entomologia* 23, 29–41.

Mifsud, D., Perez Hidalgo, N. and Barbagallo, S. (2009) Present status of aphid studies in Malta (Central Mediterranean) with special reference to tree dwelling species. *Redia* 92, 93–96.

Miller, D. (1925) *Forest and Timber Insects in New Zealand*. New Zealand State Forest Service Bulletin No. 2. Government Printer, Wellington.

Mills, N.J. (1990) Biological control of forest aphid pests in Africa. *Bulletin of Entomological Research* 80, 31–36.

Mishra, J. and Prasad, U.N. (1980) Agri-silvicultural studies on raising of oil seeds like *Sesamum indicum* Linn. (Til), *Arachys hypogea* Linn. (Groundnut) and *Glycine max* Merrill (Soybean) as cash crops in conjunction with *Dalbergia sissoo* Roxb. and *Tectona grandis* Linn, at Mandar, Ranchi. *Indian Forester* 106, 675–695.

Mishra, S.C., Veer, V. and Chandra, A. (1985) *Aristobia horridula* Hope (Coleoptera: Lamiidae) a new pest of Shisham (*Dalbergia sissoo* Roxb.) in West Bengal. *Indian Forester* 111, 738–741.

Mitchell, J.D. (2002) Termites as pests of crops, forestry, rangeland and structures in southern Africa and their control. *Sociobiology* 40, 47–69.

Mitchell, M.R. and Boland, D.J. (1989) Susceptibility to termite attack of various tree species planted in Zimbabwe. Trees for the tropics. Growing Australian multipurpose trees and shrubs in developing countries. ACIAR Monograph, pp. 215–227.

Mitchell, W.C. and Waterhouse, D.F. (1986) Spread of the Leucaena psyllid *Heteropsylla cubana* in the Pacific. *Leucaena Research Reports* 7, 6–8.

Mitter, C. and Farrell, B. (1991) Macroevolutionary aspects of insect–plant relationships. In: Bernays, E.A. (ed.) *Insect–Plant Interactions*, Vol. 3. CRC Press, Boca Raton, Florida.

Mohandas, K. (1986) *Brachymeria excarinata* Gahan (Hymenoptera: Chalcididae) as pupal parasitoid of *Calopepla leayana* Latr. in Kerala, India: a new record. *Entomon* 11, 279–280.

Moller, A.P. (1995) Leaf-mining insects and fluctuating asymmetry in elm *Ulmus glabra* leaves. *Journal of Animal Ecology* 64, 697–707.

Montagu, K.D. and Woo, K.C. (1999) Effect of two insect pests on *Acacia auriculiformis* tree growth and form in Australia's Northern Territory. *Journal of Tropical Forest Science* 11, 492–502.

Montalva, C., Rojas, E., Ruiz, C. and Lanfranco, D. (2010) The cypress aphid in Chile: a review of the current situation and preliminary data of the biological control. *Bosque* 31, 81–88.

Moog, F.A. (1992) *Heteropsylla cubana*: impact on feeding systems in southwest Asia and the Pacific. In: Speedy, A. (ed.) *Legume Trees and Other Fodder Trees as Protein Sources for Livestock*. FAO Animal Production and Health Paper No. 102, pp. 233–243.

Moore, D., Fischer, H.U. and Agounke, D. (1988) Biological control of *Rastrococcus invadens* Williams in Togo. *FAO Plant Protection Bulletin* 36, 169–174.

Moore, J.A. (1993) Rediscovery and releases of parasitoid of eucalyptus borer. *Plant Protection News* 33, 4–7.

Mopper, S. and Whitham, T.G. (1992) The plant stress paradox: effects on pinyon sawfly sex ratios and fecundity. *Ecology* 73, 515–525.

Moran, V.C., Hoffmann, J.H., Impson, F.A.C. and Jenkins, J.F.G. (1994) Herbivorous insect species in the tree canopy of a relict South African forest. *Ecological Entomology* 19, 147–154.

Morton, J.F. (1993) The black olive (*Bucida buceras* L.), a tropical timber tree, has many faults as an ornamental. *Proceedings of the Florida State Horticulture Society* 106, 338–343.

Moses, I. (2009) Interactions amongst trees and crops in taungya systems of western Kenya. *Agroforestry Systems* 76(2), 265–273.

Mrkva, R. (1994) Silver fir woolly aphid [*Adelges* (= *Dreyfusia*) *nordmannianae* Eckstein), its control and its role in fir decline. *Lesnictvi Prague* 40, 361–370.

Muddiman, S.B., Hodkinson, I.D. and Hollis, D. (1992) Legume-feeding psyllids of the genus *Heteropsylla* (Homoptera: Psylloidea). *Bulletin of Entomological Research* 82, 73–117.

Mugasha, A.G., Chamshama, S.A.O., Nshubemuki, L., Iddi, A. and Kindo, A.I. (1997) Performance of thirty-two families of *Cupressus lusitanica* at Hambalawei, Lushoto, Tanzania. *Silvae Genetica* 46, 185–192.

Mukhtar, A. (1987) Relative resistance of different clones of *Tectona grandis* to teak defoliator, *Hyblaea puera* Cram (Lepidoptera: Hyblaeidae) in South India. *Indian Forester* 113, 281–286.

Mukhtar, A., Vijayachandran, S.N. and Choudhuri, J.C.B. (1985) Biology of *Hestiasula brunneriana* Saussure (Dictyoptera: Mantidae). *Indian Forester* 111, 333–338.

Mullen, B.F. and Gutteridge, R.C. (2002) Wood and biomass production of *Leucaena* in subtropical Australia. *Agroforestry Systems* 55, 195–205.

Mullen, B.F. and Shelton, H.M. (2003) Psyllid resistance in Leucaena. Part 2. Quantification of production losses from psyllid damage. *Agroforestry Systems* 58, 163–171.

Mullen, B.F., Gabunada, F., Shelton, H.M. and Stur, W.W. (2003) Psyllid resistance in *Leucaena*. Part 1. Genetic resistance in subtropical Australia and humid-tropical Philippines. *Agroforestry Systems* 58, 149–161.

Mullen, B.F., Shelton, H.M., Gutteridge, R.C. and Basford, K.E. (2003) Agronomic evaluation of Leucaena. Part 1. Adaptation to environmental challenges in multi-environment trials. *Agroforestry Systems* 58, 77–92.

Mulrooney, J.E., Wagner, T.L. and Gerard, P.D. (2009) Fipronil: toxicity to subterranean termites and dissipation in soils. In: Peterson, C.J.S.D.M. (ed.) *Pesticides in Household, Structural and Residential Pest Management*. pp. 107–123.

Mundim, F.M., Costa, A.N. and Vasconcelos, H.L. (2009) Leaf nutrient content and host plant selection by leaf-cutter ants, *Atta laevigata*, in a Neotropical savanna. *Entomologia Experimentalis et Applicata* 130, 47–54.

Muniappan, R. (1993) Pests and diseases of *Erythrina*: a review. *Journal of Coffee Research* 23, 1–13.

Murphy, E.A. and Aucott, M. (1998) An assessment of the amounts of arsenical pesticides used historically in a geographical area. *The Science of the Total Environment* 218(2–3), 89–101.

Murphy, S.T. (1996) Status and impact of invasive conifer aphid pests in Africa. In: Nair, K.S.S., Sharma, J.K. and Varma, R.V. (eds) *Impact of Diseases and Insect Pests in Tropical Forests*. Proceedings of the IUFRO Symposium, 23–26 November 1993, Peechi, India, pp. 289–297.

Murphy, S.T. (1997) Protecting Africa's trees: status and actions for pest management in African forestry. *Proceedings XI World Forestry Congress*, Vol. 1, pp. 167–172.

Myers, N. (1996) Environmental services of biodiversity. *Proceedings of the National Academy of Sciences of the United States of America* 93, 2764–2769.

Nadel, R.L., Slippers, B., Scholes, M.C., Lawson, S.A., Noack, A.E., Wilcken, C.F., Bouret, M.J. and Wingfield, M.J. (2009) DNA bar-coding reveals source and patterns of *Thaumastocoris peregrinus* invasions in South Africa and South America. *Biological Invasions* (https://repository.up.ac.za/upspace/bitstream/2263/13690/1/Nadel_DNA(2009).pdf).

Nadel, R.L., Wingfield, M.J., Scholes, M.C., Lawson, S.A. and Slippers, B. (2011) Monitoring and control of insect pests in Southern Hemisphere forestry plantations using semiochemicals. *Annals of Forest Science*.

Naeem, S., Thompson, L.J., Lawler, S.P., Lawton, J.H. and Woodfin, R.M. (1994) Declining biodiversity can alter the performance of ecosystems. *Nature, London* 368, 734–737.

Nahrung, H.F. (2006) Paropsine beetles (Coleoptera: Chrysomelidae) in south-eastern Queensland hardwood plantations: identifying potential pest species. *Australian Forestry* 69(4), 270–274.

Nahrung, H.F. and Allen, G.R. (2003) Geographical variation, population structure and gene flow between populations of *Chrysophtharta agricola* (Chapuis) (Coleoptera: Chrysomelidae), a pest of Australian eucalypt plantations. *Bulletin of Entomological Research* 93, 137–144.

Nahrung, H.F., Schutze, M.K., Clarke, A.R., Duffy, M.P., Dunlop, E.A. and Lawson, S.A. (2008) Thermal requirements, field mortality and population phenology modelling of *Paropsis atomaria* Olivier, an emergent pest in subtropical hardwood plantations. *Forest Ecology and Management* 255, 3515–3523.

Nahrung, H.F., Waugh, R. and Hayes, R.A. (2009) *Corymbia* species and hybrids: chemical and physical foliar attributes and implications for herbivory. *Journal of Chemical Ecology* 35, 1043–1053.

Nair, K.S.S. (1988) The teak defoliator in Kerala, India. In: Berryman, A.A. (ed.) *Dynamics of Forest Insect Populations: Patterns, Causes, Implications*. Plenum Press, New York, pp. 267–289.

Nair, K.S.S. (1991) Social, economic and policy aspects of integrated pest management of forest defoliators in India. *Forest Ecology and Management* 39, 283–288.

Nair, K.S.S. (2007) *Tropical Forest Insect Pests: Ecology, Impact and Management*. Cambridge University Press, Cambridge, UK, 404 pp.

Nair, K.S.S. and Mathew, G. (1988) Biology and control of insect pests of fast-growing hardwood species. 1. *Albizia falcataria* and *Gmelina arborea*. *KFRI Research Report* No. 51.

Nair, K.S.S. and Mathew, G. (1992) Biology, infestation characteristics and impact of the bagworm *Pteroma plagiophleps* Hamps in forest plantations of *Paraserianthes falcataria*. *Entomon* 17(1–2), 179–180.

Nair, K.S.S. and Varma, R.V. (1981) Termite control in eucalypt plantations. *Kerala Forest Research Institute Research Report* No. 6.

Nair, K.S.S., Mathew, G. and Sivarajan, M. (1981) Occurrence of the bagworm, *Pteroma plagiophleps* Hampson (Lepidoptera: Psychidae) as a pest of the tree *Albizia falcataria* in Kerala, India. *Entomon* 6, 179–180.

Nair, K.S.S., Sudheendrakumar, V.V., Varma, R.V. and Chacko, K.C. (1985) Studies on the seasonal incidence of defoliators and the effect of defoliation on volume increment of teak. *Research Report, Kerala Forest Research Institute* No. 30, 78 pp.

Nair, K.S.S., Kedharnath, S., Koshy, M.P., Sudheendrakumar, V.V., Mohanadas, K., Varma, R.V., et al. (1989) Search for natural resistance to the insect pests, *Hyblaea puera*, in teak (Final report of the Research Project Entom 12/83, April 1983–December 1986). *KFRI Research Report* No. 62, vi + 32 pp.

Nair, K.S.S., Babjan, B., Sajeev, T.V., Sudheendrakumar, V.V., Ali, M.I.M., Varma, R.V., et al. (1996a) Field efficacy of nuclear polyhedrosis virus for protection of teak against the defoliator, *Hyblaea puera* Cramer (Lepidoptera: Hyblaeidae). *Journal of Biological Control* 10, 79–85.

Nair, K.S.S., Sudheendrakumar, V.V., Varma, R.V. and Chacko, K.C. (1996b) Effect of defoliation by *Hyblaea puera* and *Eutectona machaeralis* (Lepidoptera) on volume increment of teak. In: Nair, K.S.S., Sharma, J.K. and Varma, R.V. (eds) *Impact of Diseases and Insect Pests in Tropical Forests*. Proceedings of the IUFRO Symposium, 23–26 November 1993, Peechi, India, pp. 257–273.

Nair, P.K.R. (1993) State-of-the-art of agroforestry research and education. *Agroforestry Systems* 23, 95–115.

Nair, P.K.R. and Dagar, J.C. (1991) An approach to developing methodologies for evaluating agro-forestry systems in India. *Agroforestry Systems* 16, 55–81.

Napompeth, B. (1994) Leucaena psyllid in the Asia-Pacific region: implications for its management in Africa. *FAO Rapa Publication* 1994/13, 27 pp.

National Research Council (1969) *Insect Pest Management and Control. Principles of Plant and Animal Pest Control*, Vol. 3. National Academy of Sciences, Washington, DC.

Navarro, C., Montagnini, F. and Hernandez, G. (2004) Genetic variability of *Cedrela odorata* Linnaeus: results of early performance of provenances and families from Mesoamerica grown in association with coffee. *Forest Ecology and Management* 192, 217–227.

Neal, J.W. Jr, Tauber, M.J. and Tauber, C.A. (1992) Photoperiodic induction of reproductive diapause in *Corythuca cydoniae* (Heteroptera: Tingidae). *Environmental Entomology* 21, 1414–1418.

Nechols, J.R. and Seibert, T.F. (1985) Biological control of the spherical mealybug, *Nipaecoccus vastator* (Homoptera: Pseudococcidae): assessment by ant exclusion. *Environmental Entomology* 14, 45–47.

Nedved, O. and Windsor, D. (1994) Supercooling ability, fat and water contents in a diapausing tropical beetle, *Stenotarsus rotundus* (Coleoptera: Endomychidae). *European Journal of Entomology* 91, 307–312.

Nehru, C.R., Jayarathnam, K. and Karnavar, G.K. (1991) Application of entomophagous fungus *Beauveria brongniartii* for management of chafer beetle of the white grub *Holotrichia serrata* infesting rubber seedlings. *Indian Journal of Natural Rubber Research* 4, 123–125.

Nestel, D., Dickschen, F. and Altieri, M.A (1994) Seasonal and spatial population loads of a tropical insect: the case of the coffee leaf-miner in Mexico. *Ecological Entomology* 19, 159–167.

Neto, A.B.G., Felfili, J.M., da Silva, G.F., Mazzei, L., Fagg, C.W. and Nogueira, P.E. (2004) Evaluation of mahogany homogenous stands, *Swietenia macrophylla* King, compared to mixed stands with *Eucalyptus urophylla* S.T. Blake, 40 months after planting. *Revista Arvore* 28(6), 777–784.

Neumann, F.G. (1987) Introduced bark beetles on exotic trees in Australia with special reference to infestations of *Ips grandicollis* in pine plantations. *Australian Forestry* 50, 166–178.

Neumann, F.G., Morey, J.L. and McKimm, R.J. (1987) The sirex wasp in Victoria. Department of Conservation, Forests and Lands, Lands and Forest Division, Bulletin No. 29.

New, T.R. (1976) Aspects of exploitation of *Acacia phyllodes* by a mining Lepidopteran, *Acrocercops plebeia* (Gracillariidae). *Journal of the Australian Entomological Society* 15, 365–378.

Newton, A.C., Baker, P., Ramnarine, S., Mesen, J.F. and Leakey, R.R.B. (1993) The mahogany shoot borer: prospects for control. *Forest Ecology and Management* 57, 301–328.

Newton, A.C., Cornelius, J.P., Mesen, J.F., Corea, E.A. and Watt, A.D. (1998) Variation in attack by the mahogany shoot borer, *Hypsipyla grandella* (Lepidoptera: Pyralidae), in relation to host growth and phenology. *Bulletin of Entomological Research* 88, 319–326.

NFTA (1990) Leucaena psyllids – a review of the problem and its solutions. Nitrogen Fixing Tree Association, Waimanalo, Hawaii. *Tigerpaper* 17, 12–14.

Nichols, J.D., Wagner, M.R., Agyeman, V.K., Bosu, P. and Cobbinah, J.R. (1998) Influence of artificial gaps in tropical forest on survival, growth, and *Phytolyma lata* attack on *Milicia excelsa*. *Forest Ecology and Management* 110, 353–362.

Nichols, J.D., Agyeman, V.K., Agurgo, F.B., Wagner, M.R. and Cobbinah, J.R. (1999) Patterns of seedling survival in the tropical African tree *Milicia excelsa*. *Journal of Tropical Ecology* 15, 451–461.

Niemala, P., Rousi, M. and Saarenmaa, H. (1987) Topographical delimitation of *Neodiprion sertifer* (Hym., Diprionidae) outbreaks on Scots pine in relation to needle quality. *Journal of Applied Entomology* 103, 84–910.

Noack, A.E. and Rose, H.A. (2007) Life-history of *Thaumastocoris peregrinus* and *Thaumastocoris* sp. in the laboratory with some observations on behaviour. *General and Applied Entomology* 36, 27–34.

Norlander, G., Nordenhem, H. and Hellqvist, C. (2009) A flexible sand coating (Conniflex) for the protection of conifer seedlings against damage by the pine weevil *Hylobius abietis*. *Agricultural and Forest Entomology* 11, 91–100.

Noumi, Z., Touzard, B., Michalet, R. and Chaieb, M. (2010a) The effects of browsing on the structure of *Acacia tortilis* (Forssk.) Hayne ssp. *raddiana* (Savi) Brenan along a gradient of water availability in arid zones of Tunisia. *Journal of Arid Environments* 74, 625–631.

Noumi, Z., Dhaou, S.O., Abdallah, F., Touzard, B. and Chaieb, M. (2010b) *Acacia tortilis* subsp. Raddiana in the North African arid zone: the obstacles to natural regeneration. *Acta Botanica Gallica* 157(2), 231–240.

Novotny, V. and Basset, Y. (1998) Seasonality of sap sucking insects (Auchenorrhyncha, Hemiptera) feeding on *Ficus* (Moraceae) in a lowland rain forest in New Guinea. *Oecologia* 115, 514–522.

Novotny, V., Drozd, P., Miller, S.E., Kulfan, M., Janda, M., Basset, Y. and Weiblem, G.D. (2006) Why are there so many species of herbivorous insects in tropical rainforests? *Science* 313, 1115–1118.

Noyes, J. (1990) A new genus and species of encyrtid (Hymenoptera: Chalcidoidea) parasitoid of the eggs of the varicose borer, *Agrilus sexsignatus* (Fisher) (Coleoptera: Buprestidae), a pest of bagras (*Eucalyptus deglupta* Blume) in the Philippines. *Journal of Natural History* 24, 21–25.

Nyeko, P., Mutitu, E.K. and Day, R.K. (2007) Farmers' knowledge, perceptions and management of the gall-forming wasp, *Leptocybe invasa* (Hymenoptera: Eulophidae), on *Eucalyptus* species in Uganda. *International Journal of Pest Management* 53, 111–119.

Nyeko, P., Mutitu, E.K. and Day, R.K. (2009) *Eucalyptus* infestation by *Leptocybe invasa* in Uganda. *African Journal of Ecology* 47, 299–307.

O'Dowd, D.J., Green, P.T. and Lake, P.S. (2003) Invasional 'meltdown' on an oceanic island. *Ecology Letters* 6, 812–817.

Obiri, J.F. (1994) Variation of cypress aphid (*Cinara cupressi*) (Buckton) attack on the family Cupressaceae. *Commonwealth Forestry Review* 73, 43–46.

Obiri, J.F., Giathi, G. and Massawe, A. (1994) The effect of cypress aphid on *Cupressus lusitanica* orchards in Kenya and Tanzania. *East African Agricultural and Forestry Journal* 59, 227–234.

Oda, S. and Berti Filho, E. (1978) Annual volume increment of *Eucalyptus saligna* in areas with various levels of infestation by caterpillars of *Thyrinteina arnobia* (Stoll, 1782) (Lepidoptera: Geometridae). *IPEF, Piracicaba* 17, 27–31.

Odera, J.A. (1974) The incidence and host trees of pine woolly aphid, *Pineus pini* (L.), in East Africa. *Commonwealth Forestry Review* 53, 128–136.

Odera, J.A. (1991) Some opportunities for managing aphids of softwood plantations in Malawi. Assistance to Forestry Sector, MLW/86/020, Malawi, FAO, Rome.

Ofori, D.A. and Cobbinah, J.R. (2007) Integrated approach for conservation and management of genetic resources of *Milicia* species in West Africa. *Forest Ecology and Management* 238, 1–6.

Ofori, D.A., Opuni-Frimpong, E. and Cobbinah, J.R. (2007) Provenance variation in *Khaya* species for growth and resistance to shoot borer *Hypsipyla robusta*. *Forest Ecology and Management* 242(2–3), 438–443.

Ogden, J. (1995) The long-term conservation of forest diversity in New Zealand. *Pacific Conservation Biology* 2, 77–90.

Ogol, C.K.P.O. and Spence, J.R. (1997) Abundance, population dynamics and impact of the leucaena psyllid *Heteropsylla cubana* Crawford in a maize–leucaena agroforestry system in Kenya. *Insect Science and its Application* 17, 183–192.

Ohno, S. (1990a) The Scolytidae and Platypodidae (Coleoptera) from Borneo found in logs at Nagoya port I. *Research Bulletin of the Plant Protection Service, Japan* No. 26, 83–94.

Ohno, S. (1990b) The Scolytidae and Platypodidae (Coleoptera) from Borneo found in logs at Nagoya port II. *Research Bulletin of the Plant Protection Service, Japan* No. 26, 95–103.

Oliveira, A.C.D., Fonseca, E.D.P., Anjos, N.D., Santos, G.P. and Zanuncio, J.C. (1984) Resistance of *Eucalyptus* spp. (Myrtaceae) to the larval defoliator *Thyrinteina arnobia* (Lepidoptera: Geometridae). *Revista Arvore* 8, 93–103.

Oliveira, M.A., Della-Lucia, T.M.C. and Anjos, N. (1998) Occurrence and nest density of leaf-cutting ants under eucalypt plantations in southern Bahia. *Revista Brasileira de Entomologia, Sao Paulo* 42(1/2), 17–21.

Oliveira, N.C., Wilcken, C.F., Zonta-de-Carvalho, R.C., Ferreira-Filho, P.J. and Couto, E.B. (2008) Occurrence of *Pineus boerneri* Annand (Hemiptera: Adelgidae) in pine plantations in the States of Sao Paulo and Minas Gerais, Brazil. *Arquivos do Instituto Biologico, Sao Paulo* 75(3), 385–387.

Oliveira, Y.M.M., Rosot, M.A.D., Luz, N.B., Ciesla, W.M., Johnson, E.W., Rhea, R., et al. (2006) Aerial sketch-mapping for monitoring forest conditions in Southern Brazil. *USDA Forest Service Proceedings* RMRS-P-42CD, pp. 815–824.

Oliver, A.D. and Chapin, J.B. (1980) The cranberry rootworm: adult seasonal history, and factors affecting status as a pest of woody ornamentals in Louisiana. *Journal of Economic Entomology* 73, 96100.

Olkowski, W., Dietrick, E. and Olkowski, H. (1992) The biological control industry in the United States. *IPM Practitioner* 14, 1–7.

Onufrieva, K.S., Thorpe, K.W., Hickman, A.D., Leonard, D.S., Mastro, V.C. and Roberts, E.A. (2008) Gypsy moth mating disruption in open landscapes. *Agricultural and Forest Entomology* 10, 175–179.

Oogita, T., Naito, H., Soma, Y. and Kawakami, F. (1998) Effect of low dose methyl bromide on forest insect pests. *Research Bulletin of the Plant Protection Service* 34, 37–39.

Opuni-Frimpong, E., Karnosky, D.F., Storer, A.J. and Cobbinah, J.R. (2008a) Silvicultural systems for plantation mahogany in Africa: influences of canopy shade on tree growth and pest damage. *Forest Ecology and Management* 255, 328–333.

Opuni-Frimpong, E., Karnosky, D.F., Storer, A.J., Abeney, E.A. and Cobbinah, J.R. (2008b) Relative susceptibility of four species of African mahogany to the shoot borer *Hypsipyla robusta* (Lepidoptera: Pyralidae) in the moist semideciduous forest of Ghana. *Forest Ecology and Management* 255, 313–319.

Orr, A.G. and Hauser, C.L. (1996) Kuala Belalong, Brunei: a hotspot of old world butterfly diversity. *Tropical Lepidoptera* 7, 1–12.

Osland, A. and Osland, J.S. (2007) Aracruz Celulose: best practices icon but still at risk. *International Journal of Manpower* 28(5), 435–450.

Ozaki, K., Kitamura, S., Subiandoro, E. and Taketani, A. (1999) Life history of *Aulacaspis marina* Takagi and Williams (Hom., Coccoidea), a new pest of mangrove plantations in Indonesia, and its damage to mangrove seedlings. *Journal of Applied Entomology-Zeitschrift für Angewandte Entomologie* 123, 281–284.

Ozaki, K., Takashima, S. and Suko, O. (2000) Ant predation suppresses populations of the scale insect *Aulacaspis marina* in natural mangrove forests. *Biotropica* 32, 764–768.

Ozanne, C.M.P., Hambler, C., Foggo, A. and Speight, M.R. (1997) The significance of edge effects in the management of forests for invertebrate biodiversity. In: Stork, N.E., Adis, J. and Didham, R.K. (eds) *Canopy Arthropods*. Chapman and Hall, London, pp. 534–550.

Paine, T.D. and Millar, J.G. (2002) Insect pests of eucalypts in California: implications of managing invasive species. *Bulletin of Entomological Research* 92, 147–151.

Paine, T.D. and Miller, J.G. (2010) Accumulation of insects on eucalyptus in California: random process or smoking gun. *Journal of Economic Entomology* 103(6), 1943–1949.

Paine, T.D., Millar, J.G. and Hanks, L.M. (1995) Integrated program protects trees from eucalyptus longhorned borer. *California Agriculture* 49, 34–37.

Paine, T.D., Steinbauer, M.J. and Lawson, S.A. (2011) Native and exotic pests of eucalyptus: a worldwide perspective. *Annual Review of Entomology* 56, 181–201.

Palmer, B., Bray, R.A., Ibrahim, T.M. and Fulloon, M.G. (1989) The effect of the leucaena psyllid on the yield of *Leucaena leucocephala* cv. Cunningham at four sites in the tropics. *Tropical Grasslands* 23, 105–107.

Palmer, J. and Synnott, T.J. (1992) The management of natural forests. In: Sharma, N.P. (ed.) *Managing the World's Forests*. Kendall/Hunt Publishing Company, Dubuque, Iowa, pp. 337–374.

Palo, R.T., Gowda, J. and Hogberg, P. (1993) Species height and root symbiosis, two factors influencing antiherbivore defense of woody plants in East African savanna. *Oecologia Heidelberg* 93, 322–326.

Palot, M.J. and Radhakrishnan, C. (2004) First report of the infestation and epidemic outbreak of *Hyblaea puera* Cramer (Hyblaeidae: Lepidoptera: Insecta) on mangrove plant, *Avicennia*. *Records of the Zoologicial Survey of India* 103(3–4), 171–174.

Pancel, L. (1993) *Tropical Forestry Handbook*. Springer-Verlag, Berlin (2 vols).

Pandey, R.R., Sharma, G., Tripathi, S.K. and Singh, A.K. (2007) Litterfall, litter decomposition and nutrient dynamics in a subtropical natural oak forest and managed plantation in northeastern India. *Forest Ecology and Management* 240, 96–104.

Pang, Z. (2003) Insect pests of eucalypts in China. In: Turnbull, J.W. (ed.) *Eucalypts in Asia*. Proceedings of an international conference held in Zhanjiang, Guangdong, People's Republic of China, 7–11 April 2003. ACIAR Proceedings No. 111, pp. 183–184.

Paniagua, M.R., Medianero, E. and Lewis, O.T. (2009) Structure and vertical stratification of plant galler-parasitoid food webs in two tropical forests. *Ecological Entomology* 34, 310–320.

Pant, D.N., Roy, P.S. and Phounsavath, B. (2002) Monitoring and assessment of Sal forest infected by the insect, *Hoplocerambyx spinicornis*, using remote sensing and GIS. *Indian Forester* 128, 955–964.

Paquette, A., Hawryshyn, J., Senikas, A.V. and Potvin, C. (2009) Enrichment planting in secondary forests: a promising clean development mechanism to increase terrestrial carbon sinks. *Ecology and Society* 14

Parthiban, K.T., Seenivasan, R. and Rao, M.G. (2010) Contract tree farming in Tamil Nadu – a successful industrial farm forestry model. *Indian Forester* 136, 187–197.

Partridge, L.W., Britton, N.F. and Franks, N.R. (1996) Army ant population dynamics: the effects of habitat quality and reserve size on population size and time to extinction. *Proceedings of the Royal Society of London Series B-Biological Sciences* 263, 735–741.

Pascual-Alvarado, E., Cuevas-Reyes, P., Quesada, M. and Oyama, K. (2008) Interactions between galling insects and leaf-feeding insects: the role of plant phenolic compounds and their possible interference with herbivores. *Journal of Tropical Ecology* 24, 329–336.

Patel, A.N. and Patel, R.K. (2008) Biology of bark eating caterpillar, *Indarbela quadrinotata* (Walker). *Current Biotica* 2(2), 234–239.

Patil, S.U. and Naik, M.I. (1997) Evaluation of *Trichogramma chilonis* Ishii – an egg parasitoid against teak defoliator, *Hyblaea puera* Cramer. *Indian Journal of Forestry* 20, 183–186.

Pawitan, H. (1996) The use of Plantgro in forest plantation planning in Indonesia. In: Booth, T.H. (ed.) *Matching Trees and Sites*. Proceedings of an International Workshop, Bangkok, Thailand, 27–30 March, 1995. ACIAR Proceedings No. 63, pp. 97–100.

Pawson, S.M., Watt, M.S. and Brockerhoff, E.G. (2009) Using differential responses to light spectra as a monitoring and control tool for *Arhopalus ferus* (Coleoptera: Cerambycidae) and other exotic wood-boring pests. *Journal of Economic Entomology* 102(1), 79–85.

Payne, C.C. (1988) Pathogens for the control of insects: where next? *Philosophical Transactions of the Royal Society of London, Series B* 318, 225–248.

Payne, N.J. (2000) Factors influencing aerial insecticide application to forests. *Integrated Pest Management Reviews* 5, 1–10.

Peck, S.B. (1994) Aerial dispersal of insects between and to islands in the Galapagos Archipelago, Ecuador. *Annals of the Entomological Society of America* 87, 218–224.

Peeters, P.J., Sanson, G. and Read, J. (2007) Leaf biomechanical properties and the densities of herbivorous insect guilds. *Functional Ecology* 21, 246–255.

Pellmyr, O. (1992) The phylogeny of a mutualism: evolution and coadaptation between Trollius and its seed-parasitic pollinators. *Biological Journal of the Linnean Society* 47, 337–365.

Pemberton, R.W., Lee, J.H., Reed, D.K., Carlson, R.W. and Han, H.Y. (1993) Natural enemies of the Asian gypsy moth (Lepidoptera: Lymantriidae) in South Korea. *Annals of the Entomological Society of America* 86, 423–440.

Pena-Ramirez, Y.J., Juarez-Gomez, J., Gomez-Lopez, L., Jeronimo-Perez, J.L., Garcia-Shesena, I., Gonzalez-Rodriguez, J., et al. (2010) Multiple adventitious shoot formation in Spanish Red Cedar (*Cedrela odorata* L.)

cultured in vitro using juvenile and mature tissues: an improved micropropagation protocol for a highly valuable tropical tree species. *In Vitro Cellular and Developmental Biology – Plant* 46(2), 149–160.

Peng, H.Y., Chen, X.W., Jiang, Y., Shen, R.J., Zhou, X.M. and Hu, Z.H. (1998) Controlling *Dendrolimus punctatus* with *Trichogramma dendrolimi* carrying cytoplasmic polyhedrosis virus. *Chinese Journal of Biological Control* 14, 111–114.

Peng, J.W. (1989) Study on the pine caterpillars of China. Proceedings of the IUFRO Regional Workshop on Forest Insect Pests and Tree Diseases in the Northeast Asia, 28 August to 1 September, 1989, Tsukuba, Japan, pp. 92–97.

Peng, J.W., Zhou, S.J., Jiang, Y. and Deng, X.H. (1996) Improve forest structure to heighten its resistance to Masson's pine caterpillar. In: Nair, K.S.S., Sharma, J.K. and Varma, R.V. (eds) *Impact of Diseases and Insect Pests in Tropical Forests*. Proceedings of the IUFRO Symposium, Peechi, India, 23–26 November, 1993, pp. 373–378.

Pereira, F.F., Zanuncio, T.V., Zanuncio, J.C., Pratissoli, D. and Tavares, M.T. (2008) Species of Lepidoptera defoliators of *Eucalyptus* as new host for the parasitoid *Palmistichus elaeisis* (Hymenoptera: Eulophidae). *Brazilian Archives of Biology and Technology* 51, 259–262.

Pereira, J.M.M., Zanuncio, T.V., Zanuncio, J.C. and Pallini, A. (2001) Lepidoptera pests collected in *Eucalyptus urophylla* (Myrtaceae) plantations during five years in Tres Marias, State of Minal Gerais, Brazil. *Revista de Biologia Tropical* 49(3–4), 1073–1082.

Peres Filho, O. and Berti Filho, E. (2003) Thermal requirements of *Thyrinteina arnobia* (Stoll, 1782) and effects of the temperature in its biology. *Ciencia Florestal* 13, 143–151.

Peres Filho, O., Dorval, A. and Berti Filho, E. (2002) Leaf cutting ant. *Atta sexdens rubropillosa* Forei 1908 (Hymneoptera: Formicidae) preference for different species under laboratory conditions. *Ciencia Florestal* 12(2), 1–7.

Perez, J., Eigenbrode, S., Hilje, L., Tripepi, R., Elena Aguilar, M. and Mesen, F. (2010) Leaves from grafted Meliaceae species affect survival and performance of *Hypsipyla grandella* (Zeller) (Lepidoptera: Pyralidae) larvae. *Journal of Pest Science* 83, 95–104.

Perez-Salicrup, D.R. and Esquivel, R. (2008) Tree infection by *Hypsipyla grandella* in *Swietenia macrophylla* and *Cedrela odorata* (Meliaceae) in Mexico's southern Yucatan Peninsula. *Forest Ecology and Management* 255, 324–327.

Pervez, A. and Omkar (2006) Ecology and biological control application of multicoloured Asian ladybird, *Harmonia axyridis*: a review. *Biocontrol Science and Technology* 16, 111–128.

Pervez, A., Gupta, A.K. and Omkar (2006) Larval cannibalism in aphidophagous ladybirds: influencing factors, benefits and costs. *Biological Control* 38(3), 307–313.

Peshin, R., Bandral, R.S., Zhang, W., Wilson, L. and Dhawan, A.K. (2009) Integrated pest management: a global overview of history, programs and adoption. *Integrated Pest Management: Innovation-Development Process*, Vol 1, pp. 1–49.

Peters, B.C. and Fitzgerald, C.J. (2007) Developments in termite management in Queensland, Australia: life after cyclodienes (isoptera). *Sociobiology* 49, 231–249.

Pham Quang, T., Dell, B. and Burgess, T.I. (2009) Susceptibility of 18 eucalypt species to the gall wasp *Leptocybe invasa* in the nursery and young; plantations in Vietnam. *Scienceasia* 35, 113–117.

Pherobase (2011) The Pherobase (http://www.pherobase.com).

Picolo, R. and Terradas, J. (1989) Aspects of crown reconstruction and leaf morphology in *Quercus ilex* L. and *Quercus suber* L. after defoliation by *Lymantria dispar* L. *Acta Oecologica Oecologia Plantarum* 10, 69–78.

Pillai, S.R.M. and Gopi, K.C. (1990a) The bagworm *Pteroma plagiophleps* Hamp. (Lepidoptera: Psychidae) attack on *Acacia nilotica* (Linn.) Wild, ex Del. *Indian Forester* 116, 581–583.

Pillai, S.R.M. and Gopi, K.C. (1990b) Seasonal drying up of the distal shoots of neem (*Azadirachta indica* A. Juss) and important insect pests associated with it. *Myforest* 26, 33–50.

Pillai, S.R.M., Gopi, K.C., Salarkhan, A.M. and Pankajam, S. (1991) Insect pest problems of *Albizia lebbeck* (1). Benth. in nurseries and young plantations. *Indian Journal of Forestry* 14, 253–260.

Pimentel, D., McNair, S., Janecka, J., Wightman, J., Simmonds, C., O'Connell, C., Wong, E., Russell, L., Zem, J., Aquino, T. and Tsomondo, T. (2001) Economic and environmental threats of alien plant, animal, and microbe invasions. *Agriculture, Ecosystems and Environment* 84, 1–20.

Pineda, S., Smagghe, G., Schneider, M.I., Del Estal, P., Vinuela, E., Martinez, A.M. and Budia, F. (2006) Toxicity and pharmacokinetics of spinosad and methoxyfenozide to *Spodoptera littoralis* (Lepidoptera: Noctuidae). *Environmental Entomology* 35(4), 856–864.

Pinhassi, N., Nestel, D. and Rosen, D. (1996) Oviposition and emergence of olive scale (Homoptera: Diaspididae) crawlers: regional degree-day forecasting model. *Environmental Entomology* 25, 1–6.

Pinkard, E.A., Battaglia, M., Bruce, J., Leriche, A. and Kriticos, D.J. (2010) Process-based modelling of the severity and impact of foliar pest attack on eucalypt plantation productivity under current and future climates. *Forest Ecology and Management* 259, 839–847.

Plan, J. and Vennetier, M. (1998) Contribution of experimental trials from Martinique to the silviculture of the broadleaf mahogany. Special issue: Forets tropicales. *Bulletin Technique Office National des Forets* 36, 29–38.

Plath, M., Mody, K., Potvin, C. and Dorn, S. (2011) Establishment of native tropical timber trees in monoculture and mixed-species plantations: small-scale effects on tree performance and insect herbivory. *Forest Ecology and Management* 261, 741–750.

Poderoso, J.C.M., Ribeiro, G.T., Goncalves, G.B., Mendonca, P.D., Polanczyk, R.A., Zanetti, R., Serrao, J.E. and Zanuncio, J.C. (2009) Nest foraging characteristics of *Acromyrmex landolti balzani* (Hymenoptera: Formicidae) in Northeast Brazil. *Sociobiology* 54(2), 361–371.

Pomeroy, D. and Service, M.W. (1986) *Tropical Ecology*. Longman, UK.

Pong, T.Y. (1974) The termite problem in plantation forestry in Peninsular Malaysia. *The Malaysian Forester* 37, 278–283.

Porcar, M., Gomez, F., Gruppe, A., Gomez-Pajuelo, A., Segura, I. and Schroder, R. (2008) Hymenopteran specificity of *Bacillus thuringiensis* strain PS86Q3. *Biological Control* 45, 427–432.

Potter, D.A. and Hartman, J.R. (1993) Susceptibility of honeylocust cultivars to *Thyronectria austroamericana* and response of *Agrilus* borers and bagworms to infected and non-infected trees. *Journal of Environmental Horticulture* 11, 176–181.

Powell, S. and Clark, E. (2004) Combat between large derived societies: a subterranean army ant established as a predator of mature leaf-cutting ant colonies. *Insectes Sociaux* 51, 342–351.

Powell, W. (1982) Age-specific life-table data for the Eucalyptus boring beetle, *Phoracantha semipunctata* (F.) (Coleoptera: Cerambycidae), in Malawi. *Bulletin of Entomological Research* 72, 645–653.

Prasad, J., Korwar, G.R., Rao, K.V., Mandal, U.K., Rao, G.R., Srinivas, I., Venkateswarlu, B., Rao, S.N. and Kulkarni, H.D. (2011) Optimum stand density of *Leucaena leucocephala* for wood production in Andhra Pradesh, Southern India. *Biomass and Bioenergy* 35, 227–235.

Prasad, J.V.N.S., Gangaiah, B., Kundu, S., Korwar, G.R., Venkateswarlu, B. and Singh, V.P. (2009) Potential of short rotation woody crops for pulp fiber production from arable lands in India. *Indian Journal of Agronomy* 54, 380–394.

Prasad, L., Pandey, R., Ansari, I.A. and Chandra, S. (2002) Population dynamics of *Dalbergia sissoo* defoliators, *Plecoptera reflexa* and *Dichomeris eridantis*. *Indian Forester* 128, 800–812.

Prathapan, K.D. (2006) The invasive pest erythrina gall wasp, *Quadrastichus erythrinae* Kim enters Sri Lanka. *Insect Environment* 12(1), 28–29.

Preszler, R.W. and Price, P.W. (1995) A test of plant-vigor, plant-stress, and plant-genotype effects on leaf-miner oviposition and performance. *Oikos* 74, 485–492.

Prinsloo, G.L. (2004) *Oxysychus genualis* (Walker) (Hymenoptera: Pteromalidae): first record of an indigenous parasitoid of the introduced eucalyptus borer, *Phoracantha semipunctata* (Coleoptera: Cerambycidae), in South Africa. *African Entomology* 12(2), 271–274.

Prinsloo, G.L. and Kelly, J.A. (2009) The tetrastichine wasps (Hymenoptera: Chalcidoidea: Eulophidae) associated with galls on *Erythrina* species (Fabaceae) in South Africa, with the description of five new species. *Zootaxa* 2083, 27–45.

Protasov, A., Blumberg, D., Brand, D., la Salle, J. and Mendel, Z. (2007a) Biological control of the eucalyptus gall wasp *Ophelimus maskelli* (Ashmead): taxonomy and biology of the parasitoid species *Closterocerus chamaeleon* (Girault), with information on its establishment in Israel. *Biological Control* 42, 196–206.

Protasov, A., La Salle, J., Blumberg, D., Brand, D., Saphir, N., Assael, F., Fisher, N. and Mendel, Z. (2007b) Biology, revised taxonomy and impact on host plants of *Ophelimus maskelli*, an invasive gall inducer on *Eucalyptus* spp. in the Mediterranean area. *Phytoparasitica* 35(1), 50–76.

Protasov, A., Doganlar, M., La Salle, J. and Mendel, Z. (2008) Occurrence of two local *Megastigmus* species parasitic on the *Eucalyptus* gall wasp *Leptocybe invasa* in Israel and Turkey. *Phytoparasitica* 36(5), 449–459.

Putz, F.E. (2008) On the irrelevance of tropical foresters and tropical forestry. *Journal of Tropical Forest Science* 20, V–VII.

Putz, F.E. and Redford, K.H. (2009) Dangers of carbon-based conservation. *Global Environmental Change* 19, 400–401.

Putz, F.E. and Redford, K.H. (2010) The importance of defining 'forest': tropical forest degradation, deforestation, long-term phase shifts, and further transitions. *Biotropica* 42, 10–20.

Pyle, E.H., Santoni, G.W., Nascimento, H.E.M., Hutyra, L.R., Vieira, S., Curran, D.J., van Haren, J., Sateska, S.R., Chow, V.Y., Carmago, P.B., Laurance, W.F. and Wofsy, S.C. (2008) Dynamics of carbon, biomass, and structure in two Amazonian forests. *Journal of Geophysical Research-Biogeosciences* 113

Pywell, H.R. and Myhre, R.J. (1992) Monitoring forest health with airborne videography. In: Wood, G. and Turner, B. (eds) *Integrating Forest Information Over Space and Time*. IUFRO Conference 13–17 January 1992, Canberra, Australia, pp. 349–357.

Queensland Department of Forestry (1963) Techniques for the establishment and maintenance of plantations of hoop pine (*Araucaria cunninghamii*). Government Printer, Brisbane.

Rahardjo, A.W.B. (1992) Early warning system: a method to control the pest *Helopeltis* sp. on *Eucalyptus* sp. plantation. *Duta Rimba* 18(149–150), 33–35.

Raksakantong, P., Meeso, N., Kubola, J. and Siriamornpun, S. (2010) Fatty acids and proximate composition of eight Thai edible terricolous insects. *Food Research International* 43, 350–355.

Ralph, R. (1985) Occurrence of *Celosterna scabrator* var. *spinator* on *Eucalyptus*. *Journal of the Bombay Natural History Society* 82, 682–683.

Ralph, R. (1990) Infestation by the borer *Celosterna scabrator* variety *spinator* on *Acacia nilotica* in north Karnataka. *Commonwealth Forestry Review* 69, 91–94.

Ramos, L.dS., Marinho, C.G.S., Zanetti, R., Delabie, J.H.C. and Schlindwein, M.N. (2003) Impact of formicid granulated baits on non-target ants in eucalyptus plantations according to two forms of application. *Neotropical Entomology* 32, 231–237.

Ramos, V.M., Forti, L.C., Andrade, A.P.A., Noronha, N.C. and Camargo, R.S. (2008) Density and spatial distribution of *Atta sexdens rubropilosa* and *Atta laevigata* colonies (Hym., Formicidae) in *Eucalyptus* spp. forests. *Sociobiology* 51(3), 775–781.

Ramsden, M., Hemmens, T.J. and Kennedy, J.F. (2005) Laser technology enhances aerial assessment of plantation health. *International Forestry Review* 7, 168.

Ranasinghe, D.M.S.H.K. and Newman, S.M. (1993) Agroforestry research and practice in Sri Lanka. *Agroforestry Systems* 22, 119–130.

Rankin, D.M.J.M., Prance, G.T., Hutchings, R.W., Silva, M.F.D., Rodrigues, W.A. and Uehling, M.E. (1992) Preliminary results of a large-scale tree inventory of upland rain forest in the central Amazon. *Acta Amazonica* 22, 493–534.

Rao, M.S., Srinivas, K., Vanaja, M., Rao, G.G.S.N., Venkateswarlu, B. and Ramakrishna, Y.S. (2009) Host plant (*Ricinus communis* Linn.) mediated effects of elevated CO_2 on growth performance of two insect folivores. *Current Science* 97, 1047–1054.

Rawat, Y.S., Oinam, S.S., Vishvakarma, S.C.R., Kuniyal, C.P. and Kuniyal, J.C. (2006) Willow (*Salix fragilis* Linn.): a multipurpose tree species under pest attack in the cold desert of Lahaul Valley, northwestern Himalaya, India. *Ambio* 35, 43–48.

Reagan, D.P. and Waide, R.B. (eds) (1996) *The Food Web of a Tropical Rain Forest*. University of Chicago Press, Chicago, Illinois, 616 pp.

Remadevi, O.K. and Muthukrishnan, R. (1998) Incidence, damage potential and biology of wood borers of *Santalum album* L. In: Radomiljac, H.S., Ananthapadmanabho, H.S., Welbourn, R.M. and Satyanarayana, R. (eds) Sandal and its Products. Proceedings of an International Seminar, 18–19 December, 1997, Bangladore, India. ACIAR Proceedings No. 84, pp.192–195.

Remadevi, O.K., Nagaveni, H.C. and Muthukrishnan, R. (2005) Pests and diseases of sandalwood plants in nurseries and their management. *Working Papers of the Finnish Forest Research Institute* 11, 69–75.

Remaudiere, G. and Binazzi, A. (2003) The Cinara from Pakistan II. The subgenus *Cupressobium* (Hemiptera, Aphididae, Lachninae). *Revue Francaise d'Entomologie (Nouvelle Serie)* 25, 85–96.

Resende, V.F., Nogueira, P. de B., Zanuncio, J.C. and Guedes, R.N.C. (1993) Evaluation of controlled release carbosulfan for protection of eucalypt seedlings against soil termites. *Revista Arvore* 17, 10–15.

Resende, V.F., Zanuncio, J.C., Guedes, R.N.C. and Nogueira, P.B. (1995) Comparative effects of carbosulfan and Aldrin in the protection of eucalyptus seedlings against soil termites. *Anais da Sociedade Entomologica do Brasil* 24, 645–648.

Rhainds, M., Davis, D.R. and Price, P.W. (2009) Bionomics of bagworms (Lepidoptera: Psychidae). *Annual Review of Entomology* 54, 209–226.

Ribeiro, G.T., Mendonca, M.D., de Mesquita, J.B., Zanuncio, J.C. and Carvalho, G.S. (2005) Spittlebug *Cephisus siccifolius* damaging eucalypt plants in the state of Bahia, Brazil. *Pesquisa Agropecuaria Brasileira* 40, 723–726.

Ribeiro, S.C., Goncalves Jacovine, L.A., Bocchat Soares, C.P., Martins, S.V., de Souza, A.L. and Brandi Nardelli, A.M. (2009) Quantification of biomass and estimation of carbon stock in a mature forest in the municipal district of Vicosa, Minas Gerais. *Revista Arvore* 33, 917–926.

Ribeiro, S.P., Pimenta, H.R. and Fernandes, G.W. (1994) Herbivory by chewing and sucking insects on *Tabebuia ochraceae*. *Biotropica* 26, 302–307.

Richards, A., Matthews, M. and Christian, P. (1998) Ecological considerations for the environmental impact evaluation of recombinant baculovirus insecticides. *Annual Review of Entomology* 43, 493–517.

Richards, A., Cory, J., Speight, M.R. and Williams, T. (1999) Foraging in a pathogen reservoir can lead to local host population extinction: a case study of a Lepidoptera-virus interaction. *Oecologia* 118, 29–38.

Richardson, K.F. and Meakins, R.J. (1986) Inter- and intra-specific variation in the susceptibility of eucalypts to the snout beetle *Gonipterus scutellatus* Gyll. (Coleoptera: Curculionidae). *South African Forestry Journal* 139, 21–31.

Riley, M.A. and Goyer, R.A. (1986) Impact of beneficial insects on *Ips* spp. (Coleoptera: Scolytidae) bark beetles in felled loblolly and slash pines in Louisiana. *Environmental Entomology* 15, 1220–1224.

Riley, M.A. and Goyer, R.A. (1988) Seasonal abundance of beneficial insects and *Ips* spp. engraver beetles (Coleoptera: Scolytidae) in felled loblolly and slash pines in Louisiana. *Journal of Entomological Science* 23, 357–365.

Rio-Mora, A. del and Mayo-Jimenez, P. (1993) A study of the biology of, and damage by, *Megastigmus albifrons* in the Sierra Purepecha, Michoacan. *Ciencia-Forestal* 18, 101–120.

Rivera Rojas, M., Locatelli, B. and Billings, R. (2010) Climate change and outbreaks of Southern Pine Beetle *Dendroctonus frontalis* in Honduras. *Forest Systems* 19, 70–76.

Roberts, H. (1968) An outline of the biology of *Hypsipyla robusta* Moore, the shoot borer of the Meliaceae (mahoganies) of Nigeria, together with brief comments on two stem borers and one other Lepidopteran fruit borer also found in Nigerian Meliaceae. *Commonwealth Forestry Review* 47, 225–232.

Roberts, H. (1977) When ambrosia beetles attack mahogany trees in Fiji. *Unasylva* 29, 25–28.

Roberts, H. (1987) DPI Entomology Bulletin: No. 45. Forest insect pests of Papua New Guinea 1. Under-bark borers of Kamarere and Terminalia – *Agrilus* beetles. *Harvest* 12(2), 59–64.

Rodriguez, P.M. (1981) Observations on the biology and morphology of *Apate monachus* Fabricius in Cuba. *Centro Agricola* 8, 13–33.

Rogers, D. (1980) Grasshopper damage to 1–0 white pine seedlings. PCW Insects Nursery File. Morgantown Forestry, North Carolina.

Roonwal, M.L. (1990) Field-ecological observations on the *Ailanthus excelsa* defoliator, *Eligma narcissus indica* (Lepidoptera: Noctuidae), in Peninsular India. *Indian Journal of Forestry* 13, 81–84.

Roques, A., Sun, J.-H., Auger-Rozenberg, M. and Hua, O. (2003) Potential invasion of China by exotic insect pests associated with tree seeds. *Biodiversity and Conservation* 12, 2195–2210.

Rosado-Neto, G.H. (1993) Gonipterinae of *Eucalyptus*: first record of *Gonipterus scutellatus* for the State of Sao Paulo, Brazil, and some considerations on *G. gibberus* (Coleoptera: Curculionidae). *Revista Brasileira de Entomologia* 37, 465–467.

Rosales, C.J., Lobosque, O., Carvalho, P., Bermudez, L. and Acosta, C. (2008) *Glycaspis brimblecombei* Moore (Hemiptera: Psyllidae). 'Red Gum Lerp'. Nueva plaga forestall en Venezuela. *Entomotropica* 23(1), 103–104.

Ross, N.J. and Rangel, T.F. (2011) Ancient Maya agroforestry echoing through spatial relationships in the extant forest of NW Belize. *Biotropica* 43(2), 141–148.

Rouland-Lefevre, C. (2011) *Termites as Pests of Agriculture. Biology of Termites: A Modern Synthesis.* pp. 499–517.

Roy, N.J., Wane, C. and Noel, J.R. (1977) Attacks on trees by termites in the Cap Vert peninsula (Senegal). I. The case of the reafforestation of mobile dunes at Malika. *Bulletin de l'Institut Fondamental d'Afrique Noire* 39, 124–141.

Roychoudhury, N. (1997) Role of kairomones in sal borer operation. *Advances in Forestry Research in India* 17, 180–187.

Roychoudhury, N., Kalia, S. and Joshi, K.C. (1995) Pest status and larval feeding preferences of *Spodoptera litura* (Fabricius) Boursin (Lepidoptera: Noctuidae) on teak. *Indian Forester* 121, 581–583.

Roychoudhury, N., Joshi, K.C. and Rawat, P.S. (2001) *Phalantha phalantha* Drury (Lepidoptera: Nymphalidae) – a major nursery pest of poplar, *Populus deltoides* Bartr. *Indian Forester* 127(2), 252–254.

Rubinoff, D., Holland, B.S., Shibata, A., Messing, R.H. and Wright, M.G. (2010) Rapid invasion despite lack of genetic variation in the Erythrina Gall Wasp (*Quadrastichus erythrinae* Kim). *Pacific Science* 64(1), 23–31.

Ruiz-Guerra, B., Guevara, R., Mariano, N.A. and Dirzo, R. (2010) Insect herbivory declines with forest fragmentation and covaries with plant regeneration mode: evidence from a Mexican tropical rain forest. *Oikos* 119, 317–325.

Ruohomaki, K., Chapin, F.S. III, Haukioja, E., Neuvonen, S. and Suomela, J. (1996) Delayed inducible resistance in mountain birch in response to fertilization and shade. *Ecology* 77, 2302–2311.

Russellsmith, A. and Stork, N.E. (1994) Abundance and diversity of spiders from the canopy of tropical rainforests with particular reference to Sulawesi, Indonesia. *Journal of Tropical Ecology* 10, 545–558.

Ryan, M.A. (1994) Damage to pawpaw trees by the banana-spotting bug, *Amblypelta lutescens lutescens* (Distant) (Hemiptera: Coreidae), in North Queensland. *International Journal of Pest Management* 40, 280–282.

Sabiiti, E.N. and Wein, R.W. (1987) Fire and Acacia seeds: a hypothesis of colonization success. *Journal of Ecology* 75, 937–946.

Safian, S., Csontos, G. and Winkler, D. (2011) Butterfly community recovery in degraded rainforest habitats in the Upper Guinean Forest Zone (Kakum forest, Ghana). *Journal of Insect Conservation* 15, 351–359.

Sah, S.B., Ali, M.S. and Mandal, S.K. (2007) Efficacy of some conventional insecticides against *Myllocerus curvicornis* on shisham in nursery. *Asian Journal of Horticulture* 2(1), 90–91.

Sah, S.B., Mandal, S.K. and Ali, M.S. (2007) Seasonal occurrence and incidence, type of damages in seedling and sapling stages of Shisham (*Dalbergia sissoo* Roxb.) in Bihar. *Environment and Ecology* 25(1), 150–153.

Sajap, A.S. (1989) Incidence of attack by beehole borer, *Xyleutes ceramicus* Wlk., in *Gmelina arborea* Roxb. plantations in Peninsular Malaysia. Proceedings of a Regional Symposium on Recent Developments in Tree Plantations of Humid/Subhumid Tropics of Asia, Selangar, Malaysia, pp. 522–527.

Sajap, A.S. (1999) Detection of foraging activity of *Coptotermes curvignathus* (Isoptera: Rhinotermitidae) in an *Hevea brasiliensis* plantation in Malaysia. *Sociobiology* 33, 137–143.

Sajap, A.S. and Siburat, S. (1992) Incidence of entomogenous fungi in the bagworm, *Pteroma pendula* (Lepidoptera: Psychidae), a pest of *Acacia mangium*. *Journal of Plant Protection in the Tropics* 9, 105–110.

Sajap, A.S., Ling, S.Y., Wahad, Y.A. and Mardidi, A. (1996) Bionomics of *Neolithocollettis* pentadesma (Meyrick), a leafminer of angsana, *Pterocarpus indicus*. *The Malaysian Forester* 59, 153–163.

Sajap, A.S., Yaacob, A.W. and Aidah, M. (1997) Biology of *Spirama retorta* (Lepidoptera: Noctuidae), a new pest of *Acacia mangium* in Peninsular Malaysia. *Journal of Tropical Forest Science* 10, 167–175.

Sakai, I. (1989) A report on the Mulanje cedar resources and the present crisis. Forest Research Record 65, FRIM, Zomba, Malawi.

Salazar, R. (1984) Preliminary notes on the Nantucket pine tip moth, *Rhyacionia frustrana* (Lepidoptera; Tortricidae) in Costa Rica. *Turrialba* 34, 250–252.

San Juan, A. (2005) *The Lurkers Guide to Leafcutter Ants* (www.blueboard.com/leafcutters).

Sanchez, P.A. and Bandy, D. (1992) Alternatives to slash and burn: a pragmatic approach to mitigate tropical deforestation. *Anais da Academia de Brasileira de Ciencias* 64, 7–34.

Sangha, K.S. (2011) Evaluation of management tools for the control of poplar leaf defoliators (Lepidoptera: Notodontidae) in northwestern India. *Journal of Forestry Research (Harbin)* 22.

Santana, D.L.Q. and Burckhardt, D. (2007) Introduced *Eucalyptus* psyllids in Brazil. *Journal of Forest Research* 12, 337–344.

Santhakumaran, L.N., Remadevi, O.K. and Sivaramakrishnan, V.R. (1995) A new record of the insect defoliator, *Pteroma plagiophleps* Hamp. (Lepidoptera: Psychidae) from mangroves along the Goa coast (India). *Indian Forester* 121, 153–155.

Santos, G.P., Cosenza, G.W. and Albino, J.C. (1980) Biology of *Spodoptera latifascia* (Walker, 1856) (Lepidoptera: Noctuidae) on eucalyptus leaves. *Revista Brasileira de Entomologica* 24, 153–155.

Santos, G.P., Zanuncio, T.V. and Zanuncio, J.C. (2000) Development of *Thyrinteina arnobia* Stoll (Lepidoptera: Geometridae) on leaves of *Eucalyptus urophylla* and *Psidium guajava*. *Anais da Sociedade Entomologica do Brasil* 29, 13–22.

Santos, R.S. and de Freitas, S. (2008) Parasitism of *Erythmelus tingitiphagus* (Soares) (Hymenoptera: Mymaridae) in *Leptopharsa heveae* Drake & Poor (Hemiptera: Tingidae) eggs, in rubber tree plantation (*Hevea brasiliensis* Muell. Arg.). *Neotropical Entomology* 37, 571–576.

Sasakawa, M. and Kawaguchi, Y. (1987) Initial attack of the minute pine bark beetle, *Cryphalus fulvus* Niijima (Coleoptera: Scolytidae). Scientific Reports of the Kyoto Prefectural University Agriculture, Japan, pp. 12–19.

Sasidharan, K.R. and Varma, R.V. (2008) Seasonal population variations of the bark eating caterpillar (*Indarbela quadrinotata*) in *Casuarina* plantations in Tamil Nadu. *Tropical Ecology* 49(1), 79–83.
Sasidharan, K.R., Varma, R.V. and Sivaram, M. (2010) Impact of *Indarbela quadrinotata* on the growth of *Casuarina equisetifolia*. *Indian Forester* 136, 182–186.
Saur, E., Imbert, D., Etienne, J., Mian, D. and Dodd, R.S. (1999) Insect herbivory on mangrove leaves in Guadeloupe: effects on biomass and mineral content. *Hydrobiologia* 413, 89–93.
Savill, P. and Evans, J. (1986) *Plantation Silviculture in Temperate Regions*. Oxford University Press, Oxford, UK.
Savill, P., Evans, J., Auclair, D. and Falck, J. (1997) *Plantation Silviculture in Europe*. Oxford University Press, Oxford, pp. 297.
Sawyer, J. (1993) *Plantations in the Tropics: Environmental Concerns*. IUCN Forest Conservation Programme, Gland, Switzerland, pp. 83.
Schabel, H.G., Hilje, L., Nair, K.S.S. and Varma, R.V. (1999) Economic entomology in tropical forest plantations: an update. *Journal of Tropical Forest Science* 11, 303–315.
Schaefer, K.V.R., Clark, K.L., Skowronski, N. and Hamerlynck, E.P. (2010) Impact of insect defoliation on forest carbon balance as assessed with a canopy assimilation model. *Global Change Biology* 16, 546–560.
Schedl, K.E. (1962) Scolytidae und Platypodidae Afrikas. Band 2. Familie Scolytidae. *Revista de Entomologia de Mozambique* 5, 1–594.
Schlyter, F., Zhang, Q.H., Liu, G.T. and Ji, L.Z. (2003) A successful case of pheromone mass trapping of the bark beetle *Ips duplicatus* in a forest island, analysed by 20-year time-series data. *Integrated Pest Management Reviews* 6, 185–196.
Schmid, J.M., Mata, S.A. and Obedzinski R.A. (1994) Hazard rating ponderosa pine stands for mountain pine beetles in the Black Hills. USDA Forest Service, Rocky Mountain Forest and Range Experimental Station, Research Note RM 529, pp. 4.
Schoener, T.W. (1987) Leaf pubescence in buttonwood: community variation in a putative defense against defoliation. *Proceedings of the National Academy of Sciences of the United States of America* 84, 7992–7995.
Schoenherr, J. (1991) Protection of forests – two decades of research in Curitiba. O desafio das florestas neotropicais, Brasil. *Curitiba* 7, 188–203.
Schowalter, T.D., Hargrove, W.W. and Crossley, D.A. (1986) Herbivory in forested ecosystems. *Annual Review of Entomology* 31, 177–196.
Schroth, G., Krauss, U., Gasparotto, L., Aguilar, J.A.D. and Vohland, K. (2000) Pests and diseases in agroforestry systems of the humid tropics. *Agroforestry Systems* 50, 199–241.
Schuldt, A., Baruffol, M., Boehnke, M., Bruelheide, H., Haerdtle, W., Lang, A.C., Nadrowski, K., von Oheimb, G., Voigt, W., Zhou, H.Z. and Assmann, T. (2010) Tree diversity promotes insect herbivory in subtropical forests of south-east China. *Journal of Ecology* 98, 917–926.
Schulte, M.J., Martin, K., Buechse, A. and Sauerborn, J. (2009) Entomopathogens (*Beauveria bassiana* and *Steinernema carpocapsae*) for biological control of bark-feeding moth *Indarbela dea* on field-infested litchi trees. *Pest Management Science* 65, 105–112.
Schulze, C.H. and Fiedler, K. (2003) Hawkmoth diversity in northern Borneo does not reflect the influence of anthropogenic habitat disturbance. *Ecotropica (Bonn)* 9, 99–102.
Schwenke, W. (1985) Relationships between animal pests and tree diseases. *Forstwissenschaftliches Centralblatt* 104, 3–4.
Scott, A.C., Stephenson, J. and Chaloner, W.G. (1992) Interaction and coevolution of plants and arthropods during the Palaeozoic and Mesozoic. *Philosophical Transactions of the Royal Society of London B Biological Sciences* 335, 129–165.
Seagraves, M.P., Riedell, W.E. and Lundgren, J.G. (2011) Oviposition preference for water-stressed plants in *Orius insidiosus* (Hemiptera: Anthorcoridae). *Journal of Insect Behaviour* 24(2), 132–143.
Sedjo, R.A. (1987) Forest resources of the world: forests in transition. In: Kallio, M., Dykstra, D. and Binkley, C. (eds) *The Global Forest Sector: An Analytical Perspective*. John Wiley and Sons, Chichester, UK.
Sedjo, R.A. (2004) Genetically engineered trees: promise and concerns. *Resources for the Future*, pp 48.
Selander, J. and Bubula, M. (1983) A survey of pest insects in forest plantations in Zambia. Forest Department, Division of Forest Research Zambia, Research Note No. 33.
Selander, J. and Immonen, A. (1992) Effect of fertilization and watering of Scots pine seedlings on the feeding preference of the pine weevil (*Hylobius abietis* L.). *Silva Fennica* 26, 75–84.
Self, M. and Kay, C. (2005) Development of methodology for targeted post-border surveillance for exotic pest incursions. Paper presented at the Workshop on Post-border Surveillance for Exotic Pests of Plants, Canberra, Australia, 7–8 June 2005, 15 pp.

Sen-Sarma, P.K. (1987) Insect pest problems in social forestry plantations and their management. *Indian Journal of Forestry* 10, 239–244.

Sen-Sarma, P.K. and Thakur, M.L. (1983) Insect pests of *Eucalyptus* and their control. *Indian Forester* 109, 864–881.

Sen-Sarma, P.K. and Thakur, M.L. (1988) Insect factors in the management of forest resources. *Myforest* 24, 99–113.

Senthilkumar, N. and Barthakur, N.D. (2009) *Alcidodes ludificator* Faust: a serious insect pest of nursery and young plantations of *Gmelina arborea* (Roxb.) in northeastern India. *Current Biotica* 2(4), 493–494.

Sevillano, L., Horvitz, C.C. and Pratt, P.D. (2010) Natural enemy density and soil type influence growth and survival of *Melaleuca quinquenervia* seedlings. *Biological Control* 53, 168–177.

Shafiqul-Islam, M. (2011) Present status of homestead nursery of CARE-LIFT project – a case study in Patukhali. *Journal of Agriculture and Social Sciences* 7, 7–12.

Shafiullah, B. and Naik, S.T. (2003) Association of some host plants and insects with Sandal spike disease. *Indian Forester* 129, 393–400.

Shaner, P.-J.L. and Macko, S.A. (2011) Trophic shifts of a generalist consumer in response to resource pulses. *Plos One* 6

Sharaf, N.S. and Meyerdirk, D.E. (1987) A review on the biology, ecology and control of *Nipaecoccus viridis* (Homoptera: Pseudococcidae). *Miscellaneous Publications of the Entomological Society of America* 66, 1–18.

Sharma, A., Verma, T.D. and Sood, A. (2005) Some important insect pests of poplars in the western Himalayas and their management. *Indian Forester* 131(4), 553–562.

Sharma, M. and Ahmed, S.I. (2004) *Beauveria bassiana* (Balsamo) Vuillemin, a potential entomogenous fungal pathogen isolated from Marwar teak defoliator, *Patialus tecomella* Pajni, Kumar & Rose (Coleoptera: Curculionidae). *Indian Forester* 130(9), 1060–1064.

Sharma, N.P. (1992) *Managing the World's Forests: Looking for Balance Between Conservation and Development*. Kendall/Hunt Publishing Company, Dubuque, Iowa, 605 pp.

Sharma, N.P., Rowe, R., Openshaw, K. and Jacobson, M. (1992) World forests in perspective. In: Sharma, N.P. (ed.) *Managing the World's Forests*. Kendall/Hunt Publishing Company, Dubuque, Iowa.

Sharov, A.A., Liebhold, A.M. and Leonard, D.S. (2002) 'Slow the Spread': a national program to contain the gypsy moth. *Journal of Forestry*, July/August, 30–35.

Sharpe, A.L. (1983) Seed bed protection against rodent and insect damage. *Nepal Forestry Technical Bulletin NEFTIB* No. 9, 8–10.

Shea, P.J. and Neustein, M. (1995) Protection of a rare stand of Torrey Pine from *Ips paraconfusus*. In: Salom, S.M. and Hobson, K.R. (eds) *Application of Semiochemicals for Management of Bark Beetle Infestations – Proceedings of an Informal Conference*. USDA Forest Service General Technical Report INT-GTR-318, pp. 39–43.

Shearman, P.L., Ash, J., Mackey, B., Bryan, J.E. and Lokes, B. (2009) Forest conversion and degradation in Papua New Guinea 1972–2002. *Biotropica* 41, 379–390.

Sheikh, M.I. and Aleem, A. (1983) Effect of improved practices on *Dalbergia sissoo* (shisham) planted in Islamabad. *Pakistan Journal of Forestry* 33, 115–121.

Shen, G., Yu, M., Hu, X.-S., Mi, X., Ren, H., Sun, I.F. and Ma, K.P. (2009) Species–area relationships explained by the joint effects of dispersal limitation and habitat heterogeneity. *Ecology* 90, 3033–3041.

Shiddamallayya, N., Azra, Y. and Gopakumar, K. (2010) Hundred common forest medicinal plants of Karnataka in primary healthcare. *Indian Journal of Traditional Knowledge* 9(1), 90–95.

Shivayogeshwara, B., Mallikarjunaiah, H., Prasad, N.K.K. and Das, K.G.S. (1988) Incidence of babool borer *Celosterna scabrator* Fabricius (Laminae: Cerambycidae: Coleoptera) on *Eucalyptus* in Malnad Tracts of Shimoga. *Indian Forester* 114, 410–413.

Showler, A.T. (1995) Leucaena psyllid, *Heteropsylla cubana* (Homoptera: Psyllidae), in Asia. *American Entomologist* 41, 49–54.

Siaguru, P. and Taurereko, R. (1988) Line planting with striplings of *Cedrela odorata* in logged over natural forest. *Klinkii* 3, 10–16.

Siddiqi, M.R., Swarup, G. and Dasgupta, D.R. (1986) A review of the nematode genus *Fergusobia* Currie (Hexatylina) with descriptions of *F. jambophila* n. sp. and *F. magna* n. sp. In: *Plant Parasitic Nematodes of India, Problems and Progress*. Indian Agricultural Research Institute, New Delhi, India, pp. 264–278.

Siepel, H. (1996) Biodiversity of soil microarthropods: the filtering of species. *Biodiversity and Conservation* 5, 251–260.

Sileshi, G., Hailu, G. and Mafongoya, P.L. (2006) Occupancy-abundance models for predicting densities of three leaf beetles damaging the multipurpose tree *Sesbania sesban* in eastern and southern Africa. *Bulletin of Entomological Research* 96, 61–69.

Sileshi, G.W., Kuntashula, E., Matakala, P. and Nkunika, P.O. (2008) Farmers' perceptions of tree mortality, pests and pest management practices in agroforestry in Malawi, Mozambique and Zambia. *Agroforestry Systems* 72, 87–101.

Sileshi, G.W., Nyeko, P., Nkunika, P.O.Y., Sekematte, B.M., Akinnifesi, F.K. and Ajayi, O.C. (2009) Integrating ethno-ecological and scientific knowledge of termites for sustainable termite management and human welfare in Africa. *Ecology and Society* 14

Silva, J.O., Oliveira, K.N., Santos, K.J., Espirito-Santo, M.M., Neves, F.S. and Faria, M.L. (2010) Effects of landscape structure and eucalyptus genotype on the abundance and biological control of *Glycaspis brimblecombei* Moore (Hemiptera: Psyllidae). *Neotropical Entomology* 39(1), 91–96.

Silver, W.L., Brown, S. and Lugo, A.E. (1996) Effects of changes in biodiversity on ecosystem function in tropical forests. *Conservation Biology* 10, 17–24.

Silwal, B. and Mehta, J.N. (1992) Assessment of training needs for effective Forest User Group Committee development: a case study of Kaski District. *Banko Janakari* 3, 47–50.

Simmonds, N.W. (1995) Tree biology, forestry and agriculture, especially in the tropics: a personal viewpoint. *Botanical Journal of Scotland* 47, 211–227.

Simpson, J.A. and Ades, P.K. (1990) Variation in the susceptibility of *Pinus muricata* and *Pinus radiata* to two species of Aphidoidea. *Silvae Genetica* 39(5–6), 202–206.

Sinclair, R.J. and Hughes, L. (2008) Leaf mining in the Myrtaceae. *Ecological Entomology* 33, 623–630.

Singh, A.P. and Singh, K.P. (1995) Damage evaluation of *Nodostoma waterhousie* Jacoby (Coleoptera: Chrysomelidae), on different clones, provenances and species of poplar in Himachal Pradesh. *Indian Journal of Forestry* 18, 242–244.

Singh, K., Rana, B.S. and Singh, R.P. (2004) Biomass and productivity of an age series of three cottonwood clones (*Populus deltoides*) in central Himalayan tarai region, India. *Journal of Tropical Forest Science* 16, 384–395.

Singh, P. (1986) Aerial spraying of chemical for the control of teak defoliators. Proceedings of the Second Forestry Conference, FRI, Dehra Dun, 16–19 January, 1980, Vol. II, pp. 901–907.

Singh, P. and Bhandari, R.S. (1988) Insect pests of leguminous forest tree seed and their control. *Indian Forester* 114, 844–853.

Singh, P., Rawat, D.S., Mishra, R.M., Fasih, M., Prasad, G. and Tyagi, B.D.S. (1983) Epidemic defoliation of poplars and its control in Tarai Central Forest Division, Uttar Pradesh. *Indian Forester* 109, 675–693.

Singh, P., Joshi, K.C. and Gurung, D. (1988) Efficacy of some insecticides against the larvae of khasi pine shoot borer, *Dioryctria castanea* Bradley (Lepidoptera: Pyralidae). *Indian Forester* 114, 29–34.

Singh, S., Kumar, A., Senthilkumar, N. and Singh, A.N. (2006) Strategies for the management of *Craspedonta leayana* (Coleoptera: Chrysomelidae) in *Gmelina arborea*. *Indian Forester* 132, 581–588.

Singh, S.K. and Singh, G. (2003) Management of termite infestation in mango orchard by cultural practices and organic amendments. *Indian Journal of Agricultural Research* 37, 148–150.

Singh, V., Dubey, O.P., Nair, C.P.R. and Pillai, G.B. (1978) Biology and bionomics of insect pests of cinnamon. *Journal of Plantation Crops* 6, 24–27.

Sivaramakrishnan, V.R. (1976) Occurrence of Lantana lacebug, *Teleonemia scrupulosa* Std (Hemiptera: Tingidae) in South India. *Indian Forester* 102, 620–621.

Sivaramakrishnan, V.R. (1986) Note on recent outbreak of *Celosterna scabrator* Fabricius (Lamiidae: Coleoptera) on Eucalyptus in Karnatal.a. *Myforest* 22, 103–105.

Sivaramakrishnan, V.R. and Remadevi, O.K. (1996) Insect pests in forest nurseries in Karnataka, India, with notes on insecticidal control of psyllid on *Albizia lebbeck*. In: Nair, K.S.S., Sharma, J.K. and Varma, R.V. (eds) *Impact of Diseases and Insect Pests in Tropical Forests*. Proceedings of the IUFRO Symposium, 23–26 November, 1993, Peechi, India, pp. 460–463.

Slade, E.M., Mann, D.J. and Lewis, O.T. (2011) Biodiversity and ecosystem function of tropical forest dung beetles under contrasting logging regimes. *Biological Conservation* 144, 166–174.

Slaney, G.L., Lantz, V.A. and MacLean, D.A. (2009) The economics of carbon sequestration through pest management: application to forested landbases in New Brunswick and Saskatchewan, Canada. *Forest Policy and Economics* 11, 525–534.

Smith, D. (1974) Pink wax scale and its control. *Queensland Agricultural Journal* 100, 225–228.

Smith, I.W., Marks, G.C., Featherston, G.R. and Geary, P.W. (1989) Effects of inter-planted wattles on the establishment of eucalypts planted on forest sites affected by *Phytophthora cinnamomi*. *Australian Forestry* 52, 74–81.

Smith, J.J., Waage, J., Woodhall, J.W., Bishop, S.J. and Spence, N.J. (2008) The challenge of providing plant pest diagnostic services for Africa. *European Journal of Plant Pathology* 121, 365–375.

Smith, J.M.B. (1990) Drift disseminules on Fijian beaches. *New Zealand Journal of Botany* 28, 13–20.

Smith, K.D., May, P.B. and Moore, G.M. (2001) The influence of waterlogging on the establishment of four Australian landscape trees. *Journal of Arboriculture* 27, 49–56.

Snook, L.K. and Negreros-Castillo, P. (2004) Regenerating mahogany (*Swietenia macrophylla* King) on clearings in Mexico's Maya forest: the effects of clearing method and cleaning on seedling survival and growth. *Forest Ecology and Management* 189, 143–160.

Snyder, A.I., Bower, N.W. and Snyder, M.A. (2007) Susceptibility of coastal populations of Caribbean pine to attack by southern pine beetle in Belize. Ecological Society of America Annual Meeting Abstracts

Soares, M.G., da Silva, M.F.D., Fernandes, J.B. and Lago, J.H.G. (2010) Interspecific variation in the composition of volatile oils from the leaves of *Swietenia macrophylla* King (Meliaceae). *Quimica Nova* 33(5), 1141–1144.

Song, Z. (1997) Preliminary study on using *Metarhizium anisopliae* to control *Dendrolimus punctatus*. *Journal of Fujian College of Forestry* 17(2), 107–109.

Sookar, P., Seewooruthun, S.I. and Ramkhelawon, D. (2003) The redgum lerp psyllid, *Glycaspis brimblecombei*, a new pest of *Eucalyptus* sp. in Mauritius. Proceedings of the 6th Annual Meeting of Agricultural Scientists, Food and Agricultural Research Council, Reduit, Mauritius, 8–9 May, pp. 327–332.

Soto-Pinto, L., Anzueto, M., Menoza, J., Ferrer, G.J. and de Jong, B. (2010) Carbon sequestration through agroforestry in indigenous communities of Chiapas, Mexico. *Agroforestry Systems* 78(1), 39–51.

Sousa-Silva, C.R. and Ilharco, F.A. (2001) First report of *Cinara cupressi* (Lachninae : Cinarini) in Brazil. *Revista De Biologia Tropical* 49, 768–768.

Sousa, W.P., Quek, S.P. and Mitchell, B.J. (2003) Regeneration of *Rhizophora mangle* in a Caribbean mangrove forest: interacting effects of canopy disturbance and a stem-boring beetle. *Oecologia* 137, 436–445.

South, D.B. and Enebak, S.A. (2006) Integrated pest management practices in southern pine nurseries. *New Forests* 31d, 253–271.

South, D.B. and Zwolinski, J.B. (1996) Chemicals used in southern forest nurseries. *Southern Journal of Applied Forestry* 20, 127–135.

Speight, M.R. (1983) The importance of pine shoot moths (Pyralidae and Tortricidae) in tropical forestry. *Proceedings of the 10th International Conference of Plant Protection*, Brighton, UK, 20–25th November, 1983, Vol. 3.

Speight, M.R. (1996) The relationship between host tree stresses and insect attack in tropical forest plantations and its relevance to pest management. In: Nair, K.S.S., Sharma, J.K. and Varma, R.V. (eds) *Impact of Diseases and Pests in Tropical Forests*. Proceedings of the IUFRO Symposium, 23–26 November, 1993, Peechi, India, pp. 363–372.

Speight, M.R. (1997) Forest pests in the tropics: current status and future threats. In: Watt, A.D., Stork, N.E. and Hunter, M.D. (eds) *Forests and Insects*. Chapman and Hall, London, pp. 207–228.

Speight, M.R. and Cory, J.S. (2000) Integrated pest management of mahogany shoot borers. In: Floyd, R. and Hauxwell, C. (eds) *Proceedings of an International Workshop on Hypsipyla Shoot Borers of the Meliaceae, Kandy, Sri Lanka, 1996*. ACIAR.

Speight, M.R. and Speechly, H.T. (1982) Pine shoot moths in S.E. Asia I. Distribution, biology and impact. *Commonwealth Forestry Review* 61, 121–134.

Speight, M.R. and Wainhouse, D. (1989) *The Ecology and Management of Forest Insects*. Clarendon Press, Oxford, UK.

Speight, M.R. and Wylie, F.R. (2001) *Insect Pests in Tropical Forestry*, 1st edn. CAB International, Wallingford, UK, pp. 307.

Speight, M.R., Hails, R.S., Gilbert, M. and Foggo, A. (1998) Horse chestnut scale (*Pulvinaria regalis*) (Homoptera: Coccidae) and urban host tree environment. *Ecology* 79, 1503–1513.

Speight, M.R., Intachat, J., Chey, V.K. and Chung, A.Y.C. (2003) Influences of forest management on insects. In: Basset, Y., Novotny, V., Miller, S.E. and Kitching, R.L. (eds) *Arthropods of Tropical Forests*. Cambridge University Press, Cambridge, UK, pp. 380–393.

Speight, M.R., Hunter, M.D. and Watt, A.D. (2008) *Ecology of Insects: Concepts and Applications*, second edition. Blackwell Science, Oxford, UK.

Spiegel, L.H. and Price, P.W. (1996) Plant aging and the distribution of *Rhyacionia neomexicana* (Lepidoptera: Tortricidae). *Environmental Entomology* 25, 359–365.

Sreedharan, K., Balakrishnan, M.M., Samuel, S.D. and Bhat, P.K. (1991) A note on the association of wood boring beetles and fungus with the death of silver oak trees on coffee plantations. *Journal of Coffee Research* 21(2), 145–148.

Srygley, R.B., Dudley, R., Oliveira, E.G., Aizprua, R., Pelaez, N.Z. and Riveros, A.J. (2010) El Nino and dry season rainfall influence host plant phenology and an annual butterfly migration from Neotropical wet to dry forests. *Global Change Biology* 16, 936–945.

Stanaway, M.A., Zalucki, M.P., Gillespie, P.S., Rodriguez, C.M. and Maynard, G.V. (2001) Pest risk assessment of insects in sea cargo containers. *Australian Journal of Entomology* 40, 180–192.

Stape, J.L., Binkley, D. and Ryan, M.G. (2004) Eucalyptus production and the supply, use and efficiency of use of water, light and nitrogen across a geographic gradient in Brazil. *Forest Ecology and Management* 193, 17–31.

Stape, J.L., Binkley, D., Ryan, M.G., Fonseca, S., Loos, R.A., Takahashi, E.N., et al. (2010) The Brazil *Eucalyptus* potential productivity project: influence of water, nutrients and stand uniformity on wood production. *Forest Ecology and Management* 259(9), 1704–1713.

Steinbauer, M.J. and Matsuki, M. (2004) Suitability of *Eucalyptus* and *Corymbia* for *Mnesampela privata* (Guenee) (Lepidoptera: Geometridae) larvae. *Agricultural and Forest Entomology* 6, 323–332.

Steinbauer, M.J., Short, M.W. and Schmidt, S. (2006) The influence of architectural and vegetational complexity in eucalypt plantations on communities of native wasp parasitoids: Towards silviculture for sustainable pest management. *Forest Ecology and Management* 233, 153–164.

Stevens, G.C. (1992) The elevational gradient in altitudinal range: an extension of Rapoport's latitudinal rule to altitude. *American Naturalist* 140, 893–911.

Stone, C. and Coops, N.C. (2004) Assessment and monitoring of damage from insects in Australian eucalypt forests and commercial plantations. *Australian Journal of Entomology* 43, 283–292.

Stone, C. and Haywood, A. (2006) Assessing canopy health of native eucalypt forests. *Ecological Management and Restoration* 7, 24–30.

Stone, C., Simpson, J.A. and Eldridge, R.H. (1998) Insect and fungal damage to young eucalypt trial plantings in northern New South Wales. *Australian Forestry* 61, 7–20.

Stone, C., Chisholm, L. and Coops, N. (2001) Spectral reflectance characteristics of eucalypt foliage damaged by insects. *Australian Journal of Botany* 49, 687–698.

Stone, C., Matsuki, M. and Carnegie, A. (2003a) *Pest and Disease Assessment in Young Eucalypt Plantations: Field Manual for Using the Crown Damage Index* (Parsons. M., ed.) National Forest Inventory, Bureau of Rural Sciences, Canberra, Australia.

Stone, C., Wardlaw, T., Floyd, R., Carnegie, A., Wylie, R. and de Little, D. (2003b) Harmonisation of methods for the assessment and reporting of forest health in Australia – a starting point. *Australian Forestry* 66, 233–245.

Stone, C., Turner, R. and Verbesselt, J. (2008) Integrating plantation health surveillance and wood resource inventory systems using remote sensing. *Australian Forestry* 71(3), 245–253.

Stonedahl, G.M. (1991) The Oriental species of *Helopeltis* (Heteroptera: Miridae): a review of economic literature and guide to identification. *Bulletin of Entomological Research* 81, 465–490.

Stonedahl, G.M., Malipatil, M.B. and Houston, W. (1995) A new mirid (Heteroptera) pest of cashew in northern Australia. *Bulletin of Entomological Research* 85, 275–278.

Stork, N.E. (2010) Re-assessing current extinction rates. *Biodiversity and Conservation* 19, 357–371.

Strevens, C.M.J. and Bonsall, M.B. (2011) Density-dependent population dynamics and dispersal in heterogeneous metapopulations. *Journal of Animal Ecology* 80(1), 282–293.

Strong, D.R., Lawton, J.H. and Southwood, T.R.E. (1984) *Insects on Plants: Community Patterns and Mechanisms*. Blackwell Scientific, Oxford, UK.

Sudheendrakumar, V.V. (1986) Studies on the natural enemies of the teak pests, *Hyblaea puera* and *Eutectona machaeralis*. KFRI Research Report No. 38.

Sudheendrakumar, V.V., Sajeev, T.V. and Varma, R.V. (2004) Mass production of nucleopolyhedrovirus of the teak defoliator, *Hyblaea puera* Cramer using host population in teak plantations. *Journal of Biological Control* 18, 81–83.

Sugiura, S., Yamaura, Y., Tsuru, T., Goto, H., Hasegawa, M., Makihara, H. and Makino, S. (2009) Beetle responses to artificial gaps in an oceanic island forest: implications for invasive tree management to conserve endemic species diversity. *Biodiversity and Conservation* 18, 2101–2118.

Suharti, M., Irianto, R.S.B, and Santosa, S. (1994) Behaviour of the stem borer *Xystrocera festiva* Pascoe on *Paraserianthes falcataria* and integrated control. *Buletin Penelitian Hutan* 589, 1–26.

Suharti, M., Sitepu, I.R. and Anggraeni, I. (2000) Behaviour, intensity and effect to the attached of stem borer, *Indarbela acutistriata* on sengon in KPH Kediri. *Buletin Penelitian Hutan* 37–50.

Suhendi, D. (1990) Field resistance of several *Leucaena* varieties to psyllid *Heteropsylla cubana*. *Menara Perkebunan* 58, 110–114.

Sun, J.-Z., Fuxa, J.R., Richter, A. and Ring, D. (2008) Interactions of *Metarhizium anisopliae* and tree-based mulches in repellence and mycoses against *Coptotermes formosanus* (Isoptera: Rhinotermitidae). *Environmental Entomology* 37, 755–763.

Sun, W. and Hu, J.G. (1998) Study on the method of prediction of occurrence and density of *Dendrolimus punctatus*. *Journal of Zhejiang Forestry Science and Technology* 18, 44–49.

Sun, X.-L. and Peng, H.-Y. (2007) Recent advances in biological control of pests insects by using viruses in China. *Virologica Sinica* 22(2), 158–162.

Sundararaj, R. and Dubey, A.K. (2005) Potential of plant products for the management of whiteflies in nurseries. *Working Papers of the Finnish Research Institute* 11, 65–68.

Sundararaj, R., Karibasavaraja, L.R., Sharma, G. and Muthukrishnan, R. (2006) Scales and mealybugs (Coccoidea: Hemiptera) infesting sandal (*Santalum album* Linn.). *Entomon* 31, 239–241.

Sunil, T. and Balasundaran, M. (1998a) Detection of phytoplasma in spiked sandal using DAPI stain. In: Radomiljac, H.S., Ananthapadmanabho, H.S., Welbourn, R.M. and Satyanarayana, R. (eds) *Sandal and its Products*. Proceedings of an International Seminar, 18–19 December 1997, Bangladore, India, ACIAR Proceedings No. 84, pp. 182–184.

Sunil, T. and Balasundaran, M. (1998b) In situ detection of phytoplasma in spike-disease-affected sandal using DAPI stain. *Current Science* 74, 989–993.

Suratmo, F.G. (1996) Emerging insect pest problems in tropical plantation forest in Indonesia. In: Nair, K.S.S., Sharma, J.K. and Varma, R.V. (eds) *Impact of Diseases and Insect Pests in Tropical Forests*. Proceedings of the IUFRO Symposium, 23–26 November 1993, Peechi, India, pp. 502–506.

Swartz, M.B. (1998) Predation on an *Atta cephalotes* colony by an army ant, *Nomamyrmex esenbeckii*. *Biotropica* 30, 682–684.

Symonds, B.O., Orondo, O. and Day, R.K. (1994) Cypress aphid (*Cinara cupressi*) damage to a cypress (*Cupressus lusitanica*) stand in Kenya. *International Journal of Pest Management* 40, 141–144.

Takeda, S. (1992) Origins of taungya. In: Jordan, C.F., Gajaseni, J. and Watanabe, H. (eds) *Taungya Forest Plantations with Agriculture in SE Asia*. Sustainable Rural Development Series No. 1, CAB International, Wallingford, UK, pp. 9–17.

Tan, Z., Li, S., Tieszen, L.L. and Tachie-Obeng, E. (2009) Simulated dynamics of carbon stocks driven by changes in land use, management and climate in a tropical moist ecosystem of Ghana. *Agriculture Ecosystems and Environment* 130, 171–176.

Tanaka, S., Denlinger, D.L. and Wolda, H. (1987) Seasonal changes in the photoperiodic response regulating diapause in a tropical beetle, *Stenotarsus rotundus*. *Journal of Insect Physiology* 34, 1135–1142.

Tang, C., Wan, X.-J., Wan, F.-H., Ren, S.-X. and Peng, Z.-Q. (2008) The blue gum chalcid, *Leptocybe invasa*, invaded Hainan Province. *Chinese Bulletin of Entomology* (http://en.cnki.com.cn/Article_en/CJFDTOTAL-KCZS200806026.htm).

Tang, W. and Tian, Y.C. (2003) Transgenic loblolly pine (*Pinus taeda* L.) plants expressing a modified delta-endotoxin gene of *Bacillus thuringiensis* with enhanced resistance to *Dendrolimus punctatus* Walker and *Crypyothelea formosicola* Staud. *Journal of Experimental Botany* 54(383), 835–844.

Tanton, M.T. and Khan, S.M. (1978a) Effects of fenitrothion and aminocarb, at doses giving low mortality, on surviving eggs and larvae of the eucalypt-defoliating chrysomelid beetle *Paropsis atomaria* Ol. I. Methods, mortality and relative toxicity. II. Biology of survivors. III. Histological changes in treated larvae. *Australian Journal of Zoology* 26, 121–126.

Tanton, M.T. and Khan, S.M. (1978b) Aspects of the biology of the eucalypt-defoliating chrysomelid beetle *Paropsis atomaria* Ol. in the Australian Capital Territory. *Australian Journal of Zoology* 26, 113–120.

Tassin, J., Herve, C., Lesueur, D. and Riviere, J.N. (1997) Decline of filao on Reunion Island: ecological and silvicultural causes. *Bois et Forets des Tropiques* 37–46.

Taveras, R., Hilje, L., Hanson, P., Mexzon, R., Carballo, M. and Navarro, C. (2004) Population trends and damage patterns of *Hypsipyla grandella* (Lepidoptera: Pyralidae) in a mahogany stand, in Turrialba, Costa Rica. *Agricultural and Forest Entomology* 6, 89–98.

Taylor, D.E. (1981) The giant or sand cricket. *Zimbabwe Agricultural Journal* 78, 117–118.

Taylor, G.S. and Davies, K.A. (2008) New species of gall flies (Diptera: Fergusoninidae) and an associated nematode (Tylenchida: Neotylenchidae) from flower bud galls on *Corymbia* (Myrtaceae). *Australian Journal of Entomology* 47, 336–349.

Teugh-Hardi, T.W. (1995) A preventive measure against subterranean termite attack on *Acacia mangium*. *Buletin Penelitian Hutan* 29–37.

Thakur, M.L. (2000) *Forest Entomology: Ecology and Management*. Sai Publishers, Dehra Dun, India.

Thakur, M.L. and Sen-Sarma, P.K. (1980) Current status of termites as pests of forest nurseries and plantations in India. *Journal of Indian Academy of Wood Science* 11, 7–15.

Thakur, M.L. and Sen-Sarma, P.K. (2008) Termites as pests of forest nurseries and plantations. In: Jha, L.K. and Sen-Sarma, P.K. (eds) *A Manual of Forest Extension Education*. APH Publishing, New Delhi, 400 pp.

Thakur, R.K. (1992) Termites as pests of fodder and fuelwood plantations in South India. *Journal of Tropical Forestry* 8, 96–98.

Thakur, R.K. and Sivaramakrishnan, V.R. (1991) A note on the insect pest problems in Kundapur and Mangalore Forests, Kamataka (South India). *Myforest* 27, 187–190.

Thapa, R.S. (1991) Studies on widespread attack of ambrosia beetle, *Platypus* sp., in provenance trial plots of *Acacia crassicarpa* in Sipitang District of Sabah, Malaysia. Breeding Technologies for Tropical Acacias. Proceedings of an International Workshop, Tawau, Sabah, Malaysia, 1–4 July 1991. ACIAR Proceedings No. 37, pp. 31–34.

Thapa, R.S. and Singh, P. (1986) Large-scale mortality of deodar trees by the bark borer, *Scolytus major* Stebbing (Scolytidae, Coleoptera), in Kulu Forest Divisio, Himachal Pradesh. *Indian Forester* 112, 392–398.

Thapa, R.S., Carron, L.T. and Aken, K.M. (1992) Leaf-cutting ant.

Thomas, A.T. and Hodkinson, I.D. (1991) Nitrogen, water stress and the feeding efficiency of lepidopteran herbivores. *Journal of Applied Ecology* 28, 703–720.

Thomas, M.C. (1995) Invertebrate pets and the Florida Department of Agriculture and Consumer Services. *The Florida Entomologist* 78(1), 39–44.

Thomson, L.A.J. (2006) *Agathis macrophylla* (Pacific kauri), ver.1.2. In: Elevitch, C.R. (ed.) *Species Profiles for Pacific Island Agroforestry*. Permanent Agriculture Resources (PAR), Hōlualoa, Hawai'i (http://www.traditionaltree.org).

Thorpe, K.W., Tcheslavskaia, K.S., Tobin, P.C., Blackburn, L.M., Leonard, D.S. and Roberts, E.A. (2007) Persistent effects of aerial applications of disparlure on gypsy moth: trap catch and mating success. *Entomologia Experimentalis et Applicata* 125(3), 223–229.

Thu, P.Q., Dell, B. and Burgess, T.I. (2009) Susceptibility of 18 eucalypt species to the gall wasp *Leptocybe invasa* in the nursery and young plantations in Vietnam. *Science Asia* 35, 113–117.

Tilakaratna, D. (1991) Parasites of the teak defoliator, *Hyblaea puera*. *Sri Lanka Forester* 20(1–2), 23–25.

Tkacz, B.M. (2001) Risks associated with world trade in logs and unmanufactured wood. Exotic Forest Pests Online Symposium.

Toit, P.F. du (1975) Aspects of the bio-ecology and control of the barkbeetle *Hylastes angustatus* Herbst (Coleoptera: Scolytidae). *Forestry in South Africa* 17, 37–43.

Tong, P.Y. and Fang, H.L. (1989) Bionomics of *Anomala antigua* (Gyllenhal) and its control. *Insect Knowledge* 26, 150–151.

Tong, Q. and Kong, X.-B. (2010) The biological and ecological characteristics of *Dioryctria rubella* damaging the *Pinus kesiya* var. *langbianensis*. *Chinese Bulletin of Entomology* 47(2), 331–334.

Torres, J.A. (1994) Wood decomposition of *Cyrilla racemiflora* in a tropical montane forest. *Biotropica* 26, 124–140.

Torres, J.B., Zanuncio, J.C. and Moura, M.A. (2006) The predatory stinkbug *Podisus nigrispinus*: biology, ecology and augmentative releases for lepidopteran larval control in Eucalyptus forests in Brazil. *CABI Reviews: Perspectives in Agriculture, Veterinary Science, Nutrition and Natural Resources*, 1(015), 1–18.

Toy, S.J. and Newfield, M.J. (2010) The accidental introduction of invasive animals as hitchhikers through inanimate pathways: a New Zealand perspective. *Revue scientifique et technique (International Office of Epizootics)* 29(1), 123–133.

Tribe, G.D. (1990) Phenology of *Pinus radiata* log colonization by the pine bark beetle *Hylastes angustatus* (Herbst) (Coleoptera: Scolytidae) in the south-western Cape Province. *Journal of the Entomological Society of Southern Africa* 53, 93–100.

Tribe, G.D. (1991) Phenology of three exotic pine bark beetle species (Coleoptera: Scolytidae) colonising *Pinus radiata* logs in the south-western Cape Province. *South African Forestry Journal* 157, 27–31.

Tribe, G.D. and Cillie, J.J. (2004) The spread of *Sirex noctilio* Fabricius (Hymenoptera: Siricidae) in South African pine plantations and the introduction and establishment of its biocontrol agents. *African Entomology* 12(1), 9–17.

Tsankov, G. (1977) Insect pests of coniferous species in Cuba. *Gorskostopanska Nauka* 14, 54–63.

Tshernyshev, W.B. (1995) Ecological pest management (EPM): general approaches. *Journal of Applied Entomology* 119, 379–381.

Tsukamoto, J. and Sabang, J. (2005) Soil macro-fauna in an *Acacia mangium* plantation in comparison to that in a primary mixed dipterocarp forest in the lowlands of Sarawak, Malaysia. *Pedobiologia* 49, 69–80.

Tuomi, J., Niemela, P., Haukioja, E., Siren, S. and Neuvonen, S. (1984) Nutrient stress: an explanation for plant anti-herbivore responses to defoliation. *Oecologia* 61, 208–210.

Turner, I.M. (1996) Species loss in fragments of tropical rain forest: a review of the evidence. *Journal of Applied Ecology* 33, 200–209.

Uechi, N., Uesato, T. and Yukawa, J. (2007) Detection of an invasive gall-inducing pest, *Quadrastichus erythrinae* (Hymenoptera: Eulophidae), causing damage to *Erythrinae variegata* L. (Fabaceae) in Okinawa Prefecture, Japan. *Entomological Science* 10, 209–212.

Underwood, D.L.A. (1994) Intraspecific variability in host plant quality and ovipositional preferences in *Eucheria socialis* (Lepidoptera: Pieridae). *Ecological Entomology* 19, 245–256.

Underwood, R. and Nikles, D.G. (2009) The status of the domestication of African mahogany (*Khaya senegalensis*) in Australia – as documented in the CD ROM. Proceedings of a 2006 Workshop. *Bois Et Forets Des Tropiques*, 101–104.

UNEP (2009) *Vital Forest Graphics*, United Nations Environment Programme, pp. 70.

Urbas, P., Araujo, M.V., Leal, I.R. and Wirth, R. (2007) Cutting more from cut forests: edge effects on foraging and herbivory of leaf-cutting ants in Brazil. *Biotropica* 39(4), 489–495.

Utsumi, S., Ando, Y. and Ohgushi, T. (2009) Evolution of feeding preference in a leaf beetle: the importance of phenotypic plasticity of a host plant. *Ecology Letters* 12, 920–929.

Vale, C., Davila, R. and Longart, J.J. (1991) Evaluation of two mechanical seed drills and the causes of plant losses in two nurseries. *Boletin Tecnico CVG PROFORCA* 2, 1–14.

Valencia, R., Balslev, H. and Mino, C.G.P.Y. (1994) High tree alpha-diversity in Amazonian Ecuador. *Biodiversity and Conservation* 3, 21–28.

Valente, C. and Hodkinson, I. (2009) First record of the Red Gum Lerp Psyllid, *Glycaspis brimblecombei* Moore (Hem.: Psyllidae), in Europe. *Journal of Applied Entomology* 133, 315–317.

Van Bael, S.A., Aiello, A., Valderrama, A., Medianero, E., Samaniego, M. and Wright, S.J. (2004) General herbivore outbreak following an El Nino-related drought in a lowland Panamanian forest. *Journal of Tropical Ecology* 20, 625–633.

Van Den Beldt, R.J. (1995) The future role of leguminous multi-purpose trees in tropical farming systems. In: Shelton, H.M., Piggin, C.M. and Brewbaker, J.L. (eds) *Leucaena – Opportunities and Limitations*. ACIAR Proceedings No. 57, pp. 11–15.

Van Driesche, R.G., Carruthers, R.I., Center, T., Hoddle, M.S., Hough-Goldstein, J., Morin, L., et al. (2010) Classical biological control for the protection of natural ecosystems. *Biological Control* 54, S2–S33.

Van Frankenhuyzen, K. (1994) Effect of temperature on the pathogenesis of *Bacillus thuringiensis* Berliner in larvae of the spruce budworm, *Choristoneura fumiferana* Clem. (Lepidoptera: Tortricidae). *Canadian Entomologist* 126, 1061–1065.

Van Mele, P. (2008) A historical review of research on the weaver ant *Oecophylla* in biological control. *Agricultural and Forest Entomology* 10, 13–22.

Varley, G.C., Gradwell, G.R. and Hassell, M.P. (1973) *Insect Population Ecology: An Analytical Approach*. Blackwell Scientific Publishing, Oxford, UK.

Varma, R.V. (1991) Spatial and temporal distribution of *Ailanthus* pests, *Eligma narcissus* and *Atteva fabriciella* (final report of the Research Project KFRI 103/87, April 1987 to March 1990). KFRI Research Report.

Varma, R.V. (1992) Impact of *Atteva fabriciella* (Lepidoptera: Yponomeutidae) feeding on seed production in *Ailanthus triphysa*. *Indian Journal of Forestry* 15, 326–328.

Varma, R.V. and Swaran, P.R. (2007) Diversity of termites in a young eucalypt plantation in the tropical forests of Kerala, India. *International Journal of Tropical Insect Science* 27, 95–101.

Varma, R.V. and Swaran, P.R. (2009) Pest problems in root-trainer raised forest nurseries of selected species in Kerala. *Indian Journal of Forestry* 32(3), 375–377.

Varma, R.V., Sajeev, T.V. and Sudheendrakumar, V.V. (2007) Pest susceptibility of *Tectona grandis* under intensive management practices in India. *Journal of Tropical Forest Science* 19, 46–49.

Vazquez, G.J.A. and Givnish, T.J. (1998) Altitudinal gradients in tropical forest composition, structure, and diversity in the Sierra de Manantlan. *Journal of Ecology* 86, 999–1020.

Vearasilp, T., Jovanovic, T. and Booth, T.H. (1996) Plantgro soil and climate database for Thailand. In: Booth, T.H. (ed.) *Matching Trees and Sites*. Proceedings of an International Workshop, Bangkok, Thailand, 27–30 March 1995. ACIAR Proceedings No. 63, pp. 85–92.

Veenakumari, K. and Mohanraj, P. (1996) Folivorous insects damaging teak, *Tectona grandis* L. (Verbenaceae) in the Andaman Islands, Bay of Bengal, Indian Ocean. *Journal of Entomological Research (New Delhi)* 20(2), 177–178.

Veer, V. and Chandra, A. (1984) Insect pests in nurseries and plantations of *Populus deltoides* in Assam and Bengal. *Indian Forester* 110, 640–643.

Verma, A.K., Ghatak, S.S. and Mukhopadhyay, S. (1990) Effect of temperature on development of whitefly (*Bemisia tabaci*) (Homoptera: Aleyrodidae) in West Bengal (India). *Indian Journal of Agricultural Sciences* 60, 332–336.

Verma, B.R. (1991) Interception of *Merobruchus columbinus* (Sharp) (Bruchidae: Coleoptera) from saman seeds and new host record for this bruchid. *Indian Journal of Entomology* 53, 530.

Verma, B.R., Kapur, M.L. and Lai, B. (1991) A new host record for *Bruchus ervi* Froelich (Coleoptera: Bruchidae). *Indian Journal of Entomology* 53, 171–172.

Verma, M., Sharma, S. and Prasad, R. (2009) Biological alternatives for termite control: a review. *International Biodeterioration and Biodegradation* 63, 959–972.

Verma, T.D. (1988) Insect pests of agro-forestry trees and their management. In: Gupta, V.K. and Sharma, N.K. (eds) *Tree Protection*. Proceedings of a National Seminar, 10–11 March 1988, Solan, India. Indian Society of Tree Scientists, Solan, India, pp. 348–356.

Viado, G.B. (1979) Notes on insect pests of forest trees. *Sylvatrop Philippines Forest Research Journal* 4, 183–189.

Vieira-Neto, E.H.M. and Vasconcelos, H.L. (2010) Developmental changes in factors limiting colony survival and growth of the leaf-cutter ant *Atta laevigata*. *Ecography* 33, 538–544.

Viggiani, G., Laudonia, S. and Bernardo, U. (2001) The increase of insect pests in *Eucalyptus*. *Informatore-Agrario* 58(12), 86–87.

Villacarlos, L.T. and Robin, N.M. (1992) Biology and potential of *Curinus coeruleus* Mulsant, an introduced predator of *Heteropsylla cubana* Crawford. *Philippine Entomologist* 8, 1247–1258.

Villacarlos, L.T., Paglinawan, R.V. and Robin, R.P. (1990) Factors affecting leucaena psyllid populations in Leyte, Philippines. In: Napompeth, B. and MacDicken, K.G. (eds) *Leucaena Psyllid: Problems and Management*. Proceedings of an International Workshop, Bogor, Indonesia, 16–21 January 1989. Winrock International Institute for Agricultural Development, Bangkok, pp. 122–129.

Vir, S. and Parihar, D.R. (1993) Insect pests of *Acacia senegal* (L) Willd. and *Acacia tortilis* (Forsk) Hyne in the Thar Desert of India. *Myforest* 29, 105–111.

Wagner, M.R., Atuahene, S.K.N. and Cobbinah, J.R. (1991) *Forest Entomology in West Tropical Africa: Forest Insects of Ghana*. Kluwer Academic Publishers, Dordrecht, The Netherlands.

Wagner, M.R., Cobbinah, J.R. and Bosu, P.P. (2008) *Forest Entomology in West Tropical Africa: Forest Insects of Ghana*. Second edition. Springer, Dordrecht, The Netherlands, 244 pp.

Wakgari, W.M. (2001) The current status of the biocontrol of *Ceroplastes destructor* Newstead (Hemiptera: Coccidae) on *Citrus* and *Syzygium* in South Africa. *Biocontrol Science and Technology* 11, 339–352.

Waller, D.A. and Moser, J.C. (1990) Invertebrate enemies and nest associates of the leaf-cutting ant *Atta texana* (Buckley) (Formicidae, Attini). In: Vander Meer, R.K., Jaffe, K. and Cedeno, A. (eds) *Applied Myrmecology: A World Perspective*. Westview Press, Boulder, Colorado, Chapter 25.

Wallner, W.E. (1996) Invasive pests ('biological pollutants') and US forests: whose problem, who pays? *Bulletin OEPP/EPPO Bulletin* 26, 167–180.

Walmsley, J.D. and Godbold, D.L. (2010) Stump harvesting for bioenergy – a review of the environmental impacts. *Forestry* 83, 17–38.

Wang, C., Fan, M., Li, Z. and Butt, T.M. (2004) Molecular monitoring and evaluation of the application of the insect-pathogenic fungus *Beauveria bassiana* in southeast Chain. *Journal of Applied Microbiology* 96(4), 861–870.

Wang, F.G., Wu, J. and Gao, S. (1997) Studies on interpretation and orientation of the damaged pine forests by pine caterpillars with airborne video techniques. *Forest Research* 10, 270–276.

Wang, G., Liu, X., Ge, F. and Li, Z. (2005) Population dynamics of *Dendrolimus punctatus* in different pine forest. *Shengtaixue Zazhi* 24(4), 355–359.

Wang, H., Floyd, R.B., Farrow, R.A., Hong, C., Gao, C., Lin, C., et al. (1998) Insect damage on *Acacia mearnsii* in China. In: Turnbull, J.W., Crompton, H.R. and Pinyopusarerk, K. (eds) *Proceedings of the Third International Acacia Workshop*, Hanoi, Vietnam, 27–30 October 1997. ACIAR Proceedings No. 82, pp. 240–245.

Wang, H., Malcolm, D.C. and Fletcher, A.M. (1999) *Pinus caribaea* in China: introduction, genetic resources and future prospects. *Forest Ecology and Management* 117, 1–5.

Wang, N.H., Pan, H., Li, D. and Zhang, Y.R. (2009) Expert consulting service platform for forest pests and diseases prevention of rapid-growing and high-yielding forest in northeast China. In: *2009 International Conference on Artificial Intelligence and Computational Intelligence, Vol IV, Proceedings*, pp. 151–155.

Wang, Q., Millar, J.C., Reed, D.A., Mottern, J.L., Heraty, J.M., Triapitsyn, S.V., Paine, T.D. and He, X.Z. (2008) Development of a strategy for selective collection of a parasitoid attacking one member of a large herbivore guild. *Journal of Economic Entomology* 101, 1771–1778.

Wang, Y., Solberg, S., Yu, P., Myking, T., Vogt, R.D. and Du, S. (2007) Assessments of tree crown condition of two Masson pine forests in the acid rain region in south China. *Forest Ecology and Management* 242, 530–540.

Ward, S.E., Wightman, K.E. and Rodriguez Santiago, B. (2008) Early results from genetic trials on the growth of Spanish cedar and its susceptibility to the shoot borer moth in the Yucatan Peninsula, Mexico. *Forest Ecology and Management* 255, 356–364.

Wardell, D.A. (1990) The African termites: peaceful coexistence or total war. *Agroforestry Today* 2, 4–6.

Wardhaugh, C.W. and Didham, R.K. (2005) Density-dependent effects on the reproductive fitness of the New Zealand beech scale insect (*Ultracoelostoma assimile*) across multiple spatial scales. *Ecological Entomology* 30, 733–738.

Wardlaw, T., Bashford, R., Wotherspoon, K., Wylie, R. and Elliott, H. (2008) Effectiveness of routine forest health surveillance in detecting pest and disease damage in eucalypt plantations. *New Zealand Journal of Forestry Science* 38(2/3), 253–269.

Wargo, P.M. (1996) Consequences of environmental stress on oak: predisposition to pathogens. *Annales des Sciences Forestieres Paris* 53, 359–368.

Waring, G.L. and Cobb, N.S. (1992) The impact of plant stress on herbivore population dynamics. In: Bernays, E. (ed.) *Insect–Plant Interactions*, Vol. 4. CRC Press, Boca Raton, Florida, pp. 167–226.

Warner, K.D. and Getz, C. (2008) A socio-economic analysis of the North American commercial natural enemy industry and implications for augmentative biological control. *Biological Control* 45, 1–10.

Waterson, D. and Urquhart, C.A. (1995) Leaf beetles (Coleoptera: Chrysomelidae). State Forests of New South Wales, Forest Protection Research Division Series No. E6.

Watson, G.W., Voegtlin, D.J., Murphy, S.T. and Foottit, R.G. (1999) Biogeography of the *Cinara cupressi* complex (Hemiptera: Aphididae) on Cupressaceae, with a description of a pest species introduced into Africa. *Bulletin of Entomological Research* 89, 271–283.

Way, M.J. and Khoo, K.C. (1989) Relationships between *Helopeltis theobromae* damage and ants with special reference to Malaysian cocoa smallholdings. *Journal of Plant Protection in the Tropics* 6, 1–12.

Way, M.J. and Khoo, K.C. (1991) Colony dispersion and nesting habits of the ants, *Dolichoderus thoracicus* and *Oecophylla smaragdina* (Hymenoptera: Formicidae), in relation to their success as biological control agents on cocoa. *Bulletin of Entomological Research* 81, 341–350.

Way, M.J. and Khoo, K.C. (1992) Role of ants in pest management. *Annual Review of Entomology* 37, 479–503.

Way, M.J., Cammell, M.E. and Paiva, M.R. (1992) Studies on egg predation by ants (Hymenoptera: Formicidae) especially on the eucalyptus borer *Phoracantha semipunctata* (Coleoptera: Cerambycidae) in Portugal. *Bulletin of Entomological Research* 82, 425–432.

Wazihullah, A.K.M., Islam, S.S., Rahman, F. and Das, S. (1996) Reduced attack of keora (*Sonneratia apetala*) by stem borer in mixed-species plantation in coastal Bangladesh. *Journal of Tropical Forest Science* 8, 476–480.

Webb, D. van V. (1974) Forest and timber entomology in the Republic of South Africa. Department of Agricultural Technical Services, Republic of South Africa, Entomology Memoir No. 34.

Webb, D.B., Wood, P.J., Smith, J.P. and Henman, G.S. (1984) A guide to species selection for tropical and sub-tropical plantations. Tropical Forestry Papers, No. 15, Commonwealth Forestry Institute, University of Oxford, UK.

Webb, J.W. and Moran, V.C. (1978) The influence of the host plant on the population dynamics of *Acizzia russellae* (Homoptera: Psyllidae). *Ecological Entomology* 3, 313–321.

Wen, X.-S., Peng, L.-H., Wan, X.-M., Long, W., Luo J.-G. and Zhang, W. (2001) Monitoring the insecticide resistance of *Dendrolimus punctatus* and the synergism of pyrethroids by SV1. *Forest Research* 14, 141–147.

Werner, P.A. and Prior, L.D. (2007) Tree-piping termites and growth and survival of host trees in savanna woodland of north Australia. *Journal of Tropical Ecology* 23, 611–622.

Werner, P.A., Prior, L.D. and Forner, J. (2008) Growth and survival of termite-piped *Eucalyptus tetrodonta* and *E. miniata* in northern Australia: implications for harvest of trees for didgeridoos. *Forest Ecology and Management* 256, 328–334.

Wetterer, J.K. (1994) Attack by *Paraponera clavata* prevents herbivory by the leaf-cutting ant, *Atta cephalotes*. *Biotropica* 26, 462–465.

Wetterer, J.K. (1995) Forager size and ecology of *Acromyrmex coronatus* and other leaf-cutting ants in Costa Rica. *Oecologia Berlin* 104, 409–415.

Wheeler, R.A. (1988) *Leucaena* psyllid trial at Waimanalo, Hawaii. *Leucaena Research Reports* 8, 25–29.

Wheeler, R.A., Chaney, W.R., Butler, L.G. and Brewbaker, J.L. (1994) Condensed tannins in *Leucaena* and their relation to psyllid resistance. *Agroforestry Systems* 26, 139–146.

Wheeler, R.A., Chaney, W.R., Cecava, M.J. and Brewbaker, J.L. (1994) Forage yield and compositional analysis of leucaena species and hybrids adapted to cool sites. *Agroforestry Systems* 25, 263–274.

Whetten, R.W. and Kellison, R. (2010) Research gap analysis for application of biotechnology to sustaining US forests. *Journal of Forestry* 108, 193–201.

White, T.C.R. (1969) An index to measure weather-induced stress of trees associated with outbreaks of psyllids in Australia. *Ecology* 50, 905–909.

White, T.C.R. (1986) Weather, *Eucalyptus* dieback in New England (Australia), and a general hypothesis of the cause of dieback. *Pacific Science* 40, 58–78.

Whittaker, R.J. (1998) *Island Biogeography: Ecology, Evolution, and Conservation*. Oxford University Press, Oxford, UK.

Wightman, K.E., Ward, S.E., Haggar, J.P., Rodriguez Santiago, B. and Cornelius, J.P. (2008) Performance and genetic variation of big-leaf mahogany (*Swietenia macrophylla* King) in provenance and progeny trials in the Yucatan Peninsula of Mexico. *Forest Ecology and Management* 255, 346–355.

Wilcken, C.F., Raetano, C.G. and Forti, L.C. (2002) Termite pests in *Eucalyptus* forests of Brazil. *Sociobiology* 40, 179–190.

Wilcken, C.F., de Oliveira, N.C., Sartorio, R.C., Loureiro, E.B., Bezerra, N. Jr and Rosado Neto, G.H. (2008) *Gonipterus scutellatus* Gyllenhal (Coleoptera: Curculionidae) occurrence in *Eucalyptus* plantations in Espirito Santo state, Brazil. *Arqivos do Instituto Biologico Sao Paulo* 75, 113–115.

Wilcken, C.F., Soliman, E.P., Nogueira de Sa, L.A., Barbosa, L.R., Dias, T.K.R., Ferreira-Filho, P.J., Rodrigues, O. and Ricasdo, J. (2010) Bronze bug *Thaumastocoris peregrinus* Carpintero and Dellape (Hemiptera: Thaumastocoridae) on *Eucalyptus* in Brazil and its distribution. *Journal of Plant Protection Research* 50(2), 201–205.

Wiley, J. and Skelley, P. (2006) *Erythrina Gall Wasp, Quadrastichus erythrinae Kim, in Florida*. Pest Alert. Florida Department of Agriculture and Consumer Services.

Wiley, J. and Skelley, P. (2008) A *Eucalyptus* pest, *Leptocybe invasa* Fisher & La Salle (Hymenoptera: Eulophidae), genus and species new to Florida and North America. Pest Alert. Florida Department of Agriculture & Consumer Services (http://www.doacs.state.fl.us/pi/enpp/ento/leptocybe_invasa.html).

Wilf, P. (2008) Insect-damaged fossil leaves record food web response to ancient climate change and extinction. *New Phytologist* 178, 486–502.

Willig, M.R., Presley, S.J. and Bloch, C.P. (2011) Long-term dynamics of tropical walking sticks in response to multiple large-scale and intense disturbances *Oecologia* 165(2), 357–368.

Wilson, B.W. (1985) The biological control of *Acacia nilotica* indica in Australia. In: Delfosse, E.S. (ed.) *Proceedings of the 6th International Symposium on the Biological Control of Weeds*, 19–25 August 1984, Vancouver, Canada, pp. 849–853.

Wilson, L.F. (1993) China's Masson pine forest: cure or curse? *Journal of Forestry* 91, 30–33.

Wingfield, M.J. and Robison, D.J. (2004) Diseases and insect pests of *Gmelina arborea*: real threats and real opportunities. *New Forests* 28, 227–243.

Wingfield, M.J. and Swart, W.J. (1994) Integrated management of forest tree diseases in SouthAfrica. *Forest Ecology and Management* 65, 11–16.

Wingfileld, M.J., Swart, W.J. and Evans, H.F. (1994) Integrated management of forest tree diseases in South Africa. Forestry: integrated pest management programs. Papers presented at a symposium held within the XIIth International Plant Protection Congress, Rio de Janeiro, Brazil, August 1991. *Forest Ecology and Management* 65, 11–16.

Withers, T.M. (2001) Colonization of eucalypts in New Zealand by Australian insects. *Austral Ecology* 26, 467–476.

Wittenberg, R. and Cock, M.J.W. (eds) (2001) *Invasive Alien Species: A Toolkit of Best Prevention and Management Practices.* CAB International, Wallingford, UK, pp. xvii–228.

Witzgall, P., Kirsch, P. and Cork, A. (2010) Sex pheromones and their impact on pest management. *Journal of Chemical Ecology* 36, 80–100.

Wolda, H. (1992) Trends in abundance of tropical forest insects. *Oecologia* 89, 47–52.

Wolda, H., O'Brien, C.W. and Stockwell, H.P. (1998) Weevil diversity and seasonality in tropical Panama as deduced from light-trap catches (Coleoptera: Curculionoidea). *Smithsonian Contributions to Zoology* 590, 1–79.

Wood, C.R., Clark, S.J., Barlow, J.F. and Chapman, J.W. (2010) Layers of nocturnal insect migrants at high-altitude: the influence of atmospheric conditions on their formation. *Agricultural and Forest Entomology* 12, 113–121.

Wood, S.L. (1982) The bark and ambrosia beetles of North and central America (Coleoptera: Scolytidae), a taxonomic monograph. *Great Basin Naturalist Memoirs* No. 6.

Work, T.T., McCullough, D.G., Cavey, J.F. and Komsa, R. (2005) Arrival rate of nonindigenous insect species into the United States through foreign trade. *Biological Invasions* 7, 323–332.

World Bank (1996) Development indicators (http://www.worldbank.org/data).

World Bank (2010) World development indicators (http://data.worldbank.org/data-catalog/world-development-indicators).

Wormald, T.J. (1992) Mixed and pure forest plantations in the tropics and subtropics. FAO Forestry Paper, Rome.

Wright, M.S., Osbrink, W.L.A. and Lax, A.R. (2002) Transfer of entomopathogenic fungi among formosan subterranean termites and subsequent mortality. *Journal of Applied Entomology-Zeitschrift Fur Angewandte Entomologie* 126, 20–23.

Wright, M.S., Osbrink, W.L.A. and Lax, A.R. (2004) Potential of entomopathogenic fungi as biological control agents against the Formosan subterranean termite. In: Nelson, W.M. (ed.) *Agricultural Applications in Green Chemistry*, pp. 173–188.

Wright, M.S., Raina, A.K. and Lax, A.R. (2005) A strain of the fungus *Metarhizium anisopliae* for controlling subterranean termites. *Journal of Economic Entomology* 98, 1451–1458.

Wu, J.W., Fang, H.L., Yang, M.D. and Lian, Y.Y. (1988) Rearing technology of *Trichogramma dendrolimi* Matsumura for controlling pine caterpillar (*Dendrolimus punctatus* Walker) and effect of releasing *Trichogramma* to control insect pest on a large scale for sixteen successive years in Zhejiang Province. *Colloques de l'INRA* 43, 621–628.

Wyatt, T.D. (2001) *Pheromones and Animal Behaviour: Communication by Smell.* Cambridge University Press, Cambridge, UK.

Wyatt, T.D. (2003) *Pheromones and Animal Behaviour: Communication by Smell and Taste.* Cambridge University Press, Cambridge, UK.

Wylie, F.R. (1982a) Flight patterns and feeding behavior of adult *Milionia isodoxa* Prout (Lepidopterci: Geometridae) at Bulolo, Papua New Guinea. *Journal of the Lepidopterists' Society* 36, 269–278.

Wylie, F.R. (1982b) Insect problems of *Araucaria* plantations in Australia and Papua New Guinea. *Australian Forestry* 45, 125–131.

Wylie, F.R. (1989) Recent trends in plant quarantine policy in Australia and New Zealand and their implications for forestry. *New Zealand Journal of Forestry Science* 19, 308–317.

Wylie, F.R. (1992) A comparison of insect pest problems in eucalypt plantations in Australia and in southern China. Paper Presented at XIX International Congress of Entomology, Beijing, China, 28 June to 5 July, 1992.

Wylie, F.R. (1993) *Pests and Diseases of Fast-growing Hardwoods in Industrial Forest Plantations in the Philippines.* Asian Development Bank, Manila.

Wylie, F.R. (2001) Control of *Hypsipyla* spp. shoot borers with insecticides. In: Floyd, R.B. and Hauxwell, C. (eds) *Hypsipyla Shoot Borers in Meliaceae.* Proceedings of an International Workshop, Kandy, Sri Lanka 20–23 August 1996. ACIAR Proceedings No. 97, pp. 109–115.

Wylie, F.R. (2010) Surveillance for forest invasive species in Asia-Pacific: guidelines for selecting which pests to target. Paper presented at APFISN Workshop on Pathways of Biological Invasion into Forests, Thimpu, Bhutan, 8 and 11 June 2010.

Wylie, F.R. and Arentz, F. (1993) Management of pests and diseases in tropical forestry. Centre for International Forestry Research Strategic Planning Thematic Paper No.14. ACIAR, Canberra.

Wylie, F.R. and Brown, B.N. (1992) Report on Australian mission to China on *Eucalyptus* pests and diseases. Australia–China Agricultural Co-operation Agreement, Department of Primary Industries, Canberra.

Wylie, F.R. and Floyd, R. (1998) The insect threat to eucalypt plantations in tropical areas of Australia and Asia. Paper presented at IUFRO Workshop, Bangkok, Thailand.

Wylie, F.R. and Peters, B.C. (1987) Development of contingency plans for use against exotic pests and diseases of trees and timber. 2. Problems with the detection and identification of pest insect introductions into Australia, with special reference to Queensland. *Australian Forestry* 50, 16–23.

Wylie, F.R. and Peters, B.C. (1993) Insect pest problems of eucalypt plantations in Australia. 1. Queensland. *Australian Forestry* 56, 358–362.

Wylie, F.R., Johnston, P.J.M. and Eisemann, R.L. (1993) A survey of native tree dieback in Queensland. Research Paper, Department of Primary Industries, Forest Service, Queensland, No. 16, 100 pp.

Wylie, F.R., Johnston, P.J.M. and Forster, B.A. (1993b) Decline of *Casuarina* and *Eucalyptus* in the Mary River Catchment. Queensland Forest Research Institute Research Paper No. 17.

Wylie, F.R., Floyd, R.B., Elliott, H.J., Khen, C.V., Intachat, J., Hutacharem, C., et al. (1998) Insect pests of tropical acacias: a new project in southeast Asia and northern Australia. In: Turnbull, J.W., Crompton, H.R. and Pinyopusarerk, K. (eds) *Proceedings of the Third International Acacia Workshop*, Hanoi, Vietnam, 27–30 October 1997. ACIAR Proceedings No. 82, pp. 234–239.

Wylie, F.R., Peters, B., DeBaar, M., King, J. and Fitzgerald, C. (1999) Managing attack by bark and ambrosia beetles (Coleoptera: Scolytidae) in fire-damaged *Pinus* plantations and salvaged logs in Queensland, Australia. *Australian Forestry* 62, 148–153.

Wylie, F.R., Griffiths, M. and King, J. (2008) Development of hazard site surveillance programs for forest invasive species: a case study from Brisbane, Australia. *Australian Forestry* 71(3), 229–235.

Xiao, Y., Sun, X., Tang, X. and Peng, H. (2010) Propagation of *Dendrolimus punctatus* cytoplasmic polyhedrosis virus in substitutive host *Spodoptera exigua*. *Chinese Journal of Applied and Environmental Biology* 16, 84–90.

Xu, Y.-X., Sun, X.-G., Han, R.-D. and He, Z. (2006) Parasitoids of *Dendrolimus punctatus* in China. *Chinese Bulletin of Entomology* 43(6), 767–773.

Xue, X.Q. (1983) Influence of climate on *Dendrolimus punctatus* Walker in ten provinces and one autonomous region. *Journal of Nanjing Technological College of Forest Products* 3, 44–52.

Yadav, L.B. and Rizvi, S.M.A. (1994) Studies on the insect pests of wasteland plantations at Faizabad (India). *Journal of Entomological Research New Delhi* 18, 115–120.

Yamakura, T., Hagihara, A., Sukardjo, S. and Ogawa, H. (1986) Aboveground biomass of tropical rain forest stands in Indonesian Borneo. *Vegetatio* 68, 71–82.

Yamazaki, S., Taketani, A., Fujita, K., Vasques, C.P. and Ikeda, T. (1990) Ecology of *Hypsipyla grandella* and its seasonal changes in population density in Peruvian Amazon forest. *Japan Agricultural Research Quarterly* 24, 149–155.

Yang, M.M., Tung, G.S., La Salle, J. and Wu, M.L. (2004) Outbreak of erythrina gall wasp on *Erythrina* spp. (Fabaceae) in Taiwan. *Plant Protection Bulletin* 46, 391–396.

Yates, H.O. (1972) Bark beetles attacking Caribbean pine in northeastern Nicaragua. *FAO Plant Protection Bulletin* 20, 25–27.

Ye, H. (1994) Influence of temperature on the experimental population of the pine shoot beetle, *Tomicus piniperda* (L.) (Col., Scolytidae). *Journal of Applied Entomology* 117, 190–194.

Ye, W., Ma, X., Li, T., Chen, C., Wu, J. and Zhou, J. (1990) The population prediction model of the pine caterpillar. *Forest Research* 3, 427–433.

Yeo, Y.S., Chang, Y.D. and Choi, K.M. (1990) Effect of temperature on the development of *Anagrus incarnatus* Haliday (Hymenoptera: Mymaridae). *Korean Journal of Applied Entomology* 29, 217–221.

Yie, S.T., Hsu, S.J. and Chu, Y.I. (1967) Biological study of the more important insect pests attacking genus *Pinus* introduced from the USA. VI. The study on pine tip borer *Dioryctria pyreri* Ragonot. *Plant Protection Bulletin, Taiwan* 9(3–4), 1–3.

Yu, H., Gouge, D.H. and Baker, P. (2006) Parasitism of subterranean termites (Isoptera: Rhinotermitidae: Termitidae) by entomopathogenic nematodes (Rhabditida: Steinernematidae: Heterorhabditidae). *Journal of Economic Entomology* 99, 1112–1119.

Yue, Y.Z., Hao, W.Z. and Yi, H.S. (1994) Rehabilitation of eroded tropical coastal land in Guangdong, China. *Journal of Tropical Forest Science* 7, 28–38.

Zacharias, V.J. and Mohandas, K. (1990) Bird predators of the teak defoliator, *Hyblaea puera*. *Indian Journal of Forestry* 13, 122–127.

Zanetti, R., Jaffe, K., Vilela, E.F., Zanuncio, J.C. and Leite, H.G. (2000) Effect of the number and size of leaf-cutting ant nests on eucalypt wood production. *Anais da Sociedade Entomologica do Brasil* 29(1), 105–112.

Zanetti, R., Zanuncio, J.C., Vilela, E.F., Leite, H.G., Jaffe, K. and Oliveira, A.C. (2003) Level of economic damage for leaf-cutting ants (Hymenoptera: Formicidae) in Eucalyptus plantations in Brazil. *Sociobiology* 42(2), 433–442.

Zanuncio, J.C., Guedes, R.N.C., Cruz, A.P.D. and Moreira, A.M. (1992) Efficiency of aerial application of *Bacillus thuringiensis* and deltamethrin in the control of *Thyrinteina arnobia* Stoll. 1782 (Lepidoptera: Geometridae) in a eucalyptus grove in the Para region. *Acta Amazonica* 22, 485–492.

Zanuncio, J.C., Santos, G.P., Firme, D.J. and Zanuncio, T.V. (1997) Sulfluramid granulated bait (0.3%) in the control of *Atta sexdens rubropilosa* Forel, 1908 (Hymenoptera: Formicidae). *Cerne* 3, 161–169.

Zanuncio, J.C., Cruz, A.P., Oliveira, H.N. and Gomes, F.S. (1999) Control of *Acromyrmex laticeps nigrosetosus* (Hym.: Formicidae), in an *Eucalyptus* plantation in the State of Para, Brazil with granulated baits with sulfluramid or chlorpirifos. *Acta Amazonica* 29(4), 639–645.

Zanuncio, J.C., Guedes, R.N.C., Zanuncio, T.V. and Fabres, A.S. (2001) Species richness and abundance of defoliating Lepidoptera associated with *Eucalyptus grandis* in Brazil and their response to plant age. *Austral Ecology* 26, 582–589.

Zanuncio, J.C., Zanuncio, T.V., de Freitas, F.A. and Pratissoli, D. (2003) Population density of Lepidoptera in a plantation of *Eucalyptus urophylla* in the State of Minas Gerais, Brazil. *Animal Biology* 53(1), 17–26.

Zanuncio, T.V., Zanuncio, J.C., de Freitas, F.A., Pratissoli, D., Sediyama, C.A.Z. and Maffia, V.P. (2006) Main lepidopteran pest species from an eucalyptus plantation in Minas Gerais, Brazil. *Revista de Biologia Tropical* 54(2), 553–560.

Zanuncio, J.C., Pires, E.M., Almado, R.D., Zanetti, R., Monne, M.A., Pereira, J.M.M. and Serrao, J.E. (2009) Damage assessment and host plant records of *Oxymerus basalis* (Dalman, 1823) (Cerambycidae: Cerambycinae: Trachyderini) in Brazil. *Coleopterists Bulletin* 63, 179–181.

Zarbin, P.H.G., Villar, J.A.F.P. and Correa, A.G. (2007) Insect pheromone synthesis in Brazil: an overview. *Journal of the Brazilian Chemical Society* 18(6), 1100–1124.

Zhang, A.B., Wang, Z.J., Tan, S.J. and Li, D.M. (2003) Monitoring the masson pine moth, *Dendrolimus punctatus* (Walker) (Lepidoptera: Lasiocampidae) with synthetic pheromone-baited traps in Qianshan County, China. *Applied Entomology and Zoology* 38(2), 177–186.

Zhang, A.B., Tan, S.J., Gao, W., Tu, J.B., Wang, R., Hao, Q., Cheng, L.S. and Chen, L.M. (2001) Primary studies on monitoring *Dendrolimus punctatus* with sex pheromone in Qianshan county, China. *Entomological Knowledge* 38, 223–226.

Zhang, H.G. and Yang, S.J. (1994) Several hazardous termites intercepted and captured from imported logs and eradication method. *Plant Protection* 20, 38–39.

Zhang, L.W., Liu, Y.J., Yao, J.A., Wang, B., Huang, B., Li, Z.Z., et al. (2011) Evaluation of *Beauveria bassiana* (Hyphomycetes) isolates as potential agents for control of *Dendroctonus valens*. *Insect Science* 18, 209–216.

Zhang, Q.H., Byers, J.A. and Schlyter, F. (1992) Optimal attack density in the larch bark beetle, *Ips cembrae* (Coleoptera: Scolytidae). *Journal of Applied Ecology* 29, 672–678.

Zhang, Y., Ban, X., Chen, P., Feng, R., Ji, R. and Xiao, Y. (2005) *Dendrolimus* spp. damage monitoring by using NOAA/AVHRR data. *Yingyong Shengtai Xuebao* 16(5), 870–874.

Zhang, Z., Li, D. and Zha, G. (2008) Complex dynamics of mason pine caterpillar, *Dendrolimus punctatus* (Lepidoptera: Lasiocampidae) populations. *Icnc 2008: Fourth International Conference on Natural Computation, Vol 4, Proceedings*, pp. 622–629.

Zhang, Z.F., Huang, B.R., Lin, Q.Y., Cao, X.J. and Cao, G.M. (1998) A study on use of the Petrel 650B type light aircraft to control insect pests. *Journal of Nanjing Forestry University* 22, 56–60.

Zhao, D., Xu, J., Lin, M., Qiu, H., Zhong Z., Chen, M., et al. (2008) Evaluation for the growth loss of *Eucalyptus* caused by *Leptocybe invasa*. *Guangdong Forestry Science and Technology*. (http://en.cnki.com.cn/Article_en/CJFDTOTAL-GDLY200806011.htm).

Zhao, M.F., Yu, F.Q. and Li, G.C. (1998) Preliminary report on control of *Dendrolimus punctatus* Nun by trap lamp. *Journal of Zhejiang Forestry Science and Technology* 18, 49–51.

Zhao, T.-H., Chen, C.-J., Xu, J. and Zhang, Q.-W. (2004) Host range and cross infection of cytoplasmic polyhedrosis viruses from *Dendrolimus* spp. *Acta Entomologica Sinica* 47, 117–123.

Zhao, T.-H., Zhao, W.-X., Gao, R.-T., Zhang, Q.-W., Li, G.-H. and Liu, X.-X. (2007) Induced outbreaks of indigenous insect species by exotic tree species. *Acta Entomologica Sinica* 50, 826–833.

Zheng, B.Y. and Zheng, Y. (1996) Study on comprehensive control technique of *Dendrolimus punctatus* in Huangyan. *Journal of Zhejiang Forestry Science and Technology* 16, 56–62.

Zia-ud-Din, K. (1954) On the biology of the bark-eating borer. *Pakistan Journal of Science* 6(1–4), 22–30.

Zobel, B.J. (1982) The world's need for pest-resistant forest trees. In: Heybroek, H.M., Stephan, B.R. and von Weissenberg, K. (eds) *Resistance to Diseases and Pests in Forest Trees*. Proceedings of the Third International Workshop on the Genetics of Host–Parasite Interactions in Forestry, Wageningen, The Netherlands, 14–21 September, 1980. Centre for Agricultural Publishing and Documentation, Wageningen, The Netherlands, pp. 1–8.

Zobel, B.J., Wyk, G. van and Stahl, P. (1987) *Growing Exotic Forests*. John Wiley and Sons, New York, 508 pp.

Zondag, R. (1977) *Pineus laevis* (Maskell) (Hemiptera: Aphidoidea: Adelgidae), pine twig chermes or pine woolly aphid. Forest and Timber Insects in New Zealand, No. 25, New Zealand Forest Service.

Zvereva, E.L., Lanta, V. and Kozlov, M.V. (2010) Effects of sap-feeding insect herbivores on growth and reproduction of woody plants: a meta-analysis of experimental studies. *Oecologia* 163, 949–960.

Zwolinski, J.B. (1989) The pine woolly aphid, *Pineus pini* (L.) – a pest of pines in South Africa. *South African Forestry Journal* 151, 52–57.

Zwolinski, J.B., Grey, D.C. and Mather, J.A. (1988) The impact of pine woolly aphid, *Pineus pini* (L.) (Homoptera: Adelgidae), on cone development and seed production of *Pinus pinaster* in the southern Cape. *South African Forestry Journal* 148, 1–6.

Index

Abies sachalinensis 212
abiotic effects
    altitude 54–55
    canopy arthropod abundance
        relation 60
    climate change 66–67
    Coptotermes 63
    development time and rate 52, 53
    diurnal variation 54
    evolutionary process 50–51
    Heteropsylla cubana 51–52
    Holotrichia spp. cumulative
        emergence 62–63
    Hyblaea puera, seasonal incidence 49, 51
    Hypsipyla grandella population 49, 50
    insect pest species, light traps 59–60
    Ips grandicollis, rainfall pattern 61, 62
    lepidoptera biomass and damage 61, 62
    lepidoptera defoliating *Eucalyptus* 49, 50
    lethal temperature 54
    metabolic rate 51
    microclimatic humidity 58–59
    pest outbreak 49
    pest population density 52
    pest population dynamics 51
    photoperiod 58
    plant phenology, knock-on effect 58
    poikilothermic 51
    rainfall and leaves eaten area
        relation 60–61
    relative humidity 62
    sap-feeding psyllids ecology 50
    sex-pheromone baited filter paper and
        humidity 63–64
    sunlight 56, 57
    symbiotic relationship 63
    temperature-dependent development
        rate 52
    temperature, insect life cycle 52
    wind 64–66
Acacia auriculiformis 207
Acacia mangium 240
Acrocercops gemoniella 109
Acrocercops spp. 108–109
Acromyrmex landolti balzani 101
Aesiotes notabilis 210–211
Agrilus spp. 125–126
Agrotis ipsilon 187
Agrotis segetum 187
Ailanthus triphysa 234
Amblypelta cocophaga 114
Anomala antiqua 99
Anoplolepis gracilipes 114
Aonidiella orientalis 162
Apate monachus 215
APFC see Asia-Pacific Forestry Commission
Apocheima cinerarius 178
Araucaria cunninghamii 210–211
Aristobia horridula 128–129, 235
Asia-Pacific Forest Invasive Species Network
    (APFISN) 262–263
Asia-Pacific Forestry Commission
    (APFC) 262–263
Atta cephalotes 101
Atta spp. 190–191
Aulacaspis marina 115
Austrocedrus chilensis 113
Azadirachta indica 208

Bacillus thuringiensis 178, 196–197, 295
Beauveria bassiana 283–284

Beddingia siricidicola 214
Betula pendula 77
biotic effects
    beech scale fecundity 70
    density-dependent relationship 73
    functional response curve 72–73
    indigenous predators 71–72
    interspecific and intraspecific
        competition 69
    natural enemy regulation 70–71
    parasitoids 68
    pathogen-caused disease 73
    Plecoptera reflexa 71, 72
    population regulation 68–69
    predators and parasitoids action 73–74
    psyllid density and parasitoid density 71, 72
    resource limitation 69–70
Bombyx mori 260
Brachychiton rupestre 162
Brachytrupes spp. 152
Bruchidius spp. 143–144

CABI's Forestry Compendium (FC) 166, 167
Campnosperma brevipetiolata 114
Casinaria nigripes 107
Cassia siamea 210
Casuarina cunninghamiana 15
Casuarina equisetifolia 210
Cedrus deodara 215
Celosterna scabrator 129, 165
Ceroplastes spp. 113–114
Chaliopsis junodi 234
Chaliopsis (Kotachalia) junodi 157
Chilecomadia valdiviana 136–137, 260
Chrysomela populi 190
Cinara cupressi 113, 155
Cinara cupressivora 112–113
Cinara lusitanica 112–113
Cinara todocola 212
CLIMEX 166–167
Coccotrypes rhizophorae 208
Commindenrum robustum 219
Competitive exclusion principle *see* Gauses's
    axiom
Coptotermes curvignathus 156, 210
Coptotermes spp.
    C. elisae 135
    C. formosanus 135–136
    colonies and foraging 134–135
    economic loss 136
    infestation 135
    natural enemy 136
    survival *vs.* termite attack 136
Coryphodema tristis 137–138
Craspedonta leayana 109, 171
crown damage index (CDI) 243

Cryptolaemus montrouzieri 196
Ctenarytaina spatula 280–281
Ctenomorphodes tessulatus 108

Danaus plexippus 54
Democratic Republic of the Congo (DRC) 158
Dendroctonus frontalis 164
Dendroctonus ponderosae 212
Dendrolimus punctatus
    biological control 283–284
    IPM 282
    monitoring and prediction 282–283
    silvicultural technique 283
Dendrolimus punctatus Walker 106–107
Diacrisia obliqua 160
Diclidophlebia spp. 191
digital aerial sketch-mapping (DASM)
    system 246
Dioryctria rubella 156
Dioryctria spp. 140–141
Dryocosmus kuriphilus 211

ECFP *see* ecological control of forest pests
Eciton burchelli 35, 54
ecological control of forest pests (ECFP) 275
Eligma narcissus 189
Endoxyla cinerea 122
Endoxyla spp. 155
Epholcis bilobiceps 232–233
Eriococcus coriaceus 255
Essigella californica 246–247
Eucalyptus cloeziana 155
Eucalyptus grandis 97–98
Eucalyptus regnans 97–98
Eucalyptus viminalis 98–99
Eucosma glaciata 192–193
Eutectona machaeralis 170, 225
exotic pest entry pathway
    air transport 257, 259–260
    cargo container 257
    cut flower trade 256
    hitchhiking 255–256
    intentional and accidental
        transport 260–261
    logs and sawn timber 256–257
    scions and nursery stock 255
    seed 255
    timber packaging 257, 258
    wind dispersal 260

Falcataria moluccana 107, 159
Fergusonina spp. 145–146
Food and Agriculture Organization of
    the United Nations 166

forest health surveillance
    aerial and roadside inspection 244
    aerial photography 245
    aerial sketch mapping 245
    age/stocking and distance
        interaction 243–244
    airborne high-resolution imagery 246–247
    Aracruz Celulose SA system 252
    Asian gypsy moth 248
    bark beetles 250
    barrier quarantine 269–271
    brown wattle mirid 250
    CDI 243
    contingency planning 269
    controlled-environment room 242
    cost of 251
    cumulative cost *vs.* detection
        probability 251, 252
    DASM system 246
    detection percentage *vs.* observers 244, 245
    ecoclimatic zone comparison 265
    economic impact 266
    environmental impact 266
    extensive surveys and fixed-plot
        monitoring 241
    five-spined bark beetle 248–250
    forest health field form 252–254
    forest health survey 240–241
    forestry quarantine 254–255
    GIS-GPS interface tool 246
    hazard site surveillance 271–272
    human-assisted dispersal 265–266
    IFP 240
    indigenous fauna 263–264
    intensive-site ecosystem monitoring 241
    international treaties and standards 269
    invasive forest pest selection 266, 268
    landscape-scale outbreak 251
    lepidopteran defoliators 248, 249
    Lymantria dispar 242, 243
    nine-step guidelines 266, 267
    non-native species, biological invasion 261
    operational surveillance program 252
    pathway analysis 264–265
    pest awareness 262–263
    pest biology and behaviour 264
    pest eradication 266
    pest status assessment 242
    plant quarantine law 261
    post-entry quarantine 272
    probability detection *vs.* flight line
        spacing 247
    protection and harvest time 240
    quarantine focus, potential forest pest 264
    reflectance curve 246
    remote-sensing technique, cost
        comparison 251
    resource type conservation 264
    sal borer 250
    satellite technology 247
    sketch mapping and photography
        limitations 245–246
    social impact 266
    surveillance guidelines 254
    trap selection 248
    treatment 272–273
    tree health assessment 266
    visual sketch mapping 244–245
    walk-through survey 243–244
    within-country quarantine 273–274
    see also exotic pest entry pathway
Frankliniella occidentalis 192

Gauses's axiom 25
Glycaspis brimblecombei 117–118
Gmelina arborea 109, 154, 278–279
Gonipterus scutellatus 98–99, 255–256
Gryllus spp. 60

Helopeltis spp. 156
Hemiberlesia pitysophila 211
Hendersonula toruloidea 232–233
Heteronychus arator 232
Heteropsylla cubana 51–52, 157, 158
    biological control 120
    economic and social impact 119–120
    Leucaena leucocephala 118
    lifecycle 119
    spreading chronology 118–119
Hevea brasiliensis 188
Hoplocerambyx spinicornis 129–130, 214, 217
Hyblaea puera 155
Hyblaea puera Cramer 104–106
Hylastes angustatus 123, 250
Hylobius abietis 232
Hylurdrectonus araucariae 139–140
Hylurgus ligniperda 250
Hyperaspis pantherina 218
Hypomeces squamosus 190
Hypothenemus dimorphus 193
Hypsipyla grandella 160, 168–169
Hypsipyla robusta 27
    biocontrol programme 143
    life cycle 141
    Meliaceae species, damage 141, 142
    planting density and growth rate 143
    Toona ciliata 141
Hypsipyla spp. 154–155
    biological control 295
    chemical control 295
    genetics and host tree
        susceptibility 293–294

Hypsipyla spp. (*continued*)
   IPM 292–293
   monoculture plantation 294–295
   nursery stage 292
   pest management tactics 293
   shade interactions and outbreak
      intensity 293
   silvicultural technique 292
   tree manipulation 294

Indarbela quadrinotata 122–123, 235
insect-host tree interaction
   age effect 83
   chemical barrier 81
   co-evolution 89
   competitive and selective advantage 89
   double-action attempt 84–85
   DW total nitrogen 76
   evolutionary strategy 89–90
   external plant surface, defence
      system 78–79
   forest management 77
   host plant nutrients alteration 78
   host vigour and stress 86–88
   induced defence 83–84
   insecticidal property 90
   leaf chemistry 84
   leaf miners 79, 81
   leaf nitrogen effects 76
   leaf toughness 79
   natural selection, evolution 75
   NPK, tannins 85–86
   nutrient, leaf content 77, 78
   Ouratea spp., leaf damage rate 77
   physical and chemical defence 82–83
   plant secondary metabolite 81–82
   plant soluble organic nitrogen 87, 89
   recognition stimuli 82
   resin pressure 81
   trichomes 79
   trichome type and density 79, 80
   vigour promotion and tree resistance 75
   waste and energy expenditure 75–76
   watering regime 77, 78
integrated pest management (IPM) 176
   agroforestry poplars pest 285–286
   ECFP 275
   economic loss 275
   economics and impact assessment 279–280
   entomological extension service 278
   eucalyptus grasshoppers 290–292
   field survey 284–285
   forest pest management components 279, 280
   IPM toolbox 276–278
   management tactics 275–276
   monitoring 280–281

   neem-based insecticides 286
   pest's biology and crop
      situation 281–282
   Phoracantha species 287–288
   Phytolyma control 286
   psyllid control 285, 286
   silvicultural technique 287
   termites 288–290
   tropical eucalypts yeild 284
   tropical forestry, pest outbreak 276, 277
International Plant Protection Organization
   (IPPO) 262
Ips grandicollis 248–250
Ips spp.
   biological control 125
   feeding and breeding attack 124
   lifecycle 123–124
   Pinus spp. 123
   propensity and species 124
   tree attack *vs.* time 124, 125

Khaya anthotheca 208–209
Khaya ivorensis 176

Lamprosema lateritalis 190
leaf chewing insects
   Anomala spp. 99
   Atta spp. and *Acromyrmex* spp. 99–101
   Ctenomorphodes tessulatus 108
   Dendrolimus punctatus Walker 106–107
   Gonipterus scutellatus 98–99
   Hyblaea puera Cramer 104–106
   Paropsis spp. 97–98
   Paropsisterna spp. 97–98
   Perga affinis insularis 103
   Perga affinis Kirby 101–103
   Pteroma plagiophleps
      Hampson 107–108
   Thyrinteina arnobia 103–104
Lepidiota spp. 150–151
Lepidiota stigma 232
Leptocybe invasa 147–148, 173, 191
Lepto pharsa heveae 221
Leucaena leucocephala 158
lure-and-kill technique 237
Lygidolon laevigatum 250
Lymantria dispar 242, 243, 247
Lymantria xylina 236

Maconellicoccus hirsutus 274
Malacosoma incurvum 73
Manilkara hexandra 109
Manilkara zapota 160
Margalef's index 36, 37

Megastigmus spp. 144–145
Melaleuca quinquenervia 166
Melaleuca quinquinerva 109
Metarhizium anisopliae 99
Milicia excelsa 56, 207
Myllocerus spp. 190

Neoboutonia macrocalyx 60–61
Neolithocolletis pentadesma 162
Nipaecoccus aurilanatus 192
Nipaecoccus viridis 117
Nothofagus fusca 70
nursery stage
    advantages 186
    antitranspirant 203
    armyworms 187
    biological control and
        biopesticides 196–197
    botanicals 198, 201
    carbamates 202
    chemical damage 195
    chemical pesticides 198
    chlorinated hydrocarbon 201
    conifer aphids 192
    curl grubs 188
    cutworms 187–188
    defoliating caterpillars 189
    disadvantages 186–187
    folk medicine 197–198
    gall wasps 191
    grasshoppers and crickets 188–189
    inorganic insecticide 201
    insect growth regulator 202–203
    insecticide, mode of action 198–200
    leaf beetles 190–191
    leaf tiers 189–190
    mealybugs 191–192
    mirid and lygaeid bugs 192
    neurotoxicant 202
    organophosphate 201–202
    pest complex 193
    pest management cost 198, 201
    physical/mechanical control 195–196
    plantation cycle 186
    reduced risk insecticides 203
    root curling 193–194
    scale insect 192
    scarab beetles 188
    shoot and stem borers 192–193
    silvicultural control 197
    silvicultural side effects 195
    soil mixture 194
    synthetic pyrethroid 202
    termites 193
    transplant stress 194–195
    watering and shade regime 194

Nysius clevelandensis 192

Ophelimus maskelli 148–149
Orthezia insignis 218
Orthosia incerta 178
Orthotomicus erosus 164, 250
Oxydia trychiata 280

Pacific Plant Protection Organization
    (PPPO) 262
Parantbrene robiniae 195
Parlatoria oleae 52
Paropsis atomaria 52
Paropsis spp. 97–98
Paropsisterna spp. 97–98
Perga affinis insularis 103
Perga affinis Kirby 101–103
Pest risk assessment (PRA) 263–264
Phoracantha acanthocera 155–156
Phoracantha semipunctata 126–127
Phytolyma lata
    IPM 287, 288
    Milicia excelsa 286
    neem-based insecticide 286–287
    silvicultural technique 287
    tree mixtures and shade planting 286
Phytolyma spp. 146–147, 191
Pineus boerneri 111–112, 255
Pinus yunnanensis 55
planning stage
    afforestation program 153, 154
    agroforestry 157–158
    agro-silvo-pastoral system 156
    amenity planting 161–162
    CABI Crop Protection
        Compendium 179–181
    CABI's Forestry Compendium
        (FC) 166, 167
    carbon sequestration and biofuel
        production 162–163
    Cedrela attack, soil mineral 169, 170
    climatic mapping program 166–167
    commercial growing, insect
        damage 154–155
    communication and tactics 184–185
    compacted, hardpan site 165
    Craspedonta leayana, resistance levels 171
    damage score system 173–174
    decision making 182–183
    dieback incidence and relative
        abundance 177
    dry or desert site 164–165
    ECOCROP 166
    environmental factor and insect
        damage 169

planning stage (*continued*)
   eyeball score 172–173
   farm-forestry venture 163
   fodder 158
   forage yield and resistance 177
   forest health 178–179
   fuelwood 158–159
   genotype and environment interaction 168
   genotype selection 175–176
   Hypsipyla spp. distribution 183
   industrial forest plantation 154
   insect pest resistance 173
   insect-related wood defect 155
   IPM program 176
   large-scale international collaboration 183–184
   mahogany species, *Hypsipyla grandella* attack 168–169
   mean total height *vs.* mean commercial height 176
   MPTS 159–160
   pest resistance and silvicultural properties 175
   pine woolly adelgids 170–171
   psyllid resistance, *Leucaena* accession 174
   rehabilitation 160–161
   resistance and biotechnology 177–178
   sap staining 156
   seed orchards and clonal planting 157
   shade species, pest feeding guild 159
   shallow soil and stony site 165
   silvicultural characteristics and psyllid infestation 175
   silvicultural decisions 153
   silvicultural properties 172
   socio-economic model 153
   socio-economics and forest pest management 179, 181–182
   stem borer attack 155–156
   stem malformation 156
   taungya 160
   tree growing concept 153–154
   tree resistance testing 171–172
   tree selection and plant resistance 167–168
   trichomes 170
   water conservation measure 166
   waterlogged site 165–166
   wind- or frost- prone site 163–164
plantation stage
   azodrin 235
   bacteria 225, 226
   baculovirus 227
   biodiversity stability theory 206
   biological control 215–216
   carbosulfan 232
   Cecidomyiidae gall former and *Euclystis* spp. defoliator, effect 205
   chlorpyrifos 232
   cypermethrin 234–235
   diversity and stability relation 207
   enrichment planting 208–209
   entomopathogenic properties 220, 221
   forest health surveillance and monitoring 204
   forest pests and fungi target taxa 238, 239
   fungal dissemination 223–224
   fungus, biological control 222
   gas fumigation 233
   habitat modification 217
   harvesting 215
   HybCheck 229
   Hyblaea puera, productivity 229–230
   insect damage 231
   insecticidal spray application 232–233
   insecticides 230–231
   insect pest, natural enemy 216–217
   leaf-cutting ants 233
   lure-and-kill technique 237
   malathion aerial spraying 234
   mass trapping 236–237
   mating disruption 237–238
   mature trees, insecticide spraying 235–236
   Metarhizium spp. 223
   monoculture 204–205
   natural enemy *vs.* biological control program 217
   nematodes 224–225
   NPV characteristics 227–228
   NPV fate 228, 229
   pest-host plant-environment interaction 229
   pest outbreak potential 205
   pheromone efficacy 239
   pheromones 236
   α-pinene and β-pinene 238
   predatory ladybird 218–219
   predatory weaver ant 218
   prophylactic treatment 232
   pruning and brashing 210–212
   salvage and trap logs 213–214
   sanitation and hygiene 214–215
   seedling treatment 231–232
   shading and nurse crops 207–208
   spacing and thinning 212–213
   top-down and bottom-up regulation 216
   toxic baits 233–234
   transplanting tactics 209–210
   tree-husbandry technique 204
   Zeuzera conferta infestation, *Sonneratia apetala* 205–206
plant pest diagnostic services (PPDS) 15–16
Platyomopsis humeralis 207
Platypus spp. 130–131
Plecoptera reflexa 71, 72

Populus deltoides 285–286
Pseudotsuga menziesii 256
Psorocampa denticulata 248, 249
Psylla oblonga 191
Pteroma pendula 63
Pteroma plagiophleps Hampson 107–108
Pyrausta machaeralis 189

Quadrastichus erythrinae 149–150

Rhabdopterus picipes 194
Rhizophora mangle 208
Rhyacionia frustrana 274
Rhyparida limbatipennis 15
Roptrocerus xylophagorum 125
Ruta chalepensis 295

Santalum album 156
sap-feeding insect
    Amblypelta cocophaga 114
    aphids 110–111
    Aulacaspis marina 115
    Ceroplastes spp. 113–114
    Cicadas damage 110
    Cinara cupressi 113
    Cinara cupressivora 112–113
    Glycaspis brimblecombei 117–118
    Helopeltis theivora, adult female 115–116
    Heteropsylla cubana see *Heteropsylla cubana*
    Leucopsis obscum, natural enemy 112
    natural enemy, *Helopeltis* spp. 116–117
    necrosis, plant tissue 110
    Pineus boerneri 111–112
    pine woolly adelgids 111
    population fluctuation, rainfall 111
    spherical mealybug
        see *Nipaecoccus viridis*
    Thaumastocoris peregrinus 120
    Tingis beesoni 120–121
    tree vitality 110
Schizonycha ruficollis 188, 281
Scolytus major 215
Scotorythra paludicola 73
Shannon–Wiener diversity index 24
shoot boring insects
    Dioryctria spp. 140–141
    Hylurdrectonus araucariae 139–140
    Hypsipyla robusta see *Hypsipyla robusta*
    sapling attack 139
Sirex noctilio 240
    biological control 134
    integrated pest management 134
    lifecycle 132–133

pest detection, southern hemisphere 132, 133
    Pinus spp. 132
species diversity indices 23–24
Spirama retorta 217
Stenalcidia grosica 248, 249
Strepsicrates rothia Meyrick 110
Strepsicrates spp. 190
Strongylurus decoratus 151–152
Stryphnodendron microstachyum 205
Swietenia humilis 160
Swietenia macrophylla 131
Syzygium cumini 109

Tectona grandis 105, 154
termites
    control measure 289
    entomopathogenic fungi 290
    indigenous pest 288–289
    IPM 289
    plant-derived botanical insecticides 290
    termite-pathogenic fungi 290
Tetrastichus sp. 109
Thaumastocoris peregrinus 120
Thyrinteina arnobia 53, 103–104, 248, 249, 284–285
Tingis beesoni 120–121, 232–233
Tomicus piniperda 274
tropical forest
    abiotic variation 29–30
    agroforestry and silvo-pastoralism 11–12
    arthropod species composition 25–26
    artificial restoration 12
    biodiversity 23
    canopy height 30–31
    CCA 46, 47
    climatic characteristics 2–3
    complex web interaction 15
    conceptual model 46, 48
    conservation ethic 10
    conventional wisdom 24–25
    DCA 31
    distubance and productivity, species diversity 40, 41
    dung beetle abundance and species richness 31
    dung beetle population density 35–36
    dung removal 43–44
    ecological and economic importance 1–2
    ecological change 5–7
    ecological niche 26–27
    ecological pest management 46–47
    ecological stability 22, 31–32
    ecospace 26
    ecosystem disturbance 39
    edge effect 37–38
    effective pest management 47–48

tropical forest (*continued*)
    energy flow and nutrient cycling 13
    eradication programme 16
    food web 13, 14
    forest product, FAO definition 4, 5
    forest product value 3, 4, 6
    fragmentation effects 33–34
    fuelwood and charcoal 5
    fundamental conflict 5
    geographical belt 2
    global land use 1, 2
    habitats 25
    insect damage 16–19
    island biogeography theory 34
    Jaccard's coefficient 41, 43
    leaf damage 28–29
    Margalef's index 36, 37
    matrix-corridor interaction 36, 37
    microhabitats and probable functional grouping 26, 27
    native/exotic planting 44–46
    natural and plantation forest 1
    PCA 41, 43
    peatland composition 10, 12
    pest dynamic implication 40–42
    pest management components 20, 21
    phylogenetic distance *vs.* host species 28
    plant architecture 25
    plantation area 9
    plant collection and screening 8
    plant diversity and habitat heterogeneity 25
    population density and forest clearance relation 6, 7
    population density and forest cover relation 3, 5
    PPDS 15–16
    prevention 20
    primary natural forest 6
    productivity and carbon storage 13
    productivity of 10, 11
    reduced-impact logging 41, 43–44
    reforestation effects 32–33
    sap-feeding insectimpacts 16
    secondary natural forest 8–9
    selective logging 39–40
    silviculture planning and pest management 27–28
    socio-economics 3, 4
    species-area relationship 34–35
    species-area slope 35, 36
    species composition 23–24
    species diversity indices 23
    species richness and cerambycids abundance 28, 29
    sustainable-yield cycle 6, 8
    teak plantation, insect pests 20
    terrestrial ecosystem community diagram 24
    trophic interactions 13–14
    watershed protection and preservation 22
    see also tropical forest pest
tropical forest pest 14–15
    Agrilus spp. 125–126
    Aristobia horridula 128–129
    Atta, foraging trail 100
    bagworm 107–108
    bark- and wood-boring insect 121–122
    bark-eating caterpillar *see Indarbela quadrinotata*
    Brachytrupes spp. 152
    Bruchidius spp. 143–144
    Celosterna scabrator 129
    Chilecomadia valdiviana 136–137
    chrysomelid leaf beetle 97
    Coryphodema tristis 137–138
    defoliators 91–92
    effects, defoliation 92, 97
    egg mortality 103
    feeding habit and insect order 91, 93–96
    Fergusonina spp. 145–146
    forestry pests, importance 100
    Hoplocerambyx spinicornis 129–130
    Hylastes angustatus 123
    insect feeding impacts 91
    Ips spp. *see Ips* spp.
    leaf chewing insects *see* leaf chewing insects
    leaf-cutting ants 99–100
    leaf mining insect *see Acrocercops* spp.
    leaf tying/rolling *see Strepsicrates rothia* Meyrick
    Lepidiota spp. 150–151
    Leptocybe invasa 147–148
    light defoliation 97
    Megastigmus spp. 144–145
    nest density, affected ecosystem 101
    Ophelimus maskelli 148–149
    P. affinis insularis 103
    paropsine 92
    P. atomaria 98
    Perga spp. lifecycle 101–103
    Phoracantha semipunctata 126–127
    Phytolyma spp. 146–147
    Platypus spp. 130–131
    Ps. bimaculata 97
    Ps. cloelia 97–98
    Quadrastichus erythrinae 149–150
    sap-feeding insect *see* sap-feeding insect
    sawfly 101

shoot boring insects *see* shoot boring
    insects
Sirex noctilio see Sirex noctilio
sirex wasp 122
skeletonizer *see Craspedonta leayana*
snout beetle, effects 98–99
Strongylurus decoratus 151–152
sunlit leaves 97
types of pest, parts affected 91, 92
Xyleutes ceramica 138–139
Xylosandrus crassiusculus 131–132
Xystrocera festiva 127–128
Tsuga heterophylla 256

Ultracoelostoma assimile 70

Vanapa oberthuri 211

Widdringtonia whytei 155
Williams alpha diversity indices 45

Xyleutes capensis 210
Xyleutes ceramica 138–139
Xylosandrus crassiusculus 131–132
Xylotrechus spp. 60
Xystrocera festiva 127–128, 194

Zeuzera conferta 205–206
Zonocerus variegatus 188–189